31st International Congress and Exhibition on Condition Monitoring and Diagnostic Engineering Management (COMADEM 2018)

Rustenburg, South Africa
2-5 July 2018

Editors:

P. S. Heyns
G. van Schoor
P. A. van Vuuren
R. B. K. N. Rao

ISBN: 978-1-5108-7038-3

Printed from e-media with permission by:

Curran Associates, Inc.
57 Morehouse Lane
Red Hook, NY 12571

Some format issues inherent in the e-media version may also appear in this print version.

Printed by Curran Associates, Inc. (2018)

North-West University
Potchefstroom Campus
Private Bag X6001
Potchefstroom
2520
South Africa

Phone: 0860 169698 (0860 1mynwu)
Fax: (+27 18) 299-2767

studies@mynwu.info

Additional copies of this publication are available from:

Curran Associates, Inc.
57 Morehouse Lane
Red Hook, NY 12571 USA
Phone: 845-758-0400
Fax: 845-758-2633
Email: curran@proceedings.com
Web: www.proceedings.com

Proceedings of the 31st International Congress and Exhibition on Condition Monitoring and Diagnostic Engineering Management

2-5 July 2018

Sun City, Rustenburg, South Africa

COMADEM®
2018

31st International Congress and Exhibition on Condition Monitoring and Diagnostic Engineering Management

Proceedings

Editors

P.S. Heyns, P.A. van Vuuren, G. van Schoor, R.B.K.N. Rao

Published by: North-West University, South Africa
2018

Organizing committee

Prof Raj B.K.N. Rao	Comadem International	United Kingdom
Prof George van Schoor	North-West University	South Africa
Prof Stephan P. Heyns	University of Pretoria	South Africa
Prof Pieter A. van Vuuren	North-West University	South Africa
Prof Alwyn J. Hoffman	North-West University	South Africa
Prof Kenny R. Uren	North-West University	South Africa
Dr. Henri Marais	North-West University	South Africa

Technical programme committee

Prof Stephan P. Heyns	University of Pretoria	South Africa
Prof Pieter A. van Vuuren	North-West University	South Africa
Dr. Fengshou Gu	University of Huddersfield	United Kingdom
Prof David Baglee	University of Sunderland	United Kingdom
Prof Viliam Makis	University of Toronto	Canada
Dr. Nicholas West	University of the Witwatersrand	South Africa
Prof Jerry Walker	Vaal University of Technology	South Africa
Mr. Christo van der Walt	Engineering Dynamics	South Africa
Dr. Philip Loveday	Council for Scientific and Industrial Research (CSIR)	South Africa

Additional reviewers

Prof Mehdi Behzad	Sharif University of Technology	Iran
Prof Cuthbert Nyamupangedengu	University of the Witwatersrand	South Africa
Dr. Guojin Feng	Brunel University	United Kingdom
Dr. Xiange Tian	University of Portsmouth	United Kingdom
Dr. Dong Zhen	Hebei University of Technology	China
Dr. Hao Zhang	Hebei University of Technology	China
Dr. Benji Gwashavanhu	University of Pretoria	South Africa
Dr. Erfan Asaadi	University of Massachussets, Dartmouth	United States of America
Dr. Sylvester Aye	University of Pretoria	South Africa
Dr. Anriëtte Bekker	Stellenbosch University	South Africa
Dr. Stephan Schmidt	University of Pretoria	South Africa
Dr. Logan Reddy	Eskom	South Africa
Mr. Henry Omoregbee	University of Pretoria	South Africa
Mr. Dawie Diamond	University of Pretoria	South Africa
Mr. Amin Jami	University of Pretoria	South Africa

Sponsored by:

Supported by:

Table of contents

Foreword

Welcome to the 31st International Congress and Exhibition on Condition Monitoring and Diagnostic Engineering Management (Comadem 2018). In a certain sense the name says it all - this is the 31st edition of this established conference series. Since 1988 this annual conference has brought together influential role players in the field of condition monitoring to share the newest research developments, practical experience and industry offerings with the global condition monitoring community.

This year is no different with exciting technology showcased by industry and cutting edge research presented by leading experts as well as budding condition monitoring researchers. As one author of popular fiction has written: we are living in interesting times. Mankind is faced with numerous challenges in 2018 ranging from climate change, extreme poverty and hunger in developing countries to isolationism. The main theme of the 2018 edition of the Comadem conference series was therefore aptly chosen as: energy and environmental issues facing the 21st century and beyond.

This volume contains 46 accepted papers written by active researchers in the broader condition monitoring community. The authors of the accepted papers hail from ten different countries, namely: Canada, China, Germany, India, Indonesia, Iran, Kenya, Namibia, South Africa and the United Kingdom.

The papers included in the conference proceedings are original research papers that have been subjected to a double-blind peer review process under the guidance of a panel of internationally recognized experts in condition monitoring. Initially each abstract was reviewed by a member of the technical programme committee. The authors of all accepted abstracts were invited to submit full length papers, which were then reviewed by two further reviewers (once again in a double-blind manner). Eighty-five percent of all the submissions were accepted.

We want to thank each and every author, reviewer, speaker and exhibitor for their valuable contribution to this congress. A special word of thanks for the sponsors for their financial contribution. Last, but definitely not the least, thank you Andreas Alberts, Maggie Faku and Helen van Vuuren for your help without which this conference wouldn't have become a practical reality.

Pieter van Vuuren
Chair of the Comadem 2018 organizing committee

Driving the digital enterprise

Ralf Leinen[1]
[1] Siemens, South Africa

ABSTRACT

The keynote will address the theme from the 4th Industrial Revolution, the German initiative "Industry 4.0" and how Digitalization together with the cloud can bring benefits to energy generation, energy consumption by industry, energy efficiency and the effect that this will have on the environment. The five pillars of Digitalization in industry also applies to energy: Reducing Time-to-market, Enhancing Flexibility, Increasing Quality, Increasing Efficiency with Increasing Security. .

Keywords: Industry 4.0, digitalization, energy efficiency

BIOGRAPHY

Leinen is Vice President for Siemens Digital Factory & Process Industries & Drives for Southern & Eastern Africa. He studied engineering and business administration at the University of Applied Sciences Cologne (Rheinische Fachhochschule Köln) and is an experienced senior level manager in various industries related to manufacturing and processes. Leinen joined Siemens in 1990 has been in international sales and management roles at the company's headquarters and regions in Germany, the USA and South Africa. He is currently driving Digitalization in key vertical markets in both Manufacturing and Process Industries

Modelling and Simulation of a Two Stage Reciprocating Compressor for Condition Monitoring Based on Motor Current Signature Analysis

Usama Haba[1, 2], Khaldoon Brethee[1], Samir Alabied[1,2], Debanjan Mondal[1], Fengshou Gu[1] and Andrew Ball[1]

[1]Centre of Efficiency and Performance Engineering, University of Huddersfield, Queensgate, Huddersfield HD1 3DH, UK.

[2]University of Al-Jabal Al-Gharbi, Gharyan, Libya.

ABSTRACT

Reciprocating compressors (RC) are widely used in various industrial processes such as petrochemical production for compressing and moving gases. It is necessary to have accurate and cost-effective detection and diagnosis approaches for ensuring process efficiency and safety. This paper investigates into utilising motor current responses for monitoring RCs. A comprehensive numerical simulation studies were carried out when the reciprocating compressor is under a wide range of operations along with different fault cases. An extended dynamic model has been developed, which allows comment faults to be studied in five physical processes including AC induction motor, crankshaft dynamics, the mass flow rate through valves, cylinder pressure variation, and valve dynamics. The changes in motor current signatures due to the faults such as valve leakage, intercooler leakage, broken rotor bars and stator winding asymmetries can then be understood with great insights, leading to the establishment of effective and reliable diagnostic features. Simulation results show that two sideband peaks, normalised speed and current magnitude calculated using a modulation signal bispectrum are significantly correlated with compressor dynamics and can be used as feature parameters for separating between different faults over the entire operating range of the compressor. Moreover, it shows that motor current signals contain sufficient information for diagnosing different faults.

Keywords: Reciprocating compressor model; Motor current signature analysis; Modulation signal bispectrum; Condition monitoring

Corresponding author: Usama Haba (Usama.haba@hud.ac.uk)

1. INTRODUCTION

Reciprocating compressor and its driving motor is one of the most popular machine in-use in various industrial systems. This is mainly because of its simple powerful construction and flexibility which offer high value of reliability. However, many of its components work in unsuitable working conditions like high-temperature and high-pressure environment [1], which in turns reduce the compressor efficiency. Consequently, this machine subjected to different faults related to its functionalities. Such faults not only degrade performance but cause production losses and consume additional energy resulting in an additional cost and even lead to severe damages to other parts [2], resulting in a reduction of the performance of machines. Hence their conditions need to be monitored.

Damages to compressor valves are considered to be the main cause of a serious machine breakdown and led to unscheduled shutdown [2, 3]. Similarly, intercooler leakage is also considered as a common fault, caused by deterioration of pipe seals or loose joints. Additionally, faults occurring in its driving motor such as stator winding asymmetry can also affect the performance of a compressor. Increases in motor stator resistance can lead to voltage imbalances, thus the increase in motor temperature and reduction in the motor efficiency [4]. Therefore, efficient and effective condition monitoring techniques are actively studied to detect machine faults as early as possible in order to avoid catastrophic failures, reduction of maintenance cost and improve system availability [3, 5, 6].

More attention is paid to the diagnosis of compressor faults based on vibration and current analysis, in which a large amount of research work has been conducted with state of the art technologies. Gu and Ball [7] presented a new method based on the use of Wigner–Ville distribution, wavelet transform [8], support vector machines [9], induction motor current signature analysis [10] and a high-resolution spectral analysis [11] for the detection of different RC faults. Lin et al. [12] used artificial neural network (ANN) for valve diagnostic combining three indices of time, frequency and amplitude. Instantaneous angular speed [3] and dynamic cylinder pressures [13], relevant vector machines based on envelope spectrum analysis [2], and time-frequency representation under different operating conditions [14] were used as analysis tools for diagnosis of various leakages in compressor valve. The motor current signatures can also be used as a mean for detecting abnormalities in the downstream machines such as compressors systems [15]. The rotor disturbances due to mechanical problems can also be detected by analysing the induction stator current [16].

Mathematical modelling is the way to study and understand the behaviour of the machine, the interaction of its components and its performance characteristic. In order to develop more accurate fault detection and diagnosis techniques, researchers have developed different mathematical models on the compressor. The first compressor model was developed by Costagliola [17] to study the dynamics of the valve with one degree-of-freedom. In [18] different degree-of-freedom was considered in the model for valve dynamics. A similar model was developed for compressor performance prediction [19], it studies in-cylinder pressure and the effect of piston rings leakage. Similarly, Winandy et al. [20] presented a numerical model to estimate the performance of the compressor which was able to determine discharge temperature and the ambient losses considering the compression as an isentropic process. Elhaj et al. [21] developed a numerical simulation for compressor condition monitoring under different operating conditions and was validated by experimental results. In [22] a similar numerical simulation a semi-hermetic CO_2 was presented which included compressor leakages and friction loss. In all the previous studies, only a few models focused on the diagnostic of different compressor faults and their effects on compressor performance. However, most of the presented RC models have not been extended to investigate the influences of the stator winding faults on the compressor performances and have not shown the potential of using the induction motor as a mean of assessing the condition of downstream driven equipment such as reciprocating compressors.

For model development and simulation study, the mathematical model offers a number of advantages. In which, modelling provides useful fault signatures with less time and equipment, particularly in the case of simulating faults which cannot be introduced to a real compressor for the purpose of experimental studies. Secondly, it allows researchers to study different compressor performances under the same conditions [21]. The main objective of this paper is to study and analyse the dynamic responses of a compressor in the context of compressor condition monitoring. On this ground, a mathematical model is developed for RC to simulate different operating conditions. The model describes the compressor working principles, valve movements, crankshaft motions and compressor pressure for the 1st and 2nd stages. The simulation considers both healthy and faulty conditions which include valve leakage, intercooler leakage, the effects of stator asymmetry and broken rotor bar. In addition, in-cylinder pressure and motor current signals are analysed to identify the fault features. Finally, to evaluate the model and simulation results, this paper presents a comparative result of measured and predicted in-cylinder pressure signals and motor current signals.

2. EXPERIMENTAL SETUP

An experimental study was employed based on a two-stage, single acting V-shape Broom Wade TS9 reciprocating compressor. It can deliver compressed air up to 0.8MPa (120 psi) to a horizontal air tank. The compressor is driven by a three-phase KX-C184, 2.5 kw induction motor via a transmission belt as shown in Figure 1. The test data was recorded using the Power 1401 Plus CED high-speed data acquisition system with a sampling rate of 49 kHz; the data of each test was recorded for 3.6-second intervals. In addition, a shaft encoder and a pressure sensor were also used to measure the output rotating speed and in-cylinder pressure respectively.

In this work, to evaluate the performance of MCSA, the current data were collected for four working conditions performed one by one. Starting with normal condition (BL), then discharge valve leakage on the second stage (DVL) which was introduced by drilling a 2mm hole in the valve plate to simulate the leakage as shown in Figure 1(b).

Leakage on the intercooler (ICL) was introduced to the intercooler pipe by loosening the nut holding the pipeline into second stage suction valve to simulate the intercooler leakage Figure 1(c). Stator winding asymmetry (R_{fs}) introduced by phase winding by increasing resistance of 1.0 Ω using an external resistor bank, shown in Figure 1(d).

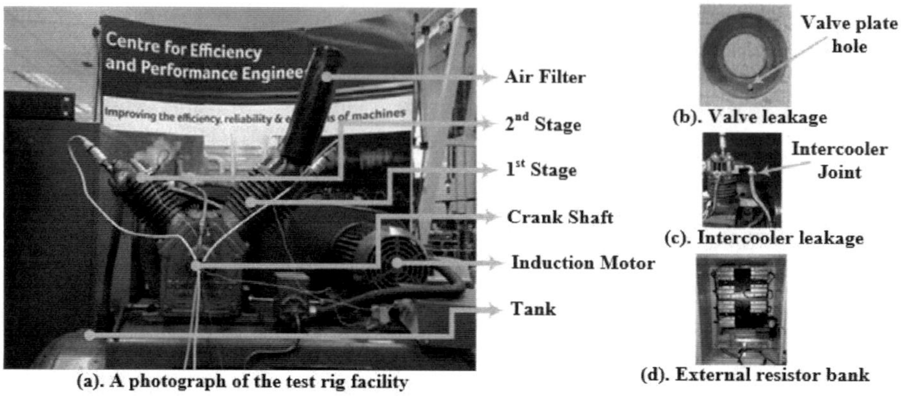

Figure 1: A photograph of the test rig facility

3. MODEL EVALUATION

Different conditions were simulated during modelling under various compressor discharge pressure to obtain features for compressor fault diagnosis at an early stage. The model dynamic responses were obtained under healthy and faulty operating conditions. Both model and experimental results are obtained and compared for in-cylinder pressure and motor current signals to evaluate the developed model with the experimental study.

3.1. Measured and predicted cylinder pressures comparison

The measured and predicted cylinder pressure for both stages is illustrated in Figure 2. The 1st stage cylinder pressure (the dashed blue line) Figure 2 clearly shows the four working processes (*expanding, suction, compression and discharge*) for each compression cycle. From the figure representation, a rapid drop in cylinder pressure can be shown during the re-expansion due to the downwards piston movement for the range of 0° to 30° crank angle allowing air enters the cylinder (*suction*) when the cylinder pressure drops sufficiently below the suction manifold, which lasts from 30° to 190°. Then the piston starts to move upwards, compressing the trapped air and pressure increases gradually up to the maximum which can be seen in the range from 190° to 285°. The discharge valve opens when the pressure exceeds the discharge manifold, allowing the compressed air to be pushed into the intercooler.

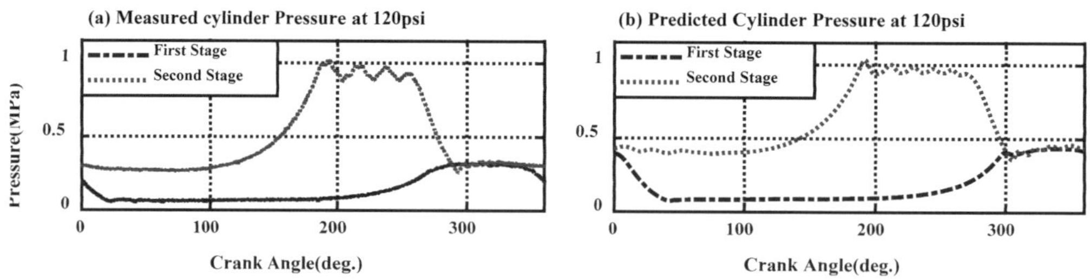

Figure 2: Pressure for low and high cylinders: (a) Measured, (b) predicted

For the 2nd stage cylinder pressure as shown by a dotted red line in Figure 2 shows the features of the compressor processes in the pressure profiles. However, each process occurs 90° earlier in terms of crank angle whereas, the 2nd stage suction process occurs within the discharge process of the first stage cylinder. Comparing both waveforms, the experimental and the predicted result generally show a good agreement for both cylinders and have the same trend which confirms that the pressure model is reliable. Yet, the measured results show a slight oscillation especially in the second stage during the discharge process and this likely caused by valve flutter.

3.2. Measured and predicted current waveform comparison

The measured and predicted motor current signal for different compressor discharge pressures are illustrated in Figure 3. The signals presented are of more than three cycles for comparison purpose. During the compressor working cycle, the motor is under a dynamic load fluctuation of about 7.3 Hz (rotation frequency of the compressor) which leads to a motor current modulation. A clear amplitude modulation (AM) effect can be clearly observed from the current waveform. Moreover, the amplitude increases as the pressure increases. Thus, the motor current signal can be used to obtain useful information from the compressor conditions. A comparison between the measured and predicted current signals shows that the basic features are the same and consistent over different compressor discharge pressures. This confirms that the model is valid for different discharge pressures and it can be used for fault simulation study.

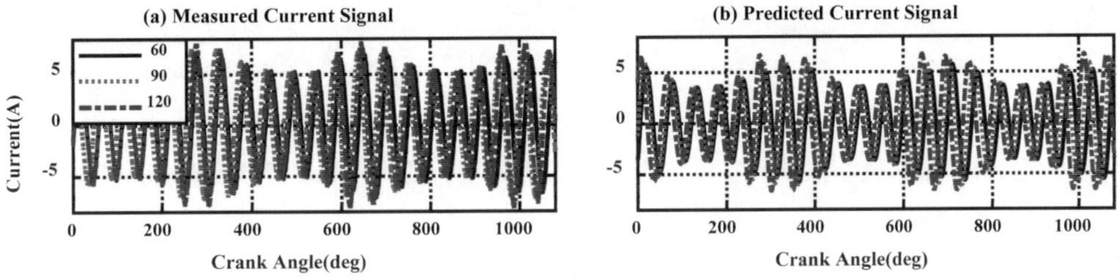

Figure 3: Current waveform for healthy compressor: (a) measured, (b) predicted

4. COMPRESSOR FAULT DETECTION

Valve leakages are considered as the most common fault in RC and the most frequent root cause of unplanned compressor shutdowns [3]. At high-pressure stages, both suction and discharge valves are exposed to a high number of impacts per second and also to high-speed flow and high temperature which cause non-uniform wear of the sealing surfaces between the valve seat and its plates which cause leakage in the valve system. Moreover, loose connection to the intercooler due to compressor vibration and pipe deterioration is also considered as a common fault affecting the compressor efficiency. Furthermore, failures to its associated motor parts such as broken rotor bar or stator winding asymmetry cause a reduction of motor efficiency. As a result, the temperature of motor goes up. The oscillatory running conditions can also affect the performance of a compressor. Therefore, this paper examines the leakage through the 2nd stage discharge valve, motor winding asymmetry and intercooler leakage. Simulation is used in each case to predict the signatures of each fault.

4.1. Valve leakage modelling

Valve leakage was simulated by drilling a small hole (2mm in diameter) into the valve plate Figure 1(b) which cover 2% of the plate. The leakage was introduced into the 2nd stage discharge valve where the valve works under high pressure and high temperature. During the simulation, valve leakage was simulated as an additional flow through an orifice in parallel to the normal flow. The cylinder pressure Equation was modified to consider the leakage through the discharge valve, in which additional term for the mass flow $\dot{m}_{vdL(L,H)}$ was included.

$$\dot{P}_{c(H)} = \frac{1}{V_{c(H)}}\left[c_{i(H)}^2 \dot{m}_{vi(H)} - c_{c(H)}^2 \dot{m}_{vd(H)} - \gamma p_{c(H)} \dot{v}_{c(H)} - c_{c(H)}^2 \dot{m}_{vdL(H)}\right] \quad (1)$$

The discharge valve leakage mass flow rate $\dot{m}_{vdL(L,H)}$ can be calculated as

$$\dot{m}_{vdL(H)} = \beta_{dL(H)} c_{dd(H)}(x).A_{dL(H)} \sqrt{2\rho_{c(H)} \left|P_{d(H)}^e - P_{c(H)}\right|} \quad (2)$$

where $\beta_{dl_H} = sign(P_{d_H} - P_{c_H})$ is the direction of the flow (+1 leakage flow from the cylinder into discharge valve and -1 flow from discharge valve into the cylinder), $A_{dL(H)}$ is the leakage flow area of the 2nd stage discharge valve.

4.2. Intercooler leakage modelling

The assumption is that there is a small leakage in the compressor's intercooler pipes due to deterioration or loose nut. This fault was simulated by loosening the nut that connects the intercooler pipe into the second stage suction valve Figure 1(c). The intercooler pressure Equation was modified to consider the leakage through the intercooler pipe in which additional term $\dot{m}_{ic}$ was included.

$$\dot{P}_{ic} = \frac{1}{V_{ic}} \left[c_{icL}^2 \dot{m}_{vdl} - c_{ic(H)}^2 \dot{m}_{vi(H)} - c_{ic}^2 \dot{m}_{ic} \right] \tag{3}$$

The intercooler leakage mass flow rate $\dot{m}_{ic}$ can be determined as

$$m_{ic} = \beta_{ic} c_{ic}(x).A_{ic} \sqrt{2P_{ic} \left| P_{ic} - P_o \right|} \tag{4}$$

where $\beta_{ic} = (P_{in} - P_o)$ is the direction of the flow and A_{ic} is the leakage flow area of the intercooler.

4.3. Motor fault modelling

Stator faults are modelled by varying the resistance matrices and inductance matrices associating with stators correspondingly. The resistance matrices in the *qd0* reference frame for the healthy motor is defined as [23]:

$$r_{sr}^{qd0} = r_{sr} \begin{bmatrix} 1 & 0 & 0 \\ 0 & 1 & 0 \\ 0 & 0 & 1 \end{bmatrix} \tag{5}$$

For the simplicity, the assumption is that the stator fault occurs only in phase *a* in the stator. Resistance matrix of the stator with shorted turns in the *qd0* reference frame can be derived as [23]:

$$r_{sr}^{qd0} = r_{sr} \begin{bmatrix} r_{s(11)} & 0 & r_{s(13)} \\ 0 & r_{s(22)} & 0 \\ r_{s(31)} & 0 & r_{s(33)} \end{bmatrix} \tag{6}$$

where r_{sf}^{qd0} is the stator winding resistance matrix for the faulty motor. Let g_{as} be the percentage of the remaining un-shorted winding in stator phase *a,* and the non-zero elements of the matrix r_{sf}^{qd0} are being defined as:

$$r_{s(11)} = \frac{1}{3}(2g_{as} + 1) \;\&\; r_{s(13)} = \frac{2}{3}(g_{as} - 1) \;\&\; r_{s(22)} = 1 \;\&\; r_{s(31)} = \frac{1}{3}(g_{as} - 1) \;\&\; r_{s(33)} = \frac{1}{3}(g_{as} + 1) \tag{7}$$

In the same way, the rotor faults (broken rotor bars) are modelled by changing the resistance matrices and inductance matrices associating with rotor [23].

5. SIMULATION AND MODEL VALIDATION

5.1. Cylinder pressures

Figure 4 illustrates both the measured and simulated cylinder pressure under 120 psi discharge pressure load, for the different operating condition. During the expansion and suction strokes, a noticeable deviation of the pressure waveform for valve leakage can be seen due to a high-pressure air backflow through the discharge valve which raises the in-cylinder pressure above the normal conditions and hence the discharge valve opens earlier compared to healthy condition. The intercooler leakage shows small decrement of in-cylinder pressure due to the fact that the pressure escapes through the intercooler leak causes a slight delay in valve opening time. It is hardly distinguishable between the stator winding asymmetry and the healthy condition. Comparing both the measured and the predicted results for high-pressure cylinder shows a good agreement and this agreement is considered to be acceptable.

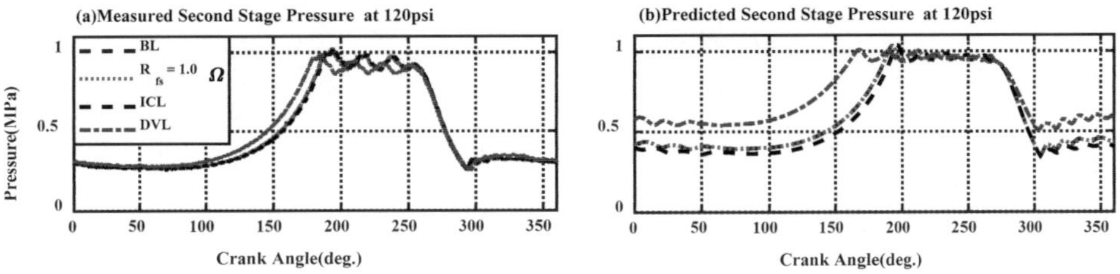

Figure 4: 2nd stage cylinder pressure for different conditions: (a) measured results, (b) predicted results

5.2. Motor current waveforms

The measured and predicted motor current results for the healthy and faulty conditions under 120 psi discharge pressure load are depicted in Figure 5. A clear amplitude modulation (AM) effect can be observed from the current waveform for both measured and predicted signals as mentioned earlier. The current amplitude for the crank angles from 250° to 350° is higher than that from 100° to 200° due to the load fluctuation caused by the increasing pressure during the compression stroke. The measured and the predicted results for current waveform also show good agreement.

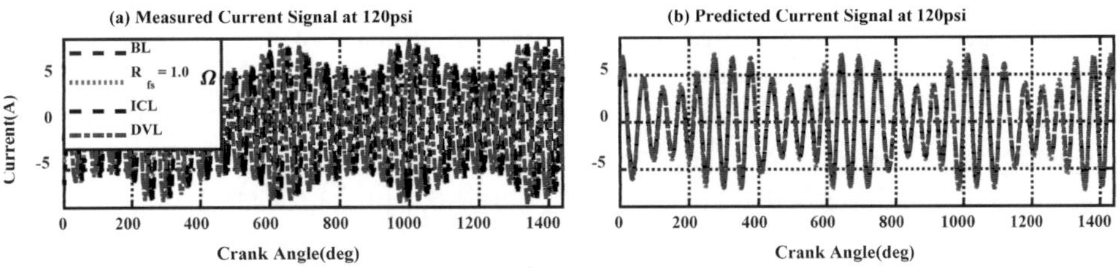

Figure 5: Current waveform for different conditions

5.3. Motor current spectra

Figure 6 depicted the measured and predicted current spectra for normal operation (healthy) and different faulty conditions under 120 psi discharge pressure load. It can be seen that signals for both measured and predicted signals exhibit a clear AM effect that causes a series of sidebands at $f_s \pm nf_r = 50 \pm n7.3Hz$. Especially, the amplitudes of sidebands show differences for different fault cases.

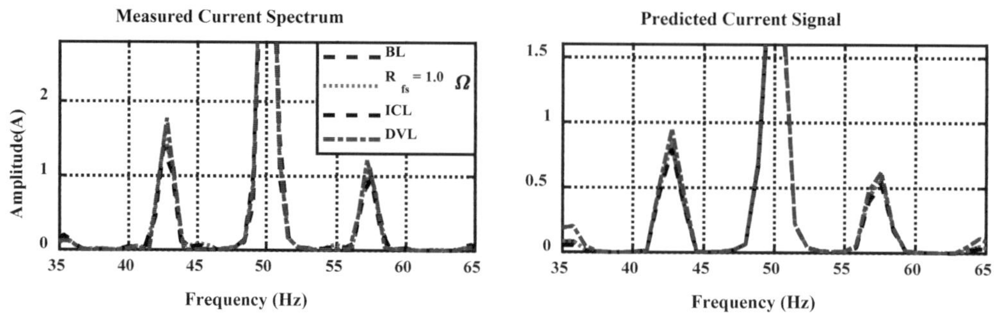

Figure 6: Current spectrum for different operating conditions

As shown in Figure 7 the sideband amplitudes increase as the compressor pressure increases. Moreover, when the sidebands amplitude for both measured and predicted signals are compared, it shows that both have the same increasing trends with discharge pressures. In addition, it also shows high consistency for different fault cases.

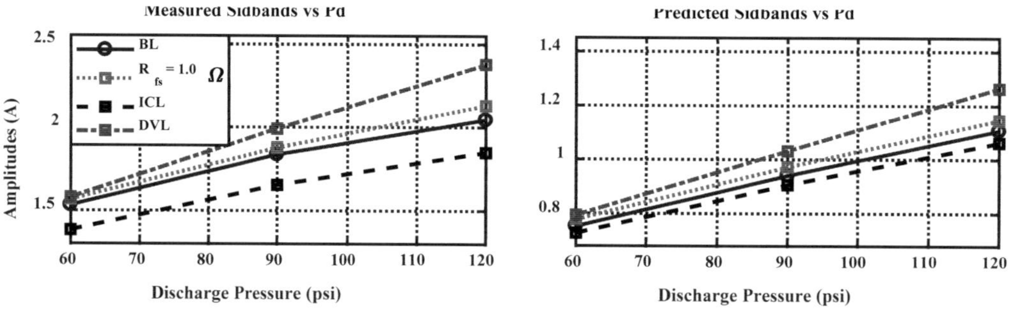

Figure 7: Measured and predicted sidebands vs pressure

6. DIAGNOSTIC FEATURES

As MSB has the unique properties of combining sidebands and noise suppression simultaneously [10, 24-26], it was applied to simulated current signals to obtain reliable diagnostic features. Figure 8 shows a typical MSB slice at supply frequency of 50 Hz. With its much simpler representation, MSB peaks can be observed clearly at various orders of harmonics associating with the rotational frequency $f_r = 7.3Hz$ of the compressor. Moreover, these peaks decrease slightly when there is a fault. In addition, the fault also induces additional frequencies components at the twice slip frequency: $f_{brb} = 2sf_s = 3.2Hz$ due to the presence of broken rotor bar (BRB) fault.

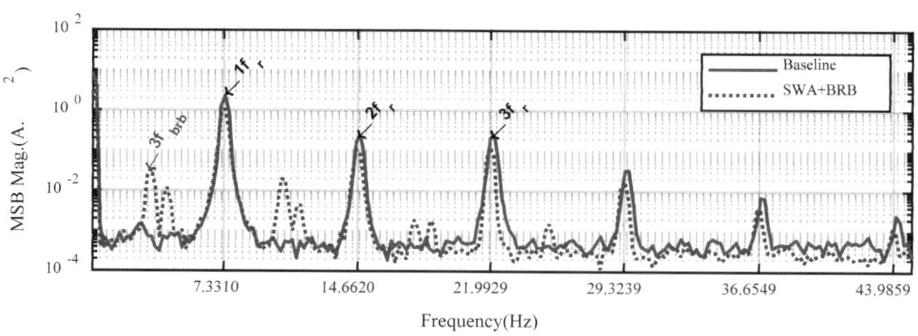

Figure 8: A typical MSB slice at 50Hz

Based on these observations, four primary diagnostic features including 1) the speed normalised by rated speed; 2) current amplitude normalised by the rated current of AC motor; and 3) two MSB peaks are extracted from the MSB slice to differentiate different fault cases, which allows diagnostics to be implemented without the need of additional measurements such as pressure, temperature, etc.

6.1.Diagnostics of compressor faults

Figure 9 shows the feature characteristics for different faults simulated on the compressor. Figure 9 (a1), (a2) and (a3) shows that all the four features exhibit distinctive differences for the inter-cooler leakage (ICL). It means that ICL cause significant influences on the pressures of both stages and hence on the dynamics of torque, and thus large deviation from the baseline operation. So this fault can be easily separated. For the suction valve leakages (SVL) and discharge valve leakage (DVL), shown in Figure 9 (b1-b3) and (c1-c3) respectively, smaller changes are induced to the pressure and hence the dynamics of torque. Consequently, only smaller deviations are shown in the current signals by the MSB peaks at $2f_r$. Nevertheless, these changes can be acceptable to make differences from the baseline and between the two faults as their changing trends are different with the normalised current values. For the losing belt transmission (LBT), the deviated speed pattern in Figure 9 (d1) can be based to separate it from others. Although the deviation values look much smaller, the high accuracy of speed estimation by higher order harmonics can ensure the diagnostic accuracy. In general, these feature parameters can be sufficient for the diagnostics of these common faults.

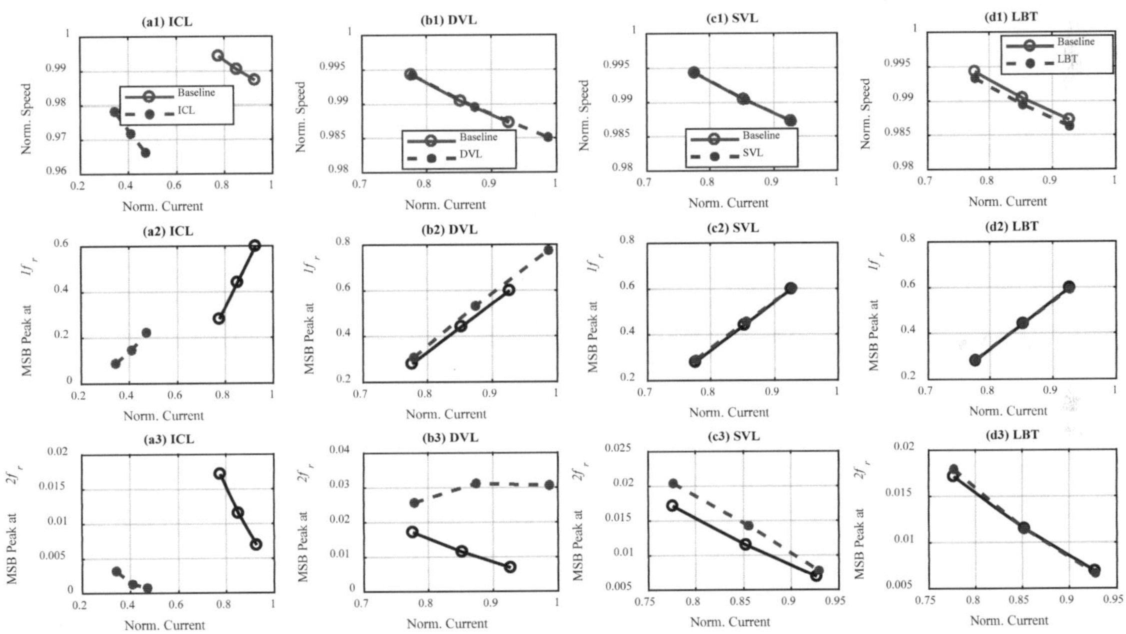

Figure 9: Characteristics of feature parameters for different mechanical faults

6.2.Diagnostics of motor faults

Figure 10 presents the feature characteristics for different faults simulated on the driving motor. For the stator winding asymmetry (SWA) which is of 10% increase of the stator phase resistance, shown in Figure 10 (a1-a3), the deviations in speed, MSB peaks are very significant and different from that of ICL in Figure 9. Therefore, this fault can be identified easily. For the BRB case (1 bar over 20 of them), only the MSB peaks at f_{brb} is significantly different from the baseline, so this unique feature can be based to separate this fault. The combination of SWA and BRB show the deviation characteristics of both, so it is can be differentiated using these features completely. In addition, as these feature patterns along with the additional one at f_{brb} are exhibit very different characteristics from that of compressor faults in Figure 9, these two types of faults can be diagnosed without any ambiguity.

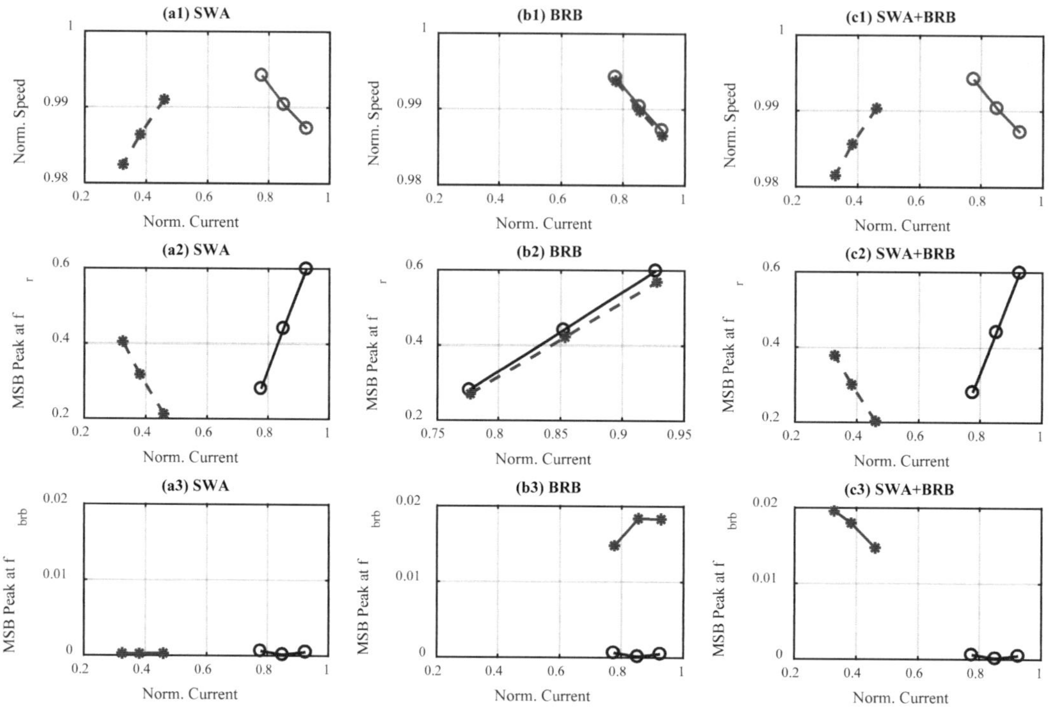

Figure 10: Characteristics of feature parameters for electrical faults

7. CONCLUSIONS

Simulation studies along with experimental verification have shown that motor current waveforms contain sufficient information and have the capability of characterising reciprocating compressor working process under various common faults for a wide range of discharge pressure. Thus, fault detection and diagnosis can be performed for reciprocating compressor condition monitoring based on the motor current analysis. It can be concluded that the agreement between the measured and predicted results for the first and second stages is considered to be acceptable. The comparative study illustrates that the diagnostic features of normalised speed, normalised current allows reflect the basic operating conditions of reciprocating compressors, and the MSB sideband peaks allow the dynamics of the torques induced by pressure changes to be accurately reflected. The combinations of these multiple current features allow various common faults (valve leakages, inter-cooler leakage, loose belt, stator winding asymmetry and broken rotor bar) to be diagnosed over a wide range of operating conditions.

REFERENCES

[1] AlThobiani, F. and A. Ball, An approach to fault diagnosis of reciprocating compressor valves using Teager–Kaiser energy operator and deep belief networks. Expert Systems with Applications, 2014. 41(9): p. 4113-4122.

[2] Ahmed, M., et al. Fault diagnosis of reciprocating compressors using revelance vector machines with a genetic algorithm based on vibration data. in Automation and Computing (ICAC), 2014 20th International Conference on. 2014. IEEE.

[3] Elhaj, M., et al., A combined practical approach to condition monitoring of reciprocating compressors using IAS and dynamic pressure. World Academy of Science, Engineering and Technology, 2010. 4(3): p. 287-293.

[4] Sridhar, S. and K.U. Rao. Detection of simultaneous unbalanced under-voltage and broken rotor fault in induction motor. in Condition Assessment Techniques in Electrical Systems (CATCON), 2013 IEEE 1st International Conference on. 2013. IEEE.

[5] N. A. Hussein, D.Y.M., E. M. Abdul-Baki, 3-phase Induction Motor Bearing Fault Detection and Isolation using MCSA Technique based on neural network Algorithm. Journal of Engineering and Development, 2012. 16(3): p. 15.

[6] Haba, U., et al. Diagnosis of Compound Faults in Reciprocating Compressors Based on Modulation Signal Bispectrum of Current Signals. in Proceedings of the 2nd International Conference on Maintenance Engineering, IncoME-II 2017,(University of Manchester, 5-6 September 2017). 2017. University of Manchester.

[7] Gu, F. and A. Ball, Use of the Smoothed Pseudo-Wigner-Ville Distribution. Maintenance & Asset Management Journal, 1995. 10(2): p. 16-23.

[8] Wang, Y., M. Liao, and T. Zhao, Application of the wavelet transform to fault diagnosis in compressor valves. Zhongguo Jixie Gongcheng/China Mechanical Engineering, 2003. 14(12): p. 1046-1048.

[9] Cui, H., et al., Research on fault diagnosis for reciprocating compressor valve using information entropy and SVM method. Journal of loss prevention in the process industries, 2009. 22(6): p. 864-867.

[10] Gu, F., et al., Electrical motor current signal analysis using a modified bispectrum for fault diagnosis of downstream mechanical equipment. Mechanical Systems and Signal Processing, 2011. 25(1): p. 360-372.

[11] Garcia-Perez, A., et al., The application of high-resolution spectral analysis for identifying multiple combined faults in induction motors. Industrial Electronics, IEEE Transactions on, 2011. 58(5): p. 2002-2010.

[12] Lin, Y.-H., H.-C. Wu, and C.-Y. Wu, Automated condition classification of a reciprocating compressor using time–frequency analysis and an artificial neural network. Smart materials and structures, 2006. 15(6): p. 1576.

[13] Pichler, K., et al. Detecting broken reciprocating compressor valves in the PV diagram. in Advanced Intelligent Mechatronics (AIM), 2013 IEEE/ASME International Conference on. 2013. IEEE.

[14] Pichler, K., et al., Fault detection in reciprocating compressor valves under varying load conditions. Mechanical Systems and Signal Processing, 2016. 70: p. 104-119.

[15] Haynes, H. and R. Kryter, Condition monitoring of machinery using motor current signature analysis. 1989, Oak Ridge National Lab., TN (USA).

[16] El Hachemi Benbouzid, M., A review of induction motors signature analysis as a medium for faults detection. Industrial Electronics, IEEE Transactions on, 2000. 47(5): p. 984-993.

[17] Costagliola, M., The theory of spring-loaded valves for reciprocating compressors. JOURNAL OF APPLIED MECHANICS-TRANSACTIONS OF THE ASME, 1950. 17(4): p. 415-420.

[18] WAMBSGANSS, M., MATHEMATICAL MODELING AND DESIGN EVALUATION OF HIGH-SPEED RECIPROCATING COMPRESSORS. 1968.

[19] Manepatil, S., G. Yadava, and B. Nakra, Modelling and computer simulation of reciprocating compressor with faults. JOURNAL-INSTITUTION OF ENGINEERS INDIA PART MC MECHANICAL ENGINEERING DIVISION, 2000: p. 108-116.

[20] Winandy, E., C. Saavedra, and J. Lebrun, Simplified modelling of an open-type reciprocating compressor. International journal of thermal sciences, 2002. 41(2): p. 183-192.

[21] Elhaj, M., et al., Numerical simulation and experimental study of a two-stage reciprocating compressor for condition monitoring. Mechanical Systems and Signal Processing, 2008. 22(2): p. 374-389.

[22] Yang, B., C.R. Bradshaw, and E.A. Groll, Modeling of a semi-hermetic CO2 reciprocating compressor including lubrication submodels for piston rings and bearings. International Journal of Refrigeration, 2013. 7(36): p. 1925-1937.

[23] Mustafa, M.O., Faults detection and diagnosis for three phase induction machines. 2012, Luleå tekniska universitet.

[24] Naid, A., et al. Bispectrum Analysis of Motor Current Signals for Fault Diagnosis of Reciprocating Compressors. in Key Engineering Materials. 2009. Trans Tech Publ.

[25] Alwodai, A., et al. Modulation signal bispectrum analysis of motor current signals for stator fault diagnosis. in Automation and Computing (ICAC), 2012 18th International Conference on. 2012. IEEE.

[26] Gu, F., et al., A new method of accurate broken rotor bar diagnosis based on modulation signal bispectrum analysis of motor current signals. Mechanical Systems and Signal Processing, 2015. 50: p. 400-413.

Experimental Study on Vibration Reduction Characteristics of a Helical Gear Coupling System Based on ISFD

Zhang Yipeng[1], He Lidong[2], Lu Kaihua[3]
Beijing Key Laboratory of Health Monitoring and Self-recovery for High end Mechanical Equipment, Beijing University of Chemical Technology, Beijing, 100029, China

ABSTRACT

In view of the characteristics of helical gear coupling systems and their complex vibration characteristics, a new type of integral squeeze film damper (ISFD) was designed to reduce the vibration transmission of the coupling process of a pair of helical gears to the bearing and foundation by squeezing film. For the experiment, two pairs of helical gears needed to be designed: one is a normal helical gear, and the other is a faulty helical gear worn on the tooth surface. The first-stage helical gear test bench was set up for the experiments. The rigid support and ISFD support were each installed on the four bearings to study the ISFD vibration reduction of the driving shaft and the driven shaft of two pairs of helical gear systems under the effects of different rotational speeds. The experimental results show that, under the same conditions, the vibration amplitude of the faulty helical gear is much larger than that of the normal gear. The ISFD can improve the vibration of the helical gear coupling system. When the system increased the damping, the resonant modulation phenomenon of the faulty helical gears was reduced. Also, ISFD has a frequency shift effect, and this can effectively shift away from the resonance region. In addition, ISFD can significantly reduce the meshing frequency of the helical gear coupling system caused by the resonance phenomenon.

Keywords: Vibration control; Helical gear; Integral squeeze film damper (ISFD); faulty gear; Time domain analysis; Frequency domain analysis .

1. INTRODUCTION

With the rapid development of modern science and technology, rotating machinery is developed quickly. Gear transmission systems are widely used in the fields of aviation, energy, petrochemicals, shipbuilding, and other fields for power and transmission in rotating machinery. However, the meshing transmission process of the gear system is complicated. The complexity is not only affected by the external factors of the prime mover and the load but also by the influence of internal excitation, such as time-varying meshing stiffness. These factors can cause strong vibration of the system, and, in severe cases, they will lead to faults in the vulnerable components, such as bearings. Helical gears are a relatively stable type of gear for transmission but vibration still exists. When the gears are worn out, glued, and fractured, the vibrations are aggravated. A large amplitude will cause fatigue damage to the helical gears. It will be transmitted to the foundation and other parts through shafts, bearings, etc., resulting in vibration of the entire mechanical system, causing safety hazards [1]. Helical gear vibration excitation is diverse and complex, mainly affecting the modulation part and the additional pulse part. The modulation part mainly for the meshing frequency and its higher harmonics, the additional pulse part for the frequency and its lower harmonic components [2].

For helical gear vibration, domestic and foreign scholars often use the following ways to control it: (1) by adjusting the structural parameters of the helical gear [3], so that the meshing situation is more reasonable, thereby reducing the excitation force and controlling the vibration source; (2) installing a damping element on a helical gear structure or system and dissipating the vibration energy through damping [4]; and (3) installing an isolation device between the vibration source and the foundation to further cut off the transmission of the vibration. This local vibration reduction measure can reduce the vibration transmitted during meshing of the helical gear along the shaft and bearing to the entire gearbox [5]. At present, the most commonly studied method can be found in (1) and (2). The method reported in (1) belongs to the active design and combines the helical gear dynamics and vibration theory to reduce the vibration by optimizing the design of the helical gear parameters. The method in (2) involves passive damping, which is commonly used when viscoelastic dampers are affected by the thickness of the damping layer [6], and the friction dampers are limited to a narrower frequency band for damping [7].

ISFD is a type of squeeze film damper (SFD) used in conjunction with bearings to reduce rotor vibration problems and improve system stability. ISFD has low stiffness and high damping. Low stiffness reduces the dynamic force that helical gears transmit to the foundation and other components through bearings. High damping dissipates vibrational energy. However, there are some problems [8-9] in SFD, such as the nonlinearity of the oil film, large occupied space, difficult installation, and nonlinear change of the stiffness and damping characteristics [10]. A new type of ISFD was designed, and it can overcome the above problems of SFD.

In this research, targeting the vibration problem of the helical gear coupling system, the new ISFD was designed to isolate the vibration source from the foundation to reduce the vibration energy transferred to the bearing and the foundation. A pair of normal helical gears and faulty helical gears was designed, and a first-order helical gear coupling test bench was set up. The vibration amplitude of ISFD was compared with that of rigid support to verify the damping effect of ISFD. A simple and convenient helical gear vibration control method was obtained.

2. HELICAL GEAR VIBRATION MODEL AND ANALYSIS

First, we establish a helical gear foundation single-degree-of-freedom vibration model, as shown in Figure 1. The helical gear is simplified as a cylindrical mass. ISFD is between the bearing and foundation, expressed in terms of the stiffness coefficient K and the damping coefficient C. According to the vibration isolation principle [11], K at the bearing is properly reduced, and C is increased; thus, the natural frequency of the system is reduced and the vibration transmission coefficient is reduced.

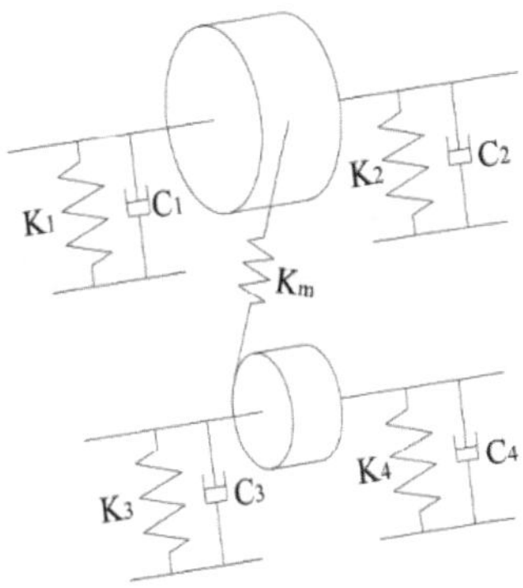

Figure 1. Single-degree-of-freedom helical gear vibration model [12].

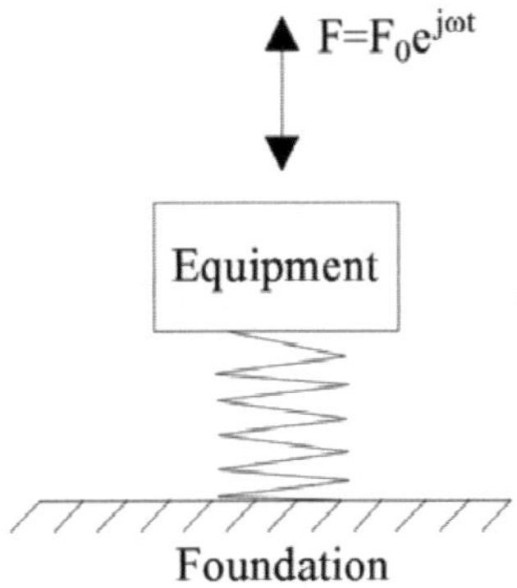

Figure 2. Single-degree-of-freedom system vibration isolation [13].

According to Fig. 1 and Fig. 2, the simplified mechanism diagram of the vibration isolation of the helical gear foundation system is analyzed. The dynamic equation of this system is:

$$M\ddot{x} + C\dot{x} + Kx = F_0 e^{j\omega t} \tag{1}$$

where *K, M,* and *C,* denote the stiffness, mass and damping of the vibration isolation system, respectively. The vibration displacement is,

$$x_0 = \left| \frac{F_0}{K} \frac{1}{(1-\eta^2 + j2\zeta\eta)} \right| \tag{2}$$

where $\omega_0 = \sqrt{K/M}$, $\zeta = C/\left(2\sqrt{KM}\right)$, $\eta = \omega / \omega_0$

The transmission force transmitted to the foundation by the vibration-isolating spring is

$$F_K = C\dot{x} + Kx \tag{3}$$

The magnitude of transmission force transferred to the foundation is given by (4)

$$F_{K0} = |j\omega Cx_0 + Kx_0| \tag{4}$$

The vibration transfer coefficient is

$$T_A = \left|\frac{F_{K0}}{F_0}\right| = \left|\frac{j\omega Cx_0 + Kx_0}{Kx_0(1-\eta^2+2j\zeta\eta)}\right| = \left|\frac{1+2j\zeta\eta}{1-\eta^2+2\zeta\eta}\right| = \frac{\sqrt{1+(2\zeta\eta)^2}}{\sqrt{(1-\eta^2)^2+(2\zeta\eta)^2}} \tag{5}$$

Through Fig. 1 and 2 and Eq. (5), it can be found that, when the vibration frequency ω is much larger than the resonance frequency ω_0 of the entire system, the vibration transmission coefficient is less than 1. The helical gear meshing frequency is higher, it is often larger than the natural frequency of the helical gear coupling system. At this point, the system has the vibration isolation function. The higher the vibration frequency is at the resonance frequency of the entire system, the smaller the vibration transmission coefficient and the better the vibration isolation effect. When reducing the stiffness in the system and increasing the damping of the system, the resonance of the system can be avoided, and the stability of the system can be improved.

3. CHARACTERISTICS OF ISFD

3.1. Structural features

ISFD adopts EDM technology to process the slits of the squeezed film to form a four area integrated structure, which can provide high damping and low stiffness for the whole system. This integrated structure has more-precise concentricity and can improve the installation accuracy. Figure 3 shows the experimental ISFD. The S-shaped spring structure can evenly distribute pressure and provide low radial static stiffness. It can also provide significant damping for the system in the process of squeezing the oil film.

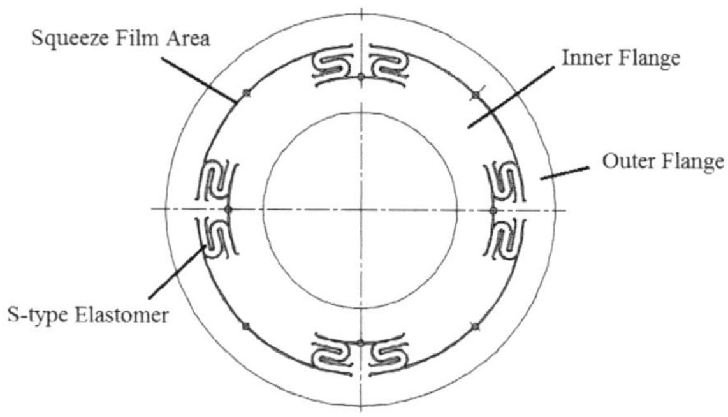

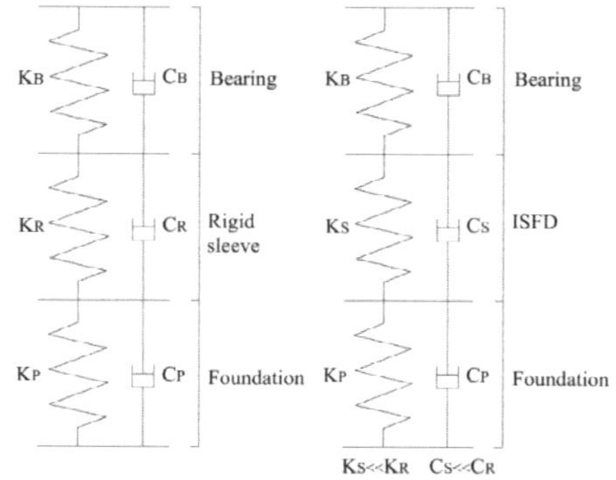

Figure 3. S-shaped integral squeeze film damper. Fig. 4. Comparison of the mechanical model between the rigid support and the ISFD.

Figure 4 is a comparison of the schematics during the experiment. The left figure shows the rigid support structure, and the right figure shows the ISFD structure. In the process of vibration transmission, the vibration deformation mainly concentrates on the elastic support part, part of the vibration energy is dissipated during the movement of the spring, and there is also the process of squeezing the oil film. The damping provided can dissipate the vibration energy in the form of thermal energy.

3.2. Energy Dissipation Mechanism

The stiffness coefficient K and the damping coefficient C in the ISFD structure are relatively independent, and the spring structure and damping can decouple. When the vibration is transmitted to the ISFD, the dynamic pressure Δp generated by the squeeze film causes the lubricant to move circumferentially along the gap. Because the ISFD has a division, it has a linear relationship and consumes energy during the flow friction. The resulting damping will effectively reduce the vibration of the helical gear system and increase the effective damping ratio of the system [14-15].

4. EXPERIMENT

To study the vibration of the normal helical gear pair and the faulty helical gear pair under the same conditions, two pairs of helical gear pair were designed: the normal helical gear pair shown in Figure 5(left one) and the faulty helical gear pair in Figure 5(right one), where the fault in the helical gear pair is mainly tooth surface wear.

Figure 5. Normal helical gear and faulty helical gear

The two types of comparative support in this experiment are rigid sleeve support and ISFD support, The rigid sleeve support consists of a rigid sleeve, rolling bearing, and foundation; ISFD support consists of the foundation, the S-shaped elastic body, the rolling bearing, the lubricating oil, the O-shaped sealing ring, and the end seal, where the O-shaped sealing ring includes the small O-shaped sealing ring between the end seal and the shaft, and the foundation and the end seal are the large O-shaped seal.

The helical gear coupled transmission test bench is shown in Figure 6. Driven by a permanent-magnet-type DC servo motor. The speed of the motor can be adjusted by the speed regulator. Before the experiment, the helical gear pair was lubricated. The photoelectric sensor was used to collect the rotational speed, and the sensor that collects the vibration was an acceleration sensor. An acceleration sensor was installed in the radial x-direction of the driving shaft and in the driven shaft. Both these sensors were connected with an LC-8008 vibration monitoring and diagnosis system. The time domain waveforms, frequency domain waveforms, and axial trajectory of the vibration could be read in real time on the computer interface.

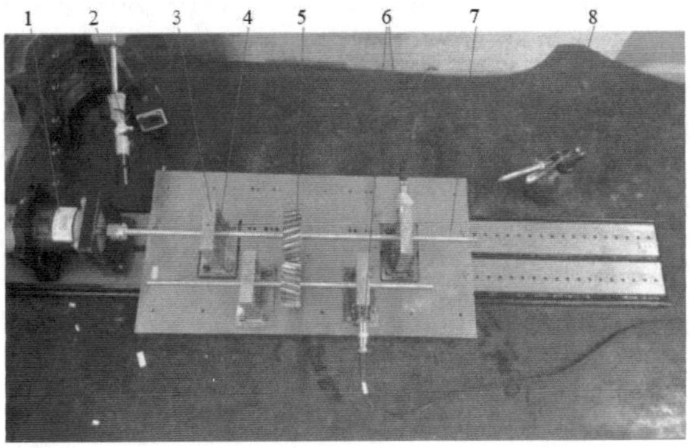

Figure 6. Experimental device

5. ANALYSIS OF RESULTS

Experimental results of the following three groups of experiments were strictly analyzed using the control variable method, and accelerometers were used to collect the acceleration and time domain signals on the master and slave axes, respectively, and the speed ranges were n1 = 300~1500 r/min. The data were recorded every 100 r/min. The LC-8008 acquisition data were set to an analysis frequency of 2 kHz and 1024 sampling points.

5.1. Experiment of Vibration of Faulty Helical Gear Pairs and Normal Helical Gear Pairs

This experiment was used to investigate the difference in vibration between the normal helical gear pair and the faulty helical gear pair under the same conditions. Figure 7 is the acceleration peak change curve of the driving shaft and the driven shaft of the normal helical gear pair and the faulty helical gear pair during the speed-increasing process.

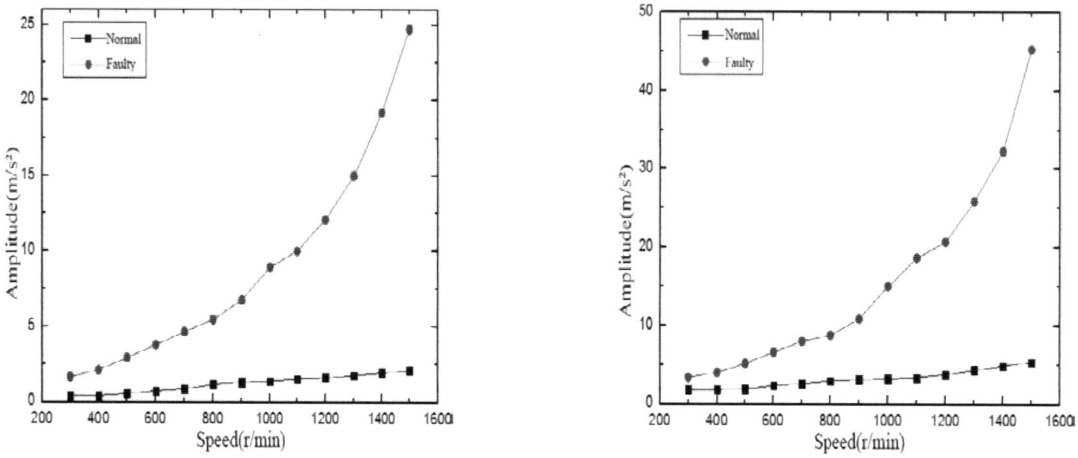

Figure 7. Curves of the amplitude at various speeds of a normal helical gear pair and a faulty helical gear pair under a rigid support

From Fig. 7, it can be observed that the vibrations of the faulty helical gear and the normal helical gear during transmission are very different, and the vibration difference at 300 r/min to 600 r/min is small, which indicates the helical gear transmission. With high stability, but with the continuous increase in speed, the faulty helical gear has a larger vibration, and the normal gear is basically maintained in a small range of vibration and is more stable. Although the smoothness of the helical gear transmission is good, the vibration value increases at high speeds.

5.2. Experiment of ISFD in Damping Test of Faulty Helical Gear pair

In Experiment 5.1, it was found that the vibration amplitude of the faulty gear increases rapidly when the speed is increased. Therefore, the following experiment was conducted to perform the vibration reduction experiment: replace the rigid sleeve for ISFD to repeat the experiment, record the data, and compare the faulty helical gear pair in the rigid sleeve Support and ISFD support. The vibration situation, as shown in Fig. 8 for the faulty helical gear pair of the DRIVING shaft and the driven shaft in the process of increasing the peak acceleration curve.

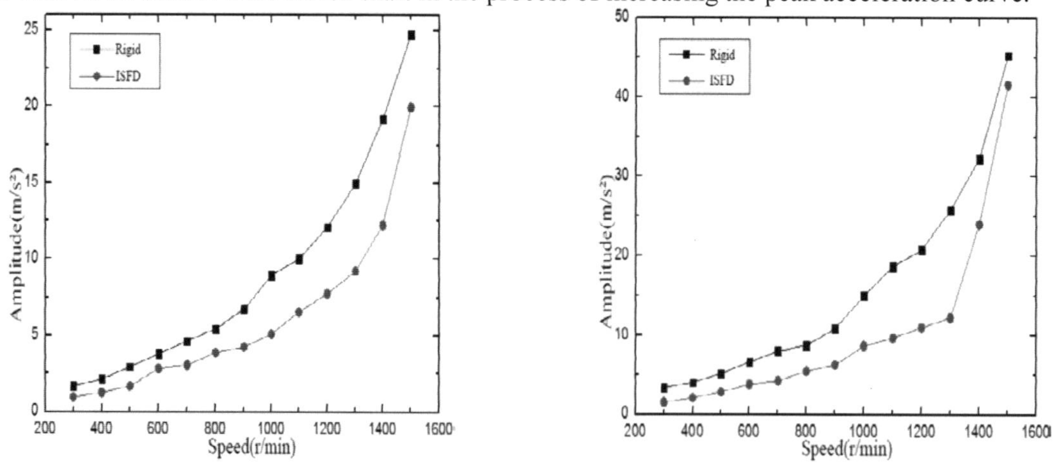

Figure 8. Acceleration curve of amplitude at various speeds with ISFD and rigid support of the faulty helical gear

From the experimental results shown in Fig. 8, the vibration values of the driving shaft and the driven shaft of the rigid sleeve support and the ISFD were reduced compared with those at the respective rotational speeds, and the highest reductions were 36.3% and 52.7%, respectively. From this, it can be concluded that the squeeze film damper can have a better damping effect on the faulty helical gear shaft.

The 1200-r/min time domain signal extraction was analyzed. Figure 9 shows the time domain signal diagram of the 1200r/min driving shaft and driven shaft. The time domain signal was converted into a frequency domain signal by fast Fourier transform, as shown in Figure 10.

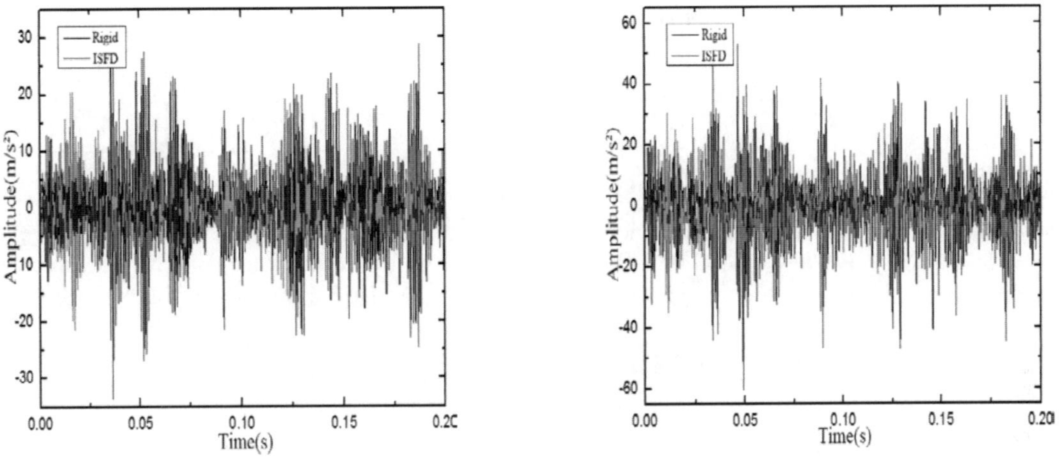

Figure 9. Comparison of two support time domain waveforms for the driving and driven shaft at 1200 r/min

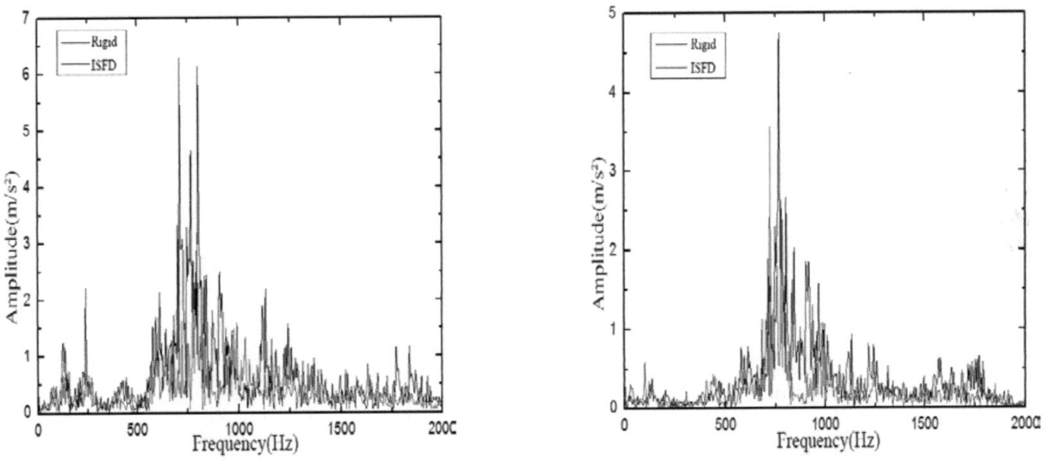

Figure 10. Comparison of two support time domain waveforms for the driving and driven shaft at 1200 r/min

From the time domain waveforms in Fig. 9, it is seen that, under the condition of a rigid sleeve, the periodic phenomenon is not obvious, and a high amplitude appears, mostly during the meshing of helical gears. Impact vibration resulted in waveform modulation. From the structural parameters of the helical gear and the analyzed speed, the meshing frequency of the helical gear pair is found to be f = f1Z1 = f2Z2 = 640 Hz. In the case of the ISFD, the vibration phenomenon of the faulty helical gear pair was significantly improved. The maximum peak value of the driving shaft was reduced from 30 m/s2 to 20 m/s2, the reduction rate was 33.3%, and the maximum peak value of the driven shaft decreased by 60 m/s2. To 30 m/s2, the drop is 50%. In Fig. 10 show that the frequency components are complex, the vibration value is relatively large near the meshing frequency, the frequency band is relatively wide, and a large vibration occurs near 765 Hz. This situation is due to frequency resonance modulation of the basic and gear shafts. According to Fig. 11 and the analysis, the vibration amplitude of the above-mentioned characteristic frequency is extracted, and the vibration reduction effects of the rigid sleeve support and the ISFD of the faulty helical gear are compared, as shown in Table 1.

Table 1: Vibration amplitude of faulty helical gear pairs under different supports

Frequency (Hz)	Acceleration amplitude of driving shaft measuring point (m/s^2)			Acceleration amplitude of driven shaft measuring point (m/s^2)		
	Rigid support	ISFD support	Decreasing amplitude (%)	Rigid support	ISFD support	Decreasing amplitude (%)
640	0.42	0.11	73.8	1.52	0.8	47.3
765	4.75	0.53	88.8	4.66	0.56	88

Table 1 shows the vibration attenuation at the measurement points of the active and driven shaft at frequencies of 640 Hz and 765 Hz. The ISFD has a good damping effect on the faulty helical gear pair, which can significantly improve the resonance modulation phenomenon caused by the characteristic frequency of the faulty helical gear.

5.3. Experiment of ISFD in Damping Test of Normal Helical Gear pair

In the same way, the experiment bench was used to perform an extensive experiment on the normal helical gear pair. As shown in Fig. 11, the amplitude of the driving shaft and the driven shaft of the normal helical gear pair.

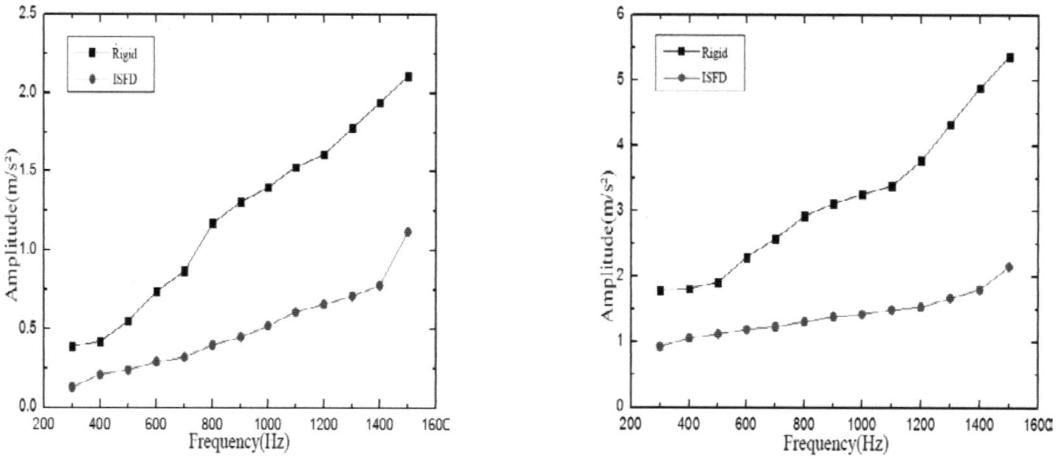

Figure 11. Acceleration curve of amplitude at various speeds with ISFD and rigid support of the normal helical gear

Figure11 shows that the vibration values of the driving shaft and driven shaft of the rigid sleeve support and the ISFD are lower than those of the ISFD at various rotation speeds, with the highest reductions reaching 65.8% and 62.9%, respectively. The squeeze film damper can have a good damping effect on the normal helical gear shaft.Table 3 shows the vibration attenuation at the measuring points of the driving and driven shaft near the three frequencies. The ISFD has a good damping effect on the normal helical gear pair and is insensitive to the type of frequency. It can achieve a good damping effect and can greatly attenuate the modulation phenomenon.

Table 3. Vibration attenuation of different normal helical gear pairs under different supports

Frequency (Hz)	Acceleration amplitude of driving shaft measuring point (m/s^2)			Acceleration amplitude of driven shaft measuring point (m/s^2)		
	Rigid support	ISFD support	Decreasing amplitude (%)	Rigid support	ISFD support	Decreasing amplitude (%)
500	0.12	0.043	64.2	0.71	0.1	85.9
640	0.064	0.015	76.5	0.22	0.03	86.3
1000	0.13	0.04	69.2	0.22	0.06	72.7

6. CONCLUSION

The above experiments showed that ISFD can effectively reduce the transmission of helical gear meshing excitation, and the following conclusions were made:
(1) The helical gear has stability. Under the same conditions, the vibration of the helical gear with low-speed rotation is not much different from the vibration of the normal helical gear, but the amplitude of the faulty helical gear rises significantly after higher-speed rotation.
(2) In the process of increasing the speed, the ISFD can produce damping effects on both the faulty helical gear and the normal helical gear.
(3) ISFD can significantly reduce the modulation of the helical gear coupling system.
(4) ISFD has the function of frequency shifting to the characteristic frequency and can effectively avoid the resonance region.
(5) ISFD is not sensitive to frequency, and it can reduce the vibration of the characteristic frequency in a wide frequency band and achieve good results.

REFERENCES

[1] Liu, F.H., Jiang, H.J., Zhang, L.. Analysis of vibration characteristic for helical gear under hydrodynamic conditions[J]. Advances in Mechanical Engineering, 2017, 9(1):1-9.
[2] Ding, K., Zhu, X.Y, Chen, Y.H.. Typical fault vibration characteristics and diagnosis strategy of gearbox[J]. Journal of Vibration and Shock, 2001, 20(3):7-12.
[3] PG Parker. Dynamic modeling and analysis of tooth profile modification for multimesh gear vibration [J]. Journal of Mechanical Design, 2008, 130(12):1500-8.
[4] RJ Wojcikowski. Gear wheel with vibration damping rings[J]. Acoustical Society of America, 1982, 72(6):2054.
[5] Wang Guozhi, Sun Peilin, Fang Kaixiang. Research on Application of Damping Vibration Reduction Technology in Transmission Gear Device[J]. Journal of Jiangsu University of Science and Technology, 2009,8(3):14-20.
[6] Huang, X.J., He, L.D., Xia, X.R.. Experimental Study on Vibration Suppression of Gear Shaft Systems with Viscous Damper Installation of Driven Shafts[J]. Mechatronic Engineering, 2014, 31(12):1551-1554.
[7] E Guglielmino, KA Edge. A controlled friction damper for vehicle applications[J]. J Control Engineering Practice, 2004, 12(4):431-443.
[8] Chen C S, Natsiavas S, Nelson H D. Coupled lateral-torsional vibration of a gear-pair system supported by a squeeze film damper[J]. Journal of Vibration and Acoustics, 1998, 120(4):860-867.
[9] Chen C S, Natsiavas S, Nelson H D. Stability analysis and complex dynamics of a gear-pair system supported by a squeeze film damper[J]. Journal of Vibration and Acoustics, 1997,119(1):85-88.
[10] K Gjika, C Groves, LS AndréS, GD Larue. Nonlinear Dynamic Behavior of Turbocharger Rotor-Bearing Systems with Hydrodynamic Oil Film and Squeeze Film Damper in Series: Prediction and Experiment[J]. Journal of Computational and Nonlinear Dynamics, 2010,5(4):2040-2049.
[11] Zhu, S.J., Lou, J.J., He, Q.W.. Vibration Theory and Vibration Isolation Technology [M]. Beijing: National Defense Industry Press, 2006.
[12] Chang, L.H.. Research on the General Modeling Method and Dynamic Incentive Influence Law of Parallel Shaft Gear Transmission System Dynamics[D]. Northwestern Polytechnical University, 2014
[13] J Vance, F Zeidan, B Murphy. Machinery Vibration and Rotordynamics[M]. Untied States:Wiley, 2010.
[14] Cai-Wan, Chang-Jian. Bifurcation and chaos analysis of the porous squeeze film damper mounted gear-bearing system[J]. Computers and Mathematics with Applications, 2012, 64(64):798-812.
[15] Santiago D O, Andrés S L. Imbalance response and damping force coefficients of a rotor supported on end sealed Integral Squeeze Film Dampers[C]. International Gas Turbine and Aeroengine Congress and Exhibition. Indianapolis, USA, 1999:1-6.

Modelling the Vibration Response of a Journal Bearing for Condition Monitoring

Osama Hassin[1], Jiaojiao Ma[2,3], Fengshou Gu[3] & Andrew D. Ball[3]
[1]University of Gharyan, Gharyan, Libya.
[2]Hebei University of Technology,China
[3]University of Huddersfield, Queens gate, Huddersfield, UK

ABSTRACT

Journal bearings are widely used to undertake heavy dynamic and static loads in many mission-critical rotating machines such as compressors, engines, turbines and centrifugal pumps. Vibration responses are often used for monitoring their healthy conditions. Vibration characteristics due to mechanical problems such as misalignment, looseness and unbalanced forces are sufficiently understood. However, fluid flow related sources such as fluid instable flows and hydrodynamic forces are less known. Especially the interaction between mechanical and hydrodynamic sources is rarely reported. Aimed at promoting fault diagnosis, this paper presents a comprehensive vibration model for studying vibration response from a journal bearing. It takes into account both conventional hydrodynamic effects and the also self-excitations of asperity collisions and fluid churns. This study successfully calculates both natural frequencies and mode shapes by considering fluid and asperity forces, which are validated based on an in-house test system.

Keywords: journal bearing, mathematical model, internal and external excitations, natural frequencies and condition monitoring.

Corresponding author: Osama Hassin (osamingo@yahoo.com)

1. INTRODUCTION

In journal bearings, the vibration response is often caused by mechanical problems such as misalignment, looseness and unbalanced forces. These problems can be easily identified at shaft frequencies and usefully indicate bearing conditions at late lifespan. To assess bearing condition at different operational phases and identify any early signs of wear or degraded lubricants, vibration mechanisms associated with hydrodynamic forces need to be understood sufficiently. Previous studies show that dynamic forces of oil film can make the shaft oscillate violently. Furthermore, asperity contact between the journal and bearing surfaces can cause random vibrations spreading over a wide frequency band. To understand the vibration phenomena in a journal bearing, a vibration model is developed to include variable excitation forces. Unbalance (F_u) creates an external excitation force. When unbalance occurs, the journal fluctuates inside the bearing which causes low frequency correlated to shaft frequency and its harmonics. The hydrodynamic (F_o) generates self-excitation force correlated to oil coefficients. Asperity contact (F_a) produces an internal self- excitation force. This random force causes corresponding high frequency vibration responses. External and internal excitation forces generate vibration of the journal. In

Figure 1, brown, blue, green and red arrows present unbalanced forces, fluid forces, asperity churn forces and asperity collision forces, respectively. The equations for small amplitude motions of the rotor-bearing system in X direction and Y direction are [1, 2].

$$m\begin{bmatrix}\ddot{x}\\ \ddot{y}\end{bmatrix} + \overbrace{\begin{bmatrix}F_{ox}\\ F_{oy}\end{bmatrix}}^{\text{Viscous Friction Force}} = \overbrace{\begin{bmatrix}F_{ex}\\ 0\end{bmatrix}}^{\text{Static Force}} + \overbrace{\begin{bmatrix}F_{ux}\\ F_{uy}\end{bmatrix}}^{\text{Unbalance Forces}} \tag{1}$$

where m is the mass of the shaft, F_{ox} and F_{oy} are oil film forces at x and y directions, F_e is static load at x direction only, F_{ux} and F_{uy} are unbalanced forces at x and y direction. By adding asperity collision forces F_{ax} and F_{ay}, Equation (1) becomes:

$$m\begin{bmatrix}\ddot{x}\\ \ddot{y}\end{bmatrix} + \overbrace{\begin{bmatrix}F_{ox}\\ F_{oy}\end{bmatrix}}^{\text{Viscous Friction Force}} + \overbrace{\begin{bmatrix}F_{ax}\\ F_{ay}\end{bmatrix}}^{\text{Asperity Friction Force}} = \overbrace{\begin{bmatrix}F_{ex}\\ 0\end{bmatrix}}^{\text{Static Force}} + \overbrace{\begin{bmatrix}F_{ux}\\ F_{uy}\end{bmatrix}}^{\text{Unbalance Forces}} \tag{2}$$

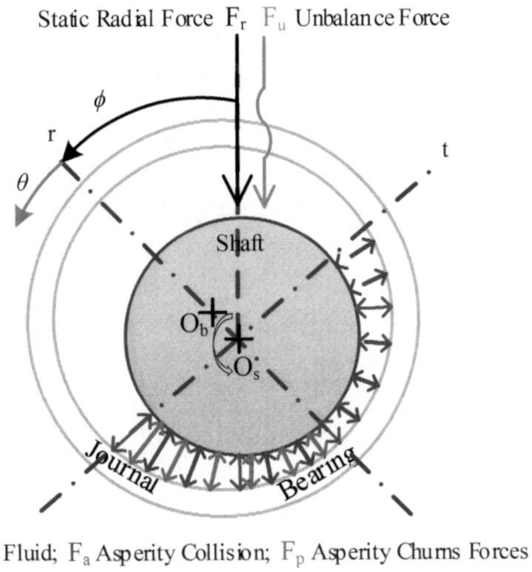

Figure 1. Free body diagram of journal bearing

2. EXTERNAL EXCITATION

Of different external excitations, unbalanced force is the most common one. As illustrated in Figure 2, unbalance can occur when the journal centre mass is not at the centre of the journal O_j.

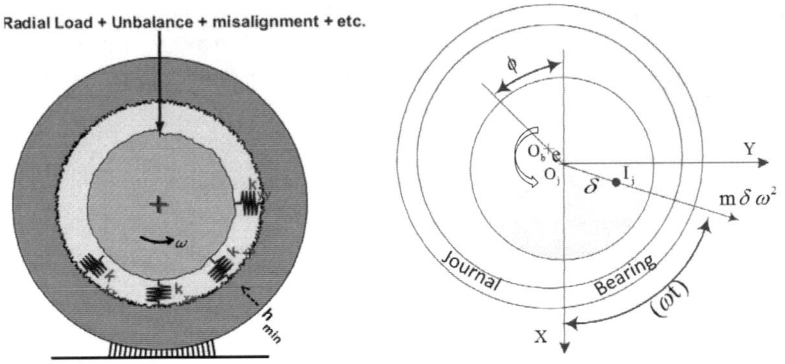

$$\begin{bmatrix}F_{ux}\\ F_{uy}\end{bmatrix} = m\delta\omega^2\begin{bmatrix}cos(\omega t)\\ sin(\omega t)\end{bmatrix} \tag{3}$$

Figure 2. Vibration caused by external excitation and out of shaft roundness [3]

3. INTERNAL EXCITATION

3.1 Fluid stiffness and damping coefficients

Fluid-film forces may be expanded in a Taylor series about the static equilibrium position[2].

$$\begin{aligned} F_x &= F_{x0} + K_{xx}\Delta x + K_{xy}\Delta y + C_{xx}\Delta\dot{x} + C_{xy}\Delta\dot{y} \\ F_y &= F_{y0} + K_{yx}\Delta x + K_{yy}\Delta y + C_{yx}\Delta\dot{x} + C_{yy}\Delta\dot{y} \end{aligned} \tag{4}$$

Fluid film bearing stiffness K_{ij} , $ij = X,Y$ and damping C_{ij} , $ij = X,Y$ force coefficients are defined as

$$K_{ij} = \frac{-\partial F_i}{\partial x_j} \qquad C_{ij} = \frac{-\partial F_i}{\partial \dot{x}_j};\quad i,j = x,y \tag{5}$$

For example, $K_{xy} = \partial F_x / \partial y$ corresponds to a stiffness produced by a fluid force in the X direction due to a journal static displacement in the Y direction. By definition, this coefficient is evaluated at the equilibrium position with other journal centre displacements and velocities set to zero [4].

3.2 Asperity collision excitation

A journal bearing is designed to operate under the hydrodynamic regime. Unfortunately, it often operates under mixed or boundary regimes due to abnormal operating conditions during transient operations and fault cases like oil leakages, oil degradations and worn surfaces [5], etc. In these two regimes asperity collisions occur and result in increased vibration. Figure 3 illustrates that asperity contact can be expressed as a source of vibrations in journal bearing. It shows that one of the asperity pairs has a larger degree of collisions, which produces higher transient vibrations; whereas the smaller contact produces lower transient responses. The stiffness of a single asperity contact pair is built and calculated based on shoulder-shoulder asperity contact [6], as shown in Figure 4.

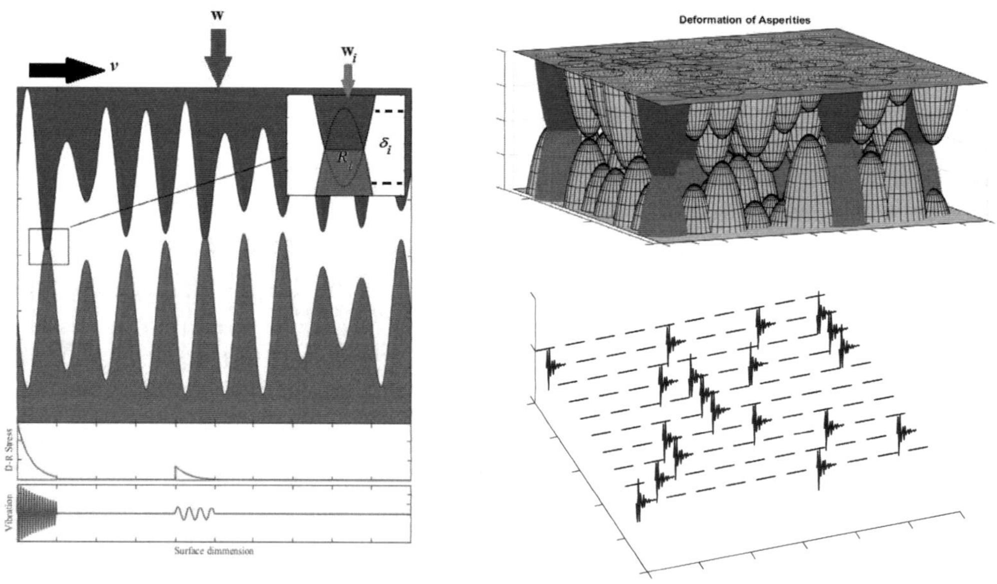

Figure 3. Asperity deformation caused by collisions [3]

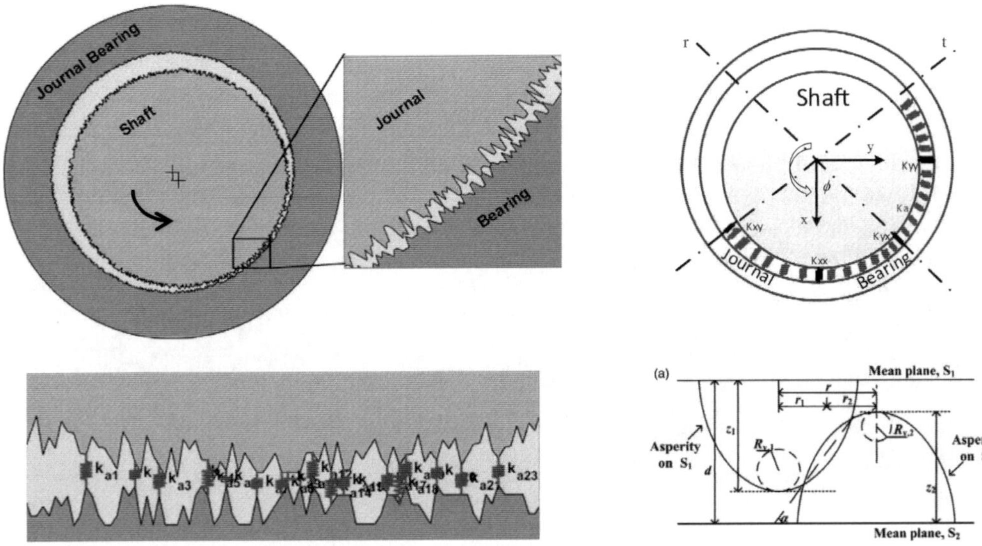

Figure 4. Vibration caused by asperity collisions [3, 6]

3.3 Asperity churn excitation

For high speeds and high oil viscosities but low radial loads, a bearing operates under hydrodynamic lubricant regime in which the bearing surfaces are separated by oil film and under this condition the wear is minimal. However, the fluid shear stress is high and consequently can cause high friction in the forms of heat generation and vibration. This type of vibration can be understood to be the effect of fluid shear stress which causes an alternation between surface asperity deformation and reclamation because the shear stress fields are not uniform i.e. random turbulent close to bearing surfaces [5]. To simplify the modelling of asperity churns, all the asperities are assumed to be as a cantilever beam. Figure 5 shows the vibration modes and maximum deflections of asperities while the oil flows at zone load.

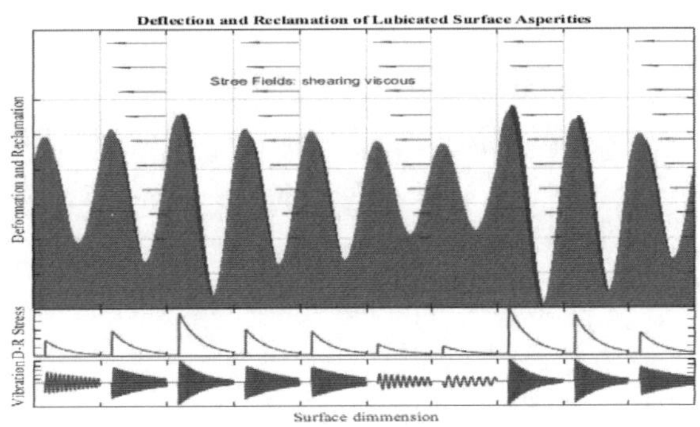

Figure 5 Asperity churns [3]

In the mixed lubrication regime, vibration is caused mainly by asperity-asperity collisions and asperity churns. The total elastic energy of asperities is gathering asperity collisions and asperity churns [3].

$$U_{total} = U_{asperity\,churn} + U_{asperity\,collisions} \tag{6}$$

4. EQUATION OF MOTION

In designed operating conditions, mainly there is not any asperity contact between the journal spins and the bearing because the oil film completely separates their surfaces. By decreasing the oil film thickness, metal to metal contact occurs and asperity stiffness affects the motion.

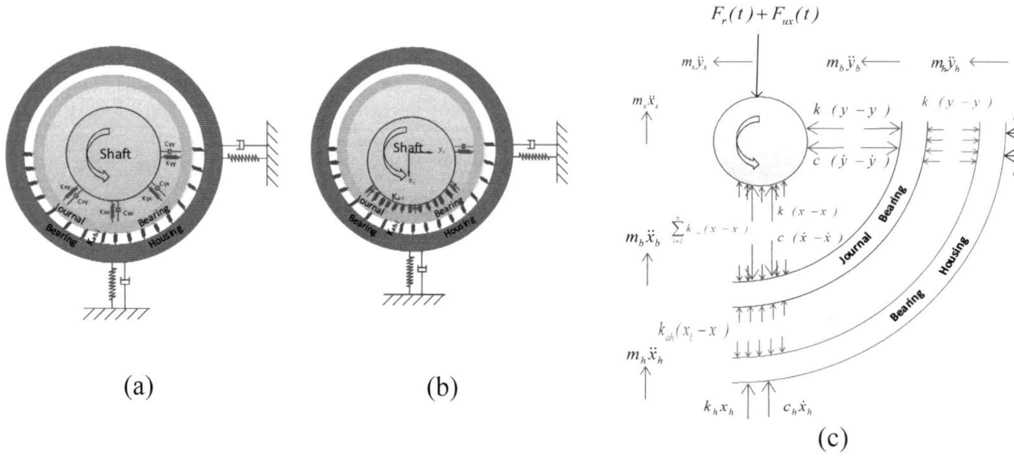

Figure 6. Vibration model. a) Considers fluid coefficients (hydrodynamic regimes). b) Considers fluid coefficients and asperity contact (boundary regimes). c) Free body diagram of a journal bearing

1 $$m_s\ddot{x}_s + c_{xx}(\dot{x}_s - \dot{x}_b) + k_{xx}(x_s - x_b) + c_{xy}(\dot{y}_s - \dot{y}_b) + k_{xy}(y_s - y_b) + \sum_{i=1}^{n} c_{a\_i}(\dot{x}_s - \dot{x}_b) + \sum_{i=1}^{n} k_{a\_i}(x_s - x_b) = F_r(t) + F_{ux}(t) \quad (7)$$

2 $$m_b\ddot{x}_b - c_{xx}(\dot{x}_s - \dot{x}_b) - k_{xx}(x_s - x_b) - c_{xy}(\dot{y}_s - \dot{y}_b) - k_{xy}(y_s - y_b) - \sum_{i=1}^{n} c_{a\_i}(\dot{x}_s - \dot{x}_b) - \sum_{i=1}^{n} k_{a\_i}(x_s - x_b) + c_{xah}(\dot{x}_b - \dot{x}_h) + k_{xah}(x_b - x_h) = 0 \quad (8)$$

3 $$m_h\ddot{x}_h - c_{xah}(\dot{x}_b - \dot{x}_h) - k_{xah}(x_b - x_h) + c_h\dot{x}_h + k_h x_h = 0 \quad (9)$$

4 $$m_s\ddot{y}_s + c_{yy}(\dot{y}_s - \dot{y}_b) + k_{yy}(y_s - y_b) - c_{yx}(\dot{x}_s - \dot{x}_b) - k_{yx}(x_s - x_b) + \sum_{i=1}^{n} c_{a\_i}(\dot{y}_s - \dot{y}_b) + \sum_{i=1}^{n} k_{a\_i}(y_s - y_b) = F_{uy}(t) \quad (10)$$

5 $$m_b\ddot{y}_b - c_{yy}(\dot{y}_s - \dot{y}_b) - k_{yy}(y_s - y_b) + c_{yx}(\dot{x}_s - \dot{x}_b) + k_{yx}(x_s - x_b) - \sum_{i=1}^{n} c_{a\_i}(\dot{y}_s - \dot{y}_b) - \sum_{i=1}^{n} k_{a\_i}(y_s - y_b) + c_{yah}(\dot{y}_b - \dot{y}_h) + k_{yah}(y_b - y_h) = 0 \quad (11)$$

6 $$m_h\ddot{y}_h - c_{yah}(\dot{y}_b - \dot{y}_h) - k_{yah}(y_b - y_h) + k_h y_h + c_h\dot{y}_h = 0 \quad (12)$$

where m_s, m_b and m_h are masses of shaft, bearing and housing, k_{xx}, k_{yy}, k_{xy} and k_{yx} are fluid film bearing stiffness coefficients, c_{xx}, c_{yy}, c_{xy} and c_{yx} are damping force coefficients of fluid film, k_{a_i} and c_{a_i} are stiffness and damping coefficients of asperity collisions between bearing and shaft, k_{xah} and c_{xah} are stiffness and damping coefficients of asperities between bearing and its house, and k_h and c_h are house stiffness and damping coefficients.

State space has been used to estimate the linear system, and its equations are presents as

$$\dot{x}(t) = Ax(t) + Bu(t) \qquad \begin{bmatrix} \dot{x}_1 \\ \vdots \\ \dot{x}_n \end{bmatrix} = A \begin{bmatrix} x_1 \\ \vdots \\ x_n \end{bmatrix} + B \begin{bmatrix} u_1 \\ \vdots \\ u_n \end{bmatrix} \qquad A = \begin{bmatrix} [0] & [I] \\ [K(-M^{-1})] & [C(-M^{-1})] \end{bmatrix}$$

$$y(t) = Cx(t) + Du(t) \qquad \begin{bmatrix} y_1 \\ \vdots \\ y_n \end{bmatrix} = C \begin{bmatrix} x_1 \\ \vdots \\ x_n \end{bmatrix} + D \begin{bmatrix} u_1 \\ \vdots \\ u_n \end{bmatrix} \qquad B = \begin{bmatrix} [0] & [0] \\ [0] & [M^{-1}] \end{bmatrix}$$

In x direction, $x_1 = x_s$, $x_2 = x_b$ and $x_3 = x_h$ are positions of shaft, bearing and house masses, respectively. And $x_7 = \dot{x}_s$, $x_8 = \dot{x}_b$ and $x_9 = \dot{x}_h$ are velocities of shaft, bearing and house, respectively. In y direction, $x_4 = y_s$, $x_5 = y_b$ and $x_6 = y_h$ are positions of shaft, bearing and house masses, respectively. And $x_{10} = \dot{y}_s$, $x_{11} = \dot{y}_b$ and $x_{12} = \dot{y}_h$ are velocities of shaft, bearing and house masses, respectively.

5. MODEL RESULTS

Based on journal bearing parameters and state space technique, eigenvalues have been calculated that determine the nature frequencies of the system. As the result of Eigen analysis, the natural frequencies of the shaft and the bearing are around 6 kHz and 13 kHz. The result bellow, in Figure 7, presents vibration response of a journal bearing under multiple operating conditions, including speed 1500 rpm, six different radial loads (1, 5, 10, 20, 40 bar) and three different oil viscosities (5, 15, 37). The vibration model has been simplified based on the value of eccentricity ration. At low eccentricity ratio, the excitation force is only correlated to fluid. At high eccentricity ratio, the excitations become more significant and correlation to both fluid and asperity forces, when the oil film thickness becomes much smaller than the size of the surface roughness.

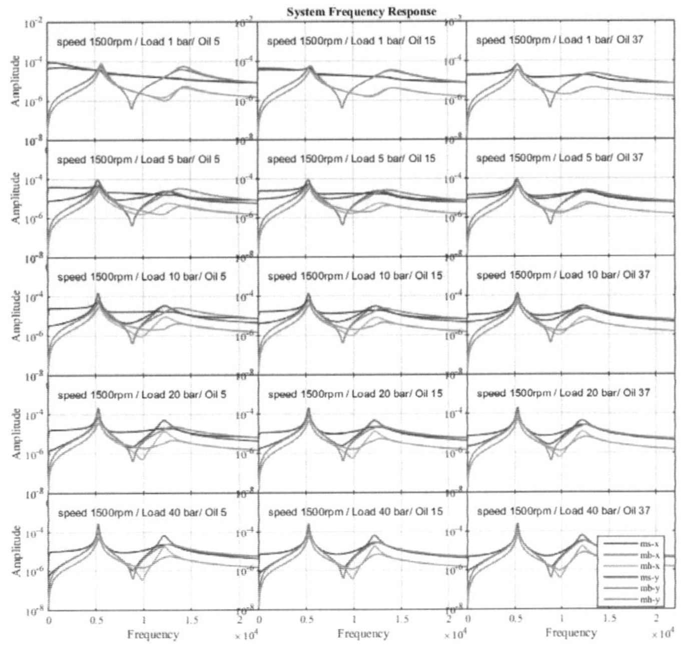

Figure 7. Natural frequency modes of multiple operating conditions

Figure 8 presents six modes, each mode shows the vibration shapes of the shaft, bearing and house masses in X and Y axis. This also presents the nature mode of vibration under different loads and oil viscosities. Figure 9 presents natural frequency for each mode under different operating conditions. Natural frequencies increase while the load increase but oil viscosity decrease.

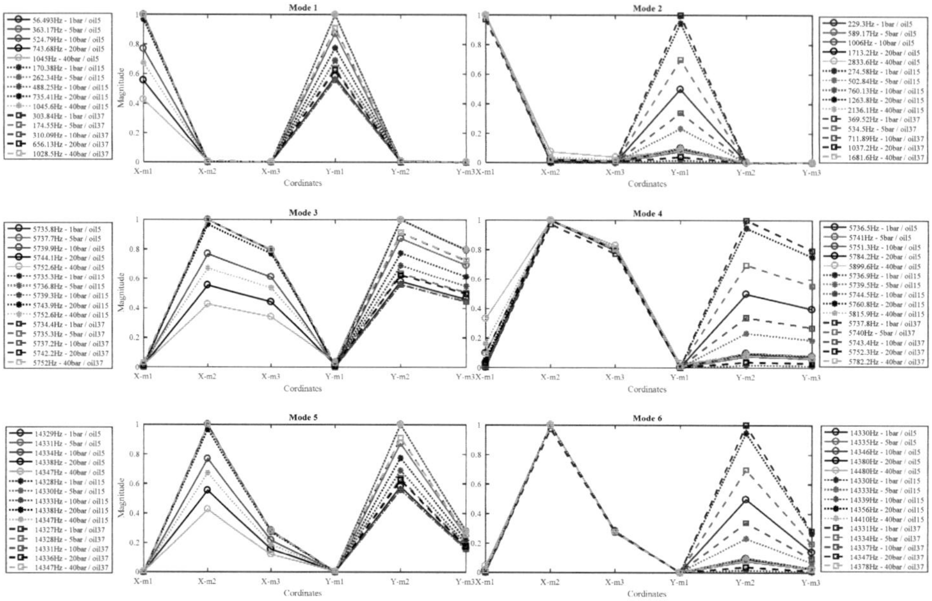

Figure 8. Natural modes of vibration

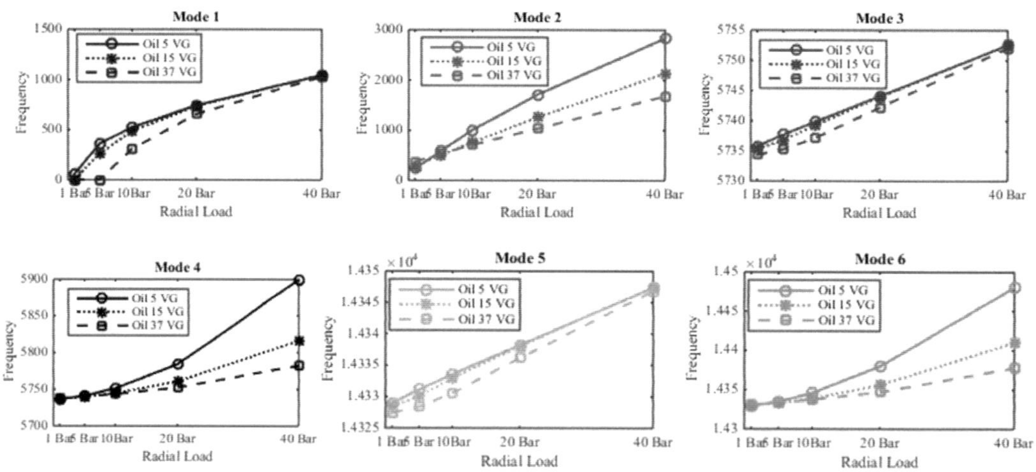

Figure 9. Natural frequency Vs radial load for three different oil viscosities

6. EXPERIMENTAL RESULTS

The test rig of journal bearing consisted two self-aligning spherical journal bearings (SA35M), electrical motor and load system exerted radial load on the shaft supported between bearings. Also measurement instrumentation transducers were mounted. Four accelerometer sensors are fixed to collect the vibration signals. And, an encoder and a pressure sensor are placed to measure the output rotating speed and radial load, respectively. The journal bearing has been tested to validate the vibration response model. The experiments were carried out to test the spectrum of three different lubrication types under different radial loads at 1500 rpm rotating speed.

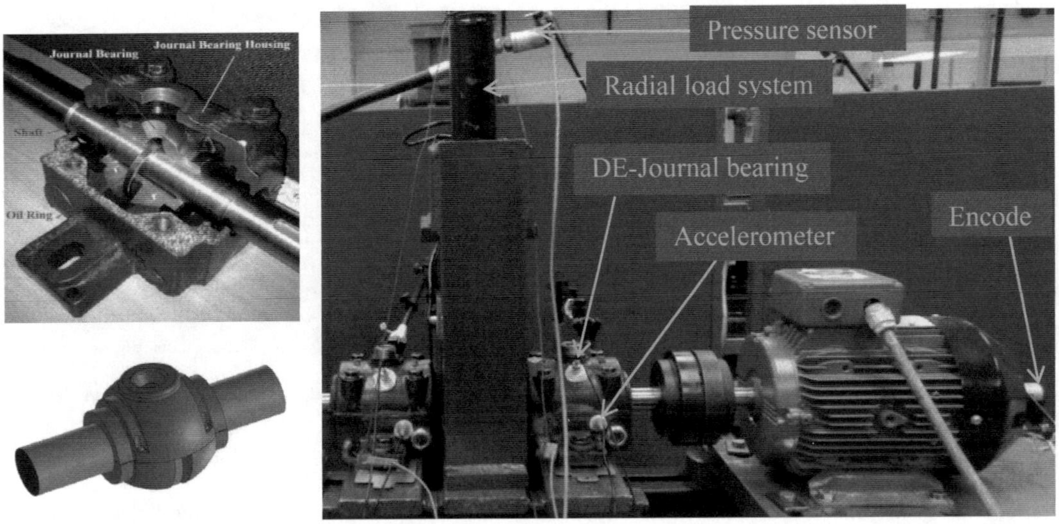

Figure 10. Self-aligning journal bearing companies and test rig [5]

Figure 11 presents spectrums of all cases of experiments. The spectrum in the frequency domain of different lubrications can be regarded to have three sub-bands: 3.5kHz-5.5kHz, 5.5kHz-7.5kHz and 7.5kHz-11kHz, according to the distinction of the spectral amplitudes. In each band, spectrum exhibit continuous profiles due to the random excitations of asperity churns and collisions. In addition, these bands are magnified by the nonlinear transfer paths of structural resonances, showing that they can be validated by mathematical model. Moreover, the frequency amplitude in each band increases with the increasing radial loads. This change clearly agrees with the vibration response model results in Figure 9.

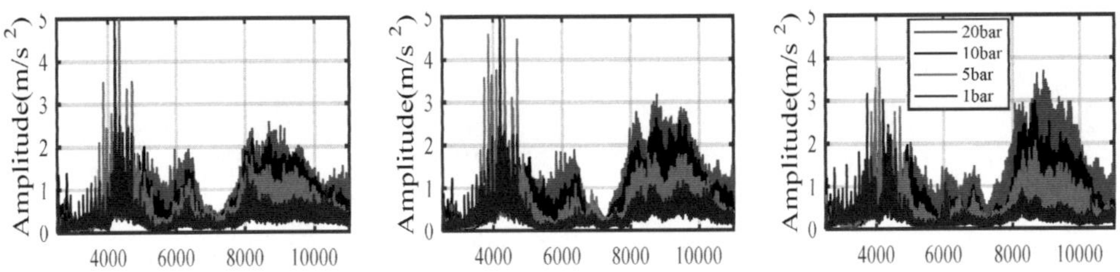

Figure 11. Experimental spectrums of three lubrications (5, 15, 37) under different radial loads

7. CONCLUSION

A mathematical model is established to carry out an in-depth bearing vibration responses. This model considers not only conventional fluid forces but also asperity collisions and churn effects, allowing explaining high frequency responses of the lubricated journal bearing to be the effect of micro asperity elastic deformations. The models with six degrees of freedom are solved by using the state space technique, each eigenvalue and volume of a certain mass in one direction present natural and shape modes. The natural frequencies of the system are found to be around 6 kHz and 13 kHz. Additionally, as a result of asperity collisions which increase with higher radial loads or lower viscous oil, the natural frequencies of the journal bearing can shift higher. An experimental comparison was made by vibration signal spectrums with different oil types and different radial loads. It has found that the structural resonances in frequency range bands can reflect the excitations of asperity churns and asperity collisions, and the vibration amplitudes exhibit significant increase. These results are validated based on lab experimental results.

REFERENCES

[1] Singhal, S., *A Simplified Thermohydrodynamic Stability Analysis of the Plain Cylindrical Hydrodynamic Journal Bearings*. 2004, Louisiana State University.

[2] Javorova, J., B. Sovilj, and I. Sovilj-Nikic, *On the derivation of dynamic force coefficients in fluid film bearings*. 2009, Monograph of FTS "Machine design", Novi Sad.

[3] Hassin, O., *Condition Monitoring of Journal Bearings for Predictive Maintenance Management Based on High Frequency Vibration Analysis*. 2017, Huddersfield.

[4] Andres, L.S., *Hydrodynamic fluid film bearings and their effect on the stability of rotating machinery*. 2006, DTIC Document.

[5] Hassin, Osama, Yao, Aiying, Wei, Nasha, Gu, Fengshou and Ball, Andrew. *Monitoring Oil Levels Of Journal Bearings Based On The Analysis Of Vibration Signals*. 2016.

[6] Bin Zhao, Song Zhang, Jia Man, Qing Zhang and Yan Chen. *A modified normal contact stiffness model considering effect of surface topography*. Proceedings of the Institution of Mechanical Engineers, Part J: Journal of Engineering Tribology, 2014: p. 1350650114558099.

A Study of the Influence of Time-varying Meshing Stiffness on Dynamic Response in Gear Transmission Systems

Yinghui Liu[1], Chen Zhang[1], Zuolu Wang[1], Dong Zhen[1,*], Hao Zhang[1], Zhanqun Shi[1]
[1]. School of Mechanical Engineering, Hebei University of Technology, Tianjin, 300140, P.R.China

ABSTRACT

Broken teeth are one of the main faults in the gear transmission system. The existing models could not accurately reflect the dynamic features of broken teeth in the fault diagnosis of gear transmission system. When broken teeth appear, the broken teeth will influence on the time-varying meshing stiffness. Therefore, optimizing the parameter of dynamics model system can improve the accuracy of the fault feature extraction for the fault diagnosis of gear transmission system. This paper mainly studies the influences of broken teeth on the time-varying meshing stiffness. Through optimizing the time-varying mesh stiffness, a gear mechanism system dynamic model is built for the simulation analysis. It aims to validate the optimized parameter setting method can more accurately reflect the characteristics of the gear fault. The analysis results show that when the failure of the broken tooth occurs, the time-varying meshing stiffness presents instantaneous shock, which is the function of the meshing frequency, and is also related to the size of the broken tooth. By studying the time-varying meshing stiffness of the broken teeth, it can provide more reliable fault information for fault diagnosis of gear transmission system.

Key words: broken gear teeth; time-varying stiffness; dynamical model; gear-mesh frequency; fault diagnosis

1. Introduction

It is well known that the undulation of the time-varying mesh stiffness is one of the main motivations that cause unwished vibration and noise in gear transmission systems. The influence of meshing stiffness on vibration has been analyzed in previous study for different types of gear system. And the influence of these factors on the vibration characteristics of the system is also studied when the nonlinearity associated with radial, dynamic backlash and time-varying meshing stiffness considered simultaneously. When the gears fail, such as broken teeth, the vibration of the system will aggravate and it will also affect the time-varying meshing stiffness. Therefore, in order to better diagnose the gear failure, it is necessary to study the influence rule on the vibration characteristics of the system at different speeds when the teeth are broken based on the consideration of time-varying meshing stiffness.

The study of the time-varying meshing stiffness is an important research field in gear system with gear failure. Many researchers have conducted various mathematical methods to calculate the time-varying meshing stiffness, and these models can be classified into four categories: finite element (FE) method, analytical method (AM), analytical -FE method and experimental method [1]. Based on the model of the time-varying meshing stiffness of the health gear system established by Yang and Lin, many time-varying meshing stiffness models for fault gear system are optimized. For the time-varying meshing stiffness of gear pair with tooth faults like root crack, spalling, breakage and pitting, Chaari et al. [2] revealed the influences of spalling and breakage on mesh stiffness and dynamic response of spur gear transmission by potential energy method. Saxena and Parey [3] revealed the influences of spalling and friction on the time-varying meshing stiffness of spur gear.

Kahraman et al.[4,5] put forward a gear system nonlinear dynamic model for the first time ,which the backlash and radial clearance were both considered, and then the model was validated by experiment, and analyzed the characteristics of the dynamic response of the gear system in the meantime. Byrtus et al. [6]. studied the influence of clearance and time-varying meshing stiffness on the vibration characteristics of the gear system. Suohuai Zhang et al. [7,8] analyzed the vibration characteristics of the gear system by using the multi-gap gear model. Yahui Cui et al. [9,10] studied the influence rule of backlash and time-varying meshing stiffness on the gear rotor system by using the experimental facility with adjustable backlash. Later, Huibo Zhang et al. [11,12] established the model of radial clearance of bearing by using the two-state model, proposed the modeling method of dynamic backlash and dynamic meshing force, and took the influence of time-varying meshing stiffness into consideration.

In order to better simulate the actual state of gear system and provide more reliable theoretical support for the fault diagnosis of gear transmission system, it is necessary to study the vibration characteristics of gear transmission system when there is a broken-tooth gear. When the gear has broken teeth, it will have a great influence on the time-varying meshing stiffness function of the gear. Because the changing of radial clearance of the bearing will affect the backlash, it is necessary to build the radial clearance model. In this study, the radial clearance model of bearing is established by using the two-state model, at the same time, the broken tooth fault is added in the model and the influence of dynamic backlash is considered.

2. **Multi-gap fault gear - rotor system dynamics model**

In the multi-clearance gear rotor system, the clearance mainly includes the backlash of gears and radial clearance of the bearing. In the actual gear transmission system, these gaps are inevitable, which can help lubricate gears and bearings, and also change the topology of the system, thus changing the vibration state of the gear system. In the actual gear transmission system, the broken teeth are a main fault, so it is necessary to study the vibration characteristics of the dynamic system when the gear tooth is broken. In this paper, both of the gear tooth failure and multi-clearance are considered. Firstly, a two-state model is used to establish the radial clearance model of the bearing and the dynamic backlash model of the gear. Finally, the dynamic model of gear rotor system considering gear tooth broken, backlash, radial clearance and time-varying meshing stiffness is established.

2.1 Bearing radial clearance model

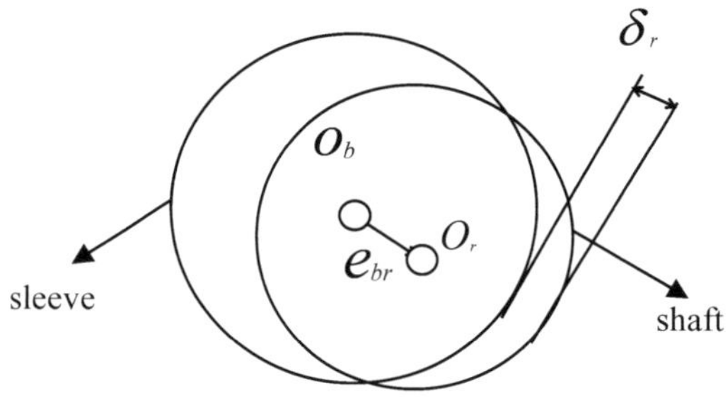

Figure 1. The bearing radial clearance model

The radial clearance of the bearing is the movement gap in the radial direction of the bearing as shown in fig. 1. In the model, it is simplified as shaft and sleeve. O_b is the center of the sleeve. O_r is the center of the shaft. e_{br} represents the vector from O_b to O_r. Supposing the radial clearance between the shaft and the sleeve is δ_r. When the gap exists, there are two states of contact and separation between the shaft and the sleeve. The radial clearance function is introduced to describe the current state of the system :

$$f_r(e_{br})=\begin{cases}|e_{br}|-\delta_r & |e_{br}|>\delta_r\ , \\ 0 & |e_{br}|=\delta_r \\ 0 & |e_{br}|<\delta_r\end{cases} \quad (1)$$

In the formula (1), e_{br} is the radial displacement between the shaft center and the sleeve center. If e_{br} is within the range of $|e_{br}|>\delta_r$, there is a free movement between the shaft and the sleeve; If e_{br} is within the range of $|e_{br}|=\delta_r$, there is a critical contact between the shaft and the sleeve. If e_{br} is within the range of $|e_{br}|<\delta_r$, there is a collision force between the shaft and the sleeve[11].

2.2 Dynamic backlash model

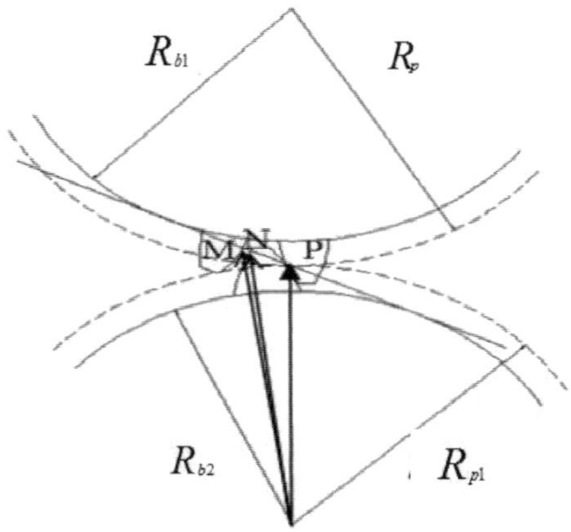

Figure 2. The backlash model

The backlash is the gap between the two gears in the direction of the meshing line when the gears mesh. When there is a radial clearance, it causes the radial run-out of the gear shaft, and the backlash will change accordingly as shown in fig. 2. When the two gears mesh at point P, the distance between M and N at the direction of the meshing line is the dynamic backlash. According to the geometrical relationship of involute, the dynamic backlash can be expressed as:

$$b_t = \left(\frac{2\pi R_p}{Z_p} - s_p - s_g\right)\cos\alpha' + b_0, \qquad (2)$$

In the formula (2), R_p represents the radius of the active gear pitch circle. Z_p represents the number of the active gear tooth. S_p and S_g represent the pitch circle tooth thickness of the active gear and driven gear respectively. b_0 is the initial backlash. α' is the actual mesh Angle and can be expressed as:

$$\alpha' = \arccos\left(\frac{A_0}{A_R}\cos\alpha_0\right), \qquad (3)$$

In the formula (3), A_0 represents ideal center distance for gear standard meshing. A_R is the actual center distance when there is a radial beat. α_0 is the presdsure angle for gear standard meshing.

Combining (2) with (3), the dynamic backlash can be written as:

$$b_t(t) = 2A_0\cos(\alpha_0)(inv(\alpha') - inv(\alpha_0)) + b_0, \qquad (4)$$

In the formula (4), inv is the involute function which can be expressed as $inv\alpha' = \tan\alpha' - \alpha'$ and $inv\alpha_0 = \tan\alpha_0 - \alpha_0$.

The relative displacement of the gear along the meshing line can be defined as:

$$g_t(t) = R_P'\cdot\theta_P - R_g'\cdot\theta_g + e_{brx}\cdot\sin\alpha' + e_{bry}\cdot\cos\alpha', \qquad (5)$$

In the formula (5), R_p' is the actual radius of the active gear pitch circle. θ_p is the active gear rotation angle. R_g' is the actual radius of the driven gear pitch circle. θ_g is the driven gear rotation angle. e_{brx} is the component on the X-axis of e_{br}. e_{bry} is the component on the Y-axis of e_{br}.

Therefore, combining (4) and (5), the function of the backlash can be expressed as:

$$f_t(t)=\begin{cases} g_t(t) & g_t(t)\geq 0 \\ 0 & -b_t(t)<g_t(t)<0 \\ g_t(t)+b_t(t) & g_t(t)\leq -b_t(t) \end{cases},\qquad(6)$$

If $g_t(t)$ is within the range of $g_t(t)\geq 0$, the two gears are in the contact state of the teeth surface, belonging to normal meshing. If $g_t(t)$ is within the range of $-b_t(t)<g_t(t)<0$, the two gears are in the separate state of the teeth surface. If $g_t(t)$ is within the range of $g_t(t)\leq -b_t(t)$, the two gears are in the contact state of the back surface of the teeth[12].

When the contact ratio is less than 2, the meshing cycle of the gears contains two teeth meshing period and single tooth meshing period. The relationship between τ_d and τ_m can be defined as:

$$\lambda=\tau_d/\tau_m,\qquad(7)$$

In the formula (7), τ_d is the two teeth meshing period and τ_m is the whole the meshing cycle of the gears.

When time is contained between 0 and $\lambda\cdot\tau_m$, the gears keep two teeth meshing. When time is contained between $\lambda\cdot\tau_m$ and τ_m, the gears keep single teeth meshing [13].

The relationship between λ and the contact ratio ε can be defined as:

$$\lambda=\varepsilon-1,\qquad(8)$$

The ratio of the time of the gear fracture part in the double tooth meshing period can be written as:

$$i=\frac{\sqrt{r_{a1}^2-r_b^2}-\sqrt{r_x^2-r_b^2}}{p_b\cdot\varepsilon\cdot(\varepsilon-1)},\qquad(9)$$

In the formula, r_{a1} is the tooth tip radius of active gear. r_b is base circle radius of active gear. r_x is broken tooth effective circle radius of active gear. p_b is the normal pitch. ε is gear coincide degree.

Therefore, the time of gear fracture part in the double tooth meshing period can be calculated by:

$$t_i=\tau_d\cdot i,\qquad(10)$$

In the process of meshing, the dynamic meshing force can be expressed as :

$$F_t(t)=K_t(t)\cdot f_t(t)+C\cdot \dot{g}_t(t),\qquad(11)$$

In the formula (11), $K_t(t)$ is time-varying meshing stiffness. C is meshing damping. $\dot{g}_t(t)$ is the relative velocity along the meshing direction.

Time-varying meshing stiffness changes along time and when there is broken teeth, $K_t(t)$ can be calculated by:

$$K_t(t)=\begin{cases} K_m+K_a\cos(\omega_n t+\varphi) & 0\leq t\leq\frac{(\varepsilon-1)\cdot 2\pi}{\varepsilon\cdot\omega_n}-t_i,\ \frac{2\pi}{\varepsilon\cdot\omega_n}\leq t\leq\frac{2\pi}{\omega_n} \\ K_m+K_a\cos(\omega_n(t+t_i)+\varphi) & \frac{(\varepsilon-1)\cdot 2\pi}{\varepsilon\cdot\omega_n}-t_i\leq t\leq\frac{(\varepsilon-1)\cdot 2\pi}{\varepsilon\omega_n} \end{cases},\qquad(12)$$

In the formula, K_m is average meshing stiffness. $K_m^{'}$ is time-varying meshing stiffness amplitude. ω_n is meshing frequency. φ is meshing phase angle. ε is the contact ratio.

2.3 Dynamic model of multi-gap gear-rotor system.

The dynamic model of multi-gap coupling gear rotor system was established by using the radial clearance model and dynamic backlash model. In the model, the active gear is connected with the motor without considering its radial run out. The function of the dynamic model of the gear rotor system can be expressed as:

$$\begin{cases} I_p \cdot \ddot{\theta}_p + \left(K_t(t) \cdot f_g(t) + C_t(t) \cdot \dot{g}_t(t)\right) \cdot R_p^{'} = T_p \\ \left(K_t(t) \cdot f_g(t) + C_t(t) \cdot \dot{g}_t(t)\right) \cdot R_g^{'} - I_g \cdot \ddot{\theta}_g = T_g \\ m_g \cdot \ddot{\delta}_x - (K_r \cdot f_{rx}(e_{brg})^n + C_r \cdot \dot{\delta}_x) = 0 \\ m_g \cdot \ddot{\delta}_y - (K_r \cdot f_{ry}(e_{brg})^n + C_r \cdot \dot{\delta}_y) = 0 \end{cases} \quad \text{..........(13)}$$

In the formula (13), θ_p and θ_g respectively represent the angular displacement of the active and driven gears. I_p and I_g respectively represent the moment of inertia of the active and driven gears. T_p is the driving torque. T_q is the load torque. $K_t(t)$ is time-varying meshing stiffness. C is meshing damping. $\ddot{\delta}_x$ and $\ddot{\delta}_y$ are the acceleration component of X and Y in the global coordinate system of the radial acceleration. m_g is the driven gear mass. $\dot{\delta}_x$ and $\dot{\delta}_y$ are the speed component of X and Y in the global coordinate system of the radial speed. K_r is the contact stiffness between shaft and sleeve. N is force index. C_r is nonlinear damping between shaft and sleeve.

3. Analysis of vibration characteristics of the gear-rotor system.

Because of the existence of the radial clearance and the backlash, the gear rotor system will produce radial vibration and torsional vibration. The torsional vibration is caused by the meshing of the gears which is related to the gear speed and the number of the gear teeth. The radial vibration is due to the collision between the shaft and the sleeve that exist the radial clearance between them. In this model, a broken tooth is added and the time-varying meshing stiffness is adopted to characterize the torsional vibration. The gear parameters used in the model is presented in table 1.

Table 1:Gear parameters

name	value	name	value
Active gear teeth	20	Active gear mass(kg)	0.784
Driven gear teeth	40	Driven gear mass(kg)	3.137
Module of gear(mm)	4	Driven gear moment of inertia(kg*m2)	0.00339
Gear pressure angle(°)	20	Active gear moment of inertia(kg*m2)	0.05123

3.1 The influence of gear speed on vibration

According to the above parameters, the dynamic model of gear rotor system with multi-gap coupling was established. But the dynamic equation is a high dimensional nonlinear equation. Therefore, the dynamic equation is solved by Newmark method. In this model, the radial clearance is 1mm, and the backlash is 0.16mm. When the active gear speed is 60r/min, 90r/min and 120r/min, the vibration acceleration spectrum of the driven gear is shown in figure 3.

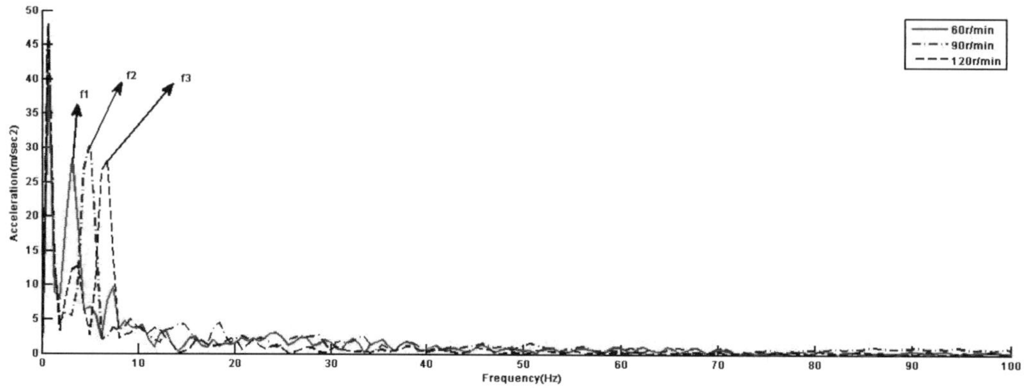

Figure 3. The vibration acceleration spectrum of the driven gear at different speed

In the vibration acceleration spectrum of the driven gear, the x-coordinate is the frequency, and the y-coordinate is the acceleration amplitude. It can be seen from the figure that the spectrum has two peaks at every speed. The first peak that caused by radial vibration of the bearing is the same. The peak f1 which appears at the speed of 60r/min is due to the broken tooth, and the corresponding frequency is 3.05 Hz. The peak f2 which appears at the speed of 90r/min is due to the broken tooth, and the corresponding frequency is 4.88 Hz. The peak f3 which appears at the speed of 120r/min is also due to the broken tooth, and the corresponding frequency is 6.71 Hz. Therefore, the frequency caused by the broken tooth is approximately proportional to the speed of the gear. But because of the low frequency resolution, the frequency and speed are not strictly proportional. In terms of numerical value of the spectrum, the amplitude of radial vibration is much larger than the vibration caused by meshing and the vibration caused by the broken tooth. The vibration caused by the broken tooth is great larger than that of the meshing vibration. Therefore, the meshing frequency and its octave are not recognized. It is indicated that the radial clearance of the bearing has the greatest influence on the vibration of the system, and the vibration of the system will increase sharply when the teeth break. And when the speed increases, the frequency caused by the broken tooth would increase correspondingly.

3.2 The influence of time-varying meshing stiffness on vibration

In order to study the influence of time-varying meshing stiffness on vibration, keep the active gear speed unchanged, and when the meshing stiffness was increased into 10^7N/m, the vibration spectrum of the driven gear was shown in figure 4.

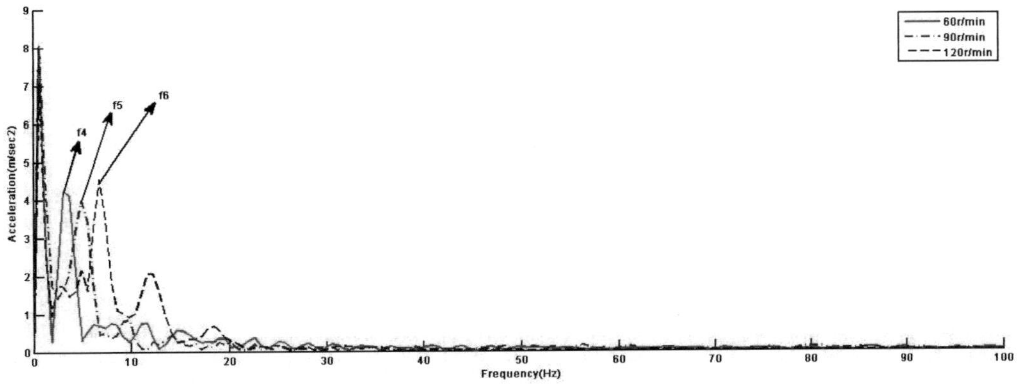

Figure 4. The vibration acceleration spectrum of the driven gear at different speed

It can be seen from the figure 4 that the spectrum has two peaks at every speed. The first peak that caused by radial vibration of the bearing is the same. The peak f4 is due to the broken tooth at the speed of 60r/min, and the corresponding frequency is 3.05 Hz. The peak f5 is due to the broken tooth at the speed of 90r/min, and the corresponding frequency is 4.88 Hz. The peak f6 is also due to the broken tooth at the speed of 120r/min, and the corresponding frequency is 6.71 Hz. When the meshing stiffness was increased, the frequency caused by the broken tooth is also approximately proportional to the speed of the gear. In terms of numerical value of the spectrum, the amplitude of radial vibration is much larger than the vibration caused by meshing and the vibration caused by the broken tooth. The vibration caused by the broken tooth is great larger than that of the meshing vibration. Therefore, the meshing frequency and its octave are not recognized. It is indicated that the radial clearance of the bearing has the greatest influence on the vibration of the system, and the vibration of the system will increase sharply when the teeth break. And when the speed increases, the frequency caused by the broken tooth would increase proportionally.

According to the figure3 and figure4, the acceleration amplitude of the driven gear will decrease when the stiffness increases.

3.3 The comparison of different mesh stiffness at the same speed

When the active gear speed is 60r/min and 90r/min, the vibration at different mesh stiffness was compared in figure5 and figure6.

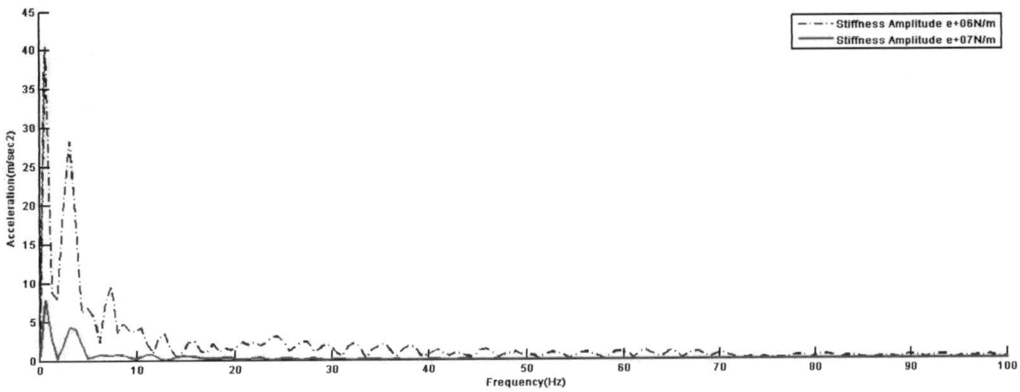

Figure 5. The vibration acceleration spectrum of the driven gear at the speed of 60r/min

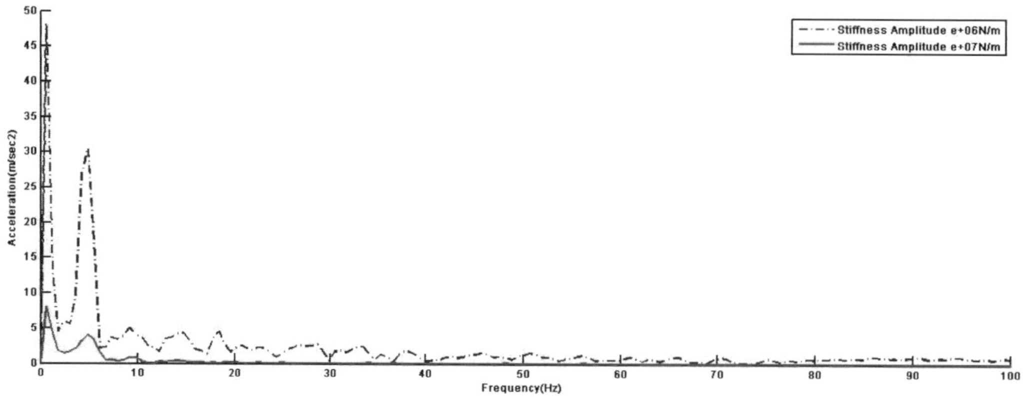

Figure 6. The vibration acceleration spectrum of the driven gear at the speed of 90r/min

It can be seen from the figure 5 and figure 6 that when the meshing stiffness is increased, the acceleration amplitude of the driven gear will decrease greatly.

4. Conclusion

The dynamic model of gear rotor system with multi-gap coupling is established in this study. In addition, the broken tooth fault is added to the system. By changing the time-varying meshing stiffness, reflect the gear meshing condition when the broken tooth exists. Then the relationship between the vibration characteristics of the fault system and the time-varying meshing stiffness is studied. By comparing the simulation results under different gear speed it can be know that there is a relationship between the fault frequency and speed. Also it shows that the time-varying meshing stiffness function is reasonable, but the parameters choice of the model is still exists room for improvement and needs further research.

ACKNOWLEDGMENT

This research was supported by the National Natural Science Foundation of China (Grant no. 51605133; 51705127), Hebei Science and Technology Projects of China (Grant no. 17394303D) and Hebei Provincial High-level Personnel Funding (Grant no. E2014100015).

REFERENCES

[1] Hui Ma, Xu Pang, Ranjiao Feng, Jin Zeng, Bangchun Wen.Improved time-varying mesh stiffness model of cracked spur gear[J].Engineering Failure Analysis,2015,55:271-287.
[2] Fakher Chaari, Walid Baccar, Mohamed Slim Abbes, et al..Effect of spalling or tooth breakage on gear mesh stiffness and dynamic response of a one-stage spur gear transmission. Eur. J. Mech. A.Solids 2008,27:691-705.
[3] Ankur Saxena, Anand Parey, Manoj Chouksey. Time varying mesh stiffness calculation of spur gear pair considering sliding friction and spalling defects. Eng. Fail. Anal. 2016,70:200-211.
[4] Kahraman A, Singh R. Non-linear dynamics of a geared rotor-bearing system with multiple clearances[J].Journal of Sound and Vibration, 1991, 144(3):469-506.
[5] Blankenship G W, Kahraman A. Steady state forced response of a mechanical oscillator with combined parametric excitation and clearance type non-linearity[J].Journal of Sound and Vibration, 1995, 185(5):743-765.
[6] Byrtus M, Zeman V. On modeling and vibration of gear drives influence by nonlinear couplings[J].Mechanism and Meshine Theory, 2011, 46:375-397.
[7] Suohuai Zhang, Yiping Li, Damou Qiu. Study on nonlinear dynamic characteristics of a geared rotor bearing system[J].Chinese Journal of Mechanical Engineering, 2001,37(9)53-61.
[8] Suohuai Zhang, Yunwen Shen, Damou Qiu.On response of mass unbalance in a geared rotor bearing system[J].Chinese Journal of Mechanical Engineering, 2002, 38(6):51-55.

[9] Yahui Cui, Zhansheng Liu, Jianhuai Ye.Dynamic response of geared roter system and the effect of clearance on jump characteristics of amplitude[J].Journal of Mechanical Engineering, 2009, 45(7):7-15.
[10] Yahui Cui, Zhansheng Liu, Jianhuai Ye, et al..Bifurcation and chaos of gear transmission system with clearance subjected to internal and external excitation[J].Journal of Mechanical Engineering, 2010, 46(11):129-136.
[11] Huibo Zhang, Jian Tian, Jun Zhou, et al..Dynamic and Experimental Investigation of Gear-rotor System with Multiple Clearances Coupled[J].Journal of Mechanical Engineering, 2017, 53(11).
[12] Huibo Zhang, Ran Wang, Zikun Chen, et al..Nonlnear dynamic analysis of a gear-rotor system with coupled multi-clearance[J].Journal of Vibration and Shock, 2015, 34(8).
[13] J.-H.Kuang, A.-D.Lin. Theoretical Aspect of Torque Responses in Spur Gearing due to Meshing Stiffness Vatriation[J]. Mechanical Systems and Processing, 2003, 17(2):255-271.

An Outer Ring Fault Quantitative Diagnosis Method of Ball Bearing Based on the Detail Impact Characteristic

Lingli Cui [1], Jinfeng Huang [2], Huaqing Wang[3], Xin Wang[2]
[1] Key Laboratory of Advanced Manufacturing Technology, Beijing University of Technology, Beijing 100124, China
[2] Beijing Engineering Research Center of Precision Measurement Technology and Instruments, Beijing 100124, China
[3] Beijing University of Chemical Technology, Beijing 100029, China

ABSTRACT

Outer ring faults that differ in size by natural multiples of the ball angular spacing will have the same time interval, which may lead to misdiagnosis. Aiming at this problem, the dynamic model of ball bearing is established in this paper. And the vibration acceleration of the bearing system with outer ring fault that differ in size by natural multiples of the ball angular spacing is simulated. Analysis of the simulation signal shows that the bearing systems with outer ring fault that differ in size by natural multiples of the ball angular spacing have the similar impact characteristics. This phenomenon will easily lead to misdiagnosis of the quantitative diagnosis method based on the time interval. Therefore, the dynamic model is optimized. In this model, the effect of the degree of freedom of balls on the vibration acceleration is considered, and then the new features which can distinguish these two faults is theoretically predicted and proposed. More importantly, experimental results show that the proposed new features can distinguish these two faults effectively, thus expanding the applicability of the diagnosis method of the fault of the bearing outer ring.

Keywords: Ball bearing; outer ring fault; quantitative diagnosis; dynamic model; impact characteristic

Corresponding author: Ling li Cui (acuilingli@163.com)

1. INTRODUCTION

Ball bearings are critical mechanical components that determine the health of machinery and its remaining lifetime in modern production machinery [1]. A defect in such a bearing, unless detected in time, causes malfunction and may even lead to catastrophic failure of the machinery [2]. From this reason, it is significant to study the fault diagnosis of a ball bearing [3].

With the development of bearing fault diagnosis, bearing fault diagnosis technology and methods have made great breakthroughs. In terms of signal processing, most researches on bearing fault diagnosis focus on the presence or absence of faults and types of faults. Zhu et al. [4] used the Laplace wavelet correlation filtering method to effectively identify the cyclic period of fault characteristic. Jia et al. [5] achieved the accurate identification of bearing fault types based on a morphological demodulation method of improved singular spectral decomposition. This is not enough for preventive and maintenance of bearings. Only by mastering the severity of the failure can it be possible to effectively prevent the occurrence of equipment downtime and save production costs and maintenance costs. The Ref. [6] also mentioned the need to achieve a breakthrough from qualitative diagnosis to quantitative diagnosis. Therefore, it is of great significance to realize quantitative diagnosis of faults.

Domestic and foreign scholars have achieved certain results in quantitative diagnosis. In order to obtain the bearing fault diagnosis mechanism, scholars established a dynamic model of the bearing, simulated the vibration response of the bearing system under different conditions, and analyzed the fault mechanism. Cao et al. [7] used the dynamic model of roller bearings to simulate the vibration response of the inner and outer rings of the bearing under different conditions. Yuan et al. [8] established a multi-body dynamics model of defective bearings, and analyzed the vibration signal of local fault and composite faults. Singh et al. [9] established the finite element model of bearing and obtained the corresponding relationship between contact force and defects. Therefore, this paper established the dynamic model of the bearing system and simulated the vibration acceleration response of the bearing system under different sizes.

2. NONLINEAR DYNAMIC MODEL OF BEARING SYSTEM

To obtain the rolling bearing vibration response characteristics in different severities and to explore its mechanism, this section from the dynamic and kinematic point of view, a dynamic model of the rolling bearing system is built. Basing on the Hertz contact theory, the contact load F_j can be derived:

$$F_j = K\delta_j^{\,n} \tag{1}$$

where K represents the load deflection factor of bearing, the contact deformation δ_j of the jth ball can be expressed as:

$$\delta_j = \begin{cases} A(\phi_j(t)) - A_0, & \delta_j > 0 \\ 0, & \delta_j \leq 0 \end{cases} \tag{2}$$

where A_0 and A are the unloaded and loaded relative displacement between the outer and inner ring groove curvature centers, respectively.

When a bearing has an rectangular outer ring fault of circumferential $\Delta\phi_f$ (as shown in Figure1), depth h, and the fault center position at ϕ_f, in this instance, the effective depth $d(\phi_j)$ for the fault can be derived as[11]:

$$d(\phi_j) = \min(h, r_b - \sqrt{r_b^2 - 0.25R_O^2(0.5\Delta\phi_f - |\phi_j - \phi_f|)^2}) \tag{3}$$

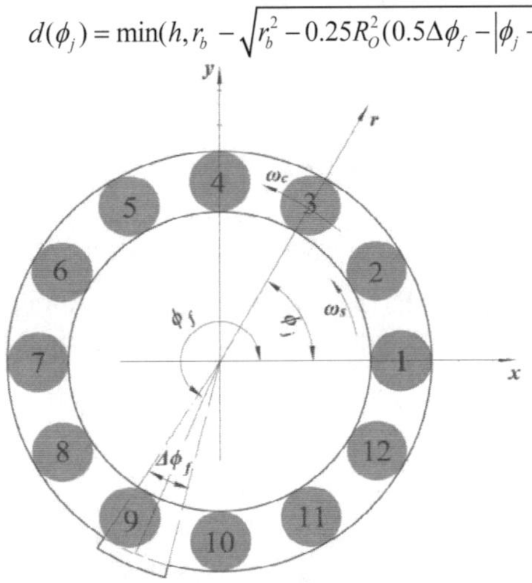

Figure1. The schematic diagram of rolling bearing with outer ring fault

where Ro is the outer ring radius, r_b is the radius of balls. The loaded relative displacement A can be expressed as:

$$A(\phi_j(t)) = \sqrt{\{A_0\cos\alpha_0 + \delta_{rj}\}^2 + \{A_0\sin\alpha_0 + \delta_{zj}\}^2} \tag{4}$$

where radial δ_{rj} and axial δ_{zj} deformations are calculated as follows:

$$\delta_{rj} = (x_i - x_o)\cos\phi_j + (y_i - y_o)\sin\phi_j - r_L - d(\phi_j)\cos\alpha_0 \tag{5}$$

$$\delta_{zj} = \delta_z + r_d(\theta_x\sin\phi_j - \theta_y\cos\phi_j) - d(\phi_j)\sin\alpha_0 \tag{6}$$

where r_l is the bearing radial clearance. (x_i, y_i) and (x_o, y_o) represent the displacement of the inner ring and outer ring, respectively. α_o is the unloaded contact angle.

The total contact forces in the x, y and z directions between balls and rings can be given by:

$$\sum_{j=1}^{N_b} F_j \begin{bmatrix} \cos\alpha_j\cos\phi_j \\ \cos\alpha_j\sin\phi_j \\ \sin\alpha_j \end{bmatrix} = \sum_{j=1}^{N_b} \begin{bmatrix} F_{xj} \\ F_{yj} \\ F_{zj} \end{bmatrix} \tag{7}$$

where the load contact angle α_j is calculated as follows:

$$\alpha_j = \arctan\left(\frac{\{A_0 \sin\alpha_0 + \delta_{zj}\}}{\{A_0 \cos\alpha_0 + \delta_{rj}\}}\right) \tag{8}$$

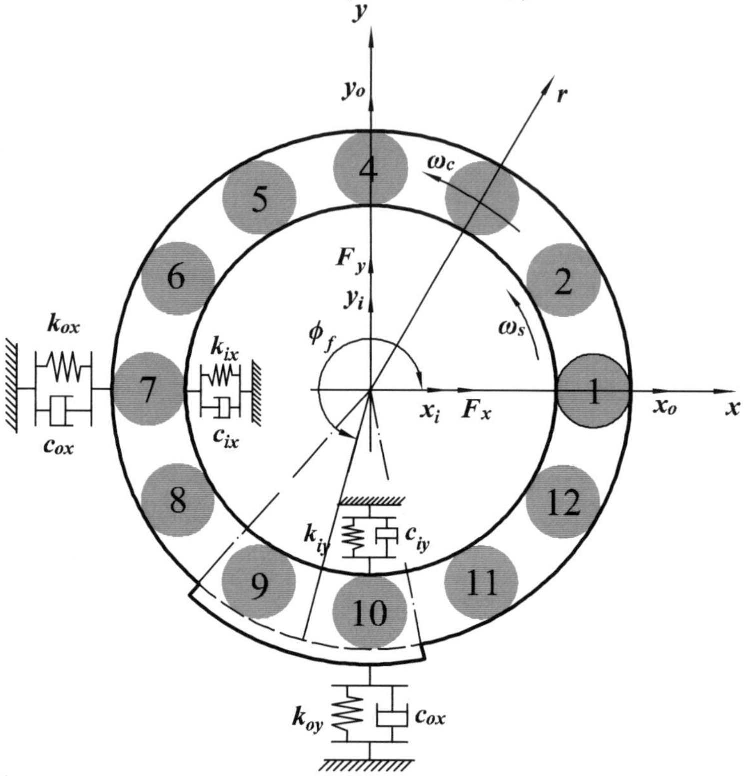

Figure 2.The nonlinear dynamic model of rolling bearing with outer ring fault

The nonlinear dynamic model of the rolling bearing system is shown in Figure 2. Based on the nonlinear dynamic model of the bearing, the dynamic differential equation of the rolling bearing is given by:

$$\begin{cases} m_o\ddot{x}_o + c_{ox}\dot{x}_o + k_{ox}x_o = \sum_{j=1}^{N_b} F_{xj} \\ m_o\ddot{y}_o + c_{oy}\dot{y}_o + k_{oy}y_o = \sum_{j=1}^{N_b} F_{yj} \\ m_i\ddot{x}_i + c_{ix}\dot{x}_i + k_{ix}x_i = F_x - \sum_{j=1}^{N_b} F_{xj} \\ m_i\ddot{y}_i + c_{iy}\dot{y}_i + k_{iy}y_i = F_y - \sum_{j=1}^{N_b} F_{yj} \end{cases} \tag{9}$$

where m_i, k_i and c_i are the mass, the stiffness and damping of inner ring and shaft, x_i and y_i are the displacements. m_o, k_o and c_o are the mass, the stiffness and damping of outer ring and bearing pedestal, x_o and y_o are the displacements.

3. VIBRATION RESPONSE OF BEARING SYSTEM

The bearings studied in this paper are NSK6308 ball bearings with eight balls. The angle extents of the outer ring faults are set to $\Delta\phi_f$ =7.4° and 52.4°, these faults differ in size by an integer number of the ball angular spacing. The Hertz contact load factor K=5453MN/m$^{1.5}$. The radial load in the horizontal and vertical directions of the bearing is F_x=0N, F_y=-100N respectively, set the shaft rotation speed f_s=7Hz and the sampling frequency F_s=65536Hz.

Figure 3 (a) and (b) shows the simulated vertical vibration response for $\Delta\phi_f$ = 7.4° and 52.4°, respectively. Combining with the theoretical analysis, it can be seen that when the ball enters or leaves the fault area, the acceleration will abruptly change and there will be shock oscillating, namely the impact 1 and the impact 2 in Figure 3(a). From the comparison of the results in Figure 3, it can be seen that the time interval between the impact 1 and impact 2 in the vibration response of $\Delta\phi_f$ =7.4° is almost equal to the time interval of $\Delta\phi_f$ =52.4°. Therefore, the two faults have the same impact characteristics in the time domain. So the quantitative diagnosis method based on the time interval may be misdiagnosed. Ref. [10] proposed a third feature method to deal with this problem. This method can effectively distinguish these faults with the same impact time interval located at 270°. Considering that in the actual engineering case, the outer ring fault angle position may appear near 270°, so this paper carries on further research to this question.

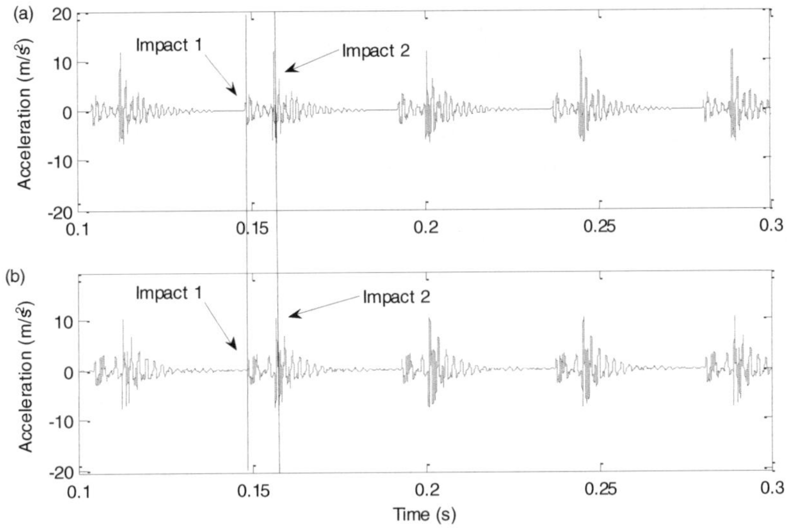

Figure 3. Simulated vertical vibration response for $\Delta\phi_f$=7.4° and 52.4°: (a) $\Delta\phi_f$=7.4°; (b) $\Delta\phi_f$=52.4°

4. MECHANISM ANALYSIS BASED ON IMPROVED BEARING SYSTEM MODEL

4.1. Analysis of bearing simulated vibration response

Based on the concern of above problem, firstly, this part will optimize the nonlinear dynamic model of bearing system shown in Figure 2. The dynamic model of Figure 2 does not consider the degree of freedom of balls, but in the actual system, the ball has independent freedom and inertia force. In Ref. [11], an improved dynamic model of the bearing system is introduced. The model is considered the degree of freedom and inertia of the ball based on the model of equation (9), and its nonlinear dynamic equation is as follows:

$$m_i\begin{bmatrix}\ddot{x}_i\\ \ddot{y}_i-g\end{bmatrix}+\begin{bmatrix}F_{i,x}+F_{di,x}\\ F_{i,y}+F_{di,y}\end{bmatrix}=\begin{bmatrix}0\\ -W\end{bmatrix}$$

$$m_o\begin{bmatrix}\ddot{x}_o\\ \ddot{y}_o-g\end{bmatrix}+\begin{bmatrix}c_{s,x}\dot{x}_o\\ c_{s,y}\dot{y}_o\end{bmatrix}+\begin{bmatrix}k_{s,x}x_o\\ k_{s,y}y_o\end{bmatrix}+\begin{bmatrix}F_{o,x}+F_{do,x}\\ F_{o,y}+F_{do,y}\end{bmatrix}=\begin{bmatrix}0\\ 0\end{bmatrix} \qquad (10)$$

$$m_b\ddot{\rho}_{i,r}-m_b\rho_{i,r}w_c^2+m_b g\sin\phi_{i,r}-\{f_r\}=0$$

where f_r is the contact force, and p_i is the radial coordinates of each ball.

Based on the optimized dynamics model of bearing system, the simulated vibration acceleration signal is shown in Figure 4. The results shown in Figure 4 indicates that the impact 1 and the impact 4 correspond respectively to the moment of the ball entry and exit fault area, and the two impacts are similar to the impact of the entry and exit fault area obtained by the equation 1. In addition, except for the two impacts, there are small impacts between the two impacts, such as the impact 2 and impact 3 shown in Figure 4, which cannot be obtained by the equation (9). Due to the existence of faults, the gap between the ball and the inner and outer raceway increases, resulting in the bouncing phenomenon of the ball in the raceway, causing the small impacts in the vibration response. This phenomenon will only appear in larger faults, while small faults such as 7.4° will not appear. For the small size fault, the ball does not contact with the bottom of the raceway in the fault area, and then rolled out the edge of the fault, so there is no bouncing phenomenon. According to the impact characteristics, it can further be used to distinguish these outer ring faults that differ in size by an integer number of the ball angular spacing.

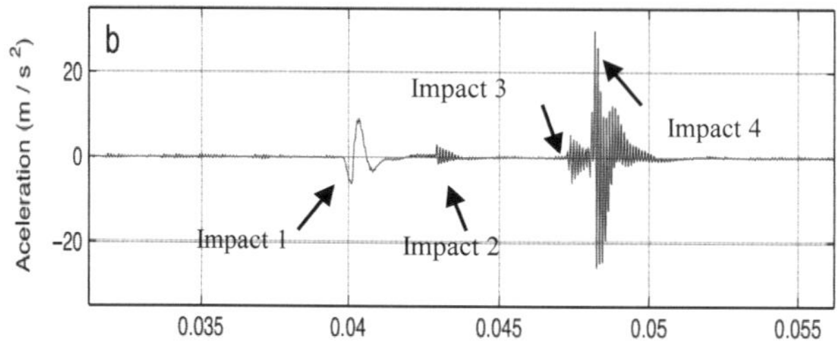

Figure 4. Simulated vibration response for Ref.[11]

4.2. Analysis of experimental data of bearing

In order to verify the feasibility of the impact characteristic proposed in this paper for distinguishing these outer ring faults that differ in size by an integer number of the ball angular spacing, the experiment was carried out. The test rig is shown in Figure 5. In this experiment, the outer ring fault of $\Delta\phi_f$ =7.4° and 52.4° in NSK6308 deep groove ball bearing is processed by wire cutting, and the fault bearing is shown in Figure 6. Rotational frequency is 7Hz, the sampling frequency is set to 65536Hz. The measurement system is NI USB4432, the sensors is HD-YD-232, and the Butterworth filter is adopted.
The experimental signals of $\Delta\phi_f$ =7.4° and 52.4° are shown in Figure 7. The results shown in Figure 7 show that only two impacts (impact 1 and impact 2 in Figure a) appear in the signal of the small size fault $\Delta\phi_f$ =7.4°, and four impacts appear in the signal of $\Delta\phi_f$ =52.4°, which are similar to the simulation conclusions in the last section. Therefore, the detail impact characteristics can be used to distinguish these outer ring faults that differ in size by an integer number of the ball angular spacing from the time domain signal.

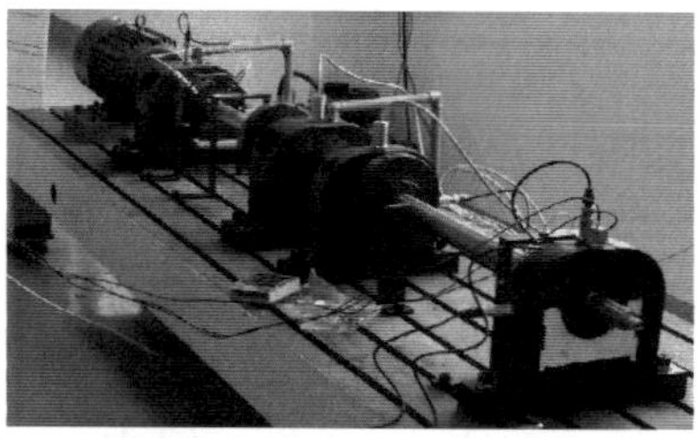
Figure 5. Photos of the test rig

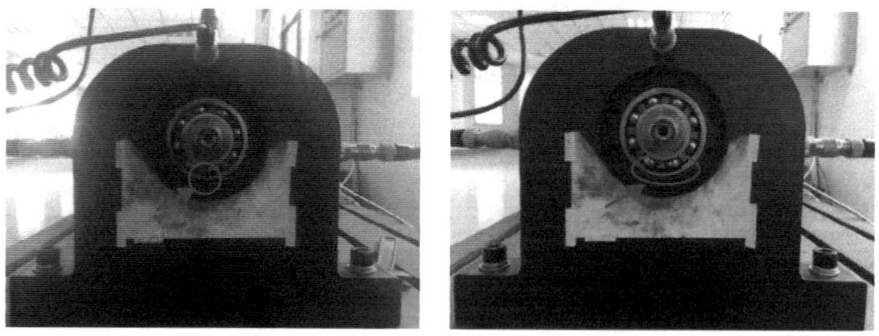
Figure 6. Defective bearing with fault size of $\Delta\phi_f$=7.4° and 52.4°: (a) $\Delta\phi_f$=7.4°; (b) $\Delta\phi_f$=52.4°

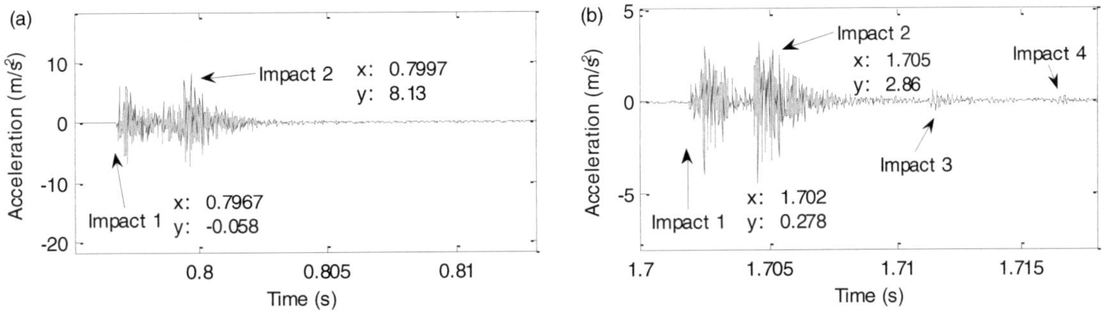

Figure 7. Measurement vibration response for $\Delta\phi_f$=7.4° and 52.4°: (a) $\Delta\phi_f$=7.4°; (b) $\Delta\phi_f$=52.4°

5. CONCLUSION

In this paper, a dynamic model of bearing system with outer ring faults is established. Based on this model, the vibration acceleration of the ball bearing system is simulated with outer ring faults that differ in size by an integer number of the ball angular spacing. Through the comparison of the simulation signals, it is known that these outer ring faults that differ in size by an integer number of the ball angular spacing have the same impact interval. This phenomenon will lead to the diagnosis method based on the impact time interval. To solve this problem, the dynamic model is further optimized. From the simulation results, it is known that there are small impacts in the vibration response of the large size fault except for the double impact, and the double impact phenomenon only occurs in the signal of small size fault. Then the new characteristics can be used to distinguish these faults that differ in size by an integer number of the ball angular spacing theoretically. The experimental results show that the new features can effectively distinguish these faults that differ in size by an integer number of the ball angular spacing.

ACKNOWLEDGMENTS

This work is supported by the National Natural Science Foundation of China (Grant No. 51575007 and No. 51675035). The authors would like to thank the editors and anonymous reviewers for their helpful comments and constructive suggestions.

REFERENCES

[1] El-Thalji I., Jantunen E., A summary of fault modelling and predictive health monitoring of rolling element bearings[J]. Mechanical Systems & Signal Processing, 2015, 60-61(1):252-272.
[2] Tandon N, Choudhury A. A review of vibration and acoustic measurement methods for the detection of defects in rolling element bearings [J]. Tribology International, 1999, 32(8):469-480.
[3] Cui L., Huang J., Zhang F., Quantitative and Localization Diagnosis of a Defective Ball Bearing Based on Vertical-Horizontal Synchronization Signal Analysis [J]. IEEE Transactions on Industrial Electronics, 2017, 64(11): 8695 - 8706.
[4] Wang S., Zhu Z., Wang A., Bearing fault feature detection based on parameter identification of transient impulse response [J]. JOURNAL OF VIBRATION ENGINEERING, 2010, 23(4):445-449.
[5] Yan X. Jia M., Morphological Demodulation Method Based on Improved Singular Spectrum Decomposition and Its Application in Rolling Bearing Fault Diagnosis [J]. Journal of Mechanical Engineering,2017, 53 (7):104-112.
[6] Wang G., He Z., Chen X., Lai Y., Basic Research on Machinery Fault Diagnosis—What is the Prescription [J]. Journal of Mechanical Engineering，2013,49(1): 63-72.
[7] Cao M., Xiao J., A comprehensive dynamic model of double-row spherical roller bearing—Model development and case studies on surface defects, preloads, and radial clearance[J]. Mechanical Systems & Signal Processing, 2008, 22(2):467-489.
[8] Xing Y., Zhang Y., Zhu Y., et al. Fault degree evaluation for rolling bearing combining backward inference with forward inference[J]. Journal of Zhejiang University, 2012, 46(11):1960-1967.
[9] Singh S., Köpke U., Howard C., et al. Analyses of contact forces and vibration response for a defective rolling element bearing using an explicit dynamics finite element model [J]. Journal of Sound & Vibration, 2014, 333(21):5356-5377.
[10] Petersen D., Howard C., Prime Z., Varying stiffness and load distributions in defective ball bearings: Analytical formulation and application to defect size estimation [J]. Journal of Sound & Vibration, 2015, 337:284-300.
[11] Ahmadi A., Petersen D., Howard C., A nonlinear dynamic vibration model of defective bearings – The importance of modelling the finite size of rolling elements [J]. Mechanical Systems & Signal Processing, 2015, 52-53(1):309-326.

Ageing Models for Lightly Loaded Distribution Power Transformers

M.Z. Mukuddem[1], J.L. Jooste [2]
[1] Department of Industrial Engineering, Stellenbosch University, South Africa
[2] Department of Industrial Engineering, Stellenbosch University, South Africa

ABSTRACT

Power transformers form an integral part of an electricity network. These assets allow for power to be transformed from different voltage levels in an efficient manner. In order to extract maximum return on investment, asset owners are required to ensure transformers are fully utilized prior to being disposed of. In line with this, there are various existing ageing models available which are used to predict the loss of life for a transformer. Some of these models include those presented by IEC [1], IEEE [2], Emsley [3], Lungaard [4] and Martin [5] These models attempt to predict loss of life based on a combination of parameters such as hot-spot temperature, moisture in paper content and oxygen content. These three parameters account for the three different degradation mechanics of transformer solid insulation (cellulose paper).

In this paper, four frequently used ageing models are applied for ten pairs of distribution power transformers, which operate at an average loading of below 50%. The data used covers a three year period. The results were verified against Furanic oil results of the transformers and it was found that the ageing models constantly produce high variations in their predicted versus measured loss of life.

To reduce the variation between predicted and measured loss of life values, three new ageing models are developed in this paper. These models are based on hot spot temperature, oxygen content in the oil and moisture content in the solid insulation. A genetic algorithm is used to determine the model parameters from available data from in-service transformers. The results show that these new ageing models produce ageing estimations with less variation between the predicted and measured loss of life in comparison to the results of the initial models investigated. The modified ageing models is shown to be easily implementable on in-service transformers, even when limited condition data is available.

Condition Monitoring, Power Transformer, Genetic Algorithm, Ageing Model, Loss of Life.

Corresponding author: Mohamed Ziyaad Mukuddem (mzmukuddem@gmail.com)

1. INTRODUCTION

Power transformers form an integral part of present day electrical networks. They allow for power to be transmitted over vast distances while minimizing losses and costs [6]. They are however one of the most expensive assets within the distribution network [7, 8, 9]. To maximize the return on investment, the transformers should operate, at minimum, for the duration of their designed life time.

Power transformers are used throughout the electricity network, as part of large power stations to small pole mount distribution transformers. The distribution networks with redundancy typically share the total load of a substation across two or more transformers [9]. Should one of the transformers not be operational (N-1 condition), the remaining transformers are still capable of supplying the full load of the substation [9]. This results in transformers operating at below full load capacity under normal network conditions [9]. A lightly loaded transformer can be considered one which operates at below 50% average loading [10]. A reduction in load factor results in a reduction of hot-spot temperature below the rated operating hot-spot temperature [1].

In the South African context distribution power transformers are commonly built with network redundancy. This typically results in two transformers sharing a total load which can be supplied by each individual unit. As a result, these transformers operate at below 50% loading. The operating hot spot temperature which they operate at is therefore below that of its designed rating.

A transformer consists of two main components, namely the winding and the core. The core is covered with solid insulation. Kraft paper is the most commonly used solid insulation. The loss of life of a transformer is directly affected by the condition of the solid insulation within the transformer windings [1, 2], since it is not ordinarily feasible to replace the solid insulation during the life of a transformer. The strength of solid insulation is measured via the degree of polymerization of the paper (DP), which is directly related to the physical strength of the solid insulation. Ageing mechanics such as hydrolytic degradation, oxidative degradation and thermal degradation degrade the strength of the solid insulation inside a transformer. A transformer degrades as a result of these mechanics until it reaches it end of life criteria. A widely accepted physical end-of-life criteria is when the degree of polymerization of the paper reaches 200 or less [1, 2, 3, 4, 5].

There are various models for calculating this rate of loss-of-life in solid insulation [1, 2, 3, 4, 5]. These models use different permutations of the parameters, hot-spot temperature, moisture and oxygen content, to calculate degradation. The accuracy of the ageing models is important for the effective management of transformers. The norm is to use a reference hot-spot temperature of 98°C for calculating loss-of-life of kraft paper insulated transformers. Transformers in distribution networks operating below average loading, therefore also operate below this reference temperature. The problem is that it is unknown how accurate the existing ageing models are for estimated loss-of-life of lightly loaded transformers. The need is therefore to identify a suitable ageing model which is sufficiently accurate to effectively manage lightly loaded transformers throughout their lifetime.

2. METHODOLOGY

The data available in support of this study is limited to transformer loading and oil sampling data. The study is therefore approached in three phases to cater for the data limitations. The existing ageing models require hot-spot temperature data for estimating loss-of-life. Hot-spot temperature data therefore needs to be firstly derived from the loading data. Secondly, the data is applied to the existing ageing models to estimate the loss-of-life and to compare the accuracy of the estimates. Finally, three new ageing models are derived and evaluated to improve on the accuracy of loss-of-life estimation for lightly loaded transformers.

Since the data available is limited to transformer loading data, an algorithm is required to simulate the operating hot-spot temperature of the transformers. IEC [1] provides an algorithm, which is capable of using loading and ambient temperature data for determining the theoretical hot-spot temperature.

The algorithm is developed in Matlab and validated in two ways. Against a dataset provided by IEC [1] and against data of a pair of in-service transformers which have digital temperature probes installed. The validation results produce a deviation of 0.11% against IEC [1]. The second validation against in-service transformers produces an average variation of 6%. The variation between the measured and calculated values can be attributed to the lack of ambient temperature which is not available in the given dataset.

The algorithm is expanded for calculating the estimated loss of life based on the existing ageing models. The algorithm logic is presented in Figure 1. For each time interval the hot-spot temperature and ageing is calculated until the full time period is reached, followed by the reporting of the total loss of life for the transformer over the full time period.

The existing ageing models investigated as part of this study are the models by IEC [1], Emsley and Stevens [3], Lungaard et al. [4] and Martin et al. [5]. These ageing models vary from single to multi-parameter models. The simplest model [1] only uses the hot-spot temperature of the transformer to determine the loss of life, while the most complex model [5] utilizes hot-spot temperature, oxygen content and moisture content.

The ageing models under investigation come in two forms, either in relative ageing rate form or in Arrhenius form. For simplification the Arrhenius form models [3, 4, 5] are converted to the relative ageing rate form by using equation (1):

$$\frac{A_T}{A_{RT}} \cdot e^{\left(\frac{E}{R(RT+273)} - \frac{E}{R(T+273)}\right)}, \quad (1)$$

where A_T is a constant at the temperature, moisture and oxygen for a specific interval, A_{RT} is a constant at the reference temperature, moisture and oxygen, E is the activation energy, R is the ideal gas constant, 8.314 J mol-1

K-1, T is the hot-spot temperature (°C), and RT is the reference temperature (°C). The parameters values used for the conversion are provided in Tables 1 to 3.

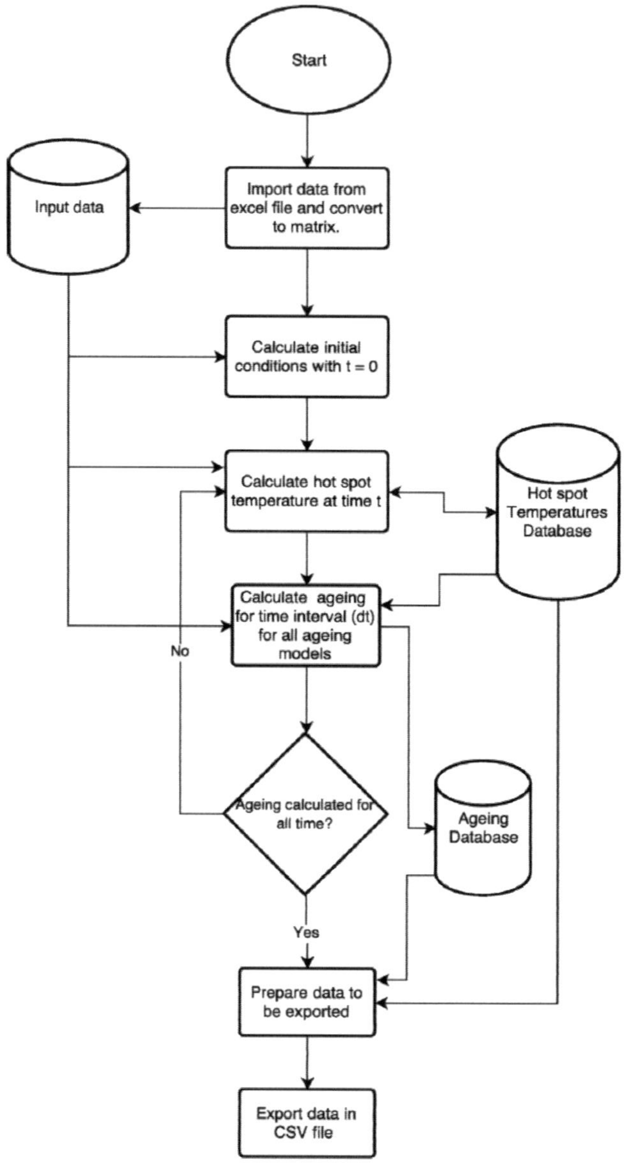

Figure 1: Graphical representation of algorithm used to determine hot-spot temperature and transformer ageing

Table 1: Constant, A_{RT}, for the ageing models

Ageing Model	**A_{RT} (Kraft paper, Moisture < 0.5%, Oxygen < 7000 ppm)**
Emsley & Stevens [3]	1.07×10^8
Lungaard, et al. [4]	2.00×10^8
Martin, et al. [5]	$0.5^2 \times (1.78 \times 10^{12}) + 0.5 \times (1.10 \times 10^{10}) + 5.28 \times 10^7$

Table 2: Constant, A_T, for the Emsley and Stevens [4] and Lungaard et al. [5] models

Ageing Model	**Kraft Paper + 1% Moisture**	**Kraft Paper + 2% Moisture**	**Kraft Paper + 3% Moisture**	**Kraft Paper + 4% Moisture**
Emsley & Stevens [3]	3.5×10^8	7.8×10^8	35×10^8	35×10^8
Lungaard, et al. [4]	6.2×10^8	6.2×10^8	21×10^8	21×10^8

Table 3: Constant, A_T, for the Martin et al. model, were w is the moisture content as a percentage

Martin, et al. [5]	A_T
Low oxygen (<7000ppm)	$(1.78\times10^{12})w^2 + (1.10\times10^{10})w + 5.28\times10^7$
Medium Oxygen (7 000 – 14 000 ppm)	$(2.07\times10^{12})w^2 + (5.61\times10^{10})w + 2.31\times10^8$
High oxygen (> 14 000 ppm)	$(2.29\times10^{12})w^2 + (9.78\times10^{10})w + 3.86\times10^8$

3. LOSS OF LIFE ESTIMATION WITH EXISTING AGEING MODELS

Furanic oil sampling is non-intrusive and a universally accepted method for determining the DP value of a given transformer. The results from the ageing models provide a loss-of-life of the transformer in minutes. To compare the models to the actual Furanic oil samples of the twenty transformers under investigation, a loss of degree of polymerization needs to be determined. The conversion from loss of life minutes to degree of polymerization is derived from IEC [1] and IEEE [2]:

$$DP\ loss = 7.41\times10^{-5}\times loss\ of\ life\ (in\ minutes), \quad (2)$$

To compare these values a variation factor is calculated, which indicate how closely the ageing models compare to the measured Furanic oil sample results. A variation percentage of zero indicates that the ageing model and the measured results are identical. A value less than zero indicates an underestimation, and values greater than zero indicate overestimation. Table 4 presents a scale used to classify variation as low, medium and high and Figure 2 shows the frequency plot for the variation categories.

Table 4: Variation classification scale in percentage

Var ≤ -100%	-100% < Var < -20%	-20% ≤ Var ≤ 20%	20% < Var < 100%	Var ≥ 100%
High	Medium	Low	Medium	High

From the results it is evident that the ageing models produce results which tend to underestimate the loss of life of the sample of transformers. This could be due to the low operating temperature which the transformers operate at. The transformers operate at low hot-spot temperatures due to shared load and therefore could explain the underestimation. The shared load is due to the distribution network configuration which caters for contingencies, where two transformers share a load, although one transformer is able to handle the full load, if required. Based on the ageing models, for approximately every decrease of 6 °C of the hot-spot temperature, the ageing rate of paper halves. The highest average temperature from the sample of transformers is 66°C. This is well below the rated operating temperature of 98°C. Based on the relationship above, the ageing of the paper would be at 0.03125 of the rated ageing.

The results in Figure 2 show that the ageing models investigated produce results of predominantly medium variation compared to the actual Furanic oil sample values. If these models are used by assets owners who use a shared load distribution network configuration, the transformers would be replaced prematurely. To improve on the large variations, three modified ageing models are proposed and assessed against the existing models' result in the next section.

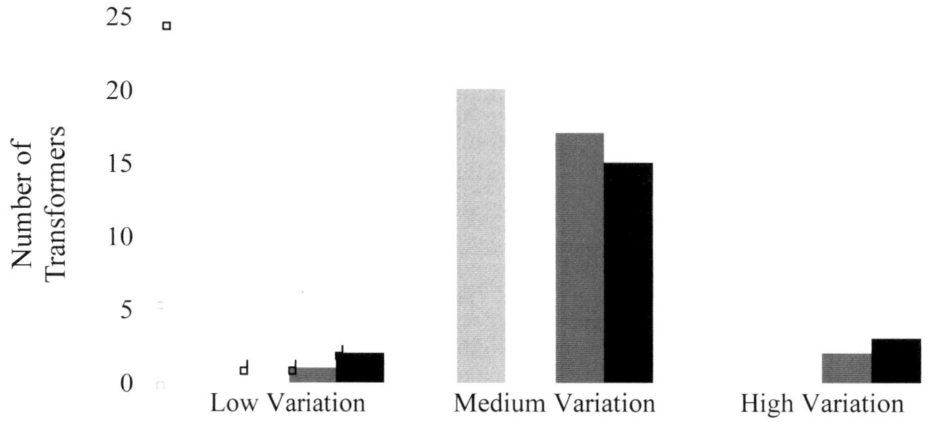

	Low Variation	Medium Variation	High Variation
IEC [1]	0	20	0
Esmley [3]	1	16	3
Lungaard [4]	1	17	2
Martin [5]	2	15	3

Figure 2: Frequency plot of the classifications for the sample of transformers

4. MODIFIED AGEING MODELS & RESULTS

The ageing models presented previously have three variables which can be selected by the user, namely, the activation energy (E), pre-exponent constant (A) and the reference temperature for ageing. The activation energy has been accepted by several authors as 111 000 J/Mol and is independent of the quantity of moisture, oxygen or hot-spot temperature [1, 2, 3, 4, 5]. Thus, the aim is to determine the value of the pre-exponent constant, A. The constant, A, is used to account for accelerated ageing depending on the moisture and oxygen content within the transformer. The second variable which is investigated is the reference temperature (RT). In the case of kraft paper, the reference temperature is 98°C, which increases to 110°C for thermally upgraded paper. To improve the accuracy of the ageing models two variables, namely the reference temperature, RT, and the constant, A, are investigated.

The process involves the investigation of three alternative cases of ageing models. Case 1 only considers the hot-spot temperature, Case 2 only oxygen content, and Case 3 only moisture content. The same basic form of the ageing model as in (1) is used for the investigation.

Case 1 aims to identify whether the reference ageing temperature should be altered to decrease the percentage variation of the loss of life estimate. In Case 1 a single variable is considered for optimization, namely the reference temperature, RT. The parameters, A_{RT} and A_T, are kept constant at a value of one for this case, reducing equation (1) to equation (3). The genetic algorithm previously presented is used to solve for variable RT which produces the most accurate ageing estimation.

$$e^{\left(\frac{E}{R(RT+273)} - \frac{E}{R(T+273)}\right)}, \quad \text{.........} \quad (3)$$

Case 2 incorporates oxygen content. The reference temperature in this case is kept constant at 98°C. This case uses a form similar to Martin et al. [5]. The values for the pre-exponents presented by Martin et al. [5] are based on a combination of moisture and oxygen content. In this case, the pre-exponent, A_T, is based only on the oxygen content during a given time period. The values for A_T and the limits used are presented in Table 5. For Case 2 equation (1) is used and the three variables, A_A, A_B, and A_C, is optimized. A_{RT} is kept constant at a value of one which is the reference for this case.

Case 3 examines the moisture content in isolation to the reference temperature and oxygen content. The reference temperature in this case is held constant at 98°C. This case uses a similar approach to Emsley & Stevens [3] and Lungaard et al. [4]. Case 3 considers the optimisation of four pre-exponent variables and are based on the moisture content of the paper. The moisture limits are specified in Table 5. For Case 3 equation (1) is used and the four variables, A_w, A_x, A_y and A_z, is optimized. A_{RT} is kept constant at a value of one which is the reference for this case.

Table 5: Pre-exponents, A_T, used in Case 2 with their oxygen limits and Case 3 with their moisture limits

Case 2 **Oxygen Content [PPM]**	A_T
<1000	A_{RT}
>1000; <= 7000	A_A
>7000; <=16500	A_B
> 16500	A_C

Case 3 **Moisture in paper [%]**	A_T
<1%	A_{RT}
>=1%; <2%	A_w
>=2%; <3%	A_x
>=3%; <4%	A_y
>4%	A_z

The optimization is performed with the premise that the ageing models are black-box problems. Values for the variables is passed into the ageing model and the model produces an accuracy value. This accuracy value is used as the cost function for the optimization model. The number of variables to be determined for each of the three cases are, one, three and four, respectively. An approximate optimization method is utilized as the problem does not exist within the scope of a simple mathematical model.

In order to carry out the optimization, the algorithms for Cases 1, 2 and 3 are passed through the genetic algorithm. The genetic algorithm is able to vary the specified number of variables in each case algorithm. The genetic algorithm evaluates the accuracy of the model at each iteration to determine its next step and terminates when it finds a solution which satisfies the exit requirements. This algorithm operates with the specified cost function in (4) and it aims to minimize the cost function to a specified fitness limit.

$$Cost = |DP_{Predicted} - DP_{Actual}|, \quad (4)$$

A genetic algorithm by nature may produce slightly different result each time it is executed. To overcome this, the genetic algorithm is executed five times for each case. The results are averaged to obtain a final value.

Case 1 was executed five times with the genetic algorithm. The output of each iteration produced a reference temperature for the transformers which produce a predicted DP value equal to the measured DP value. These results are averaged out across all the iterations to produce a specific reference temperature for each transformer.

The two variables, average reference temperature and average operating hot-spot temperature, are analyzed using correlation analysis to determine if the two variables are correlated. The results from the correlation indicate a positive correlation with a value of 0.912, which is indicative that the reference temperature for ageing can be based on the average operating hot-spot temperature. A three-year average is used to carry out regression analysis for defining the function between the variables. Hypothesis testing, at 95% significance, for the slope as well as the intercept of the regression line concludes that equation (5) is an acceptable representation of the relationship. Equations (3) in conjunction with (5) forms the modified ageing model for Case 1.

$$R_T = [1.682 \times (Average\ historic\ operating\ hotspot\ temperature)] - 19.996, \quad (5)$$

Case 2 was executed five times using the genetic algorithm. Correlation analysis revealed no significant correlation (at 95% significance) between the variables A_A, A_B and A_C and the input parameters; load, hot-spot temperature or moisture. A weak correlation is found between A_A, A_C and average oxygen content. To utilize these results correctly, the minimum and maximum oxygen values for each transformer is derived. The minimum and maximum value is used to classify which results from the genetic algorithm should be accepted. If the oxygen content of a transformer never enters any of the three categorization limits, the results are neglected. If a transformer operates within the range of a category, a one is attributed, otherwise a zero is assigned. The complete set of results, in conjunction with the operating oxygen range, is used to determine the final variables A_A, A_B and A_C. To determine the value of these variables, the following steps are carried out: The average variable value for each of A_A, A_B and A_C are calculated per transformer from the five iterations' outputs. The overall average of the variable is calculated if the transformer operates within the given categories and finally the results are substituted into the (1). The final variable values for Case 2 is A_A = 159.47, A_B = 235.27 and A_C = 128.63. These variables in conjunction with (1) forms the modified ageing model for Case 2.

The results for Case 3 were obtained in a similar manner than for Case 2. For this case, the correlation analysis also did not show significant correlation. The final variable values obtained for Case 3 are: A_w = 243.15, A_x = 133.08, A_y = 143.13, and A_z = 1058.60. These variables in conjunction with (1) forms the modified ageing model for Case 3.

The three modified ageing models are executed through the same loss of life algorithm presented in section 2. The results are compared with the results of the initial four ageing model in section 3 to establish whether the modified ageing model produce more accurate loss of life estimations.

From the results in Figure 3 it is evident that the modified ageing models produce more results in the low variation range in comparison to the initial four models used. Case 1 and Case 2 present the highest number in this category, accounting for 25% of the sample, respectively. Case 3 also have results in this category which account for 20% of the sample. Case 1 does however have more results within the medium variation category in comparison to Case 2. In addition, the new modified ageing models have a higher tendency to overestimate the age in comparison to the original four models which predominantly underestimated ageing.

To gain further insight, the results are categorised. The categories consider whether the predicted value is an over or under estimation of the measured value. The categories are defined based on the percentage variation, where less than -50% is considered an underestimation, between -50% and 50% is considered acceptable and more than 50% is considered an overestimation. The categorized results are shown in Table 7.

Table 7: Categorised ageing model results

Variation	IEC[1]	Esmley [3]	Lungaard [4]	Martin [5]	Case 1	Case 2	Case 3
Underestimation	19	14	17	10	2	2	1
Acceptable	0	3	1	5	12	7	8
Overestimation	1	3	2	5	6	11	11

When considering the three modified ageing models, Case 1 produces the largest improvement in the age estimation results. Its improvement lies in the number of prediction in the low variation category when compared to the IEC [1], Emsley and Stevens [3], Lungaard et al. [4] and Martin et al. [5] ageing models. In addition, the ageing model described by Case 1 is the simplest of the three modified models. It only accounts for hot-spot temperature, but it however reduces the reference temperature for ageing. This makes the model easily implementable in practice, with the equation using the reduction of the reference ageing temperature to account for moisture and oxygen content.

5. CONCLUSIONS

The Case 1 ageing model is recommended to be better suited for lightly loaded transformers when compared to the original four ageing models investigated. The ageing estimation are to be used in the asset management of power transformers. It is evident that while IEC, Emsley and Stevens and Lungaard et al. primarily produce results which underestimate the ageing of a transformer, Case 1 shifts the results into the categories of reasonable and overestimation. In terms of asset management, an underestimation may result in no action being taken on the specified asset which may result in an unplanned breakdown. This can be due to lack of action as it is assumed the asset does not require any intervention. An overestimation, will however start off a further investigation process. In terms of power transformers, further investigation can entail oil and paper Furanic tests. This will confirm whether action is required or not.

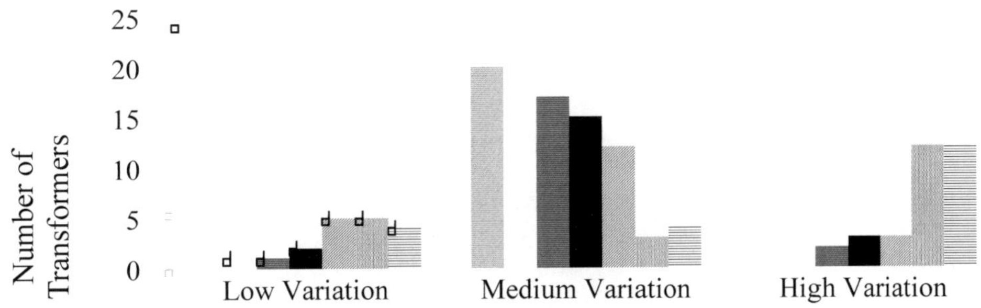

	Low Variation	Medium Variation	High Variation
IEC [1]	0	20	0
Esmley [3]	1	16	3
Lungaard [4]	1	17	2
Martin [5]	2	15	3
Case 1	5	12	3
Case 2	5	3	12
Case 3	4	4	12

Figure 3: Frequency plot of the classifications for the sample of transformers including modified ageing models

This study set out to investigate the accuracy of ageing models when applied to lightly loaded distribution power transformers. The results indicate that the available ageing models produce results with a high degree of variation between predicted and measured DP values, which is an indication of loss of life. Three cases for modified ageing models is proposed, which generate ageing estimations with a lower variation rate in comparison to the existing models available. The modified ageing model which produces the best improvement results is based only on the operating hot-spot temperature of a transformer.

Opportunities for future research is to investigate the existing and modified models with a larger sample of data, to determine what the optimal time period for historic hotspot temperature should be, and to use technology to digitally record hotspot temperature, moisture and oxygen measurements more accurately.

REFERENCES

[1] IEC (2005). Power Transformers – Part 7: Loading Guide for oil-immersed power transformers

[2] IEEE (2011). IEEE Guide for Loading Mineral Oil-Immersed transformers and step-voltage regulators,

[3] Emsley, A. M. & Stevens, G. C. (1994). Review of chemical indicators of degradation of cellulosic electrical paper insulation in oil-filled transformers. IEE Proc-Sci, Volume 141.

[4] Lundgaard, L., Hansen, W., Linhjell, D. & Painter, T. (2002). Ageing of oil-impregnated paper in power transformers. IEEE Transactions on Power Delivery, 19, pp.230-239

[5] Martin, D. et al. (2015). An updated model to determine the life remaining of transformer insulation. IEEE Transactions on Power Delivery, Volume 30(1).

[6] Heathcote, M. J. (2007). The J & P Transformer Book. 13 ed. Oxford: Elsevier.

[7] Abu-Elanien, A. E. & Salama, M. M. (2010). Asset Management Technique for Transformers. Electric Power Systems Research, Volume 80, pp. 456-464.

[8] Brandtzaeg, G. (2015). Health Indexing of Norwegian Power Transformers, Masters Thesis, Norwegian University of Science and Technology, Trondheim, Norway.

[9] Eskom Holdings Ltd and ABB Powertech, 2008. Theory, Design, Maintenance and Life Management fo Power Transformers. Johannesburg: Y-Land Design, Print and Promotions.

[10] Felber Engineering GMBH, 2016. Transformer Design and Technology Training Course, Weiz: Felber Engineering.

Measurement and visualization of the corona inception gradient on an HVDC transmission line under the effect of wet conditions

J. S. Djeumen[1*], J. J. Walker[1] and N. J. West[2]
[1]Department of Electrical Engineering, Vaal University of Technology, Private Bag X021,
[1]Andries Potgieter Blvd, Vanderbijlpark, South Africa, 1900
[2]School of Electrical. and Info. Eng. University of the Witwatersrand, Johannesburg Private Bag 2, 2050
[2]Jan Smuts Avenue, Braamfontein, Johannesburg, 2000
[1]*julesd@vut.ac.za; [1]jerryw@vut,ac,za and [2]nicholas.west@wits.ac.za

ABSTRACT

Due to the increasing demand to transmit a high amount of electric power over very long distances, High Voltage Direct Current (HVDC) transmission is gaining increasingly in importance across the world. South Africa has important coal resources. Moreover, with the implementation of the Cohora Bassa to Apollo line transmission project and many others, transmission lines will inevitably pass through high altitude regions with complex atmospheric conditions. These complex atmospheric conditions are one of the main causes of power loss on transmission lines resulting in adverse economic and environmental effects. Therefore, understanding the atmospheric conditions is crucial in designing HVDC transmission lines. Typical environmental conditions include temperature, wind or air pressure and rain. This paper focuses on the measurement and the visualization of the corona inception voltage in a corona cage at Vaal University Technology high voltage laboratory for the positive and negative HVDC applications on the transmission line under wet conditions. The focus of these measurements and visualizations were to investigate and present an overall picture of corona discharge performance and the influence of the water mist on the transmission line in the corona cage at ±100 kV with the variation of the room's humidity and the temperature.

Keywords: *Corona Discharge, CoroCam8, Partial Discharge, HVDC, Fault detection*

1. INTRODUCTION

In modern developed or developing societies and in South Africa in particular, due to the growth of the industrial sector and world population, the demand for electricity is ever increasing. A large amount of power is generated, transmitted and distributed. As an example, in South Africa, the total official unit sentout in 2010 was 237,073,293 MWH and reached 216,473,767 MWH in 2017 [1, 2]. The centres of power generation are distant from each other, thus extensive transmission systems are needed to transmit energy over long distances. Transmission systems are the backbone of the power system that carries the large amount of power at high voltage, very high voltage and ultra-high voltage.

South Africa's climatic conditions generally range from Mediterranean in the southwestern corner of South Africa to temperate in the interior plateau, and subtropical in the northeast. These diverse climatic conditions pose a problem for the power system transmission networks as lines often have to travel for many kilometers and cross-different climatic regions. ESKOM, the South African power utility, is planning in the near future to build new HVDC transmission lines, convert some of the existing HVAC lines into HVDC as well as moving towards hybrid AC-DC towers. This will only be possible of course, with a good capital expenditure plan, which is the key of investments, and with specific environmental studies [1].

Weather conditions such as rain, temperature and snow easily cause the corona loss rather than normal air weather condition [3-4]. In addition, several natural factors such as weather, conductor size, location, altitude above mean sea level and wet atmospheric conditions affect corona discharge phenomena on transmission lines. In this study, the influence of moisture on DC corona discharge pulses was measured and visualized experimentally while varying the applied voltage and the atmospheric conditions in the room. Artificial rain which aimed at increasing humidity and dropping the temperature in the HV laboratory room was introduced. A corona camera was used to visualize the corona discharge patterns and distribution. A digital partial discharge detection system through its control screen, enabled measurement of the intensity of the corona discharge.

2. FACTORS AFFECTING AND CORONA LOSS EQUATION

The corona inception voltage is defined as the lowest voltage at which continuous corona discharge of a specified pulse amplitude occurs as the applied voltage is gradually increased [5]. Corona discharge is one of the main problems encountered in the design and operation of transmission lines. This is a well-known phenomenon that can cause damage and losses in a transmission or power system [5-6]. Corona loss can be described as the ionization of air surrounding the power transmission line [7]. Consequently, the corona becomes non-uniform along the transmission line; for instance, if the transmission line is rough, then the glow will be relatively brighter. In addition, positive DC corona will be observed as having a 'smooth' glow while negative DC corona will tend to appear erratic.

2.1. Factors affecting Corona discharge

MATLAB simulations have shown that under wet conditions, there is more corona loss than during the dry season [7]. Besides the wet conditions as factors affecting the corona loss, the surface roughness of the conductor can also increase corona loss as the roughness of the surface decreases the value of breakdown voltage [6]. Among the factors, which affect the corona loss mentioned [8], the atmospheric conditions were one of the parameters emphasized. From Peek's equation, the conductor surface field intensity is proportional to the size of conductor. Also, parameters such as the supply frequency, size of conductor bundles, spacing between conductors, mean load current at sea level, conductivity of the atmosphere and temperature of the conductor from load current may have an effect on cause corona loss.

2.2. Corona Loss Equation

According to Peek's equation [7], the power loss due to corona under fair weather conditions can be expressed as:

$$P_l = \frac{K}{\delta}(f + 25)\sqrt{\frac{R}{d}}\left(\frac{V_L}{\sqrt{3}} - E_d\right)^2 \times L \times 10^{-5} \quad kW/Km/phase \qquad (1)$$

The normalized air density correction factor, δ can be elaborated as:

$$\delta = \frac{0.386 \times P}{273 + t} \qquad (2)$$

Where:

P_l: Corona loss under fair weather conditions
t: Temperature of the surroundings
E_d: Disruptive critical voltage
R: Radius of the conductor
K: Fixed Constant
V_L: Line to line voltage
***f*:** Frequency
L: Length of the conductor
δ: Air density correction factor
P: Atmospheric pressure
d: Conductor spacing

3. EXPERIMENTAL PROCEDURE AND SETUP

3.1. Corona Cage Experimental setup

The corona cage used for this research has a diameter of 1.5 m, a height of 2.4 m and is 4.7 m in length, with two outer or protective segments of 0.6 m each and one inner segment of 1.2 m usually called measured segment. The whole structure of the cage has a total length of 4.7 m. A single TERN aluminium conductor with a diameter of 30 mm runs through the centre of the cage. The conductor was energised with a HVDC modular system power supply with a maximum output of 270 kV. The shield segments are electrically isolated from the measuring segment(s) to measure the corona discharge on the conductor' surface and to eliminate the corona discharge on the cage [9-10]. A high conductor surface electric field is generated when a sufficient voltage is applied between the conductor and the cage.

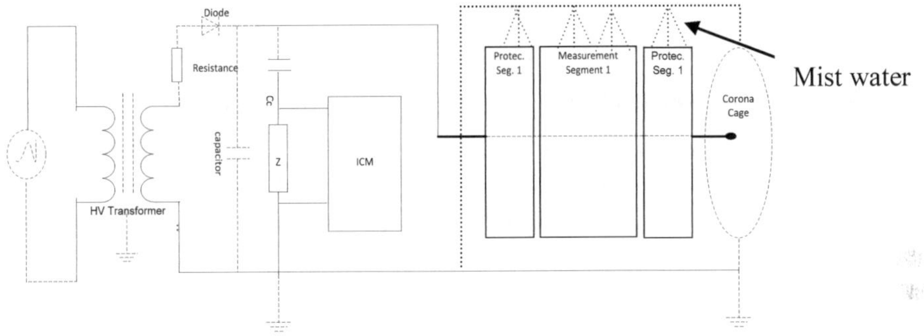

Figure 1: Test system configuration

3.2. Experimental setup

The atmospheric condition, which in this case was the "mist", has been simulated by a mist cooling system. This system was installed just outside of the cage's surface on the circular mesh of the cage in order to prevent corona discharges being generated on the mist nozzles The water were sprayed through the outer mesh electrode on to the inner energized conductor. Preassembled flexible tubing was installed around the cage, one on top of the conductor and two on the sides of the conductor. The tubing was attached to the water supply system, making it possible to control the flow of water during the measurements. The mist spray system also helps to reduce the ambient temperature around the conductor up to 15°C [11, 12, 13] have also studied the influence of humidity on the first corona inception. It can be observed that the humidity is a parameter to consider during the design of the transmission lines.

Figure 2: The installation of the mist cooling in the cage

4. TEST ARRANGEMENT AND INCEPTION VOLTAGE MEASUREMENT

4.1. Test Arrangement

The positive and negative corona inception voltage of the aluminum ACSR stranded conductor (TERN) is obtained by using the corona cage test platform as shown in figure 1 above. The conductor is placed concentrically inside the corona cage. With a total length of approximatively 4.7 m, the cross section of the cage is circular in its dimension. The inception is affected by many factors, such as the temperature, the air pressure and the relative humidity.

4.2. Inception Voltage measurement

The positive inception voltage of the test system configuration was measured using the partial discharge detector with the setting at the normal condition of 20.5°C and 41% relative humidity. The background noise was 0.87 pC and the system was calibrated at 5 pC and the scale of the monitor was 12 pC/Div. The positive inception voltage was observed around a voltage of +24 kV, the first pulses appeared on the negative half cycle of the phase resolved-pattern and the negative inception voltage was found to be -19.5 kV. The temperature in the room at the time of measurement was 20°C and the humidity 49%. The same calibration of 5 pC was used and a background noise of 0.59 pC was measured. Here the first appearance of corona pulses on the pattern was observed on the positive half cycle of the phase-resolved pattern with the aid of the ICM monitor PD detector. Figure 3 and 4 below show the measurement and magnitudes of the positive and negative inception voltages respectively.
It must be noted that the sine wave do not represent the supply voltage as this stage we are using a DC supply, but the sine wave only represent the internal voltage used for the triggering purposes.

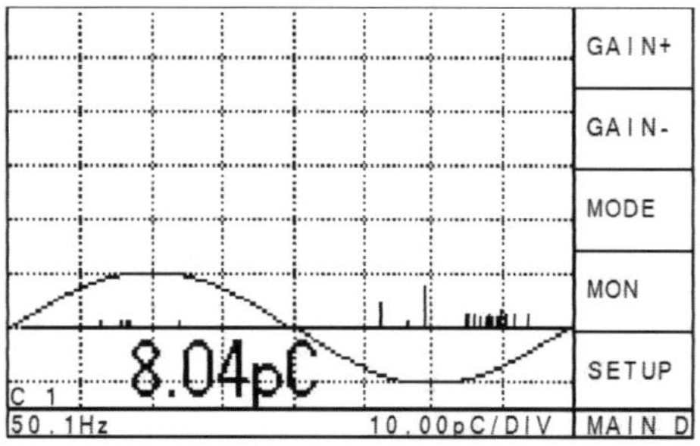

Figure 3. Inception voltage at +24 kV for Positive voltage.

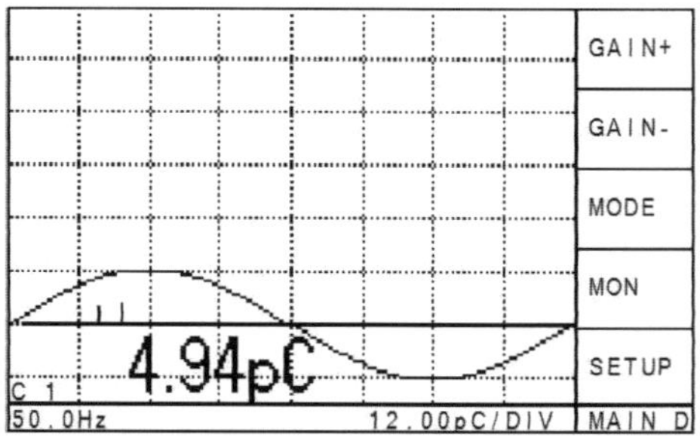

Figure 4. Inception voltage at -19.5 kV for Negative voltage.

As can be seen in Figure 3 and Figure 4, the magnitude of the positive inception voltage, which is 8.04 pC at 10 pC/DIV, is greater than the negative inception's magnitude, which is 4.94 pC at a voltage of -19.5 kV with a 12 pC/DIV.

4.3. Visualization with CoroCam8

Before the measurements were performed visual detection has been used to determine if the cage was corona free. The CoroCam8 has been used to visualize the entire cage and the sharp edges. The CoroCAM8 is a combined thermal IR, solar blind corona and video camera. The combination of these three components makes it possible to do both thermal IR and UV inspections at the same time, saving time and effort. The co-location of electrical discharges and hot spots gives the inspector more insights into the cause of a fault. The CoroCam8 can be used to inspect HV infrastructure operating at 3.3 kV and above. Figure 5 below displays the visualization of an artificial corona between the ring and the stand from transparent figure 5a to UV visualization only figure 5d.

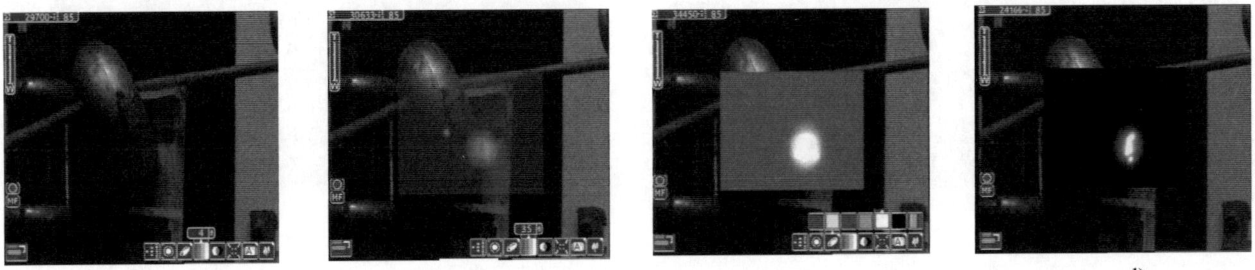

a) b) c) d)

Figure 5. Process from visible to UV visualization of corona.

5. EXPERIMENTAL RESULTS, VISUALIZATION AND MEASUREMENTS

5.1. Virtualization results with CoroCam8

The first step of the experiment consisted of visualizing the entire corona cage using the corona camera to make sure that the cage was corona free. Some discharge sources were found on the cage. After investigation, the main source of the discharge was found and fixed. The second step was the visualization of the corona on the conductor. Four different scenarios were observed: Dry, wet with one tap (top spray nozzles), two taps (top and one side spray nozzles) and all three taps (top and both sides spray nozzles) open. The supply voltage was constant at 100 kV DC.

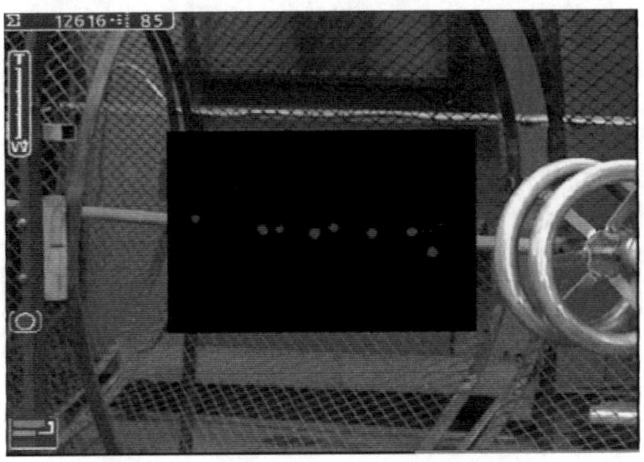

Figure 6. Visualization at normal condition -100 kV

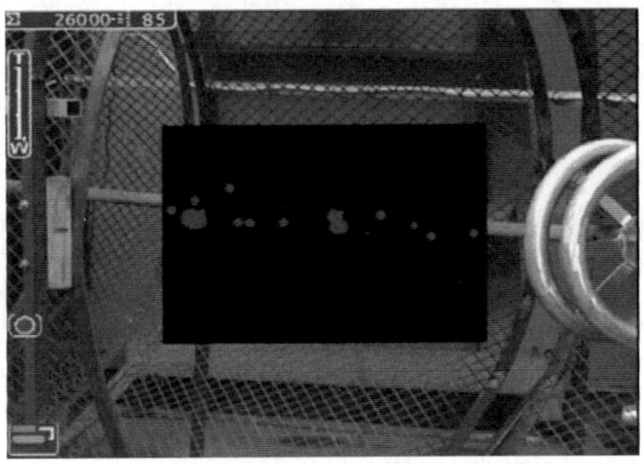

Figure 7. Visualization with one pipe on at -100 kV

Figure 6 and 7 represent the case without mist water with the ambient temperature at 23˚C, 41% humidity and "the top spray nozzles" on with 63% humidity and the ambient temperature at 19˚C respectively. With the opening of the left side spray nozzles, the humidity increased to 78% and the temperature dropped to 18˚C and when the right side spray nozzles was opened, the humidity still increased to 85% and the temperature dropped to 17˚C. For good accuracy in the results, the corona camera was set at a fixed position during the visualization at a distance of 3.5 m from the corona source and the test voltage at – 100 kV. Figure 8 and 9 below illustrate the two above scenarios where the mist water was having more pressure than the conditions in figures 6 and 7.

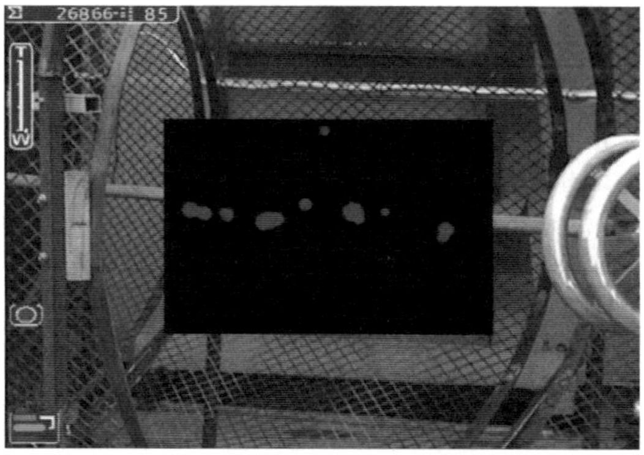

Figure 8. Visualization with two pipes on at -100 kV

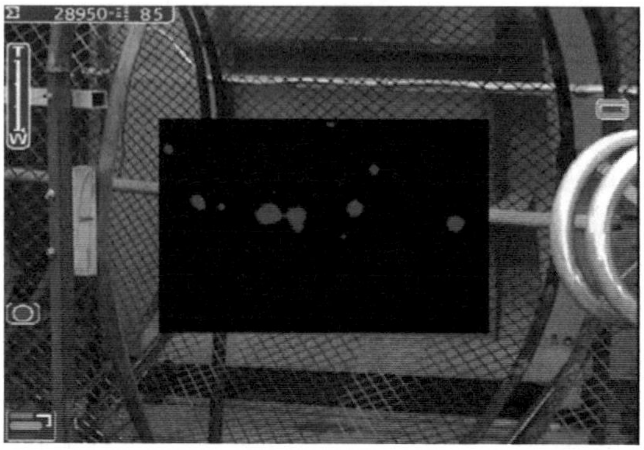

Figure 9. Visualization with three pipes on at -100 kV

When the voltage polarity was reversed (+100 kV), the partial discharge unit was again calibrated to 5.0 pC. The same set of four experiments were conducted. The temperature and humidity readings were 21°C, humidity of 45% and 19°C, humidity of 55% respectively. Figure 10 and 11 display the presence of corona during the measurements.

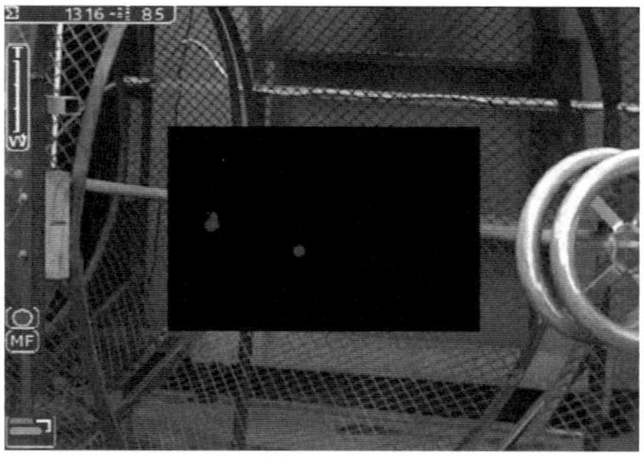

Figure 10. Visualization at normal condition +100 kV

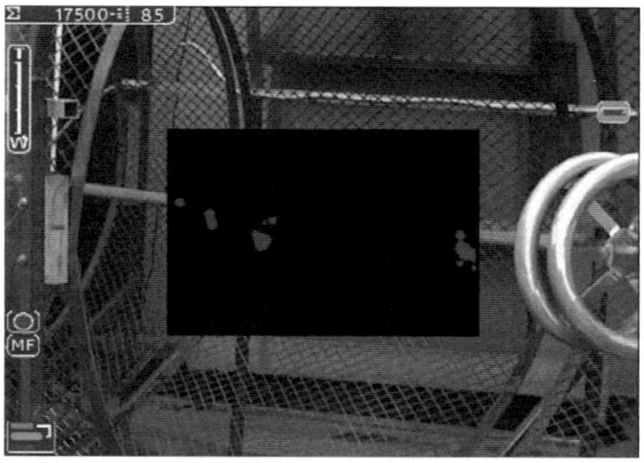

Figure 11. Visualization with one pipe on at +100 kV

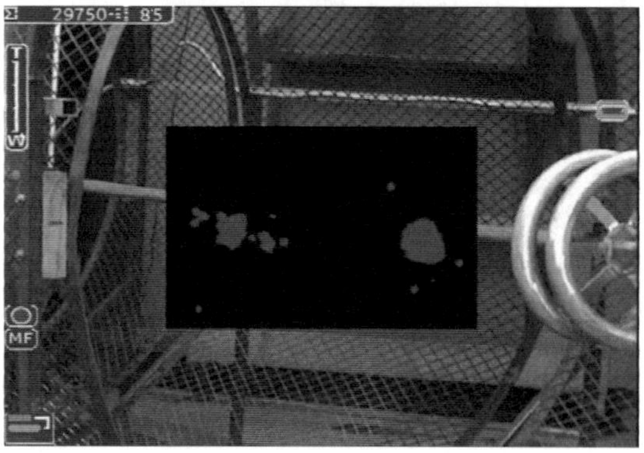

Figure 12. Visualization with two pipes on at +100 kV

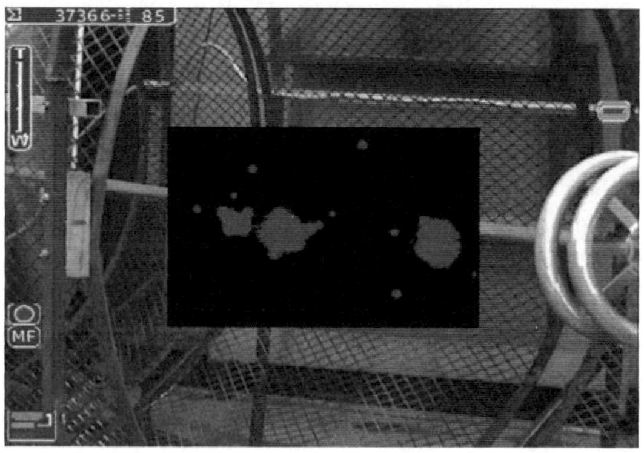

Figure 13. Visualization with three pipes on at +100 kV

5.2. Measurement results with ICM monitor

The third step of the experiment was the measurement of the corona discharge using the partial discharge detector at similar atmospheric conditions as mentioned earlier above. The corona magnitude was measured at negative supply voltage with the sensitivity of the instrument at 200 pC/DIV (figures 14 and 15) with a change in humidity and temperature from 63% to 85% and from 19°C to 17°C respectively In addition, the magnitude of the corona discharge increased from 562.9 pC to 851.1 pC.

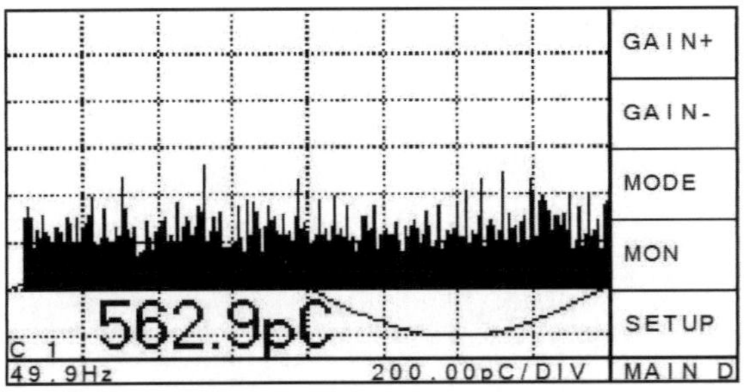

Figure 14. Measurement with two pipes on at -100 kV

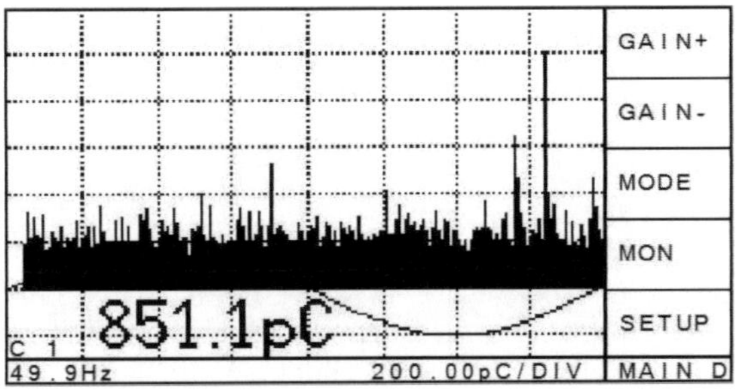

Figure 15. Measurement with three pipes on at -100 kV

The corona magnitude was measured at positive voltage with the sensitivity of the instrument at 2 nC/DIV. Figure 17 and 18 display the patterns of the similar conditions mentioned above. An increase in the magnitude was observed from 8.54 nC to 9.39 nC with a variation in the atmospheric conditions.

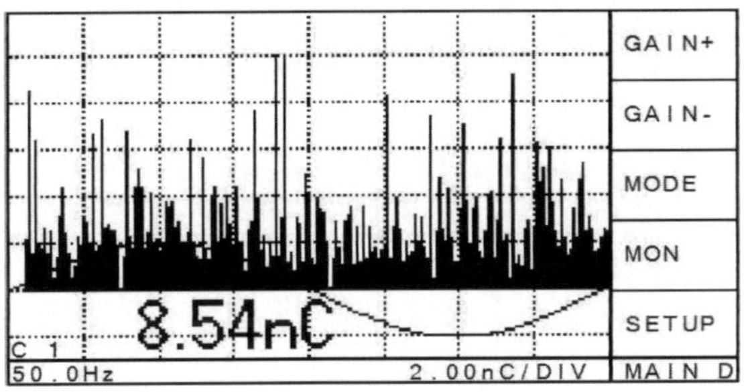

Figure 16. Measurement with two pipes on at +100 kV

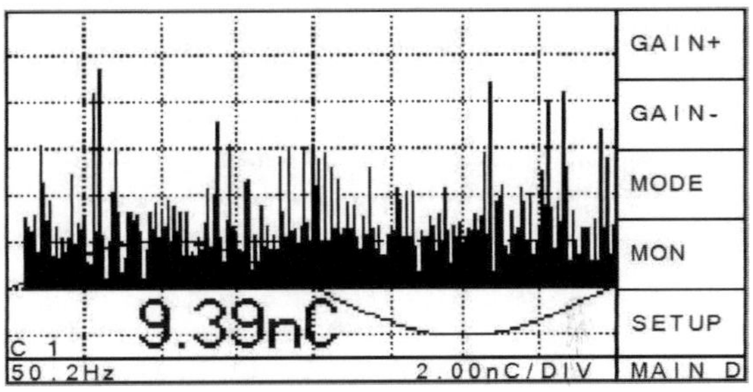

Figure 17. Measurement with three pipes on at +100kV

5.3. Summary of the measurement results

The summary of the results can be summarized into table 1 below, by bringing the sensitivity of the instrument at 200 pC/DIV for all the measurements, table 1 below clearly show the difference between the positive and negative corona at the wet condition.

Table 1: Results summary at different wet condition level

	Inception Voltage		**T=20°C & H=45%**	**T=19°C & H=63%**	**T=17°C & H=85%**
Voltages Value	+24 kV	-19.5 kV	50 kV	100 kV	100 kV
Positive Voltage	8.04 pC		1800 pC	8540 pC	9390 pC
Negative Voltage		4.17 pC	15.7 pC	562.9 pC	851.1 pC

6. CONCLUSION

In conclusion, the main objective of this experiment was to study corona on the conductor by understanding the physical mechanisms involved in corona discharge as well as the resulting corona effects under wet conditions. This study presented the corona test results obtained during the visualization and measurement of the inception voltage using the conventional method of apparent charge measurement in a corona cage having three sections under positive and negative voltages. The results show that corona inception gradient under the effect of wet conditions on HVDC Transmission line, visualization and patterns magnitude of the corona discharge increase very little with the increase of the humidity and the decrease of the temperature under the DC voltage supply. Moreover, the negative inception voltage is lower than the positive inception voltage.

REFERENCES

[1] T. Govender. (2015) *Transmission development plan 2016-2025. ESKOM TDP.* Pp.1-103.

[2] S. Bisnath, A. Britten, D. Cretchley, D. Muftic, T. Pillay, and R. Vajeth, The planning, design and construction of overhead power lines. Johannesburg: Crown Publication cc, 2005.

[3] L. Youwei, L. Jihong, and L. Bin, *Influence of air density and humidity on the corona performance of conductor, Power System Technology*, vol. 14(4), pp 46-50, 1990.

[4] P. S. Maruvada, *Corona Performance of High-Voltage Transmission Lines.* India: Narinder Kumar Lijhara for Overseas Press India Private Limited, 2006, p. 310.

[5] J.S. Djeumen & J.J. Walker, "*Environmental influence on corona inception with HVDC application*", International Symposium on High Voltage Engineering, (ISH 2017), Buenos Aires, Argentina, pp. 1-6, 2017.

[6] E.A. Yahaya, T. Jacob, M. Nwohu, A. Abubakar, "*Power Loss due to Corona on High Voltage Transmission Lines*," Journal of Electrical and Electronics Engineering (IOSR-JEEE), vol. 8, no. 3, pp. 14-19, Dec. 2013.

[7] M. C. Anumaka, "*Influence of Corona and Skin Effect on the Nigerian 330 kV Interconnected Power System*," IJRRAS, vol. 12, no. 2, pp. 328-333, August 2012.

[8] J.S. Djeumen & J.J. Walker, "*Visualization of Positive and Negative Corona under Environment Influence with HVDC Application*", South African Universities of Power Engineering Conference, (SAUPEC 2017), Johannesburg, South Africa, pp. 1-6, 2017.

[9] R. Urban, H. Reader, and J. P. Holtzhausen, "*AC transmission line corona noise issues in a small corona cage*," in Africon Conference in Africa, 2002. IEEE AFRICON. 6th, 2002, pp. 639-644 vol.2.

[10] Y.-p. Liu, L.-j. Ren, y. Chen, Q.-f. Wan, S.-h. You, and W. Chen, "*Research on High Altitude Corona Loss Measurement System Based on Small Corona Cage,*" in High Voltage Engineering and Application, 2008. ICHVE 2008. International Conference on, 2008, pp. 144-147.

[11] P. Ortega, R. Dıaz, F. Heilbronner and F Ruhling, "*Influence of negative ions on the humidity effect on the first corona inception*", Journal of Physics D: Applied Physics, vol. 40, pp. 7000–7007, 2007.

[12] P. A. Calva and F. Espino , "*Effect of the humidity in the ionic mobility in reduced air-density*," in Annual Report Conference on Electrical Insulation and Dielectric Phenomena (Cat. No.98CH36257), pp. 508-511 vol. 2, 1998.

[13] T. Soban, R. Ohyama, "*A Study on the Humidity Effect of AC Corona Discharge for a Thin-Wire Electrode Arrangement*," in Proceedings of the School of Engineering Tokai University Tokyo, Japan, 2016, pp. 1-7.

A New Fuzzy Logic-based Approach to Predict Fault in Transformer Oil Based on Health Index using Dissolved Gas Analysis

Kharisma Utomo Mulyodinoto[1,2], Suwarno[2,3] A. Abu-Siada[3,4]
[1]PT PLN (Persero), Jl. Trunojoyo Blok M I/135 Kebayoran Baru, 12160, Jakarta, Indonesia
[2]Senior Member, IEEE
[3]School of Electrical and Informatics, Institut Teknologi Bandung, Indonesia
[4]Electrical and Computer Engineering Department, Curtin University, Australia

ABSTRACT

Dissolved gas-in-oil analysis has been widely accepted as a powerful technique to detect such incipient faults. While the measurement of dissolved gases within transformer oil samples has been standardized over the past two decades, analysis of the results is not always straight forward as it depends on personnel expertise more than mathematical formulas. IEEE C57.104-2008 standards has classified the health condition of the transformer based on the absolute value of individual dissolved gases along with the Total Dissolved Combustible Gas (TDCG) within transformer oil into 4 conditions. While the technique is easy to implement, it is considered as a very conservative technique and is not widely accepted as a reliable interpretation tool. Moreover, measured gases for the same oil sample can be within various conditions limits and hence, misinterpretation of the data is expected. To overcome this limitation, this paper introduces a fuzzy logic approach to predict the health condition of the transformer oil based on IEEE C57.104-2008 standards along with Roger ratio and IEC ratio-based methods. DGA results of 31 chosen oil samples from 469 transformer oil samples of normal transformers and pre-known fault-type transformers that were collected from Indonesia Electrical Utility Company, PT. PLN (Persero), from different voltage rating: 500/150 kV, 150/20 kV and 70/20 kV; different capacity: 500 MVA, 60 MVA, 50 MVA, 30 MVA, 20 MVA, 15 MVA and 10 MVA; and different life span, are used to test and establish the fuzzy logic model. Results show that the proposed approach is of good accuracy and can be considered as a platform toward the standardization of the dissolved gas interpretation process.

Keywords — **Dissolved gas analysis, Transformer, Roger ratio method, IEC ratio method, Fuzzy logic, Health Index, IEEE C57.104-2008**

I. INTRODUCTION

Power transformers are critical components in any transmission and primary distribution networks. Effective monitoring and diagnostic techniques must be adopted to decrease possibility of transformer catastrophic failures and increase the reliability of this asset. Transformer dielectric oil is considered as a key source to detect incipient and fast developing faults within a transformer [1].

There are several chemical and electrical diagnostic techniques to assess the health condition of transformer [2]. Dissolved gas analysis (DGA) is one of the most effective techniques to detect incipient and fast developing faults within power transformers. This technique uses the concentration of various key gases which are produced due to the decomposition of oil and paper insulation [3]. Gases such as Hydrogen (H_2), Methane (CH_4), Acetylene (C_2H_4), and Ethane (C_2H_6) are mainly generated due to oil decomposition [4]-[6]. On the other side, Carbon oxides (CO and CO_2) may be generated due to paper decomposition, oil oxidation or due to atmospheric leak. Faults that can be detected using this technique include thermal, arcing and partial discharge faults. These faults produce particular characteristic gases, which can be used to assess and estimate the health index of the transformer [7], [8]. H_2 and CH_4 gases are generated due to low level energy discharge / partial discharge fault while arcing and high level energy discharge generate all gas including traceable amount of C_2H_2. C_2H_4 and C_2H_6 gases are often generated as a result of high thermal faults [4]. While the fault type can be identified based on the amount and type of dissolved gases, the analysis may lead to misinterpretation since there is a possibility of multiple faults cases [9].

From the expert personal experience, it is difficult to determine whether a transformer operates normally or not if it does not have history of dissolved gases along with a trending of each gas. Four criteria as given in Table 1 have been identified in the IEEE C57.104-2008 standards to classify the risk of a transformer without DGA history [12]. Based on individual gases concentrations, the transformer can be grouped into one of the four conditions listed in Table 1. The limits for all key gases in each condition group are listed in Table 2 [12].

Table 1. IEEE C57.104-2008 standard [12].

LEVEL 1	TDCG below this level indicates that the transformer is operating properly. If any of the gas value exceeds this level limit, investigation must be done immediately
LEVEL 2	TDCG at this level indicates the level of flammable gas has exceeded the normal limit. If one of the gas values exceeds these levels, it should be investigated immediately. Possible, fault has occured.
LEVEL 3	TDCG at this level indicates a high level of decomposition. If any of the gas values exceeds this level limit, an investigation must be done immediately. Possible, fault has occured.
LEVEL 4	TDCG exceeds this value limit indicates a very high level of deterioration. Continuing the transformer operation may result in damage to the transformer.

Table 2. Key gases limits for each condition [12]

Limit Concentration of Dissolved Gas (ppm)		STATUS			
		Condition 1	Condition 2	Condition 3	Condition 4
Hidrogen	(H_2)	100	101-700	701-1800	>1800
Methane	(CH_4)	120	121-400	401-1000	>1000
Acetylhene	(C_2H_2)	1	2-9	10-35	>35
Ethylene	(C_2H_4)	50	51-100	101-200	>200
Ethane	(C_2H_6)	65	66-100	101-150	>150
Carbon Monoxyde	(CO)	350	351-570	571-1400	>1400
Carbon Dioxyde	(CO_2)	2500	2500-4000	4001-10000	>10000
TDCG		720	721-1920	1921-4630	>4630

Unfortunately, this technique can only identify the health condition of the transformer without identifying the exact type of fault e of fault within the transformer. Therefore, ratio-based methods such as Roger and IEC were developed to overcome this limitation. However, ratio-based methods cannot be used unless the concentration of the gases used in the analysis is more than certain limits [13-21].

This paper presents a fuzzy logic-based approach that combines the features of IEEE C57.104-2008 standard, Roger and IEC Ratio-based methods.

II. PROPOSED SCORING INDEX AND FUZZY LOGIC MODEL

The proposed DGA-based health scoring index comprises 3 scoring indexes: good with 9 scoring index, fair with 6 scoring index, and poor with 1 scoring index as shown in Table 3. The scoring index reflects the health condition of a transformer based on the measured individual dissolved gases in the oil sample. The simple workflow of this method is shown in Figure 1. The measured concentration of each gas is to be compared with the limits in Table 2. If the health index is good, no further investigation will be required. On the other hand, if the health index is fair or poor, further investigation using Roger and IEC ratio methods are used to accurately identify and quantify the fault type and level.

Table 3. Proposed transformer dissolved gas health index

Condition	Scoring Index	Explanation
Good	9	Indicates that transformer is working and operating properly
Fair	6	Indicates that transformer oil should be investigated immediately using dissolved gas analysis interpretation techniques such as ratio methods (Doernenberg, Roger ratio, IEC), Duval triangle method to accurately identify the fault type. The oil must be re-tested and re-sampled to obtain the trending and dissolve combustible gas growth rate.
Poor	1	Indicates that the transformer have problems especially with oil decomposition and high level of deterioration. The transformer should be investigated immediately using dissolved gas analysis interpretation techniques such as ratio methods (Doernenberg, Roger ratio, IEC), Duval triangle method to accurately identify the fault type. The transformer's operation must be stopped for further analysis.

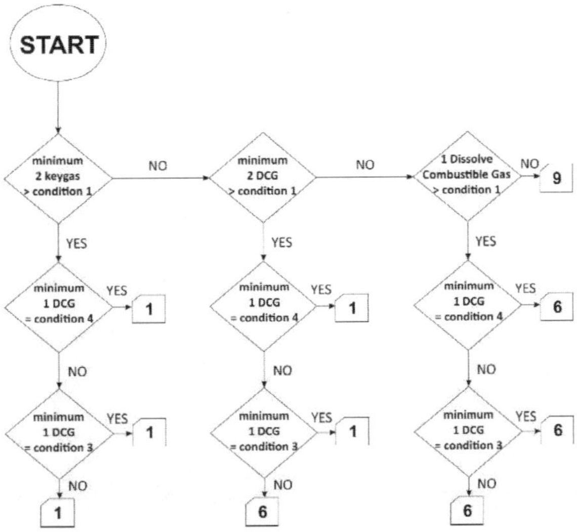

Figure 1. Flowchart of first assessment using IEEE C57.104-2008

The health index score should be 1 (Poor) in the following cases:

-If there are minimum of two key gases (H_2, CH_4, C_2H_4, and C_2H_2) not in condition 1 and at least one gas (H_2, CH_4, C_2H_4, C_2H_2, C_2H_6, and CO) within conditions 2, 3 or 4.

-If there are minimum of two gases ((H_2, CH_4, C_2H_4, C_2H_2, C_2H_6, and CO) not in condition 1 and at least one gas (H_2, CH_4, C_2H_4, C_2H_2, C_2H_6, and CO) within condition 3 or 4.

The health index score should be 6 (Fair) in the following cases:

-If there are minimum of two dissolved combustible gases (H_2, CH_4, C_2H_4, C_2H_2, C_2H_6, and CO) not in condition 1 and at least one gas (H_2, CH_4, C_2H_4, C_2H_2, C_2H_6, and CO) within condition 2.

-If there are minimum of one gas (H_2, CH_4, C_2H_4, C_2H_2, C_2H_6, and CO) not in condition 1 and at least one gas (H_2, CH_4, C_2H_4, C_2H_2, C_2H_6, and CO) within reach condition 2, 3 or 4.

The health index score should be 9 (Good) in the following cases:

-If all of dissolved gases (H_2, CH_4, C_2H_4, C_2H_2, C_2H_6, and CO) in condition 1.

Fuzzy logic rules have been developed using MATLAB fuzzy logic toolbox based on the above as listed in Table 4.

Table 4. The logic bit for the input and output of the proposed fuzzy logic approach

Logic Bit Input					Output of Fuzzy Logic	Output Result Inter-pretation
2 keygas >condition 1 state	2 dissolved combustible gas >condition 1 state	1 dissolved combustible gas >condition 1 state	1 dissolved combustible gas condition 4 state	1 dissolved combustible gas condition 4 state		
1 (High)	1 (High)	1 (High)	0 (Low)	0 (Low)	1	POOR
1 (High)	1 (High)	1 (High)	1 (High)	1 (High)	1	POOR
1 (High)	1 (High)	1 (High)	0 (Low)	1 (High)	1	POOR
0 (Low)	1 (High)	1 (High)	0 (Low)	0 (Low)	6	FAIR
0 (Low)	1 (High)	1 (High)	1 (High)	1 (High)	1	POOR
0 (Low)	1 (High)	1 (High)	0 (Low)	1 (High)	1	POOR
0 (Low)	0 (Low)	1 (High)	0 (Low)	0 (Low)	6	FAIR
0 (Low)	0 (Low)	1 (High)	1 (High)	1 (High)	6	FAIR
0 (Low)	0 (Low)	1 (High)	0 (Low)	1 (High)	6	FAIR
0 (Low)	0 (Low)	0 (Low)	0 (Low)	0 (Low)	9	GOOD

The fuzzy logic output of the IEEE C57.104-2008 comprises three triangle membership functions that reflect the health index of the transformer based on Table 4.

Among the ratio-based methods, Roger ratio method (RRM) and IEC ratio method (IRM) are most widely used by utilities companies, including the Indonesian State Own Electric Utility Company, PT. PLN (Persero). Tables 5 shows proposed fuzzy logic outputs of these two methods. The output membership functions are shown in Figure 2.

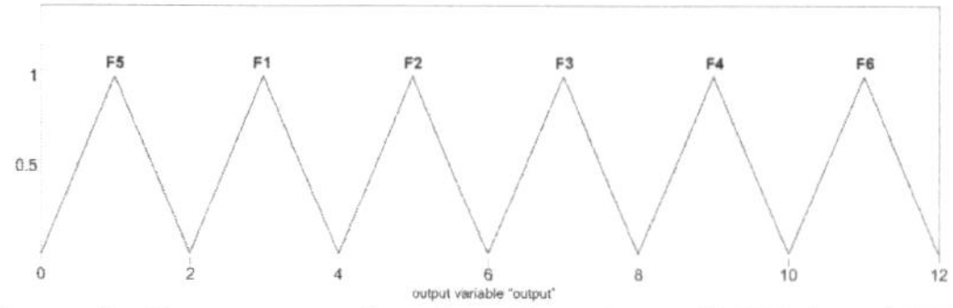

Figure 2. Output membership function of RRM and IRM

Table 5. Fuzzy logic output of RRM and IRM

Method	F1 Thermal in Cellulose Fault	F2 Thermal in Oil Fault	F3 Partial Discharge Fault	F4 Arcing Fault	F5 (Normal condition)	F6 (Out Of Code Condition)
Roger Ratio Method	- Low Temperature Thermal	- Thermal Fault <700°C - Thermal Fault >700°C	Low Energy Density Arcing	High Energy Discharge (Arcing)		
IEC Ratio Method	- Thermal Fault <300°C	- Thermal Fault between 300°-700°C - Thermal Fault >700°C	Low Energy Discharge (Partial DIscharge)	High Energy Discharge (Arcing)		

Inputs of RRM fuzzy logic model are the three ratios CH_4/H_2, C_2H_2/C_2H_6 and C_2H_4/C_2H_6 with member ship functions based on Table 6. Inputs of IRM are the same ratios as in RRM with different ranges and fault types as per Table 7.

Table 6. Fault Diagnosis Interpretation of Roger Ratio Method [4]

RATIO			FAULT DIAGNOSIS
R2 C_2H_2 / C_2H_4	R1 CH_4 / H_2	R5 C_2H_4 / C_2H_6	
< 0,1	0,1 - 1	< 1	0 : Normal
< 0,1	< 0,1	< 1	1 : Partial Discharge
0,1 - 3	0,1 - 1	> 3	2 : Arcing - high energy discharge
< 0,1	0,1 - 1	1 – 3	3 : Low temperature thermal
< 0,1	> 1	1 – 3	4 : Thermal < 700°C
< 0,1	> 1	> 3	5 : Thermal > 700°C

Table 7. Fault Diagnosis Interpretation based on IEC 60599 [22]

RATIO			FAULT DIAGNOSIS
R1 CH_4 / H_2	R2 C_2H_2 / C_2H_4	R5 C_2H_4 / C_2H_6	
< 0,1	NS	< 0,2	PD : Partial Discharge
0,1 – 0,5	> 1	> 1	D1 : Low Energy Discharge
0,1 - 1	0,6 – 2,5	> 2	D2 : High Energy Discharge
> 1 (NS)	NS	< 1	T1 : Thermal fault < 300°C
> 1	< 0,1	1 – 4	T2 : Thermal fault 300°C-700°C
> 1	< 0,2	> 4	T3 : Thermal fault > 700°C

All fuzzy logic models are developed in accordance to the fuzzy inference flowchart shown in Figure 3.

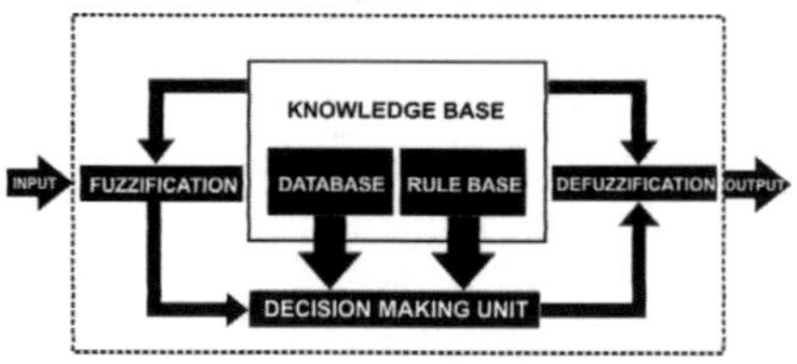

Figure 3. Fuzzy Logic Inference Flowchart

The overall proposed fuzzy logic model is a combination of the three methods discussed above as shown in in Figure 4. The logic table of this model is given in Table 8.

Table 8. Inputs and outputs of the proposed overall model

Individual Concentration Limit of TDCG output	Total Dissolved Combustible Gas (TDCG) value	Roger Ratio Method (RRM) output value	IEC Ratio Method (IRM) output value	Output
9 (Good)	All Condition	All Condition	All Condition	Normal
6 (Fair)	Condition 1	normal	not identified	Normal
		not identified	not identified	
6 (Fair)	Condition 2 / Condition 3 / Condition 4	normal	not identified	Discharge Thermal Fault
		not identified	not identified	
6 (Fair)	Condition 1 / Condition 2 / Condition 3 / Condition 4	not identified	Partial Discharge	Partial Discharge Fault
		not identified	Arcing	Arcing Fault
		Partial Discharge	not identified	Partial Discharge Fault
		Arcing	not identified	Arcing Fault
		normal	Partial Discharge	Partial Discharge Fault
		normal	Arcing	Arcing Fault
		Partial Discharge	Partial Discharge	Partial Discharge Fault
		Partial Discharge	Arcing	Arcing Fault
		Arcing	Partial Discharge	Arcing Fault
		Arcing	Arcing	Arcing Fault
		not identified	thermal in cellulose fault	Thermal in Cellulose Fault
		not identified	thermal in oil fault	Thermal in Oil Fault
		thermal in cellulose fault	not identified	Thermal in Cellulose Fault
		thermal in oil fault	not identified	Thermal in Oil Fault
		normal	thermal in cellulose fault	Thermal in Cellulose Fault
		normal	thermal in oil fault	Thermal in Oil Fault
		thermal in cellulose fault	thermal in cellulose fault	Thermal in Cellulose Fault
		thermal in cellulose fault	thermal in oil fault	Thermal in Oil Fault
		thermal in oil fault	thermal in cellulose fault	Thermal in Oil Fault
		thermal in oil fault	thermal in oil fault	Thermal in Oil Fault
		Partial Discharge	thermal in cellulose fault	Discharge Thermal Fault
		Partial Discharge	thermal in oil fault	
		Arcing	thermal in cellulose fault	
		Arcing	thermal in oil fault	
		thermal in cellulose fault	Partial Discharge	
		thermal in oil fault	Partial Discharge	
		thermal in cellulose fault	Arcing	
		thermal in oil fault	Arcing	
1 (Poor)	Condition 1 / Condition 2 / Condition 3 / Condition 4	not identified	Partial Discharge	Partial Discharge Fault
		not identified	Arcing	Arcing Fault
		Partial Discharge	not identified	Partial Discharge Fault
		Arcing	not identified	Arcing Fault
		normal	Partial Discharge	Partial Discharge Fault
		normal	Arcing	Arcing Fault
		Partial Discharge	Partial Discharge	Partial Discharge Fault
		Partial Discharge	Arcing	Arcing Fault
		Arcing	Partial Discharge	Arcing Fault
		Arcing	Arcing	Arcing Fault
		not identified	thermal in cellulose fault	Thermal in Cellulose Fault
		not identified	thermal in oil fault	Thermal in Oil Fault
		thermal in cellulose fault	not identified	Thermal in Cellulose Fault
		thermal in oil fault	not identified	Thermal in Oil Fault
		normal	thermal in cellulose fault	Thermal in Cellulose Fault
		normal	thermal in oil fault	Thermal in Oil Fault
		thermal in cellulose fault	thermal in cellulose fault	Thermal in Cellulose Fault
		thermal in cellulose fault	thermal in oil fault	Thermal in Oil Fault
		thermal in oil fault	thermal in cellulose fault	Thermal in Oil Fault
		thermal in oil fault	thermal in oil fault	Thermal in Oil Fault
		not identified	not identified	Discharge Thermal Fault
		normal	not identified	
		Partial Discharge	thermal in cellulose fault	
		Partial Discharge	thermal in oil fault	
		Arcing	thermal in cellulose fault	
		Arcing	thermal in oil fault	
		thermal in cellulose fault	Partial Discharge	
		thermal in oil fault	Partial Discharge	
		thermal in cellulose fault	Arcing	
		thermal in oil fault	Arcing	

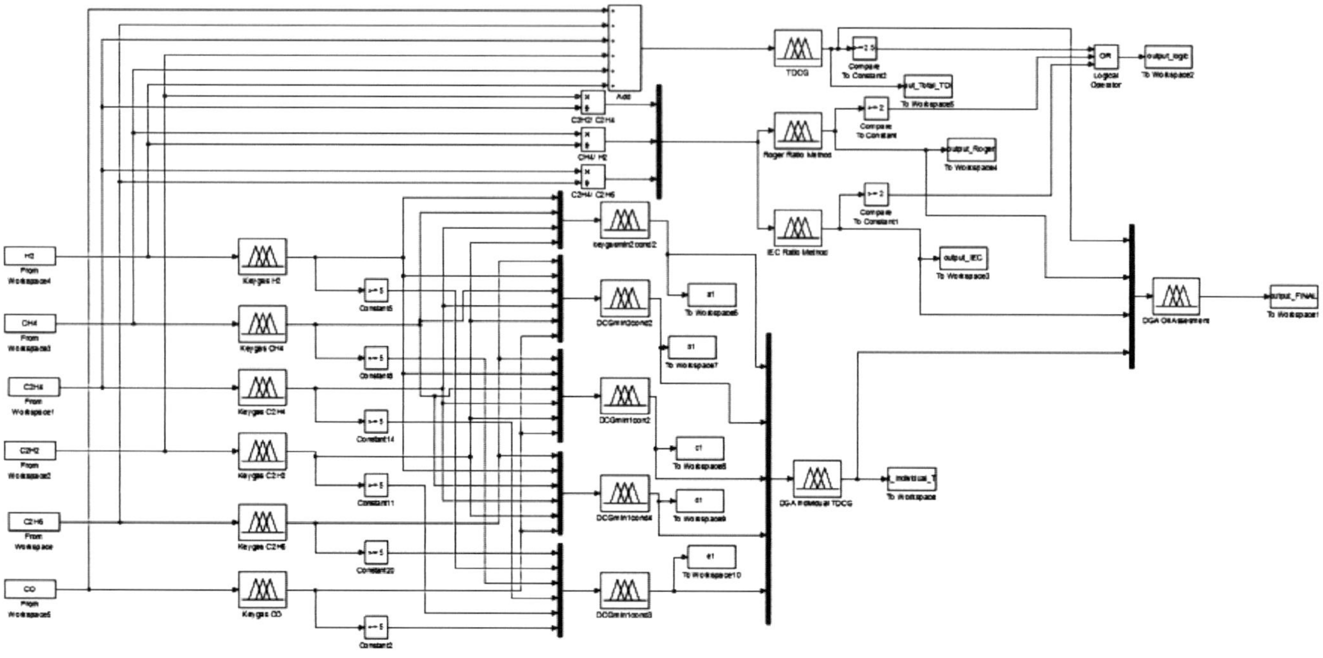

Figure 4. Overall fuzzy logic-based model

Table 9. Data Simulation Result of Dissolved Gas Transformer Oil Samples

Sample Number	Power	Voltage Rating	Operational Ages (Years)	(CO)	(H2)	(CH4)	(C2H4)	(C2H6)	(C2H2)	Simulation Individual gas TDCG	Simulation TDCG each gas Result	Simulation TDCG Result	Simulation Roger Ratio Result	Simulation IEC Ratio Result	Output Proposed Fuzzy	Output Result	Real Condition
1	500 MVA	500/150 kV	7	0	96,896	84,038	3,675	3,675	0	9	GOOD	CONDITION 1	Thermal Fault in Cellulose	Not Identified	10,99813	Normal	Normal
2	20 MVA	150/20 kV	29	244,946	42,664	18,543	1,717	37,789	0	9	GOOD	CONDITION 1	Normal	Thermal Fault in Cellulose	10,99846	Normal	Normal
3	60 MVA	150/20 kV	12	16,36	36,24	6,45	4,26	12,36	1,42	9	GOOD	CONDITION 1	Not Identified	Thermal Fault in Cellulose	10,99846	Normal	Normal
4	20 MVA	150/20 kV	35	91,16	47,49	0	3,75	6,89	0	9	GOOD	CONDITION 1	Partial Discharge	Not Identified	10,99846	Normal	Normal
5	60 MVA	150/20 kV	6	357,855	156,3	10,72	8,9	5,21	0	6	FAIR	CONDITION 1	Not Identified	Not Identified	10,99846	Normal	Normal
6	30 MVA	150/20 kV	25	124,03	481,82	9,03	5,72	37,04	0	6	FAIR	CONDITION 1	Partial Discharge	Partial Discharge	4,998462	Partial Discharge Fault	Fault, High Temperature in Winding
7	50 MVA	150/20 kV	12	492,34	20	0	1,2	3,19	0	6	FAIR	CONDITION 1	Partial Discharge	Not Identified	4,998462	Partial Discharge Fault	Fault, Partial Discharge
8	30 MVA	150/20 kV	8	36,92	20	0	2,23	46,34	14,3	6	FAIR	CONDITION 1	Not Identified	Partial Discharge	4,998462	Partial Discharge Fault	Fault, Partial Discharge
9	60 MVA	150/20 kV	12	412,8	54,1	63,8	77	60,9	0	6	FAIR	CONDITION 1	Thermal Fault in Oil	Thermal Fault in Oil	7,001538	Thermal in Oil Fault	Fault, overheat in oil
10	60 MVA	150/20 kV	25	128,212	51,026	74,558	60,841	62,067	1	6	FAIR	CONDITION 1	Arcing	Thermal Fault in Cellulose	1,004054	Discharge and Thermal Fault	Fault, arcing and overheat
11	10 MVA	150/20 kV	17	0	20	50,32	4,67	154,93	32,63	1	POOR	CONDITION 1	Not Identified	Thermal Fault in Cellulose	9	Thermal in Cellulose Fault	Fault, overheat-high temperature
12	60 MVA	150/20 kV	15	154,665	79,16	0	78,825	106,898	0	1	POOR	CONDITION 1	Partial Discharge	Not Identified	4,998462	Partial Discharge Fault	Fault, Partial Discharge and overheat
13	30 MVA	150/20 kV	22	0	20	223,26	98,05	5,67	236,6	1	POOR	CONDITION 1	Not Identified	Not Identified	1,001538	Discharge and Thermal Fault	Fault, Partial Discharge and overheat
14	30 MVA	150/20 kV	21	16,41	20	0	4,94	206,71	14,67	1	POOR	CONDITION 1	Not Identified	Partial Discharge	4,998462	Partial Discharge Fault	Fault, Partial Discharge
15	500 MVA	500/150 kV	7	640	46	21	13	5	1	6	FAIR	CONDITION 2	Thermal Fault in Cellulose	Not Identified	9	Thermal in Cellulose Fault	Fault, overheat-high temperature
16	60 MVA	150/20 kV	19	623,379	67,019	19,603	49,183	1	1	6	FAIR	CONDITION 2	Not Identified	Not Identified	1,001538	Discharge and Thermal Fault	Fault, Partial Discharge and overheat
17	30 MVA	150/20 kV	23	883,07	20	22,55	42,37	195,19	0	1	POOR	CONDITION 2	Not Identified	Thermal Fault in Cellulose	9	Thermal in Cellulose Fault	Fault, overheat-high temperature
18	30 MVA	150/20 kV	15	689,093	201,78	30,81	89,921	89,921	0	1	POOR	CONDITION 2	Thermal Fault in Cellulose	Not Identified	9	Thermal in Cellulose Fault	Fault, High Temperature in Winding
19	30 MVA	150/20 kV	12	869,921	177,47	10,945	146,379	146,379	0	1	POOR	CONDITION 2	Not Identified	Not Identified	1,002611	Discharge and Thermal Fault	Fault, Partial Discharge and High Temperature in Winding
20	50 MVA	150/20 kV	12	673,88	183,54	184,39	34,09	382,27	0	1	POOR	CONDITION 2	Arcing	Thermal Fault in Cellulose	1,001527	Discharge and Thermal Fault	Fault, arcing and overheat
21	50 MVA	150/20 kV	23	556,64	478,42	14,74	2,56	35,39	21,29	1	POOR	CONDITION 2	Not Identified	Partial Discharge	4,998462	Partial Discharge Fault	Fault, Partial Discharge
22	60 MVA	150/20 kV	22	2354,91	20	21,42	2,27	6,28	0	6	FAIR	CONDITION 3	Not Identified	Thermal Fault in Cellulose	9	Thermal in Cellulose Fault	Fault, overheat-high temperature
23	50 MVA	150/20 kV	23	0	2230,9	10,78	6,89	4,61	0	6	FAIR	CONDITION 3	Not Identified	Not Identified	1,001538	Discharge and Thermal Fault	Fault, Partial Discharge and overheat
24	60 MVA	150/20 kV	22	0	36,98	2943,7	34,99	2,97	0	6	FAIR	CONDITION 3	Thermal Fault in Oil	Thermal Fault in Oil	7,001538	Thermal in Oil Fault	Fault, overheat in oil
25	15 MVA	150/20 kV	32	3300	112	73	64	26	0	1	POOR	CONDITION 3	Thermal Fault in Cellulose	Not Identified	9	Thermal in Cellulose Fault	Fault, overheat-high temperature
26	30 MVA	150/20 kV	20	1589,71	522,08	35,88	32,85	18,61	187,19	1	POOR	CONDITION 3	Not Identified	Not Identified	1,001538	Discharge and Thermal Fault	Fault, Partial Discharge and High Temperature in Oil
27	30 MVA	150/20 kV	8	749,56	1322,2	627,77	1032,55	158,48	1139,65	1	POOR	CONDITION 4	Arcing	Arcing	3	Arcing Fault	Fault, Arcing in Oil
28	50 MVA	150/20 kV	12	0	1454	1179,9	440,45	642,27	1028,94	1	POOR	CONDITION 4	Not Identified	Thermal Fault in Cellulose	9	Thermal in Cellulose Fault	Fault, High Temperature in Winding
29	60 MVA	150/20 kV	18	473,41	2591,1	907,02	971,03	639,85	2058,19	1	POOR	CONDITION 4	Not Identified	Partial Discharge	4,998462	Partial Discharge Fault	Fault, Partial Discharge
30	20 MVA	70/20 kV	26	0	247,34	216,59	3547,55	1036,51	0	1	POOR	CONDITION 4	Not Identified	Not Identified	1,001538	Discharge and Thermal Fault	Fault, arcing and high temperature
31	60 MVA	150/20 kV	13	3255,27	3324,9	378,33	6,39	80,92	0	1	POOR	CONDITION 4	Normal	Thermal Fault in Cellulose	9	Thermal in Cellulose Fault	Fault, overheat-high temperature

3. VALIDATION OF PROPOSED FUZZY LOGIC MODEL

To assess the accuracy of the proposed model in identifying and quantifying various faults based on the three combined DGA interpretation techniques, 31 dissolved gas transformer oil data were collected from Indonesia National Electrical Utility Company, PT. PLN (Persero) Jawa, Bali, and Sumatera. To design the fuzzy logic model, 469 transformer oil samples that were collected from different voltage rating: 500/150 kV, 150/20 kV, and 70/20 kV; different capacity: 500 MVA, 60 MVA, 50 MVA, 30 MVA, 20 MVA, 15 MVA, and 10 MVA; and different life span from 7 years to 35 years operational ages, are used to test and establish the fuzzy logic model. Normal and pre-known fault types transformers are tested using the proposed fuzzy logic model as shown in Table 9. From 31 transformer oil samples, there are 5 transformers in good condition: sample number 1 to 5; and 26 transformers in faulty condition: sample number 6 to 31. The normal transformers that were chosen in table 9 had different life span and operational ages: 2 transformer with long operational ages, 1 transformer with medium operational ages and 2 transformer with early operational ages. The data samples of faulty transformers that were chosen also reflect that 3 condition of operational ages: long operational ages, normal operational ages, and early operational ages. Each samples also have unique combination of each dissolved gas analysis interpretation method so it can check the result of proposed fuzzy logic model with the real fault of transformers. Results in Table 9 reveal a high agreement of the model output with the real fault. It proves that proposed fuzzy logic approach which combine several methods: Individual Dissolved Combustible Gas concentration called Health Index, TDCG, Roger Ratio Methods, and IEC Ratio Method, can be used and can be applied to different voltage rating, different power rating, and different lifespan operational ages of transformer with satisfying result. As such, it can be concluded that the proposed model can be used to identify the type of internal faults within a transformer based on dissolved gas analysis data with high degree of accuracy.

IV. CONCLUSION

This paper presents a fuzzy logic-based technique that combines the feature of three dissolved gas analysis interpretation techniques; IEEE C57.104-2008, Roger and IEC ratio methods. The proposed technique is aiming at overcoming the limitations involved in these methods. This includes the conservative nature of the IEEE method and the inability of the ratio methods to diagnose faults when gases of low concentrations exist within the transformer oil samples. Accuracy of the proposed technique to identify various faults has been assessed by comparing the output of the developed model with the pre-known faults of 31 samples collected from PT. PLN (Persero) Jawa, Bali, and Sumatera with different voltage rating, different power rating, and different lifespan operational ages. Results reveal the high agreement between the model's output and the real fault types. The proposed expert system can be used by non-experienced persons to help them gaining an experience in dissolved gas analysis.

REFERENCES

[1] A. Abu-Siada and S. Islam "A Novel online technique to detect Power Transformer Winding Fault." IEEE Trans Power Delivery, Vol. 27 No 2, pp. 849-857, 2012.
[2] T. K. Saha. "Review of Modern diagnostic techniques for assessing insulation condition in aged transformer Dielectric and Electrical Insulation, IEEE Transaction on Vol. 10, pp 903-917, 2013.
[3] N. A. Bakar, A. Abu Siada, and S. Islam. "A review of dissolved gas analysis measurement and interpretation techniques." *IEEE Electrical Insulation Magazine*, Vol. 30, No. 3, pp 39-49, 2014.
[4] "IEEE Guide for the interpretation of Gases Generated in Oil Immersed Transformers—Redline." IEEE Std. C57.104-2008, pp, 1-45, 2009.
[5] "IEEE Guide for the Detection and Determination of Generated Gas in Oil-Immersed Transformer and the relation to Serviceability of the Equipment." ANSI/IEEE C57.104-1978 p. 0-1, 1978.
[6] "IEEE Guide for the Interpretation on Gases Generated in Oil-Immersed Transformer." IEEE Std. C57.104-1991, p 0-1, 1992.
[7] M. Duval and A. dePabla,"Interpretation of gas in oil analysis using new IEC Publication 60599 and IEC TC 10 databases." *IEEE Electrical Insulation Magazine*, Vol. 17, pp. 31-41, 2001.
[8] H-C Sun, Y-C Huang, and C-M. Huang. "A Review of Dissolved Gas Analysis in Power Transformer." *Energy Procedia*, Vol. 14, pp. 1220-1225, 2012.

[9] A. Abu-Siada and S. Islam." A new approach to identify power transformer and asset management decision based on dissolved gas in oil analysis." *IEEE Transaction on Dielectric and Electrical Insulation*, vol 19, pp. 1007-1012, 2012.

[10] H. Li and M. Gupta, Fuzzy Logic and Intelligent Systems, International Series in Intelligent Technologies, Kluwer Academic Publisher, 1995.

[11] A. Abu-Siada, S. Hmood, and S. Islam. "A New Fuzzy Logic Approach for Consistent Interpretation of Dissolved Gas-in-Oil Analysis." *IEEE Transactions on Dielectrics and Electrical Insulation*, Vol. 20, No. 6, pp. 2348, December 2013.

[12] "IEEE Guide for the Interpretation on Gases Generated in Oil-Immersed Transformer." IEEE Std. C57.104-2008, p 9-10, 2008.

[13] M. Arshad and S. Islam,"A Novel Fuzzy Logic Technique for Power Transformer Asset Management," in *Industry Application Conference, 2006, 41st IAS Annual Meeting Conference Record of the 2006 IEEE,* 2006, pp 276-286.

[14] M. Arshad, S. Islam, and A. Khaliq," Fuzzy logic approach in power transformer management and decision making,"*IEEE Transactions on Dielectric and Electrical Insulation,*vol 20, pp. 2343-2354, 2014.

[15] A. Abu-Siada, M. Arshad, and S. Islam, "Fuzzy logic approach to identify transformer critically using dissolve gas analysis," *in Power and Energy Society General Meeting, 2010, IEEE*, pp. 1-5, 2010.

[16] S. Mofizul Islam, T. Wu, and G. Ledwich," A novel fuzzy logic approach to transformer fault diagnosis," *IEEE Transaction on Dielectric and Electrical Insulation,* vol.7, pp.177-186, 2000.

[17] S. Hmood, A. Abu Siada, Mohammad A.S. Masoum amd Syed M. Islam. "Standardization of DGA Interpretation Techniques using Fuzzy Logic Approach." *2012 IEEE Int'l Conf. on Condition Monitoring and Diagnosis, 23-27 September 2012, Bali, Indonesia*. Pp. 929-932, 2012.

[18] A. Abu-Siada and S. Hmood, "Fuzzy Logic Approach for Power Transformer Asset Management Based on Dissolved Gas-in-Oil Analysis", *Chemical Engineering Transactions*, Vol 33, No 2, pp. 997-1002, September 2013.

[19] N. A. Muhammad, B.T. Phung, and T. R. Blackburn, "Comparative study and analysis of DGA methods for mineral oil using fuzzy logic", *Internasional Conf. Power Engineering (IPEC)*, pp. 1301-1306, 2007.

[20] M. Hongzhong, L. Zheng, P. Ju, H. Jingdong, and Z. Limin, "Diagnosis of power transformer faults on fuzzy three-ratio method", 7th International Conference Power Engineering (IPEC), pp. 1-4, 2005.

[21] S. M. Islam, T. Wu, and G. Ledwich," A novel fuzzy logic approach to transformer fault diagnosis," *IEEE Transaction Dielectric Electrical Insulation*, Vol. 7, pp. 177-186, 2000.

[22] "International Standard Mineral Oil-Filled Equipment in service – Guidance on the interpretation of dissolved and free gases analysis." IEC 60599 Edition 3.0, p 14-15, 2015.

Condition monitoring thermal properties of a 20A hydraulic-magnetic MCB

R. Kleynhans[1], P.A. van Vuuren[1], R. A. Thomas[2]
[1] School of Electrical, Electronic and Computer Engineering, North-West University, South Africa
[2] School of Management, Swansea University, Bay Campus, Swansea, Wales, UK

ABSTRACT

Heat is one of the major causes of degradation which can cause a reduction in performance, reliability and life span of miniature circuit breakers (MCBs) [1]. In order to diagnose the condition of electrical equipment, the thermal stress of that equipment should be known [2]. Due to the compact design of MCB's it is important to minimize or dissipate the heat that builds up inside MCBs during normal use. With time electrical components and contact surfaces heats up and begin to deteriorate, this increase in heat is often a product of increased electrical resistance which is a result of degradation [3]. Infrared thermography is a condition monitoring technique that can be applied to electrical distribution boards but because of design limitations MCBs are stacked next to each other on a rail in a panel allowing a direct view only from the front. Is the surface temperature of the MCB a reflection of the internal degradation or are there other factors that need to be considered to accurately diagnose the MCB?

The following methodology for the exploratory study was followed to determine the heat profile after degradation, a 20A hydraulic-magnetic MCB was switched 9000 times at rated voltage and current. The volt drop across the conduction path as well as thermal images of the front and sides were taken after every 500 switching operations, after the MCB was allowed to heat up for 1 hour at rated current. When comparing the surface temperature rise above ambient of the three sides the results shows that the number of operations does not influence the surface temperature of the front and the left (hydraulic magnetic unit) as much as the right (conduction path).

The aim of this paper is to use a lumped model and compare with empirical infrared thermographic results relating to the condition of MCBs and to:

Define the MCB surface heat distribution after a number of switching operations at rated current.
Determine whether the frontal surface temperature of the MCB is an accurate indicator of internal degradation.
Determine a field methodology to determine degradation and to anticipate failure of MCBs?

Due to the construction of the MCB the front, left and right sides give different heat patterns that complicate matter with regards to determining the correct internal health state of the MCB.

Keywords: condition monitoring, thermal imaging, miniature circuit breaker, degradation, hydraulic magnetic

1. INTRODUCTION

An electrical installation protection system is the most important component affecting the performance and safety of personnel under abnormal conditions. Protection devices such as miniature circuit breaker (MCB) are designed to detect electrical faults and to promptly disconnect the fault component, in order to reduce the effect of the fault on the rest of the installation. The speed at which the MCB disconnect the fault is dependent on the type of MCB and can be affected by temperature, for example, fatigue failure. The importance of protection devices such as circuit breakers in a LV (low voltage) network are often overlooked and may lead to unnecessary damage to expensive equipment [3]. A small fraction of downtime during a system failure is spent to maintain or repair the faulty components; however, during this downtime 80% is spent trying to locate the source of the fault [4].

In terms of electrical failure the same natural laws i.e. Stefan-Boltzmann law, Ohm's law, Lenz's law, Newton's, Faraday's law etc. are applied in normal operation of equipment. Electrical component failure is thus obeying natural laws when it fails. Electrical degradation processes may include electrical heating, thermal degradation and contact resistance [5]. The uncontrolled flow of electrical energy is generally the result of misapplication, misuse or accident and manifests itself in the form of electrical fault currents either in intended or unintended electrical circuits or paths [6]. Thermal imaging has proven itself as a useful condition monitoring tool as part of an intelligent asset management program to prevent unplanned breakdowns and to increase plant availability [2].

Due to the internal mechanical connections of MCB's which can radiate heat under load, it is not always straightforward to pin point the root cause of the excessive heat due to resistance. Early detection and analysis will lead to better prediction of remaining useful life estimates. Prognostics and health management for MCB's will detect, isolate and predict the onset and source of degradation as well as time to failure [7]. Misdiagnosed thermal anomalies can result in wasted time and money [8]. As heat is a product of normal energy associated with normal operations, heat inside an electrical system can be seen as wasted energy and also contributes to the wear of components whether it is insulation material or mechanical parts. The thermal performance of MCB's is dependent on the geometry of their construction, materials used, dynamics of operation and the aerodynamics in the cabinet. A hydraulic magnetic trip unit uses the induced magnetic flux to separate the contacts and not heat due to current flow. Hydraulic magnetic thermal MCB's are usually temperature stable and aren't affected by ambient changes as the case is with bimetallic trip units.

A lumped model is useful to determine the thermal distribution inside a MCB often in a topographic way which could be compared to the thermographic results. The difficulty in this is that the heat generation mechanisms are often tightly coupled with the electrical operation of the circuit. In order to create a lumped parameter model of the MCB the individual segment resistance of the current path should be determined. Due to these segments being arranged in series a simple network model of the whole circuit can be built. The resistance across the terminals will then give an indication of the combined resistance of the conduction path. The electrical notations for "current", "voltage" and "electrical resistance" can be replaced by "heat flux (power)", "temperature" and "thermal resistance" in a lumped parameter model [9]. Modeling a thermal equivalent circuit consist of three steps:

- Identification of heat transfer mechanism.
- Definition of thermal parameters of each part of the MCB.
- Validation of the model [10].

The rest of this paper is structured as follows: Section 2 provides an overview of related work. Background information on the construction and operation of MCBs is given in Section 3. Section 4 describes the methodology that was followed to prepare the exploratory practical experiments. Section 5 summarizes the results and Section 6 discusses those results. Lastly, a conclusion is drawn in Section 7.

2. RELATED WORK

Siegel *et al.* concluded that contact points inside a MCB influences the electrical and thermal resistance. The electrical resistance is responsible for terminal temperature rise and the thermal resistance the temperature of the internal hot spot [11].

The majority of MCB simulations found to validate experimental results were completed to determine the forces and arc behavior inside a MCB. Arulanantha, 2015, validated that an inverse problem can be used to model the complex arcing phenomenon, which is governed by coupled physics of electromagnetism and thermal dynamics [12]. Stammberger *et al.* discussed lumped network models for calculating short circuit current, sequential coupling of finite element codes and tight coupling of magnetic field and fluid dynamics that run simultaneously. Proving that calculating the thermal behavior is complex due to the combination of separate finite element and finite volume calculations where boundary conditions and source terms are exchanged manually [13]. Barcikowki, 2002, performed comprehensive analysis on heat transportation in switchgear and came to the conclusion that convection and radiation inside a device has a negligible effect on the resulting temperature distribution. Gregory B McIntosh, 2013, discussed various considerations for assessing the condition of electrical connections using infrared thermography revealing a thermal anomaly but limited his work to electrical connections.

3. BACKGROUND ON THE CONSTRUCTION AND PRINCIPLE OF OPERATION OF A HYDRAULIC MAGNETIC MCB

MCB construction is simple but very precise. With no replacement parts the unit is replaced when it fails to operate as per manufacturer's chart [14]. LV MCBs must be able to carry rated current, interrupt an overload current, successfully interrupt short circuits and enable manual on/off operations within prescribed temperature and humidity ranges [15].

In order for a MCB to carry out the four functions mentioned above it needs to consist of five sub-systems either independent or as a combination:

I. **Conduction path** – Carrying the rated current and consist of conductors and contact tips. Since the conduction path is the only heat source it is primarily responsible for the temperature rise of the MCB [15]. The conduction path starts with the line terminal, through the moving contacts into the top and bottom frame, through the sensing unit and out the load terminal.

II. **Sensing unit** – Detect the overload and initiate opening of the moving contacts. A hydraulic magnetic MCB operates by using the magnetic force produced by the current flowing through a coil where the coil and the load are in series. The coil is wound around a hermetically sealed tube containing a mild steel core, a spring and silicon oil as a hydraulic damping fluid.The silicon oil slows the movement of the core's velocity to create a controlled time delay. The magnetic flux created by the coil during an overcurrent produces sufficient pull on the core to start movement towards the pole piece. With a persistant overload the core reaches the pole piece and by doing so the reluctance of the magnetic circuit drops considerably. The armature is attracted to the pole piece with sufficient force to collapse the latching mechanism [15].

   During a short circuit the magnetic force produced by the coil is sufficient to attract the armature to the pole piece and trip the MCB. During this operation the core has not moved [15].

III. **Arc Chamber** – This is where the arc is formed during contact opening and is controled and extinguished to interrupt the current. When the contacts open a magnetic force blows the arc along the arc runners and into the arc grids. Erosion is reduced if the arc is removed from the contact area.The purpose of the arc grids is to split the arc into a number of shorter arcs and to cool the arc. This will raise the impedance and hence increase the arc voltage drop. The rapid rise in arc voltage limits the current. Current ceases to flow and the arc is extinguished when the current is forced to zero. The damage to the MCB is determined by the amount of energy let-through and is determined by the opening time of the moving contacts and efficiency of the magnetic force which blows the arc off the contact tips and into the arc grids [15].

IV. **Mechanism** – Open automatically when activated by the sensing unit or open and close the moving contacts manually [15]. The speed is dependent on the type of MCB.

V. **Molded case** – The housing of the MCB which is made out of [15].

Figure 1 shows a photograph of the components and configuration inside an MCB.

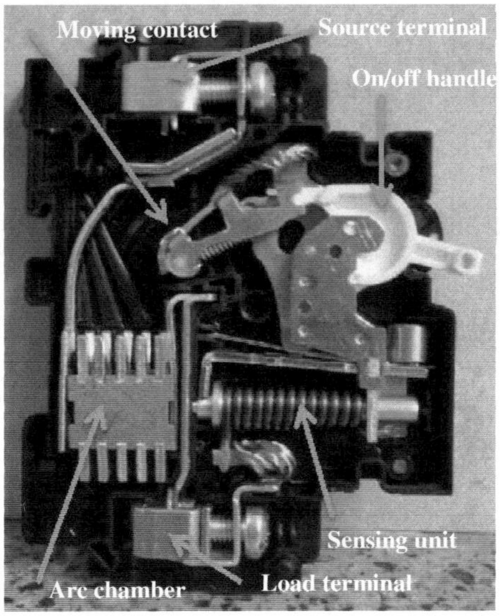

Figure1: Photograph showing the typical components of a miniature circuit breaker.

4. METHODOLOGY

An electrical lumped parameter model of the main heat generating components in a MCB is as follows:

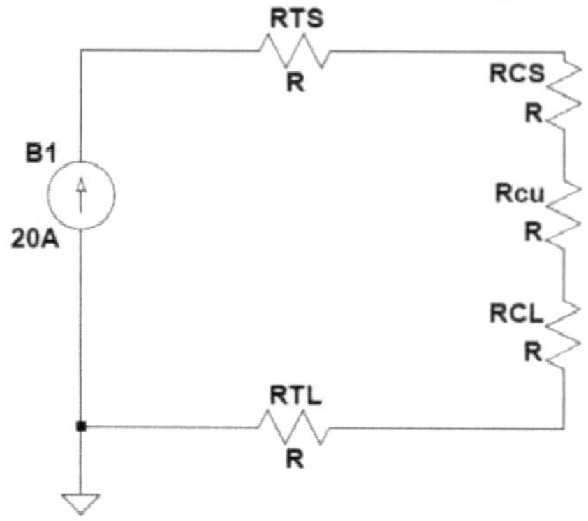

Figure 2: Circuit diagram of the electrical domain of an MCB

Where:

- I_{source} represents the external DC power supply
- R_{TS} represents the resistance at the terminal contact between the source and the MCB
- R_{CS} represents the resistance at the contact point between the switch and the source
- R_{cu} represents the copper resistance of all conductors in the system (lumped together)
- R_{CL} represents the resistance of the connection point between the switching mechanism and the load
- R_{TL} represents the resistance of the terminal contact between the MCB and the load

The electrical resistance of a length of conductor is calculated from:

$$R = \frac{\rho l}{A}, \qquad (1)$$

With:

- ρ = resistivity [Ω.m]
- l = length of conductor [m]
- A = cross section area [m^2]

The resistivity in turn is temperature dependent:

$$\rho - \rho_0 = \rho_0 \alpha (T - T_0)\,, \qquad (2)$$

With:

- ρ_0 = resistivity at reference temperature T_0 [Ω.m]
- T = current temperature [°C]
- ρ = resistivity at the current temperature [Ω.m]
- α = temperature coefficient of resistance [1/°C]

For the practical experiment a commercially available 20A curve 2 single pole MCB was used to perform an accelerated endurance test at ambient temperature which are aligned with the IEC 60947-2 international standard. The aim of an endurance test is to check the electrical and mechanical performance after a number of switching operations at rated voltage and current [16]. The following steps were followed:

Step 1: A never operated MCB was placed on a static test bench where it was allowed to heat up at rated current, 20A, for 1 hour at ambient temperature, i.e. until steady state has been reached. The load simulation was carried out with a DC power source to maintain a constant current, to limit the additional heat generation, due to the influence of alternating current, like eddy currents and core losses [11]. Infrared thermal images were taken of the front, left and right side of the MCB with a FLUKE Ti400 radiometer (7.5μm to 14μm, ≤0.05°C at 30°C target temp). It was assumed that the emissivity of the MCB housing was 0.96. The ambient temperature was recorded with a Kestrel 3000 pocket weather meter to determine the temperature rise above ambient. The volt drop over the terminals was measured to determine the combined resistance of the conduction path. The total resistance of the conduction path is the sum of the inherent resistances of the components as well as the sum of the connection resistances. The thermal images as well as the resistance values obtained were used as reference values.

Step 2: The MCB was moved to the endurance test bench and positioned in the off position for 9 seconds, then switched to the on position for 1 second. This test was performed at rated current, 20A, and voltage, 250V. The test frequency was 6 switching cycles per minute. After 500 switching cycles the MCB was moved from the endurance test bench back to the static test bench.

Step 3: The MCB was placed on the static test bench where it was allowed to heat up at rated current, 20A, for 1 hour. After steady state has been reached thermal images of the front and sides of the MCB were taken to evaluate the temperature rise after some degradation has set in and to compare the temperature rise to the reference values obtained in step 1. The ambient temperature was recorded to determine the temperature rise above ambient. The volt drop over the terminals was also measured to determine the combined resistance of the conduction path to record the change in resistance after multiple switching operations at rated current and voltage.

Step 2 and *step 3* was repeated until the MCB has reached 9000 operations.

5. RESULTS

A graphical representation in Figure 3 illustrates the surface temperature rise above ambient of the highest temperatures recorded on the front, left and right sides of the MCB starting with the reference values up to 9000 switching operations. The conducted and/or convected temperatures on the outer shell of the MCB of interest on the right is that of the top frame and on the left that of the hydraulic magnetic unit. During the experiment the ambient temperature fluctuated by 3.9°C. The ΔT for the rise above ambient on the front, left and right sides were 1.58°C, 3.91°C and 11.13°C between 0 and 9000 operations.

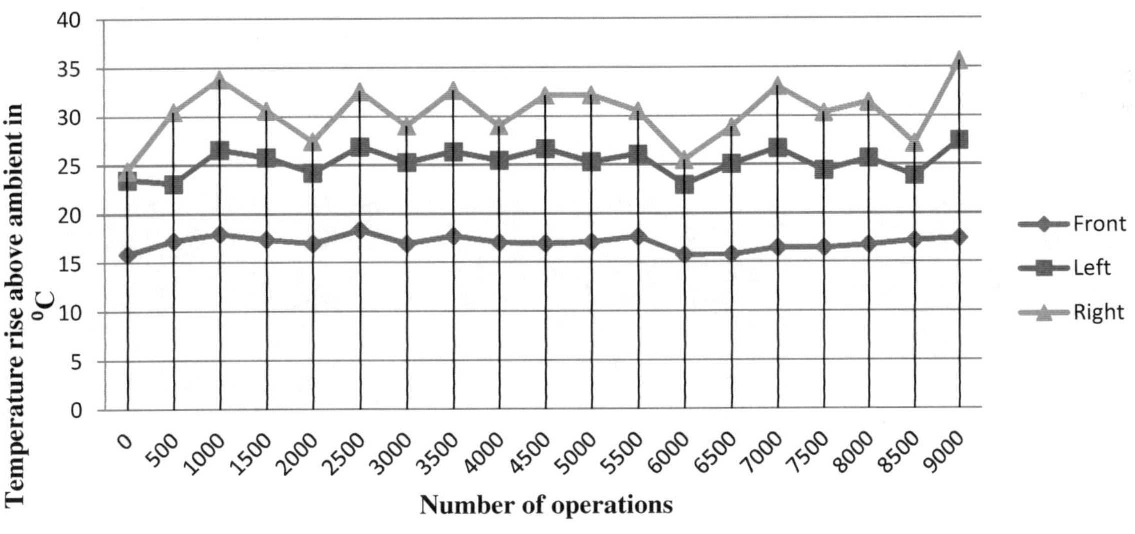

Figure 3 Temperature rise above ambient of a 20A MCB after multiple switching cycles at full load

The change in the measured resistance (black line) in Figure 4 shows the conduction path after multiple operations at rated current and voltage. The blue line represents the calculated resistance value of the conduction path over the switching period. The fact that it is much lower than the measured resistance can be contributed to the fact that the exact transfer mechanism as well as the exact parameters of each part of the MCB was not known. The temperature rise on the right side (red line) of the MCB is also plotted on the same graph fitted with a quartic to make the trend clear. The electrical resistance of the components in the current path is dependent on its geometry and the specific electrical resistance of the material used. The specific electrical resistance is temperature dependent [11]. Oxidation and area of contact of the moving contacts is a significant contributor to the overall measured resistance of the conduction path.

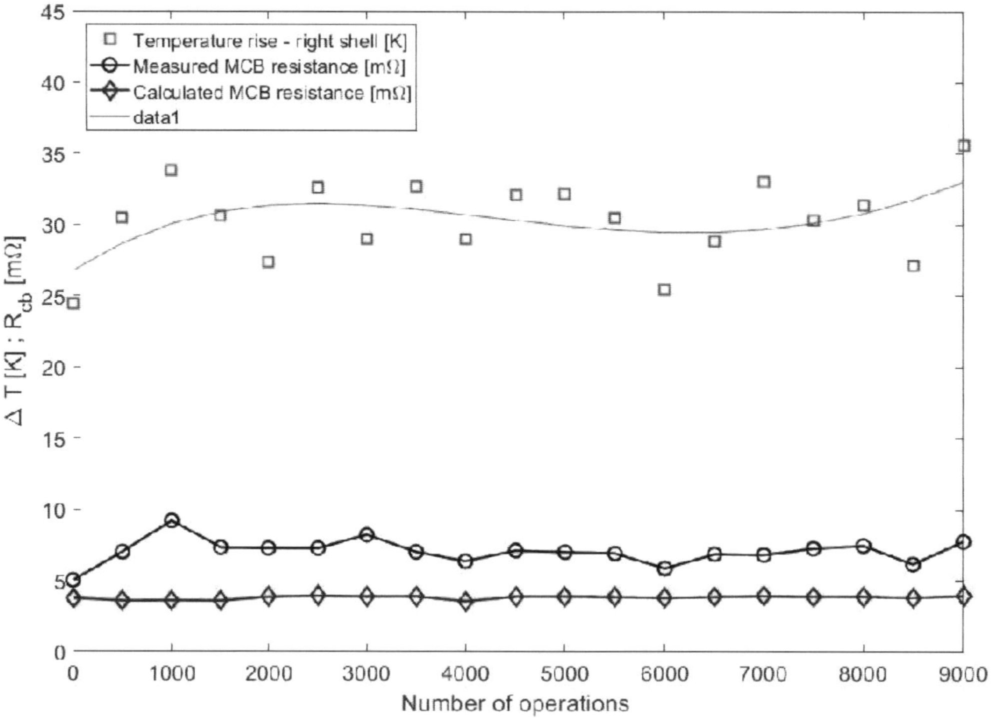

Figure 4 Measured and calculated resistance of conduction path after multiple operations and temperature rise on right side

Figure 5 shows the reference front (a), right (b) and left(c) thermal view of the MCB before any operations but after it was heated for 1 hour at rated current, 20A, as described in step 1.

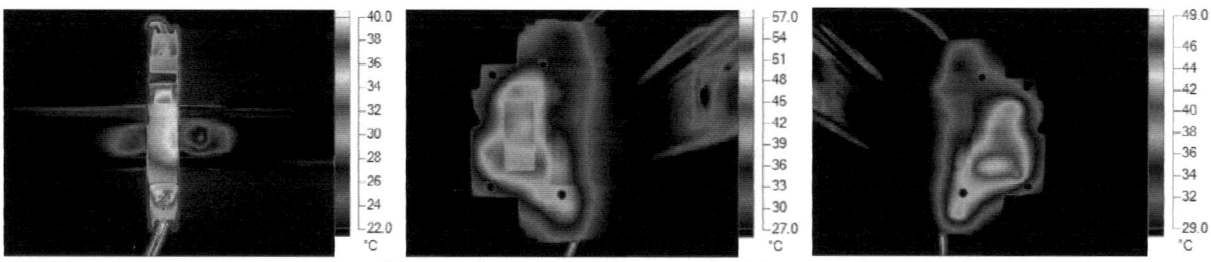

(a) (b) (c)

Figure 5 Thermal images of the front (a), right (b) and left(c) of the MCB before any operations

Figure 6 shows the front (a), right (b) and left(c) thermal view of the MCB after 9000 operations and heated for 1 hour at rated current, 20A, as described in step 2 and step 3.

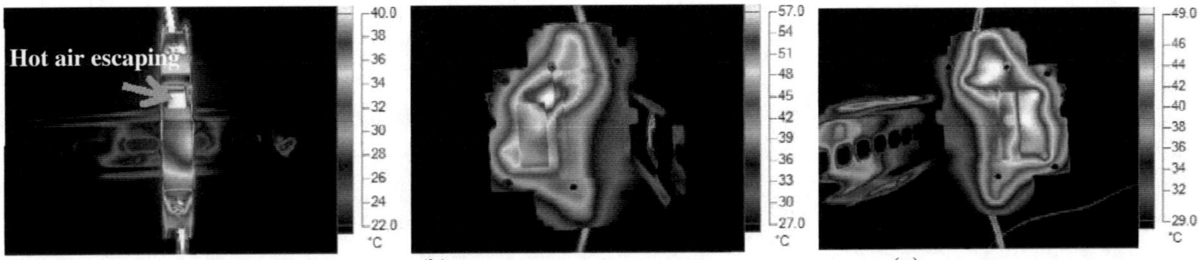

(a) (b) (c)

Figure 6 Thermal images of the front (a), right (b) and left(c) of the MCB after 9000 operations.

6. DISCUSSION OF THE RESULTS

Due to the construction of the MCB used for this paper the front face shows a lower temperature rise above ambient as the sides with the right side being the highest. The higher temperature of the right is due to the electrical current that flows down the conduction path in the bottom frame which is in contact with the right shell of the MCB and thus releasing heat to the housing by conduction and convention. This heat on the sides of the MCB housing will result in conduction of the heat to adjacent MCBs when stacked in a distribution board (DB). This is also known as the proximity effect. The effect of this heat conduction due to the proximity effect is outside the scope for this paper but could be further investigated using infrared thermography due to the high emissivity of MCB casing and with an infrared window could take into consideration the internal temperatures.

The change in surface temperature after multiple switching operations is due to the changing resistance of the conduction path as can be seen from Figure 4. The increase in voltage drop of figure 4 is according to Ohm's law caused by the increase of the electrical resistance of mainly the switching contacts. The electrical resistance is determined by the contact area and is among other things influenced by the contact force, the moduli of elasticity and the Poisson's ratio. During switching, the contact surface changes due to oxidation. [17]. There is an initial rise in temperature up to 1000 operations where it stabilizes. This can be due to the erosion of the silver plated contact tips and a thin layer of oxidation that is formed on the contact tips. The small deviations in temperature thereafter can be contributed to how the contacts make contact in the conduction path, the so called "a-spots" [17]. The drop in resistance over the course of the experiment may be due to wiping of the oxidation each time the contacts are opened and closed. Thereafter, increases are a result of the oxide build up on the contacts. The heat flow through the current path is affected by the additional or reduced thermal resistance and electrical resistance.

The thermal resistance is assumed from the electrical resistance and is subject to the average thermal conductivity of the two different materials [9]. The surface temperature observed from the thermal imager is a result of heat convection externally around the MCB, metallic conduction and radiation from the emissive outer surface. The first law of thermodynamics states that the total energy of an isolated system is constant where the internal energy is equal to the amount of heat supplied to the system minus the amount of work done by the system on its surroundings. Energy can be transformed from one to another but cannot be created nor destroyed. Considering the experiment the first law of thermodynamics can be represented as:

$$Pin = \frac{Tr}{Rth} + Cth\frac{dTr}{dt}, \quad \text{......(3)}$$

Where:
$Pin = I^2Rcb$ = power dissipated in the total circuit breaker resistance, Rcb, by the current, I = 20A
Tr = temperature of the right shell
Cth = thermal capacitance of the circuit breaker and cables
Rth = thermal resistance between the heat source in the breaker and the right shell
t = time

Cth and *Rth* are the two critical components in the equation to accurately calculate the internal heat of the MCB.

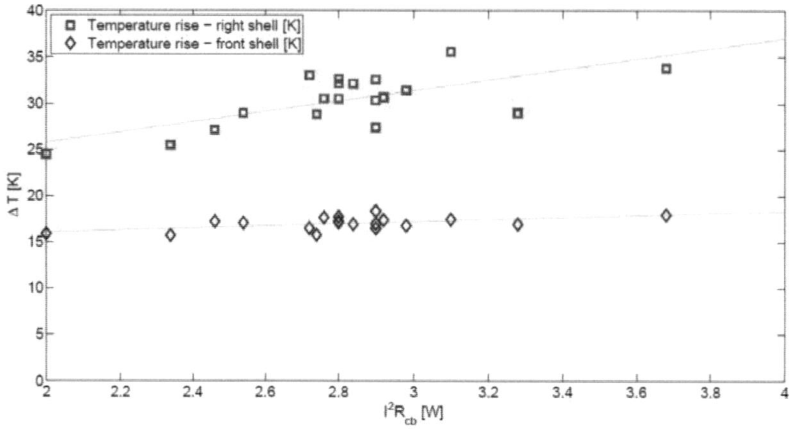

Figure 7 Temperature rise versus power dissipated

Figure 7 shows the graph of ΔT versus I^2R where the slopes of the fitted lines are the thermal resistances between the heat source in the MCB and outer shell on the right and front respectively. The almost linearity of the lines are dependent on the total resistance of the circuit breaker. This verifies that the surface temperature of both the front and right sides of the MCB correlate and there is a possibility that a prediction of the inside condition and hence the state of health of the MCB can be developed. The front face shows a flatter slope compared to the right shell suggesting a less accurate predictor of the overall MCB resistance.

The calculated resistance of the MCB is much lower than the actual measured value as can be seen in Figure 4. This can be contributed to the fact that the exact transfer mechanism as well as the exact parameters of each part of the MCB was not known. The internal temperature of the MCB due to current flow and degradation of internal components is much higher than the measured surface temperature. With the internal temperature of the individual components known a more accurate resistivity of each of the components can be calculated. The true effect of contact contamination due to oxidation is also unknown.

When examining the thermal images of Figure 5 it is seen that after heating the MCB at rated current for one hour without it performing a single operation, the most heat is concentrated around the hydraulic magnetic unit due to the coil in the sensing unit. This change as the unit is switched at rated current and the contacts start to degrade due to the erosion of the silver-plated contact tips, formation of oxidation and subsequent resistance increases as can be seen from Figure 4. With every switching operation under load an arc is formed. The high temperature of the arc will vaporize a percentage of the contact material which will lead to contact erosion.

7. CONCLUSION

Different MCBs will have different thermal and electrical resistances due to manufacturing diversities that result in unique signatures. Thus, comparing the surface temperature rise above ambient after various operating cycles is not a true reflection of the temperature profile for all MCB types but can be used as a comparative reference. From the results it is seen that after 9000 operations at rated current and voltage degradation has not really set in due to the small change in temperature and measured resistance. International standard IEC 60947-2 does only make provision for temperature rise limits for specified parts at 1500 operations at rated current. More testing post 9000 operations is required in order to see the thermal properties after degradation has set in and how the heat generated due to the rise in resistance is conducted and confected to the housing. These results will assist to determine a field methodology to anticipate failure of MCBs.

When comparing the surface temperature rise above ambient of the 3 sides it is clear from Figure 3 that the number of operations does not influence the front as much as the conduction path on the right side/shell. This can be contributed to the airgap between the components and the front shell. The heat inside rises due to buoyancy and is dissipated through the gap between the housing and the on/off handle. Resulting in less heat transferred to the front face due to convection. This is illustrated in Figure 6(a). Due to changes in resistance the temperature rise above ambient is more for the moving contacts. The temperature difference as well as the slopes in Figure 7 between the front and right sides suggests that the front surface temperature only is not a true reflection of the internal temperature due to degradation.

A certified thermographer will only have access to the front view of a MCB when inspecting distribution boards. The surface temperature rise above ambient from the results of the experiment obtained of the front, left and right sides, clearly show that the rise on the left side is on average 8.15°C and the right side 12.53°C higher than the front. This value may change as degradation sets in and the product near failure. Although the front face is an indicator as indicated on Figure 4, it is displaying a lower surface temperature due to the construction of the MCB and requires a correction factor to accurately determine the state of health of the MCB to avoid misinterpreting the state of health of the MCB. It is proposed that MCB infrared thermography comparative severity criteria be investigated/developed as regions of external interest both normal and abnormal to determine the internal condition of the MCB.

8. REFERENCES

[1] N. Gorjian, L. Ma, M. Mittinty, P. Yarlagadda and Y. Sun, "A Review On Degradation Models In Reliability Analysis," in *Proceedings of the 4th world congress on engineering asset management*, Athens, Greece, 2009.

[2] S. Bagavathiappan, B. Lahiri, T. Saravanan, T. Jayakumar and J. Philip, "Infrared Thermography for Condition Monitoring - A Review," *Infrared physics & Technology,* vol. 60, pp. 35-55, 2013.

[3] V. Cohen, Application Guide for the Protection of L.V. Distribution Systems, Second ed., Johannesburg: Creda Communications, 1998.

[4] R. L. Kegg, "On-line machine and process diagnostics," *Annals of the CIRP,* vol. 32, no. 2, pp. 469-573, 1984.

[5] A. Hattangadi, Failure Prevention of Plant Machinery, New York: McGraw Hill Education Private Limited, 2004.

[6] V. Cohen, "Overcurrent Protection in Low Voltage Electrical Circuits - The Perils of Adjustment," CBI, Johannesburg, 2012.

[7] C. S. Kulkarni, J. R. Celaya, G. Biswas and K. Goebel, "Prognostics of Power Electronics, Methods and Validation Experiments," in *48th Systems Readiness Technology Conference*, Anaheim, 2012.

[8] W. C. Garber, "Resistance in Circuit Breakers," Industrial Electric Testing, Jacksonville, 2010.

[9] P. U. Frei and H. O. Weichert, "Advanced Thermal Simulation of a Circuit Breaker," in *22nd International Conference on Electrical Contacts*, Seattle, 2004.

[10] M. Rasid, A. Ospina, K. El Kadri Benkara and V. Lanfranchi, "A Thermal Study on Small Synchronous Reluctance Machine in Automotive Cycle," in *25th IEEE International Symposium on Industrial Electronics*, Santa Clara, 2016.

[11] D. Siegel and M. Anheuser, "Significance of resistances of switching contacts for the temperature rise of LV circuit breakers," in *27th International Conference on Electrical Contacts*, Dressden, Germany, 2014.

[12] A. Prabu, "An Inverse Problem Approach to Modeling of Circuit Breaker Arc Voltage," in *2014 International Symposium on Fundamentals of Electrcal Engineering*, Bucharest, Romania, 2015.

[13] H. Stammberger, C. Rumpler and A. Zacharias, "Interaction in low-voltage switchgear - a staggered simulation approach," 2001.

[14] D. Pop, L. Neamt, R. Tirnovan and D. Sabou, "3D Finite Element Analysis of a Miniature Circuit Breaker," in *The 8th international symposium on advances topics in electrical engineering*, Bucharest, Romania, 2013.

[15] E. G. van Niekerk, "Fundamentals of low voltage circuit breaker operation," Private, Johannesburg, 2002.

[16] E. Larsen, M. Valdes, G. H. Fox, K. Rempe and C. G. Walker, "IEEE3004.5 Recommended Practice for the Application of Low-Voltage Circuit Breakers in Industrial and Commercial Power Systems," 2016, 2016.

[17] T. E. Browne, Circuit interruption:Theory and techniques, Taylor & Francis, 1984.

A Comparative Study of Partial Discharge Measurement using Electrical and Inductive Coupling Sensors

IK Kyere[1], TR Becker[2], JJ Walker[3]
Vaal University of Technology, Vanderbijlpark, South Africa
[1]isaack@vut.ac.za, [2]209034092@edu.vut.ac.za, [3]jerrywalker@walmet.co.za

ABSTRACT

Since the acceptance of partial discharge measurements as a tool used for diagnostic condition assessment of high voltage insulation systems, various equipment and methodologies are employed to acquire and analyse the data obtained, be it in the factory or field environment. Some of the equipment and methods can only detect the presence of partial discharges and cannot be used to evaluate the true condition of the insulation by determining the apparent charge magnitude. Advancement in signal processing technology allowed manufacturers to design digital acquisition equipment with much better diagnostic applications. Consideration must therefore be given to the type of sensor used and its application in the measuring circuit.

The aim of this study is to compare the results obtained by using two different sensing techniques on two different defects, namely, an artificial defect in the form of a void as well as corona discharge in point-to-plane configuration. A method published and recommended in IEC 60270 makes use of a coupling capacitor with associated accessories to determine the apparent charge (in coulomb/pico coulomb) which includes a measuring impedance and preamplifier to measure within a specific frequency spectrum. The results are then compared to a different decoupling method by means of high frequency current transformers, having different frequency spectrums and a preamplifier with the same frequency response as the coupling capacitor technique. Sensitivity assessments and merits of applying each different coupling techniques are presented and discussed.

Keywords: Partial discharge, Electrical, Diagnostic, Inductive, Sensor.

1. INTRODUCTION

A partial discharge (PD) is generally a highly localised or bound electrical discharge in an insulation medium between two conductors, and at times PD activity is the forerunner to an entire electrical breakdown. The occurrence of PD can lead to electrically-induced ageing of certain insulating materials, for example, by formation of corrosive gaseous by products, surface erosion, and tree development [1].

Internal discharges occur within the dielectric medium and can include various cavities of different sizes and geometries and the discharge occurs when the critical inception field of the void is reached, granted that a free electron is available. Surface discharge ensues remotely along the surface of the insulation, while a corona discharge occurs externally in non-uniform gaps and at sharp conductive edges [2].

The impact that partial discharge activity has on different insulation materials and the ability of the insulation to sustain this activity for prolonged period is directly influenced by the nature of the insulation material. Epoxy mica insulation used in rotating machines has the capability of withstanding the presence of partial discharge activity for a longer period compared to the very sensitive polymeric insulation systems used in cables [3].

Coupling sensors are fundamental components in the partial discharge measurement circuit. They are required to capture high frequency PD signals originating from the object under test. Field testing or routine tests in a factory environment determines the choice and application of coupling sensors. IEC 60270 recommends standard test circuits and the use of different sensing techniques to capture the current impulses. The arrival of the current pulse at the terminals of the chosen sensor is attenuated due to a change in impedance and/or reflections within the object and the location of the defect. Therefore, the sensitivity and the sensors impact on the overall test circuit is crucial to reduce any signal distortion [4]. Various techniques have been used for partial discharge measurement with most of them focusing on wideband measurements with various kinds of sensors [5, 6, 7]. High Frequency Current Transformer (HFCT) partial discharge detection systems cannot verify apparent charge by calibration procedure, however, there is the need for sensitivity checks and their performance to ensure validity of non-conventional partial discharge measurement [8].

In this paper, the partial discharge (PD) measurements using two different sensing techniques are presented. Inductive sensing using two high frequency current transformers (HFCT) with different frequency responses and a capacitive coupling device with measuring impedance as published in IEC 60270, was used in the laboratory, to investigate the difference in sensitivity in the discharge pattern and PD inception voltage in defects.

The partial discharge data was presented in the form of Φ-q-n pattern where Φ is the phase angle of partial discharge occurrence, q the PD magnitude and n is the PD pulse number.

2. THE MEASUREMENT SYSTEM

Figure 1 shows the experimental setup of the measuring circuit; the coupling capacitor is in series with the measuring impedance and the HFCT's are placed in the earth leg of the test object. During the first experiment the partial discharge phase-resolved measurement method has been applied for pattern acquisition, in accordance with IEC 60270 standard [9] consisting of 50 Hz a.c. power supply, the test sample (*Ck*), the Coupling Capacitor (*Cc*), and the measuring impedance (*Zm*). In the following experiments the outputs of the HFCT sensors was connected to the digital measuring PD detector.

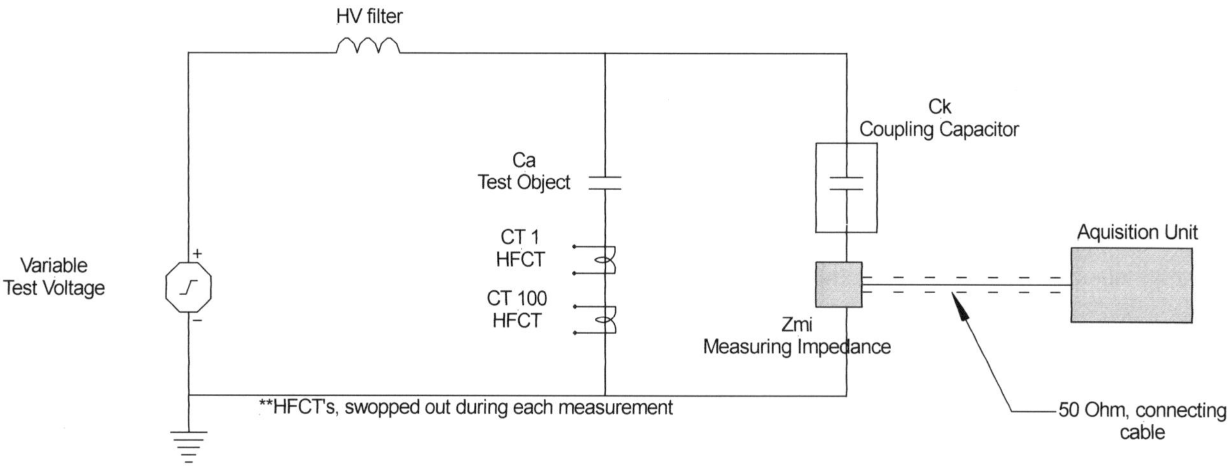

Figure 1. Partial discharge measuring circuit

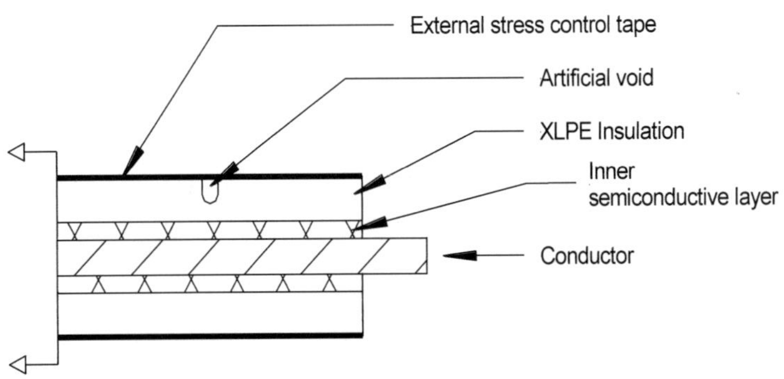

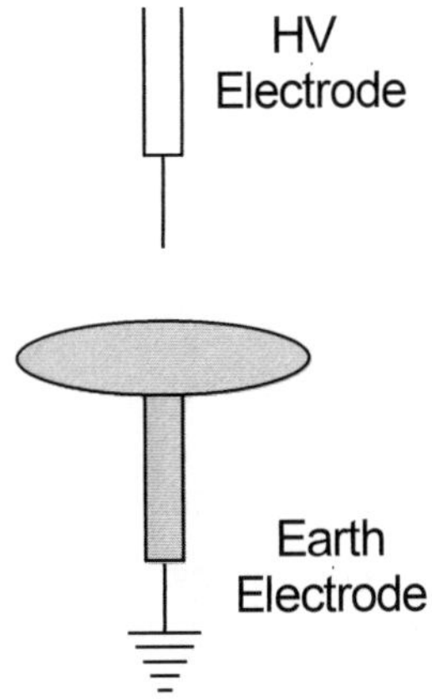

Figure 2. (a) Artificial void (b) Point-to-plane configuration

Figure 2(a) shows the defect (cavity) that was artificially created by drilling a hole of 1,5 mm diameter 1 mm deep into the XLPE insulation of a cable. In addition to that, as shown in Figure 2(b), corona is obtained by using a point-to-plane electrode configuration in air. The high voltage electrode is a steel needle and the earth electrode is the flat plane. The distance between them is 60 mm.

The measurement circuit was calibrated according to the procedure described in IEC 60287 before voltage application. A 5 pC calibration pulse was applied across the coupling capacitor leg, as close as possible to the test object.

3. COUPLING SENSORS

Choosing an efficient way to sense the pulses depends on a number of aspects ranging from sensitivity, type of test object and field environment. Consideration as to calibration efficacy must also be taken into account. Other methods of detecting include electromagnetic (antennas), acoustic and optical detection.

Since sensors should be adapted to the test object and the test environment it is important to state that the detecting method has an impact on the signal transmission.

3.1. Capacitive coupling

a) PD measurement using coupling capacitor and measuring impedance

A coupling capacitor is a capacitive sensor, in combination with a measuring impedance and should be discharge free at the test voltage as well as providing a path that captures PD pulses.

3.2. Inductive coupling

Using a ring type of ferrite core, the basic structure of HFCT consists of several turns of copper wire over the ring core. Ferrites being ferromagnetic Ceramics with very high resistivity and permeability are the most attractive materials for high frequency applications [11]. The sensitivity of HFCT's is increased due to the ferrite material being used as core material [12] and can detect the PD signals in a wide frequency band [13].

b) PD measurement using CT1 in earth lead

CT1 is a high frequency current transformer with a window of 15 mm, and for the experiments it was placed in the earth leg of the test object. It has a -3dB bandwidth of 0.5 - 80 MHz and a -6dB bandwidth of 0.3 - 100 MHz.

c) PD measurement using CT100 in earth lead

CT100 is a clamp–on high frequency current transformer with a window of 100 mm which is also placed in the earth leg of the test object. CT100 can therefore be applied without any disconnection of earth leads or interruption of the supply. It has a -3dB bandwidth of 2 - 25 MHz and a -6dB bandwidth of 1.2 - 40 MHz.

4. COMPARISON OF RESULTS AND EXPLANATION

The performance of the sensors used in this paper are validated by the measurement using different partial discharge sources. PD activity in the test samples are detected to obtain representative patterns for each individual sensor. Initial tests were performed using a point plane PD source. Figures 3 – 5, including table 1 include the results obtained for PD inception voltage condition. A further increase beyond inception is depicted in figures 6 – 8 followed by the corresponding results in table 2.

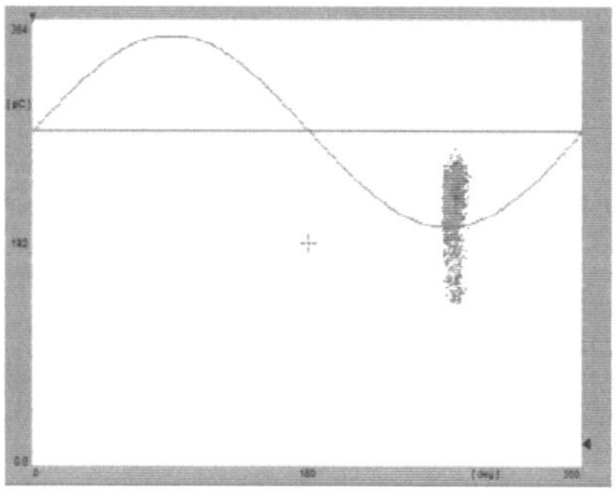

Figure 3 CC, gain setting 3.50 pC/DIV

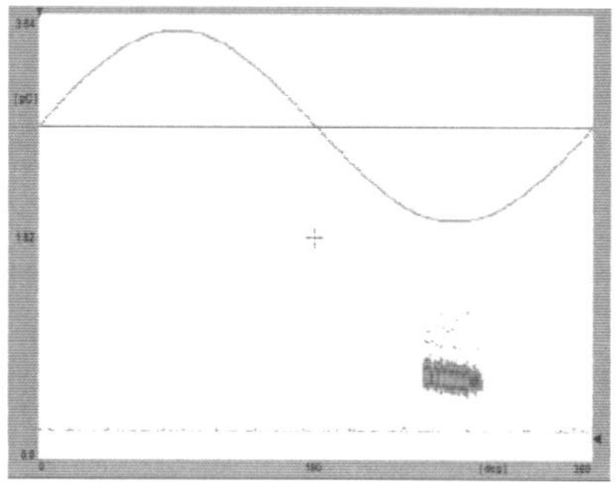

Figure 4 CT 1, gain setting 0.70 pC/DIV

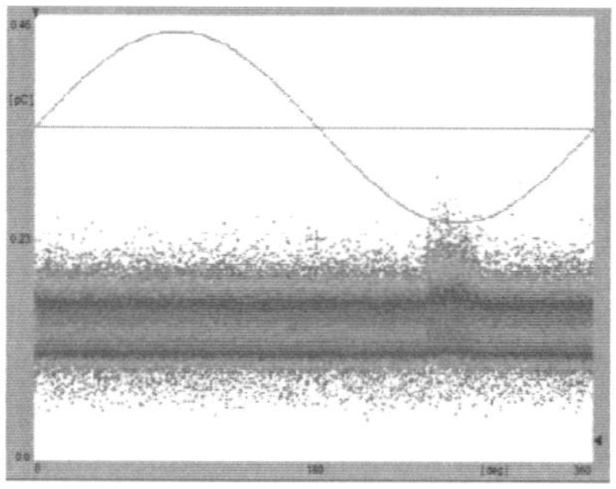

Figure 5 CT 100, gain setting 0.09 pC/DIV

Table 1 Summary of results with different coupling sensors for a point plane discharge at inception voltage.

	Coupling capacitor	CT 1	CT 100
Q Magnitude (pC)	138.6	0.84	0.21
Voltage (kV)	6.2	6.2	6.2
Total counts	2 549	22 208	59 1640

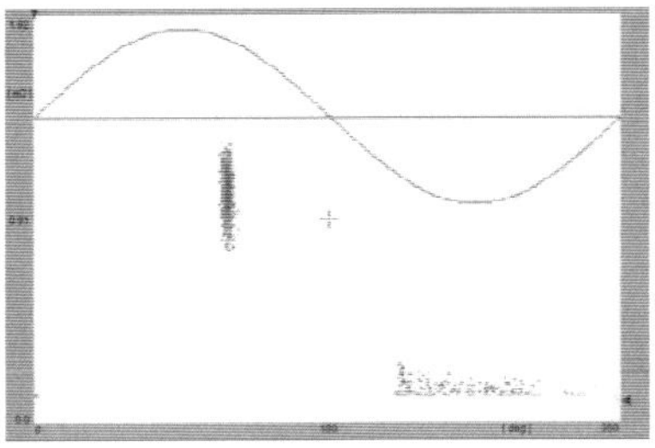

Figure 6 CC, gain setting 350 pC/DIV

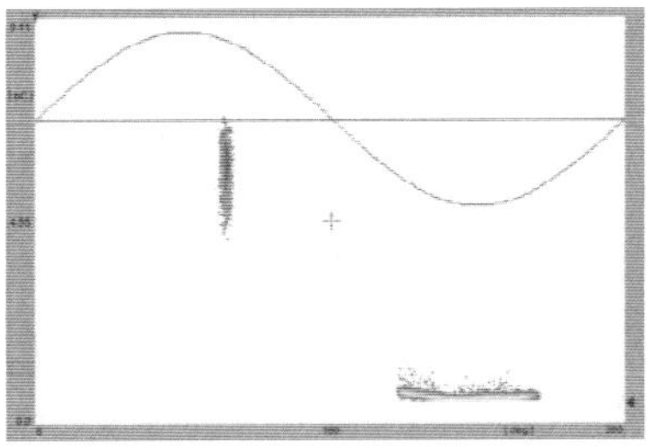

Figure 7 CT 1, gain setting 1.50 pC/DIV

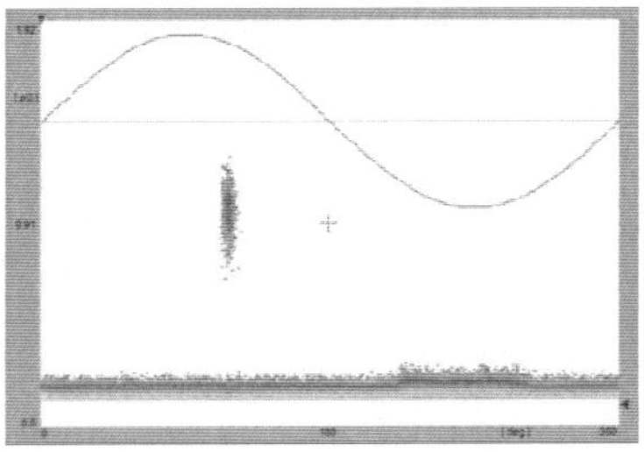

Figure 8 CT 100, gain setting 0.35 pC/DIV

Table 2: Summary of results with different coupling sensors for a point plane discharge above inception (streamer discharge)

	Coupling capacitor	CT 1	CT 100
Q Magnitude (pC)	1130	6.25	1.25
Voltage (kV)	9.4	9.4	9.4
Total counts	23 364	121 273	542 116

Below are the patterns obtained for an artificial void using different coupling sensors. Figure 9 indicates the pattern using the coupling capacitor followed by figure 10 – 11 using HFCT's.

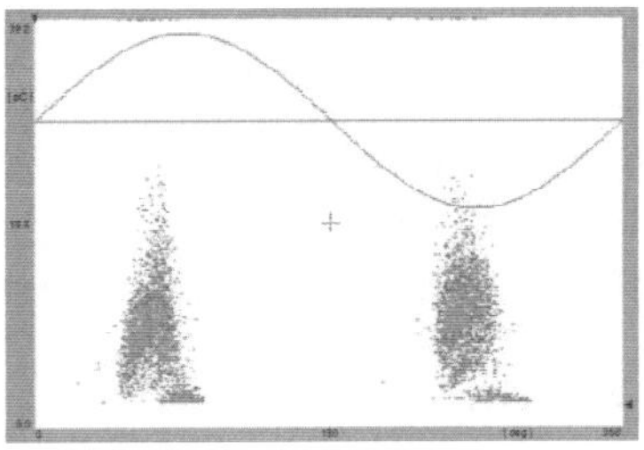

Figure 9. CC, gain setting 7pC/DIV

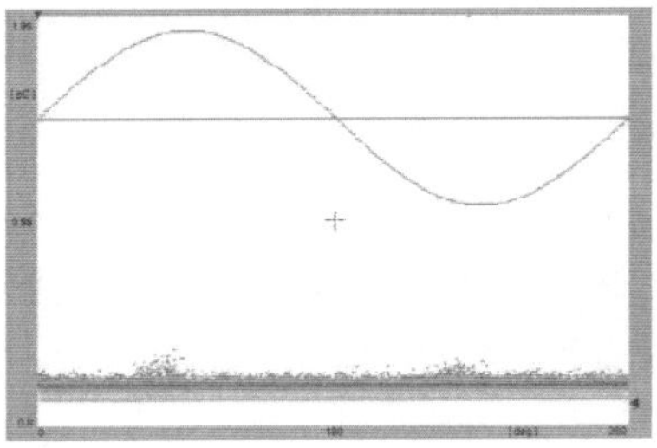

Figure 10. CT 1, gain setting 0.35 pC/DIV

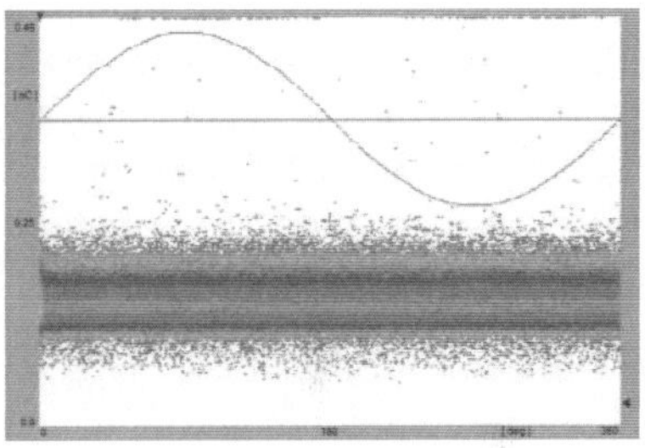

Figure 11. CT 100, gain setting 0.09 pC/DIV

Table 3: Summary of results with different coupling sensors for an artificial void defect at inception.

	Coupling capacitor	CT 1	CT 100
Q Magnitude(pC)	13.23	0.23	0.22
Voltage (kV)	14.5	14.5	14.5
Total counts	6 312	57 3424	59 5258

Referring to Figures 3 – 5 and Table 1(Point to Plane), the PD pattern measured using coupling capacitor are greater in magnitude than the pulses measured with the HFCTs. The same can be explained for pulses measured with CT1, also have greater magnitude compared to the results with CT100. Although the signature of the pattern acquired for each sensor is indicative of a corona discharge on the high voltage electrode the pulse amplitude and height distribution are different for each sensor. The amplification factor was also adjusted for each measurement, the gain setting was adjusted to 3.50 pC/DIV for the coupling capacitor however the CT 100 amplification factor was set to 0.09 pC/DIV. A further increase in voltage beyond inception from 6.2 kV to 9.4 kV showed similar characteristics in gain settings and pulse amplitude – phase height distribution as with patterns acquired at inception.

Aside from point plane, an artificial void was used in subsequent tests as a PD source. Only the patterns and results at inception voltage were reported as illustrated in Figures 9 – 11 as well as Table 3. The issue of sensitivity can be highlighted by a visible pattern seen in figure 9 which does not exist in figures 10 – 11. From the pattern seen in figure 9 a diagnosis of the defect can be explained as a combination of a surface and void discharge which cannot be said for the remaining results as only the noise flow is visible.

5. CONCLUSION

The measurement of partial discharge activities with two detection techniques has been undertaken. The phase – resolved partial discharge pattern in different cases obtained from HFCT measurement are compared with that from the conventional IEC method. From the measurement results obtained it's clear that the performance of the sensors employed in this study can be distinguished. The coupling capacitor/measuring impedance sensor has a better sensitivity than HFCTs. However, CT1 with a smaller window and different bandwidth is more sensitive than CT100 with the larger window. The results show that the HFCT can detect the discharge pattern but from a diagnostic point of view analysis could not be performed. Comparing to the conventional method, the commercial HFCT cannot detect the partial discharge at the inception voltage due to its poor sensitivity.

The HFCT's do not detect some PD pulses with some amount of energy as shown in Figures 4, 5, 10 and 11. The differences in the measurement results can be attributed to the size of the air gap between the conductor and the CT core and the difference in frequency bandwidth of the current transformers. The capacitance of the transformer winding also play a significant role in the measurement with the high frequency current transformers. This can occur when the frequency bandwidth of the PD pulses is low [13].

The practical implications of the measurements show that it is imperative to determine the correct coupling method based on the application of the partial discharge measurements and finally how the results will be assessed. The sensitivity can be determined based on whether the application is for monitoring, diagnostic or on site testing applications and keeping in mind the nature of the test object.

REFERENCES

[1] Fruth,. B.A, Gross,. D.W., (1994) Phase resolving partial discharge pattern acquisition and frequency spectrum analysis. Proceedings of the 4th international conference on properties and application of dielectric materials.

[2] Fruth,. B.A, Gross,. D.W., (1996) Partial discharge signal conditioning techniques for on line noise rejection and improvement of calibration. Conference record of IEEE international symposium on electrical insulation. Montreal, Quebec, Canada.

[3] Gross, D., (2016) Acquisition and location of partial discharge – esp. in transformers.

[4] Bergius, O. (2011) Implementation of On-Line Partial Discharge Measurements in Medium Voltage Cable network. Master thesis, Tampere University of Technology, pp.33-34.

[5] Tian Y., Lewin P., Davies A. (2002) Comparison of On-line Partial Discharge Detection Methods for HV Cable Joints. IEEE Trans. Dielectr. Electr. Insul. ;9:604–615. doi: 10.1109/TDEI.2002.1024439.

[6] Rodrigo A., Llovera P., Fuster V., Quijano A. (2011) Influence of High Frequency Current Transformers Bandwidth on Charge Evaluation in Partial Discharge Measurements. IEEE Trans. Dielectr. Electr. Insul. ;18:1798–1802. doi: 10.1109/TDEI.2011.6032852

[7] Rodrigo A., Llovera P., Fuster V., Quijano A. (2012) Study of Partial Discharge Charge Evaluation and the Associated Uncertainty by Means of High Frequency Current Transformers. IEEE Trans. Dielectr. Electr. Insul. ;19:434–442. doi: 10.1109/TDEI.2012.6180236.

[8] Cho, S.I. (2011) On – Line PD (Partial Discharge) Monitoring of Power System Components. Msc Thesis. Aalto University.

[9] IEC60270 (2000) Standard. High – Voltage Test Techniques – Partial Discharge Measurements.

[10] Mallikarjunappa, K.; Ratra, M.C. (1990) Detection of partial discharges in power capacitors using high frequency current transformers,‖ Electrical Insulation and Dielectric Phenomena, 1990. Annual Report. Conference on, Page(s): 379 – 384.

[11] Bhosale, N., Kothoke, P., Deshpande, A and Cheeran, A. N. (2013) Analysis of Partial Discharge Using Phase-Resolved (Φ-Q) and (Φ-N) Statistical Techniques. International Journal of Engineering Research & Technology (IJERT), Vol. 2 Issue 5, Pp 1960-1965.

[12] Luo, G and Zhang, D. (2010) Study on Performance of HFCT and UHF Sensors in Partial Discharge Detection. In proceedings of IPEC 2010 conference, 27-29 Oct., Singapore, pp. 630-635.

Review of Transmission Line Fault Location using Travelling Wave Method

Gunawan Wijaya[12], Suwarno[1], A. Abu-Siada[3]
[1] PT PLN, Indonesian Electricity Company
[2] School of Electrical and Informatics, Institut Teknologi Bandung, Indonesia
[3] Electrical and Computer Engineering Department, Curtin University, Australia

ABSTRACT

This paper presents a state of the art review for travelling wave-based fault locating techniques. The application of these techniques to estimate the fault location in various configurations of transmission lines is also presented. The presented configurations include series capacitor compensated transmission line, tapped transmission line, and hybrid transmission line. Both one-ended and two-ended synchronized methods are explained and compared.

Keywords: Fault location, Transmission lines, Travelling-waves

Corresponding author: Gunawan Wijaya (awan8019@yahoo.com)

1. INTRODUCTION

Several studies on traveling-wave-based fault location in long transmission lines (TLs) can be found in the literatures [1]–[9]. Faults generate voltage signals that propagate through the TL and reflect back at the two ends of the line. By identifying these waves and measuring its arrival time toward the ends, the fault location can be identified.

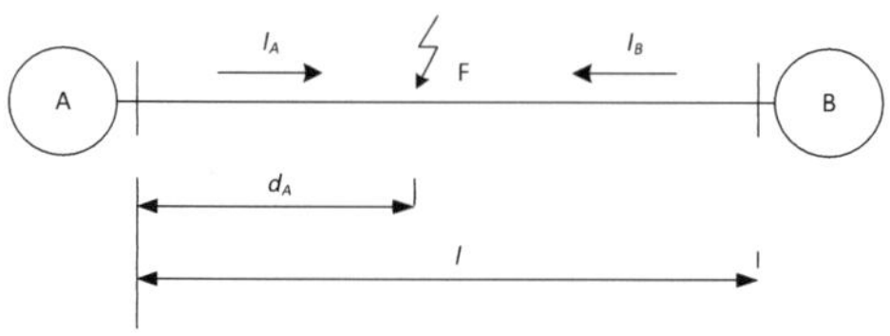

Figure 1. Fault on a TL fed from both ends

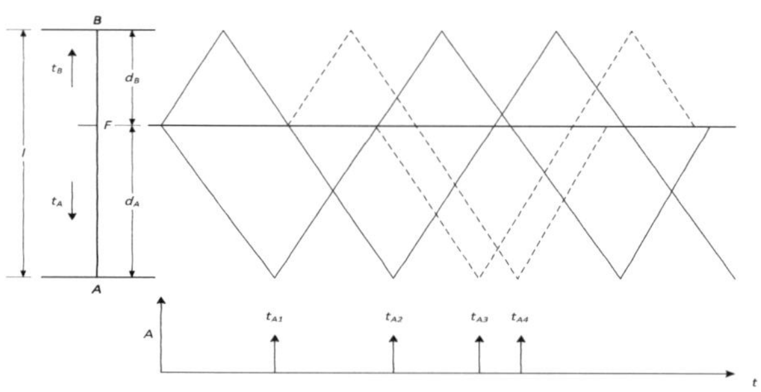

Figure 2. Bewley lattice diagram for the TL in Figure 1

It was reported that this technique is independent on power swing, current transformer saturation, fault type, fault resistance, fault-inception angle and system parameters [10]. The application of the travelling waves can be extended to identify fault location within under-ground cables. Figs. 1 and 2 show schematics for a fault in a TL and the corresponding Bewley lattice diagram. Once the arrival time for the fault travelling waves is recorded, it can be used to identify the fault location.

Data can be collected from one end of the line or at both ends. Fault location based on data collected from one end is categorized as one-ended algorithm, while using data from both ends is referred to two-ended algorithms. The later employs time-synchronization technique between both ends of the line.

Although basic calculation of travelling wave-based fault location is similar for all fault types, a fault occurs on a series capacitor compensated line, hybrid lines (mix of overhead lines and underground cables) and tapped line differ in characteristics. For example, a series capacitor in compensated lines introduces non-linearity to the system. In the long TL and underground cables, capacitive current can be considerably large in comparison to fault current [11]–[13].

Other techniques based on impedance measurements were presented in the literature to overcome the limitation of impedance-based methods in identifying faults in a compensated line. Sadeh et. al. presented an impedance based algorithm to identify fault locations in a series capacitor-compensated TL [14].

Huang et. al. modelled series capacitor in the middle of a long TL and measurements were captured at one-end of TL [15]. Their wavelet analysis showed that the current travelling wave is more accurate than the voltage travelling wave in locating faults in such configuration. The main difficulty is on distinguishing the source of the reflected waves of the same polarity as it may be due to a reflection from the series capacitor or a remote bus. Spoor et. al. used the polarity of the arrival waves to distinguish reflected signals from the fault point and the opposing terminal [7]. Abedini et. al. combined the two methods reported in [15] and [7] with a fault locator installed in the middle of a series capacitor compensated line [8].

The fault location on tapped lines or on three terminal lines can be figured using Bewley lattice diagram as in Fig. 2. Analyzing measured data using wavelet was introduced by Magnago and Abur [16] that was applied for a tapped line as reported in [3]. Hamidi et. al. used wavelet to analyse the travelling wave signals on multi hybrid terminals [4], [5]. Though this algorithm is able to estimate fault location, its accuracy decreases as faults occurred close to the joint points and end-terminals. For this kind of TLs, identification of first and consecutive traveling wave is the major challenge of wavelet-based fault location.

Wavelet has been widely used to process traveling waves when a fault occurs on a TL [2]–[8], [15], [16] and in TL protection [15], [17], [18]. Wavelet-based digital signal processing has been compared with Artificial Neural Networks (ANN) and Fourier Transformation (FT) [1]. Results show the higher accuracy of the wavelet in identifying the location of phase to ground faults when compared to ANN and FT. On another research, Hilbert-Huang Transform has been utilized [19]. However, this approach calls for additional Rogowski coil to measure the voltage at both terminals of the line and also in the middle of TL.

2. TRAVELLING WAVE THEORY

When a fault occurs on TL, current will be drawn at the fault location from both terminals. Voltage waves occurred at the fault point travel toward the ends of the line. The fault will divide the TL into two sections at the location of fault as shown in Fig. 2. The wave arrival times to both ends of the line are tA1, tA2, tB1, tB2, as shown in the Bewley lattice diagram of Fig. 2. These waves propagate along the line until they find a discontinuity point such as joints, buses, or another fault point. Then these waves undergo reflections and refractions. This process continues until the wave energy is completely dissipated along the line [10]. The wave arrival time on an end-terminal depend on fault point and propagation velocity of the line medium. Although several wavelet-based algorithms are proposed in the literatures, the basic principle of fault location has the same major steps [2]–[4], [6]–[8], [16], [17]. The first step is transforming the voltage or current signal to modal mode. For this purpose, Clarke Transformation Matrix is used. The second step is applying Discrete Wavelet Transform (DWT) to the modal voltages or current. Squaring the wavelet transform coefficients (WTC2) will determine the maximum value of the wave due to the signal energy reaches its maximum value. Many researches utilize Daubechies family as the mother wavelet due to its simplicity that reduces the computation process and the fewer resource it requires when implemented [8]. The final step is to observe WTC2 to determine the fault location. This step varies to specific cases and will be explained below.

2.1. Modal Analysis

Clarke's Transformation Matrix is used to transform a signal into several independent modes of signal propagation. The matrix will transform a three-phase system into a three-mode signals known as mode 0, 1, and 2. For fault location purpose, these mode have different function. The zero-mode, depends on the frequency. At this mode, wave travels at slower velocity due to the line high attenuation. The other modes, modes 1 and 2, are known as independent aerial modes. At these modes, wave travels at a speed close to the speed of light [7]. All these modes can be calculated from signal measurement following a general equation.

$$S_{mode} = TS_{phase} \quad \text{,...} \quad (1)$$

where Smode is modal-transformed voltage or current signals, Sphase is the signal phase and T is Clarke's Transformation Matrix for fully transposed line that is defined as [6], [7], [15], [16]

$$T = \frac{1}{3}\begin{bmatrix} 1 & 1 & 1 \\ 2 & -1 & -1 \\ 0 & \sqrt{3} & -\sqrt{3} \end{bmatrix} \quad \text{,...} \quad (2)$$

2.2. Discrete Wavelet Transform

DWT analysis is widely used in power systems because of its capability to capture both frequency and time simultaneously. Wavelet analysis can improve the analysis of signals with localized-impulses and also oscillations. DWT for a signal f(k) with respect to a "mother" wavelet ψ(k) is given by [7], [16]:

$$DWT(m,n) = \frac{1}{\sqrt{a_0^m}} \sum_k f(k)\, \psi\left(\frac{n-ka_0^m}{a_0^m}\right) \quad \text{,................................} \quad (3)$$

Where a_0^m is a scaling factor and ka_0^m is a shifting factor, k is an integer variable and m is the level of decomposition.

Wavelet analysis is dividing the signal into two-separate blocks. The signal will be shifted and scaled based-on a power of 2 as seen to Fig. 3. Unlike Continuous Wavelet Transform (CWT), DWT determines the coefficients on the two blocks. The higher scale applied, the lower frequency will be obtained. It means that scale factor has an inverse relationship to the frequency, and the resulting series of wavelet coefficients is referred to as the wavelet series decomposition [7].

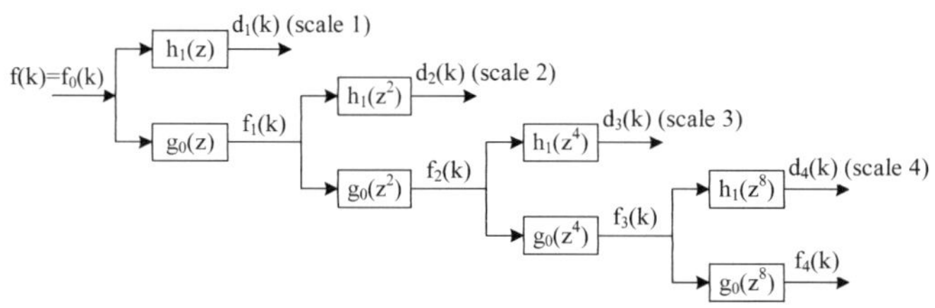

Figure 3. Structure of the algorithm for signal wavelet at 4th level of decomposition [15]

The pseudo-frequency (Fig. 4), corresponding to a particular level of decomposition or scale, can be determined from [7], [18]

$$frequency = T_s F_c / a_0^m \quad \text{,...} \quad (4)$$

where T_s is the sampling period of input signal, F_c is the wavelet function, and a_0^m is scaling factor.

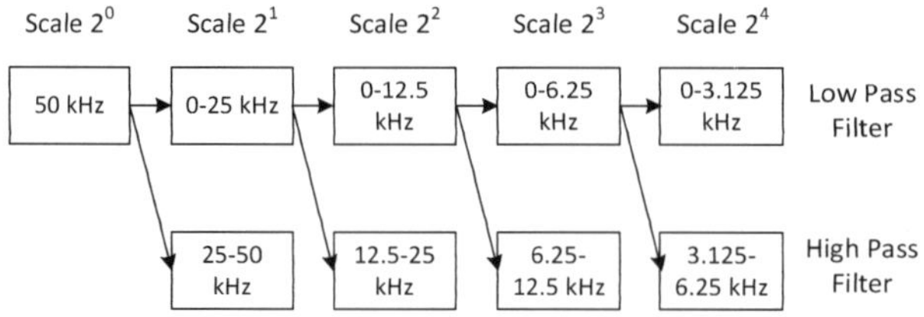

Figure 4. Pseudo-frequency of 4th scale of wavelet transform [18]

3. FAULT LOCATION AND ERROR CALCULATION

Once a fault is detected and recorded, its fault point can be estimated. After calculating the fault location, the error in the location can be calculated based on [13]. The error is represented as a percentage of TL length. Assuming FLcalc, FLexact, and l respectively as calculated fault location, actual fault location and TL length, the error is then calculated as :

$$Error\ (\%) = \frac{(FL_{cal} - FL_{exact})}{l}\ x\ 100\% \quad (5)$$

3.1. One-ended Measurements

In this technique, the recording unit is installed at one or two terminals without mutual communication. Based on Fig. 2, when conventional approach is considered [2], [3], [6]–[8], [16], which uses propagation velocity (v), the fault location on the first section of the line from bus A (dA) is calculated by (6). If the fault occurs on the second section of the line then dA is calculated by (7). As seen from (6) and (7), speed of wave propagation, as calculated in (8), is required to calculate the fault location.

$$d_A = \frac{v.(t_{A2} - t_{A1})}{2} \quad (6)$$

$$d_A = l - \frac{v.(t_{A2} - t_{A1})}{2} \quad (7)$$

$$v = \sqrt{\frac{1}{L'C'}} \quad (8)$$

where L' and C' are respectively the line inductance and capacitance per unit length.

Another fault locating technique is presented in [2] without using of the velocity of wave propagation. This technique uses three consecutive traveling waves appearance instead of two traveling waves as per the equations below. Fault location can be calculated at first section of the line using (10) if (9) is valid. Otherwise, (11) is used to calculate fault location. It is shown on [2] that traveling wave at tREF has higher magnitude compared to the reflected one from the fault point. Fault location within the second section of the line can be calculated using (12) if (9) is valid. Otherwise, (13) is used.

$$t_{A2} - t_{A1} = t_{A3} - t_{A2} \quad (9)$$

For fault occurred within the first section :

$$d_A = \frac{(t_{A2} - t_{A1}).l}{(t_{A2} - 2t_{A1} + t_{REF})} \quad (10)$$

$$d_A = \frac{(t_{A2} - t_{A1}).l}{(t_{A2} - 2t_{A1} + t_{A3})} \quad (11)$$

For fault occurred within the second section :

$$d_A = \frac{(t_{REF} - t_{A1}).l}{(t_{A2} - 2t_{A1} + t_{REF})} \quad (12)$$

$$d_A = \frac{(t_{A3} - t_{A1}).l}{(t_{A2} - 2t_{A1} + t_{A3})} \quad (13)$$

3.2. Two-ended Measurements

In this technique, a recording unit should be installed at both ends and should be synchronized through global positioning system (GPS). In this method, fault signals are captured and recorded at both terminals by the two recording units using the same time-synchronized reference. In synchronized measurements, the delay time td between the fault detection time at both ends can be determined [16] and the distance to the fault point measured from bus A can be calculated by:

$$d_A = \frac{l-(v_m.t_d)}{2} \quad ,\ldots\ldots \quad (14)$$

where $t_d = t_{B1} - t_{A1}$, and v_m is velocity of traveling wave for mode m.

Another method, reported in [2], utilizes the first signal received from the opposite terminal without using the signal velocity in calculating the fault location. Compared to one-ended algorithm, information from both terminals can calculate the fault location using the local measurement as primary data as in (15). Equations (16) and (17) could be used to calculate the fault location dA, while (18) and (19) for the other section measured from the opposite terminal (1-dA) or dB.

$$\begin{array}{ll} t_{A1} < t_{B1} & , fault\ on\ 1^{st} section \\ t_{A1} = t_{B1} & , fault\ on\ center \\ t_{A1} > t_{B1} & , fault\ on\ 2^{nd} section \end{array} \quad ,\ldots\ldots \quad (15)$$

Fault location dA for fault occurred at first section:

$$d_A = \frac{(t_{A2}-t_{A1}).l}{2(t_{A2}-2t_{A1}+t_{B1})} \quad ,\ldots\ldots \quad (16)$$

For fault occurred at second section:

$$d_A = \frac{(t_{B1}+t_{B2}-2t_{A1}).l}{2(t_{B2}-t_{A1})} \quad ,\ldots\ldots \quad (17)$$

Fault location (1-dA) or dB for fault occurred at first section:

$$d_B = \frac{(t_{A2}-3t_{A1}+2t_{B1}).l}{2(t_{A2}-2t_{A1}+t_{B1})} \quad ,\ldots\ldots \quad (18)$$

Fault location (1-dA) or dB for fault occurred at second section:

$$d_B = \frac{(t_{B2}-t_{B1}).l}{2(t_{B2}-t_{A1})} \quad ,\ldots\ldots \quad (19)$$

All these equations can be applied for fault location calculation on the opposite terminal.

3.3. Series compensated line

Series capacitors compensation on a TL affects the line impedance. Therefore, fault location using impedance-based calculation needs an information about the degree of series compensation and location of the capacitor [11], [14], [20]. By using traveling wave based methods, fault location on compensated TL is not affected by the presence of the series capacitors since the impedance modification at high frequencies due to the series capacitor will be negligible [16]. Fault location series capacitor compensated line can be calculated using one or two of the above equations of (6) – (19).

3.4. Hybrid line

Fault location that uses wave propagation velocity on hybrid type line (Bewley lattice diagram of such line is shown in Fig. 5) is more prone to large errors introduced by changing parameter over its age such as relative permittivity [9]. Therefore, in order to use fault location technique accurately, it is suggested the analysis should not take into account the wave propagation velocity [5], [9]. The use of modal transformation analysis on hybrid type TLs mitigates the effect of cross-bonding. Therefore, travelling wave-based algorithms is preferred due to its less sensitivity to fault type, non-linear fault impedance, and transmission loading conditions [4]. Fault location on this line can be calculated using one or two of equations of (14) – (19).

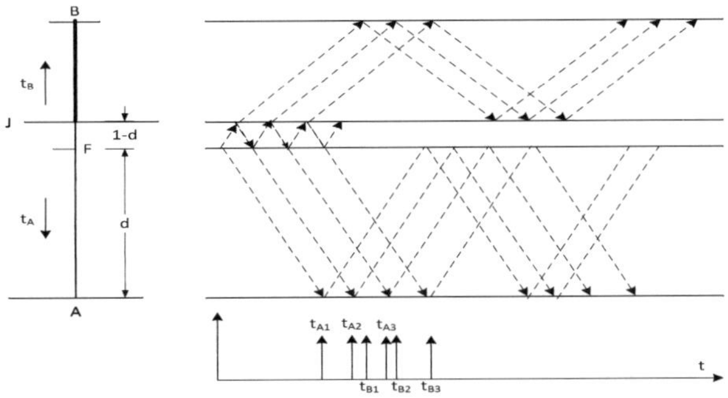

Figure 5. Bewley lattice diagram of a fault on hybrid line

3.5. Tapped line

On a tapped line, fault detection has to be done before fault location calculation [3]–[6]. Obtaining high accuracy is done by synchronizing measurement from all terminal. The measurement or protection device at joint point will increase the error of fault location.

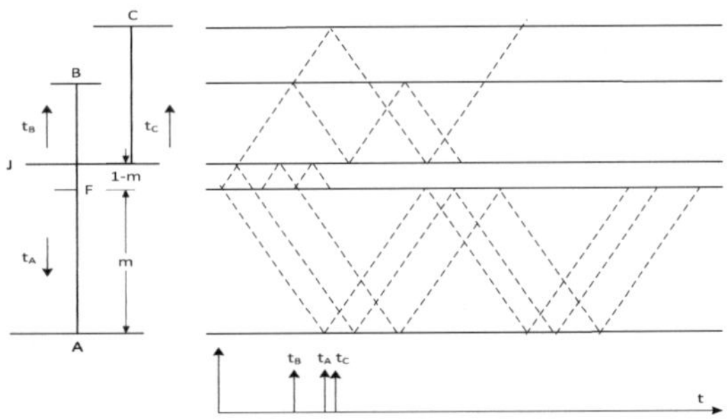

Figure 6. Bewley lattice diagram of a fault on tapped line

Fig. 6 shows the Bewley diagram for a tapped line where first-wave arrival time at all terminals are defined as tA, tB, and tC. Travelling times for segments AJ, BJ, and CJ can be defined as τAJ, τBJ, and τCJ. The relationships between these variables are defined by the following equations [4]:

$$t_A = m.\tau_{AJ} + t_F \qquad ,\ldots\ldots \quad (20)$$
$$t_B = (1-m).\tau_{AJ} + \tau_{BJ} + t_F \qquad ,\ldots\ldots \quad (21)$$
$$t_C = (1-m).\tau_{AJ} + \tau_{CJ} + t_F \qquad ,\ldots\ldots \quad (22)$$

where t_F is the time when fault occurred and m is the ratio of fault location to faulty segment length (dA/lAJ). Subtracting arrival times one another will eliminate t_F. For fault on segment AJ, eliminating equations (20) – (22) each other will result in first-wave arrival time differences at three terminals [4]

$$t_{AB} = |t_B - t_A| = |(1-2m).\tau_{AJ} + \tau_{BJ}| \qquad ,\ldots\ldots \quad (23)$$
$$t_{AC} = |t_C - t_A| = |(1-2m).\tau_{AJ} + \tau_{CJ}| \qquad ,\ldots\ldots \quad (24)$$
$$t_{BC} = |t_C - t_B| = |\tau_{CJ} - \tau_{BJ}| \qquad ,\ldots\ldots \quad (25)$$

Similarly, for fault occurred on segment BJ and CJ, (25) can be represented as [4]

$$t_{AC} = |t_C - t_A| = |\tau_{CJ} - \tau_{AJ}| \qquad ,\ldots\ldots \quad (26)$$
$$t_{AB} = |t_B - t_A| = |\tau_{BJ} - \tau_{AJ}| \qquad ,\ldots\ldots \quad (27)$$

Equations (25) – (27) will result in a unique difference for each segment if it is subtracted to propagation time between two terminals. This unique value can be used to recognize the faulty segment [4].

$$\Delta_{ij} = |t_{ij} - \tau_{ij}|, \qquad \forall i^*j^*: \Delta_{i^*j^*} = 0 \quad ,\ldots\ldots \quad (28)$$

where Δ_{ij}'s are differences between measured first-wave arrival times t_{ij} and pre-calculated wave travelling time along two segments τ_{ij}. The zero or approaching zero value of Δ_{ij} identify the faulty segment. The approaching zero value could be happen due to low sampling rate of measurement [4]. Having calculating it, a statistical mean and standard deviation can be used to find the outlier value which represent the faulty segment. The general form of faulty segment is defined as [4]

$$O_\Delta = \{x \mid x \in \Delta_{ij}'s,\ x \notin [\mu_\Delta - \sigma_\Delta , \mu_\Delta - \sigma_\Delta]\} \quad (29)$$

where O_Δ represents the outlier showing the faulty segment, while μ_Δ and σ_Δ respectively are population mean and standard deviation of $\Delta_{ij}'s$. Fault location from corresponding terminal is done by using one of the following equations [4].

$$O_\Delta = \Delta_{BC}\ ,\ x = \left(\frac{1}{2} - \frac{|t_{AB} - \tau_{BJ}|}{2\tau_{AJ}}\right) L_{AJ},\ if\ AJ\ is\ faulty \quad (30)$$

$$O_\Delta = \Delta_{AC}\ ,\ x = \left(\frac{1}{2} - \frac{|t_{AB} - \tau_{AJ}|}{2\tau_{BJ}}\right) L_{BJ},\ if\ BJ\ is\ faulty \quad (31)$$

$$O_\Delta = \Delta_{AB}\ ,\ x = \left(\frac{1}{2} - \frac{|t_{AC} - \tau_{AJ}|}{2\tau_{CJ}}\right) L_{CJ},\ if\ CJ\ is\ faulty \quad (32)$$

Where L_{AJ} , L_{BJ} , and L_{CJ} are length of segment AJ, BJ, and CJ, respectively. Summary of travelling wave-based fault locating technique is provided in Table I.

4. CONCLUSION

This paper presented a literature review for fault locating techniques based on travelling waves. The technique can be implemented as one-ended or two-ended measurements. Travelling wave techniques are insusceptible to the fault type. Traveling wave fault location is not affected by the line parameters variation such as line impedance, cable permittivity, as well as line configuration modification. On the other hand, traveling wave-based fault location techniques have to be used in pair with wide range band of frequency measurement, including high sampling frequency measurement unit, optical cable for measurement unit, and also higher processing recording unit. Absence of one of these requirements may lead to less accuracy of such techniques. The technique is relatively expensive and of low accuracy for faults close to the terminals of the line.

Table 1 : Travelling wave based algorithms characteristic

Characteristics on several TLs	One-ended measurement		Two-ended measurement (sync'd)	
	Velocity based	Time based	Velocity based	Time based
Complexity[1]				
Simple	low	fair	low	fair
CC	low	fair	low	fair
Hybrid	fair	high	low	high
Tapped	fair	high	low	high
Key features[2]				
Simple	Fast calculation	Fair accuracy and/or calculation	Fast calculation	High accuracy
CC	Fast calculation	Fair accuracy and/or calculation	Fast calculation	High accuracy
Hybrid	Fast calculation and/or difficult FSD	Fair calculation and/or hard FSD	Fast calculation	Fair accuracy
Tapped	Fast calculation	Fair calculation	Fast calculation	High accuracy and/or slow calculation

Limitation				
Simple	• Higher error on faults close to terminals • Higher error on short transmission			
CC	• Higher error on faults close to terminals • Higher error on short transmission			
Hybrid	• Higher error on faults close to terminals and joints • Higher error on short transmission • Difficulties on distinguishing reflected wave from fault point and joint • Difficulties on fault close to terminals and joints			
Tapped	• Higher error on faults close to terminals and joints • Higher error on short transmission • Difficulties on distinguishing reflected wave from fault point and joint • Difficulties on fault close to terminals and joints			
Improvement to accuracy (conventional substation)	• Higher frequency sampling • Higher processing unit • Wider frequency-band measurement sensor (optical sensor) • Lower measurement-cable impedance (optical cable)			
Cost[3]				
Simple	low - medium	low - medium	medium	medium
CC	low - medium	low - medium	medium	medium
Hybrid	low - medium	low - medium	medium	medium
Tapped	low - high	low - high	high	high

These characteristics are qualitatively measured compared to wavelet-based algorithm on simple TLs.
[1]Based on its filter processing, Fault Section Detection (FSD), FL formula
[2]Based on FL calculation processing, accuracy, and FSD
[3]Estimation costs on new installation compared to simple TLs one-ended algorithm and limited data from [21], including sensors, GPS, communication module, number of Input/Output (I/O), and Central Processing Unit (CPU).

5. REFERENCES

[1] A. Abdollahi and S. Seyedtabaii, "Transmission line fault location estimation by Fourier & wavelet transforms using ANN," *Power Eng. ...*, no. June, pp. 23–24, 2010.

[2] O. Altay, E. Gursoy, and O. Kalenderli, "Single end travelling wave fault location on transmission systems using wavelet analysis," in *2014 ICHVE International Conference on High Voltage Engineering and Application*, 2014, pp. 1–4.

[3] C. Y. Evrenosoğlu and A. Abur, "Travelling wave based fault location for teed circuits," *IEEE Trans. Power Deliv.*, vol. 20, no. 2 I, pp. 1115–1121, 2005.

[4] R. J. Hamidi and H. Livani, "A travelling wave-based fault location method for hybrid three-terminal circuits," *IEEE Power Energy Soc. Gen. Meet.*, vol. 2015–Septe, pp. 1–5, 2015.

[5] R. J. Hamidi and H. Livani, "Traveling-Wave-Based Fault-Location Algorithm for Hybrid Multiterminal Circuits," *IEEE Trans. Power Deliv.*, vol. 32, no. 1, pp. 135–144, 2017.

[6] K. Andanapalli and B. R. K. Varma, "Travelling wave based fault location for teed circuits using unsynchronised measurements," *2013 Int. Conf. Power, Energy Control*, pp. 227–232, 2013.

[7] D. Spoor and J. G. Zhu, "Improved single-ended traveling-wave fault- location algorithm based on experience with conventional substation transducers," *IEEE Trans. Power Deliv.*, vol. 21, no. 3, pp. 1714–1720, 2006.

[8] M. Abedini, A. Hasani, A. H. Hajbabaie, and V. Khaligh, "A new traveling wave fault location algorithm in series compensated transmission line," *2013 21st Iran. Conf. Electr. Eng.*, pp. 1–6, 2013.

[9] O. Altay, E. Gursoy, A. Font, and O. Kalenderli, "Travelling Wave Fault Location on Hybrid Power Lines," pp. 1–4, 2016.

[10] M. Saha, J. Izykowski, and E. Rosolowski, *Fault Location on Power Networks*. Springer, 2010.

[11] J. Sadeh *et al.*, "Accurate Fault Location Algorithm for Series Compensated Transmission Lines," vol. 0, no. c, pp. 46–53, 2013.

[12] S. Kapuduwage, "Fault Location on the High Voltage Series Compensated Power Transmission Networks," 2006.

[13] P. System, R. Committee, I. Power, and E. Society, *IEEE Guide for Determining Fault Location on AC Transmission and Distribution Lines IEEE Power and Energy Society*, vol. 2014. 2014.

[14] J. Sadeh, N. Hadjsaid, A. M. Ranjbar, and R. Feuillet, "Line Parameter-Free Fault Location Algorithm for Series

Compensated Transmission Lines," *IEEE Trans. Power Deliv.*, vol. 15, no. 3, 2000.
[15] Z. Huang, Y. Chen, and Q. Gong, "A protection and fault location scheme for EHV line with series capacitor based on travelling waves and wavelet analysis," *PowerCon 2002 - 2002 Int. Conf. Power Syst. Technol. Proc.*, vol. 1, pp. 290–294, 2002.
[16] F. H. H. Magnago and A. Abur, "Fault location Using Wavelets," *IEEE Trans. Power Deliv.*, vol. 13, no. 4, pp. 1475–1480, 1998.
[17] B. Bhalja, R. P. Maheshwari, E. Engineering, and I. I. T. Roorkee, "Wavelet Transform Based Differential Protection Scheme for Tapped Transmission Line Crossleyiand," *2006 IEEE Int. Conf. Ind. Technol.*, pp. 1004–1008, 2006.
[18] W. Chen, O. P. Malik, X. Yin, D. Chen, and Z. Zhang, "Study of Wavelet-Based Ultra-High-Speed Directional Transmission Line Protection," *IEEE Power Eng. Rev.*, vol. 22, no. 11, p. 54, 2002.
[19] D. Wang, H. L. Gao, S. B. Luo, and G. B. Zou, "Travelling wave fault location principle based on Rogowski coil ' s differential output and Hilbert-Huang transform," pp. 1–6, 2016.
[20] S. Kapuduwage and M. Al-Dabbagh, "Development of Efficient Algorithm for Fault Location on Series Compensated Parallel Transmission Lines," *TENCON 2005 - 2005 IEEE Reg. 10 Conf.*, pp. 1–5, 2005.
[21] PLN, *Fault Recorder Contract*. 2014.

Off Line Partial Discharge as an essential Condition Based Monitoring (CBM) Tool for busbar insulation monitoring on Air Insulated Switchgear (AIS)

J.V. Bissett[1], P.A. van Vuuren[2]
[1]Erongo Regional Electricity Distributor, Namibia
[2] School of Electrical, Electronic and Computer Engineering, North-West University, South Africa

ABSTRACT

This paper investigates the effectiveness of current test methods and procedures associated with insulation testing on the busbar of Air Insulated Switchgear (AIS) during scheduled planned maintenance. An improved test method proposes combining traditional AC voltage withstand test with Partial Discharge (PD) measurement. Three PD sensors were considered, namely Airborne Acoustic (AA), Transient Earth Voltage (TEV) and High Frequency Current Transformer (HFCT). The results confirm that HFCT is most sensitive and revealed significant difference at higher voltages between the six AIS samples. The TEV results produced consistent readings with smaller differences at higher voltages. The Ultra-sound instrument was found to be extremely sensitive but lack the capability to produce reliable measurement readings. Peak TEV measurements indicate a noticeable difference between older and newer samples.

The paper is supported by a practical case study.

Keywords: Air Insulated Switchgear (AIS), Remaining Useful Life (RUL), Condition Based Monitoring (CBM), Partial Discharge (PD), On Line Partial Discharge (OLPD)

Corresponding author: Vermaas Bissett (vbissett@erongored.com.na)

1. INTRODUCTION

The field of Partial Discharge (PD) is enormous and extensive research during the last few decades produced excellent results and value to the industry. The methods and techniques of partial discharge assessment using non-intrusive test equipment appears to be more comprehensive in HV and EHV switchgear, with fewer studies on MV switchgear [1].

In Electricity Distribution Networks, AIS forms a critical node in the electricity network to distribute energy from transmission levels to the end customers. The electricity utility must operate and maintain the AIS in the most effective and efficient manner to ensure compliance to statutory requirements. Whilst many of the AIS are approaching the end of lifetime phase, most utilities are not in a position to replace the obsolete AIS substations due to financial and technical constraints. Failure of AIS have enormous impact on business integrity and should be avoided by the utility at all cost.

AIS consists of various components that are assembled in a compartmentalized manner and includes the circuit breaker-, busbar-, current transformer- and cable compartments. The most under-maintained component in electrical distribution systems is the main busbar insulation of medium voltage switchgear [2]. Busbar failure is catastrophic and results in switchgear failure and unavailability of the substation. The busbar is not easily accessible and requires a power outage for inspection and maintenance. During preventative maintenance, a very elementary Alternating Current (AC) voltage withstand test is recommended by Original Equipment Manufacturers (OEM) in accordance with the International Electrotechnical Commission (IEC62271-200) [3][4]. The final result of this prescribed test reveals either a pass or fail outcome and does not necessarily provide additional information on the condition of the insulation.

PD phenomena has many macroscopic-physical effects as illustrated in Figure 1. The measurement of PD in electrical apparatus can be divided into the conventional (IEC60270) [5] and unconventional (IEC62478) [6] measurement methods. This paper will only consider unconventional measurement by electromagnetic and acoustic methods. Unconventional methods are the preferred type for on-site assessment because switchgear can remain in service while conducting the assessment with portable hand held devices. The investment cost is relatively low and most utilities can easily accommodate this budget requirement.

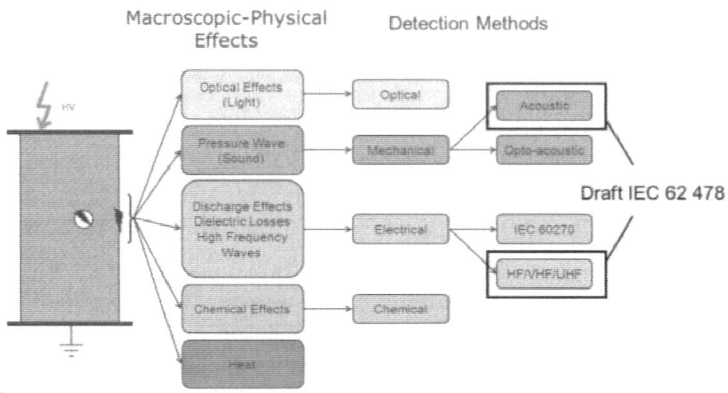

Figure 1: Macroscopic-physical effects of PD with detection methods

The measurement of PD is a suitable means of detecting certain defects in the equipment under test and is a useful complement to dielectric tests. Experience shows that PD may lead in particular arrangements to a progressive degradation in the dielectric strength of the insulation, especially of solid insulation. It is not yet possible to establish a reliable relationship between the results of PD measurement and the life expectancy of the equipment owing to the complexity of the insulation systems used in metal enclosed switchgear [7].

This paper proposes a new test method by obtaining PD measurements during over voltage withstand test across a range of voltages up to the maximum rating of the equipment. The test can be used for on-site testing before and after major maintenance has been completed, but can also be considered for Factory- and Site Acceptance Tests. The method provides more insight into leakage current discharge when performing normal AC voltage withstand testing. The results are used for further interpretation of insulation condition and RUL.

AIS suffer from several insulation failure processes, namely electrical tracking, inadequate clearance between phases and earth potential, as well as electrical treeing on voltage- and current transformers, all of which produce undesirable PD. Poor workmanship, inferior insulation material and sub-standard manufacturing processes, as well as insulation degradation due to normal aging, all contributes to this phenomena.

PD assessments is currently one of the best methods that are used for asset management on AIS. Both On Line and Off Line PD monitoring systems have been installed in the electricity industry with great success [8]. Considerable work was done to determine the most appropriate sensors applicable to electrical assets, as well as PD test techniques and the applications thereof [9]. No literature could be found that made reference to the proposed new test method.

Research has been done to simulate accelerated aging of switchgear by changing the environmental conditions [10][11]. Other research focused on the extension of the lifetime of the switchgear by applying certain maintenance techniques to eliminate discharge in MV AIS [12]. Degradation analysis of epoxy resin surfaces exposed to partial discharge was also popular research and some new work was recently completed by Tokyo City University [13]. Good research was also done on deterioration of insulating materials by internal discharge where life versus voltage characteristic were measured [14].

Asset management remains a core responsibility of electricity utilities. Risk and health indices are highlighted as important components of effective condition assessment of MV switchgear [15][16]. The utilisation of Off Line Partial Discharge measurements on AIS as a CBM tool is furthermore recommended as an essential tool for distribution utilities. Monitoring of the busbar and other compartments with various PD sensors are presented. Detailed results provide insight into the condition of insulation with opportunities to determine RUL and to extend the lifetime of the equipment.

No research could however be found to quantify Remaining Useful Life of AIS, either by extrapolating information or trending data, or life expectancy formula or health indices, or by interpretation of PD results.

2. QUESTION

Is the power frequency withstand voltage test adequate for insulation testing and what additional information on the condition of the insulation could be obtained by taking a variety of PD measurements during the over voltage withstand test across a range of voltages up to the maximum rating of the equipment?

3. CURRENT INSULATION TEST METHODS AND PROCEDURES

The current test method recommended by IEC (International Electrotechnical Commission) for metalclad switchgear prescribes applying a 50Hz power frequency test voltage to prove integrity of the insulation. The test is typically done during Factory Acceptance Test (FAT) and repeated on site (Site Acceptance Test) at a lower test voltage [17].

Milli Ampere reading and in some cases also Ohm readings are displayed on the test equipment. The interpretation of the test results could be debated as high milli Ampere reading does not necessarily indicate a breakdown in the insulation. Likewise, high Ohm readings cannot reliably indicate good insulation. Traditionally, 2.5 mA per panel was considered acceptable and used as a bench mark [18].

More recently, conventional PD testing during FAT has been conducted provided that the manufacturer's facility can accommodate this test. Maximum permissible PD quantity is the apparent charge that is usually expressed in picocoulombs (pC) [4]. The standard requires that at the correctly rated test voltage permissible discharge quantity shall be less than 100pC. This is a conventional test method with directly connected circuit to the test object. Due to many constraints this test cannot easily be repeated on-site.

After completion of on-site maintenance the switchgear will be tested at a reduced test voltage, typically 75% - 80% of the original test voltage, to safeguard equipment against unwanted stresses caused by over voltage. The equipment is considered to have passed the test if the test voltage is maintained for 1 minute without flashover. This test can be seen as formality and will seldom fail the voltage withstand test. The main purpose of this test as part of normal maintenance practices is to prove that it is safe to energise the switchgear after completion of all work and that no major defects are present on the switchgear.

The shortcomings of the existing test methods and general industry practice could be summarized as follows:

(a) Limited guidelines on the interpretation of ohm (MΩ) and leakage current (mA) readings.
(c) The voltage withstand test lacks information on the condition of the insulation.
(d) Inconsistency on test voltage compensation based on environmental conditions (both temperature and relative humidity) to avoid discrepancies when comparing results of previous tests.
(e) The voltage withstand test provides limited information on how effective the maintenance activities were executed.

4. IDENTIFICATION OF ALL AIS ELEMENTS TO BE MONITORED WITH EMPHASIS ON INSULATION

Equipment breakdown on the primary parts or components of AIS is depicted in Figure 2. Advice and recommendation from OEM was considered to identify all the critical areas of inspection. Location of PD sensors and positioning thereof is absolutely critical for the successful recording of PD pulses [9]. As metal enclosed switchgear is compartmentalized it is important to understand the coverage of the sensors on the entire board.

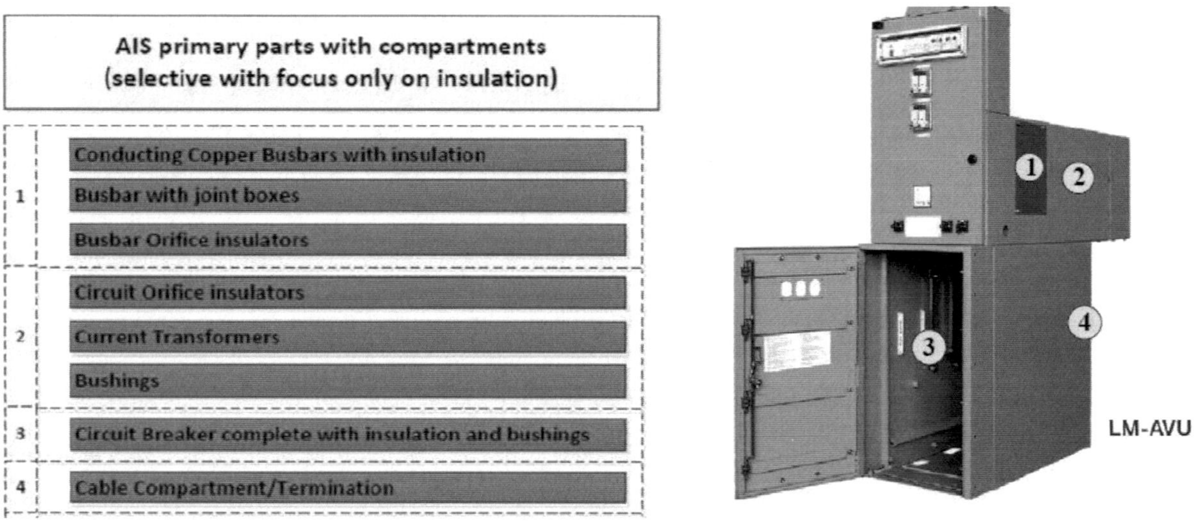

Figure 2: Typical component breakdown of AIS with compartments

5. PD APPLICATION ON AIS AND ASSET MANAGEMENT

Already from the early years of PD study, researchers discovered the potential of PD assessment on insulation material to determine the health condition of the apparatus. In recent years more emphasis was on relating the assessment with health indices as well as condition risk assessment [19]. The outcome of assessment values could be used as input drivers for assessment management. There is already On Line monitoring systems available that provide continuous figures on health statuses of the entire plant in real time.

During Factory Acceptance Test (FAT) it has lately become customary for clients to specify PD measurement on newly manufactured switchgear in accordance with IEC standards [4]. It is typically a single measurement value that should be below 100pC. Although FAT is assumed to represent the actual and final site configuration, it should be noted that the main busbar assembly has not been completed and this PD value is only reflecting on the combination of all the individual bays including the circuit breakers. Repeatability of the same FAT test for PD measurement on site after final commissioning and during major maintenance are therefore not possible, unless it is specifically prescribed by the engineer.

Three PD sensor technologies are currently available for AIS assessment and will furthermore be evaluated in this paper, namely Airborne Acoustic (AA), Transient Earth Voltage (TEV) and High Frequency Current Transformer (HFCT).

6. PD ASSESSMENT METHODS CURRENTLY DEPLOYED WITIN THE INDUSTRY

Table 1 provides comparison criteria between OLPD and Off Line PD solution for in-service AIS. PD Maintenance Testing or Site Acceptance Test (SAT) is introduced as a new proposed method for consideration.

Table 1: Comparison of PD measurement methods

Comparison Criteria	On Line Partial Discharge - OLPD Continuous	On Line Partial Discharge - OLPD Periodic	Off Line Partial Discharge	Off Line PD Maintenance Testing (Site Acceptance Test)
Standard/Measurement type	•IEC62478/Unconventional	•IEC62478/Unconventional	•IEC270/Conventional	•IEC62478/Unconventional
Monitoring	•In service •Continuous	•In serice •Periodic	•Out of service •Scheduled according to need	•Out of service •Scheduled according to need
Voltage source	•3ϕ grid, fixed •System voltage	•3ϕ grid, fixed •System voltage	• External power source	•External power source •1ϕ, variable •1 - 24kV
Assessment areas/Compartments	•Fixed •Dependent on sensor location	•Dependent on sensor positioning •All compartments	• Specific, one/all compartment	•Flexible •Dependent on sensor positioning
Panel & Compartment Assessment	•Limited compartments	•All Compartments	•Limited compartments as specified	•All Compartments •Individual Phases
Benefits/Advantages	•No need to isolate the circuit •Cicuit loaded when tested •Economical & non-invasive •Test asset under normal operating conditions •Continuos recording •Relate PD to work cycle	•No need to isolate the circuit •Cicuit loaded when tested •Economical & non-invasive •Test asset under normal operating conditions	•Proven technology •Better sensitivity	•Combine PD sensor measurements •PD assessment at higher voltages •Measure individual compartments
Shortcomings/Disadvantages	•Data interpretation can be difficult •Earthing pre-requisites	•Snapshot only at instance of time •No cyclic environmental information	•Circuit not loaded during testing •Outage required •Expensive and time consuming	•Circuit not loaded during testing •Outage required •No cyclic environmental information

There is currently not an acceptable standard for interpretation of PD levels for unconventional measurements. Manufacturers and industry users have over the years developed own benchmark levels as guidelines for assessment of AIS. Based on field experience many utilities have developed their own PD level guidelines. The guideline from HVPD which is based on similar switchgear within the UK industry and was found that this guideline with four level indication can successfully be used and trusted with high confidence [20].

7. PROPOSED NEW TEST METHOD BY COMBINING PD MEASUREMENT DURING OVER VOLTAGE WITHSTAND TEST

Unconventional test methods are sometimes not preferred by industry users because the interpretation of the physical discharge quantity cannot be determined with high accuracy and confidence.

The conventional test methods were originally developed to determine true discharge quantities and connects directly to the test object with suitable capacitor. This preferred test is normally performed in controlled laboratory condition where noise levels can be controlled and reduced to minimum levels. This test procedure is difficult to perform on-site due to the uncontrolled environment and availability of test equipment.

The unconventional (IEC62478) measurement method has lately become the preferred method and all PD equipment manufacturers have released several new Hand Held Devices as well as OLPD systems.

A test method that has not yet been considered by the industry proposes to combine the functionality and measuring capabilities of PD instrumentation while conducting off line the power frequency voltage withstand test over a range of voltages. A considerable amount of additional information can be captured during this test because measuring at higher level frequency components are possible.

8. PRACTICAL TESTS AND ASSESSMENT ON AIS SAMPLE PANELS AND THE INTERPRETATION OF RESULTS

8.1. Selection of most suitable sensor and measurement instrument

The sensor type, instrument capabilities and responsiveness to increasing AC voltages were used as selection criteria to identify the most suitable instruments. The PDS Insight from HVPD (with AA, TEV, HFCT) and Ultra TEV from eaTechnology (with contact probe AA, TEV) were selected as most appropriate and the preferred instruments.

Capturing measurements at the higher test voltages were problematic as the continuous display of readings on the ultrasonic instrument changes rapidly. A 10 second measuring time was allowed for the ultrasonic reading. The 5 second recording capability of the HVPD Insight was an advantage. Repeatability was proven by conducting the same test more than once at the same voltage. The results were acceptable and within a 10% discrepancy between readings.

It should be noted that measurements were taken at the same voltage with each separate instrument and sensor, but not at the exact same instance of time. Although there is good correlation between the various measurements, the values because it is not taken at the exact same instance of time.

8.2. Selection of panel samples

The selection of AIS panel samples consisted of 6 individual Reyrolle/ABB panel types, namely: LMT, LMS and LMR [21]. The LMT type was followed by two improved designs, namely LMS and LMR, all being fully compatible with one another.

Equipment from a redundant substation became available for research and tests and a sample from the LMT, LMS and LMR were taken to focus on equipment that has been in service for a long duration, respectively 52, 24 and 14 years. The selected panels were labelled Sample #1, #2 and #3 respectively.

It was observed that all the compartments were covered with some dust deposits and that the insulation was not fully clean and free from dirt. The condition basically resembles typical substation environment. It was decided not to immediately clean the switchgear (Sample #1, #2 and #3) before testing, but to rather look at the impact of this undesirable condition on the measurement results, and correlate the test results with visual inspection. The same test procedure will be followed once the switchgear has been properly cleaned and restored to satisfactorily condition. This test will be conducted at later stage and identified as future work.

Table 3: Detail on panel samples used during the tests.

Sample Nr	Type	Group	Manufacturer	Year	In service years	Remarks
Sample #1	LMT	A	Reyrolle	1965	52	First generation of LM switchgear and oldest sample
Sample #2	LMS	A	Reyrolle	1993	24	Second generation of LM switchgear
Sample #3	LMR0	A	ABB	2003	14	Third generation of LM switchgear
Sample #4	LMR1	B	ABB	2003	3	Relatively new sample in service for 3 year
Sample #5	LMR2	B	ABB	2007	1	Relatively new sample in service only for 1 year
Sample #6	LMR2	B	ABB	2007	0	New sample that was never in service

A total of 6 samples were therefore selected that were used during the practical test and assessment. Table 3 provides for summary of information on the different panel samples. The samples can be divided into two distinct groups, namely an older Group A (Samples #1, #2 and #3) and a newer Group B (Samples #4, #5 and #6).

8.3. Test procedure and set-up

The approach on how the assessment and tests were conducted can be summarized as follows:-

a) Same test method and approach on LMT, LMS and LMR panel types.
b) Test only the primary busbar compartment, with all phases connected.
c) Take HFCT, TEV and Ultrasonic measurements during the test.
d) Increase the applied test voltage from 12kV to 21kV in 3kV steps.
e) Cleaning of insulation was only done on the Group B samples.

8.4. Conducting Practical Tests

A systematic approach was followed by conducting the practical tests to arrive at three distinctive objectives:-

<u>Objective A:</u> Voltage and current, as well as PD measurements will be taken during all tests. This is to confirm the correlation between leakage current and PD to prove in general that the test set-up and procedure is valid and acceptable.

<u>Objective B:</u> A variety of PD sensors are available for AIS assessments. It is important to identify the most suitable sensors applicable to the type of asset assessment, in this case AIS.

<u>Objective C:</u> Determine the sensitivity of sensor and measurement equipment by conducting before and after tests on specific samples. Group B (Samples #4, #5 and #6) was tested before and after cleaning of insulation.

8.5. Observation, interpretation and comparison of the results

a) There is excellent correlation between the leakage current (mA) at different voltages, as well as between the different panel types. The overall increase in leakage current between lowest and highest voltage readings are 3.3, 6 and 3.6 times respectively for LMT (Sample #1), LMS (Sample #2) and LMR (Sample #3) panels. This is typical and acceptable results for AIS.

b) It is noticeable that the leakage current for the newer Group B (Samples #4, #5 and #6) are higher than Group A. It was expected to achieve lower readings because the insulation is new.

c) There is excellent correlation between peak HFCT PD measurements at different voltages, for each individual panel, as well as between the different panel types. The overall increase in peak HFCT PD between lowest and highest voltage readings are 398, 178 and 153 times respectively for LMT (Sample #1), LMS (Sample #2) and LMR (Sample #3) panels.

d)

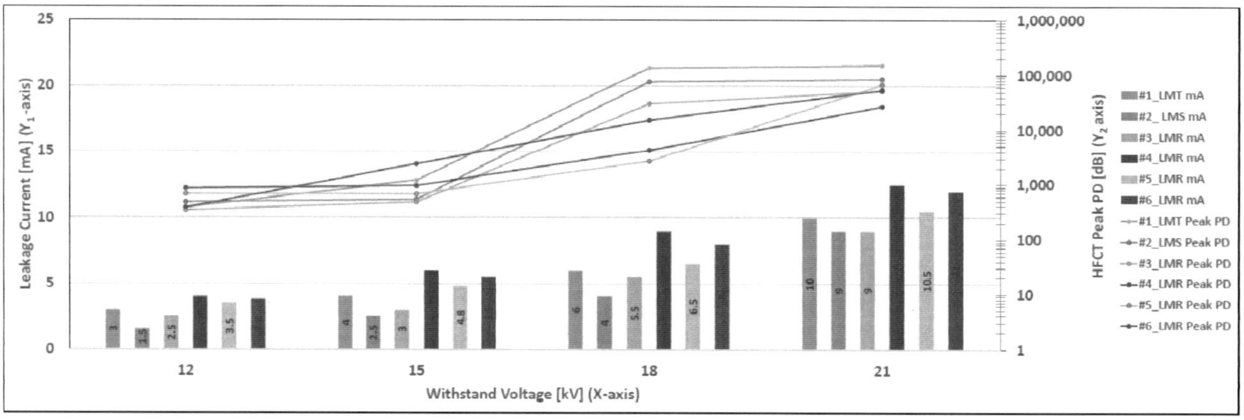

Graph 1: Combined HFCT Peak PD and Leakage Current measurements for all 6 samples "Withstand Voltage (X - axis) versus Leakage Current (Y_1 axis) & HFCT Peak PD (Y_2 axis)"

e) There is good correlation between peak TEV PD measurements at different voltages, for each individual panel, as well as between the different panel types. The overall increase in peak TEV PD between lowest and highest voltage readings are around 10 times for all 3 panels. The measurements appear to be less sensitive compared to HFCT.

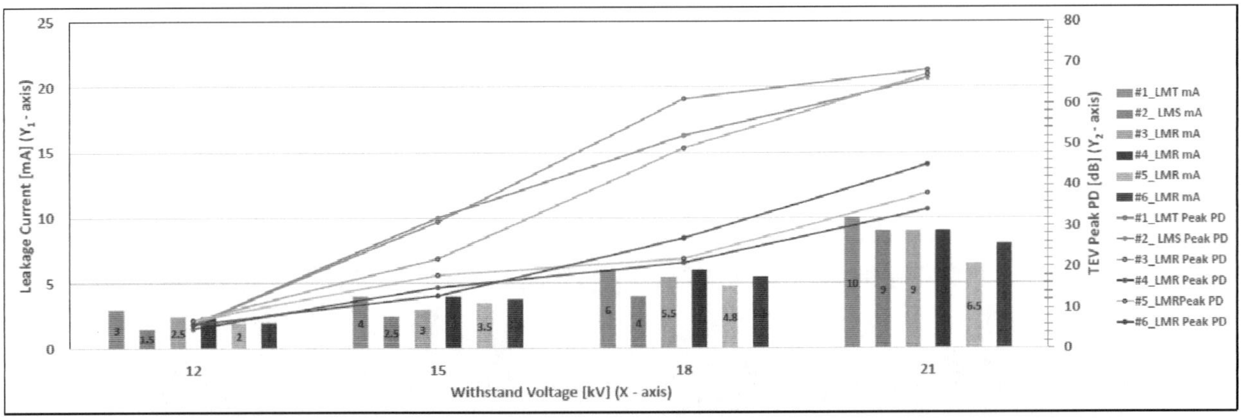

Graph 2: Combined TEV Peak PD and Leakage Current for all 6 samples
"Withstand Voltage (X_1) versus Leakage Current (Y_1) & TEV Peak PD (Y_2)"

f) Sensitivity of measurement and test set-up was proven by first obtaining TEV (Graph 3) and HFCT (Graph 4) measurements on samples 4, 5 and 6, for Group B. The test was repeated after the insulation was thoroughly cleaned. Results of all three tests indicate that both TEV and HFCT measure significantly lower levels of partial discharge. For illustration purposes, only results of Sample #6 is shown which is a good representation of the entire sample group. Measurements are consistently lower for the cleaned insulation, except for the first two TEV readings which are approximately the same value.

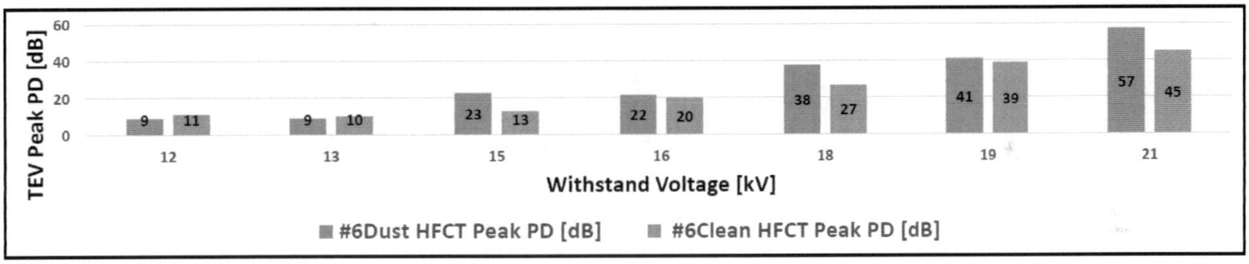

Graph 3: Withstand Voltage versus TEV Peak PD for dust and clean conditions for sample 6 only

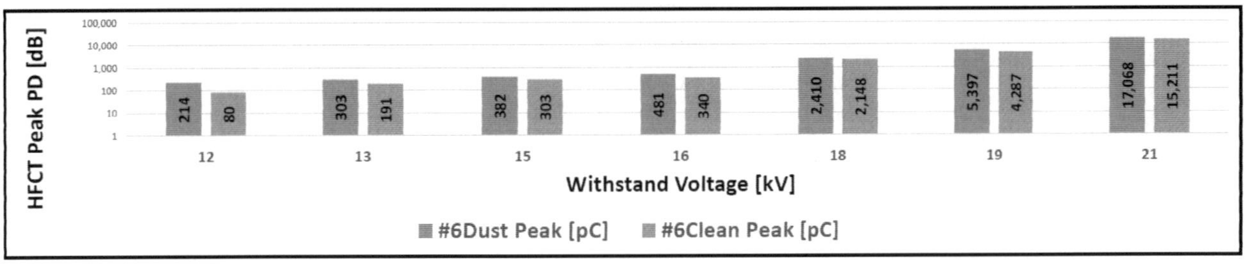

Graph 4: Withstand Voltage versus HFCT Peak PD for dust and clean conditions for sample 6 only

g) Graph 5 and 6 indicate HFCT Peak PD and TEV Peak PD for all 6 samples. There is a clear distinction between the two sample groups, namely the older Group A and newer Group B. This is depicted on the graph by the blue and green ovals, respectively for the older Group A and newer Group B. It is more predominant on the TEV readings where the newer sample group measures consistently lower levels at three of the four data points. It is only at the 12kV data point where no distinction is visible. For the HFCT measurements, this phenomena is less prevailing although also clearly visible at the higher voltages at 18kV and 21kV.

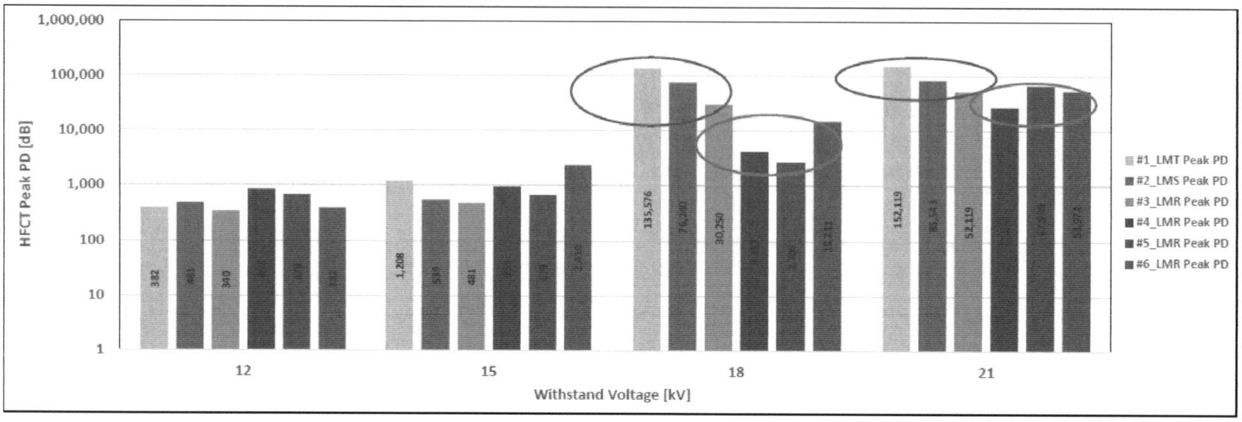

Graph 5: Withstand Voltage versus HFCT Peak PD for all 6 samples

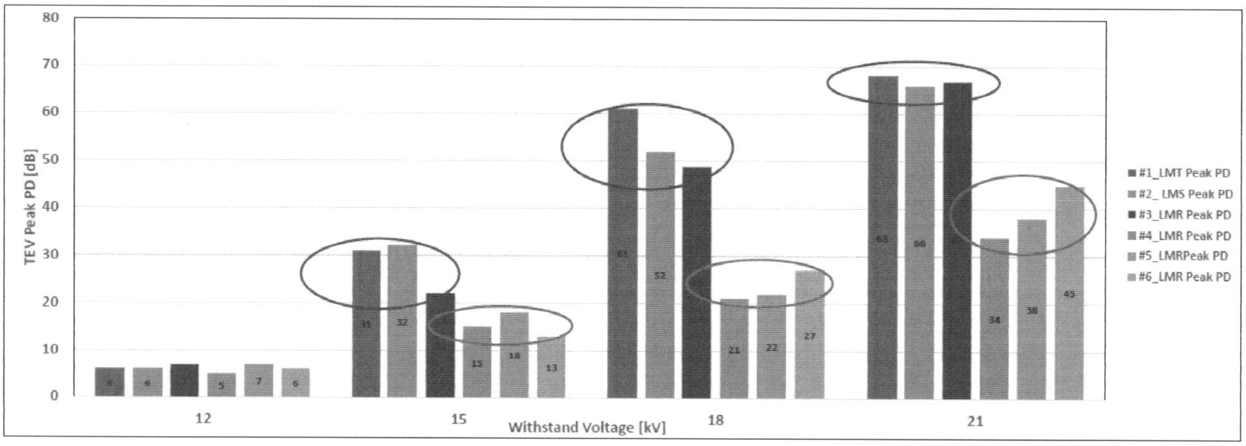

Graph 6: Withstand Voltage versus TEV Peak PD for all 6 samples

9. CONCLUSIONS

Off Line PD measurement combined with power frequency withstand voltage test during planned maintenance as Condition Based Monitoring (CBM) tool can provide additional information for asset management systems. Peak PD measurements were found to be a good indicator for analysis purposes.

It was confirmed that there is excellent correlation between leakage current and both HFCT and TEV PD measurements. Both HFCT and TEV sensors and measurement capabilities proved to be sensitive enough to distinguish and differentiate between deliberate surface discharge and cleaned insulation. The detection method is therefore ideal to differentiate between good insulation and deteriorated insulation.

Although HFCT, TEV and ultra-sonic methods were found to produce repeatable results, HFCT appears to be the most sensitive. TEV produced consistent results although not as sensitive compared to HFCT. HFCT measurement results is most sensitive and revealed significant increases in PD peak levels at higher voltages. Ultrasonic instrument could not reliable display and record the measured values and was therefore not included in this paper.

It appears that the busbar insulation design, development and manufacturing from the Reyrolle switchgear range improved, as expected, from the older LMT to the newest LMR type as reflected by significantly lower PD measurements. The proposed new method could therefore be considered for the developing of health indices for AIS.

10. FUTURE PRACTICAL WORK AND TESTS

This paper has not considered the full value of count and activity measurements which is typically available on some HHD instruments. It is believed that these parameters could also contribute towards assessment methods.

Conduct conventional PD test on Sample #1, #2 and #6 to compare conventional measurement results with unconventional results. The test can be conducted at the Rotek facility at Eskom in Johannesburg, South Africa.

The improved test method can provide information on the condition of the insulation in order to determine Remaining Useful Life (RUL) of the insulation by making use of TEV measurements. Develop a proposal to determine or calculate RUL for AIS insulation by theoretic and practical model.

Both HVPD and eaTechnology recently released advanced Phase Resolved Partial Discharge HHD with full recording and sophisticated noise filtering capability. This will improve integrity of field assessments and provide more credible measurements.

11. REFERENCES

[1] G. Dennis, "Non-intrusive techniques for detecting partial discharge in HV and EHV switchgear," University of Bath, Oct 2008.

[2] G. Paoletti, M. Stephens, G. Herman, M. Whitehead, "The most ignored maintenance electrical item in the plant electrical power distribution system and practical solutions."

[3] IEC 60060, "High Voltage test techniques, Part 1, General definitions and test requirements," 1989.

[4] IEC62271-200, "International Electrotechnical Commission, High Voltage Switchgear and controlgear," 2011.

[5] IEC60270, "International Standard, High Voltage Test Techniques: Partial Discharge Measurement," Third Edition, 2000

[6] IEC62478, "International Standard Technical Specification, High voltage test techniques - Measurement of partial discharges by electromagnetic and acoustic methods," Edition 1.0, 2016

[7] J.E. Smith, G. Paoletti, I. Blokhintsev, "Experience with On Line Partial Discharge Analysis as a tool for predictive maintenance for medium voltage (MV) Switchgear Systems," IEEE Petroleum and Chemical Industry Committee, 2002.

[8] L. Renforth, R. Giussani, M. Seltzer-Grant, M. Marcarelli, M. Foxall, "On Line and Off Line PD monitoring: The experience in testing MV and HV apparatus from an industrial point of view"

[9] M. Muhr, T. Strehl, E. Gulski, K. Feser, E. Gockenbach, W, Hauschild, E. Lemke, "Sensor and sensing used for non-conventional PD detection," Cigré, 2006

[10] D. Koenig, P. Roesch, "Contribution on the specification of an aging test for the insulation of enclosed, air insulated switchgear," IEEE Transactions on Electrical Insulation, 1987

[11] S. Keim, D. Koenig, "Standardised and non-standardised aging tests for indoor switchgear and switchgear components under severe climatic conditions," IEEE International Symposium on Electrical Insulation , 2002

[12] X. Liu, X. Xiaohui, Z, Suo, L. Luo, "Maintenance techniques to eliminate partial discharge in MV air insulated switchgear"

[13] Y. Ehara, K. Aono, S. Koshio, "Degradation analysis of epoxy resin surfaces exposed to partial discharge," IEEE, 2016

[14] H. Okamoto, M. Kanazashi, T. Tanaka, "Deterioration of insulating materials by internal discharge," IEEE Transactions on Power Apparatus and Systems, 1977.

[15] C. Lowsley, N. Davies, D. Miller, "Effective condition assessment of MV switchgear," eaTechnology.

[16] ABB, "Air insulated switchgear Life Cycle Services," 2011.

[17] IEEE Standard C37, "Standard for Metal-Clad Switchgear," 2015.

[18] R.H. Goodwin, "A handy guide to the safe on-site testing of electrical equipment," H.V. Test (Pty) Ltd.

[19] G. Pudlo, "Advanced PD monitoring of power transformers supporting asset management," Doble, Germany, 2009.

[20] HVPD, "Guidelines for TEV PD Measurements for MV Switchgear," High Voltage Partial Discharge Ltd, May 2009.

[21] ABB, "Retrofitting of Reyrolle type LMT switchgear," Technical Data Sheet, http://www.abb.com/heritage.

A Condition Based Reliability Simulator Framework based on a Heuristic Fault Model

Dr. H.F. Swanepoel[1*] & Prof. J.H. Wichers[2]

[1]Department of Mechanical and Nuclear Engineering
North West University, South Africa
swanephf@eskom.co.za

[2]Department of Mechanical and Nuclear Engineering
North West University, South Africa
Harry.Wichers@nwu.co.za

ABSTRACT

There are significant concerns as well as remedial efforts by the SA Power Utility to improve the generation plant performance that showed a significant decline over the past number of years. There is general consensus on the significant opportunity for the power utility to leverage Condition Monitoring (CM) and Advanced Analytics (AA) technologies to assist in the turn–around of technical performance of the power station fleet. Over the years, extensive online-monitoring technology and predictive condition monitoring capability have been introduced on high-value assets like the generator, turbine, transformer and boiler; with much less focus on Balance of Plant (BoP) equipment and machine trains.

The research study developed a Condition Based Reliability Simulator (CBRS) Framework that uses a heuristic fault and failure analytical model based on the Design Basis of plant systems and equipment. By developing integrated advanced analytical and predictive fault models based on a thorough understanding of critical Design Basis parameters and typical failure behaviors, it is possible to better identify impending failures and apply the most appropriate CM technology at the right time. The research considered operating philosophy and maintenance strategy impacts and demonstrated how the CBRS Framework and its associated heuristic fault model approach can significantly enhance predictive capability and assist with asset management activities to achieve the lowest total asset cost (Figure 1).

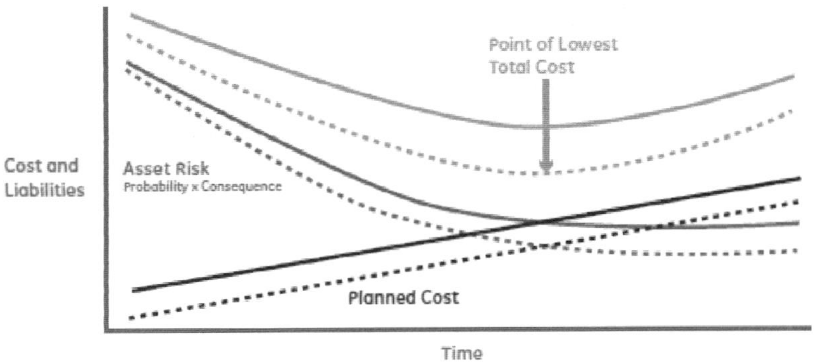

Figure 1: Optimising Assets towards the point of lowest total cost [2]

[1] The author is enrolled for a Ph.D (Mechanical) degree in the Department of Mechanical and Nuclear Engineering, North West University.
[2] The author is Director of the Department of Mechanical and Nuclear Engineering, North West University

Due to the generic nature of BoP equipment and machine trains, it would have application to the wider process plant industry that uses similar equipment. The successful and practical application of this CBRS Analytical Framework and Fault Model development and implementation methods are discussed in this paper.

Keywords: Condition Monitoring; Pattern recognition; Fault detection and localization; Diagnosis; Industrial asset management.

1. INTRODUCTION

Modern design practices for plant components use the "Design for Reliability (DfR)" principle creating a highly reliable asset base where high levels of availability and reliability should be the norm - and if proactive steps are taken to monitor, evaluate and optimise operations, plant asset life can be maximized to achieve and even exceed design life.

The current performance of South African power generating plants is no longer on par with average to top quartile global performers, while in the 1990's the utility consistently performed in the top quartile[3]. Most power stations in the Utility have a formal Condition Monitoring program in place, but still experience unacceptably high levels of UCLF/PCLF (which measures unplanned outages and extended maintenance outage durations – see Figure 2).

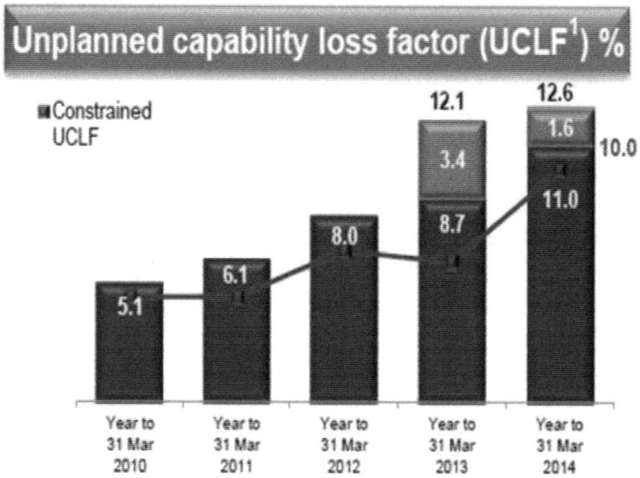

Figure 2: Increased Unavailability (UCLF) in the Power Station Fleet [3]

The research study confirmed that balance of plant (BoP) equipment failures significantly contributes to partial load losses and overall plant reliability [10]. As such, a BoP machine train became the focus of this experimental research study and basis for the heuristic fault model development aspects related to the Condition-based Reliability Simulator (CBRS) Analytical Framework. The BoP areas are generally serviced with hand-held condition monitoring devices and a condition monitoring program that is frequency based. Fault condition analysis and prediction is left to the vibration analyst, assuming a high level of skill, competency and in-depth plant Design Basis knowledge (which is many times not the case).

Further evaluation of the current CM program in the Utility shows a blanket implementation of the ISO standards related to condition and vibration monitoring practices, with limited consideration of the Design Base of the systems and equipment involved. A significant challenge of CM program efficiency is the amount of "standards" and "guidelines" provided for health assessments, many times contradicting each other. Even the International Standards Organisation [4] acknowledges that they can *at best* provide guidelines.

A further challenge is that research and practical experience show that age–related failures is the minority and that "Random Failures" makes up 89% of typical failures [5]. Generic failure modes also do not consider cases where specific plant design or operational conditions imposes very specific stresses on general machine condition and operational capabilities.

This has been pointed out in a number of research studies, for example Sondalini that describes the risk when acting on CM results without considering operational and production delivery stresses. He discusses how incorrect life prediction and failure curves can add to the challenges of ensuring an effective CM Program and points out how a Predictive Maintenance strategy without economic considerations is often the root cause for early abandonment of a CM Program.

Incorrect CM program frequencies and overly specified monitoring requirements also result in some CM programs becoming cost inhibitive and eventually abandoned [6].

An Emerson study states overwhelmed operators and complex plant operations as one of the biggest challenges in process plants that demand a new approach to managing plant more efficiently and predictively [7]. Their view is that abnormal situation prevention is one of the biggest potential productivity gains in the process plant operating space.

By developing integrated advanced analytical and predictive fault models based on a thorough understanding of critical Design Basis parameters and typical failure behaviors, it is possible to better identify impending failures and also better utilise the most appropriate CM technology at the right time.

The researcher postulated the theory that effective and efficient decisions regarding plant condition can only be made taking into consideration all aspects, including Plant Design Basis, equipment design, plant age, process design and influences; and doing this in a consistent manner using a standardised framework to guide decisions.

The primary aim and outcome of the experimental research study were to establish a Condition Based Reliability Simulator (CBRS) for power stations based on a standardised Advanced Analytics Framework, and which includes a Fault Model development and implementation methodology.

2. THE CBRS ANALYTICAL FRAMEWORK AND FAULT MODEL

Since 2013, there has been significant movement and maturity of technology elements considered to be fundamental building blocks for advanced analytics capability development (i.e. Big Data, Content Analytics, Prescriptive analytics, predictive analytics, augmented reality, etc.) [8].

2.1. The Advanced Analytics Framework

The research study developed the CBRS Analytical Framework based on the MIMOSA OSA–CBM® Specification [9], as well as analytical framework recommendations by Angeli [1].

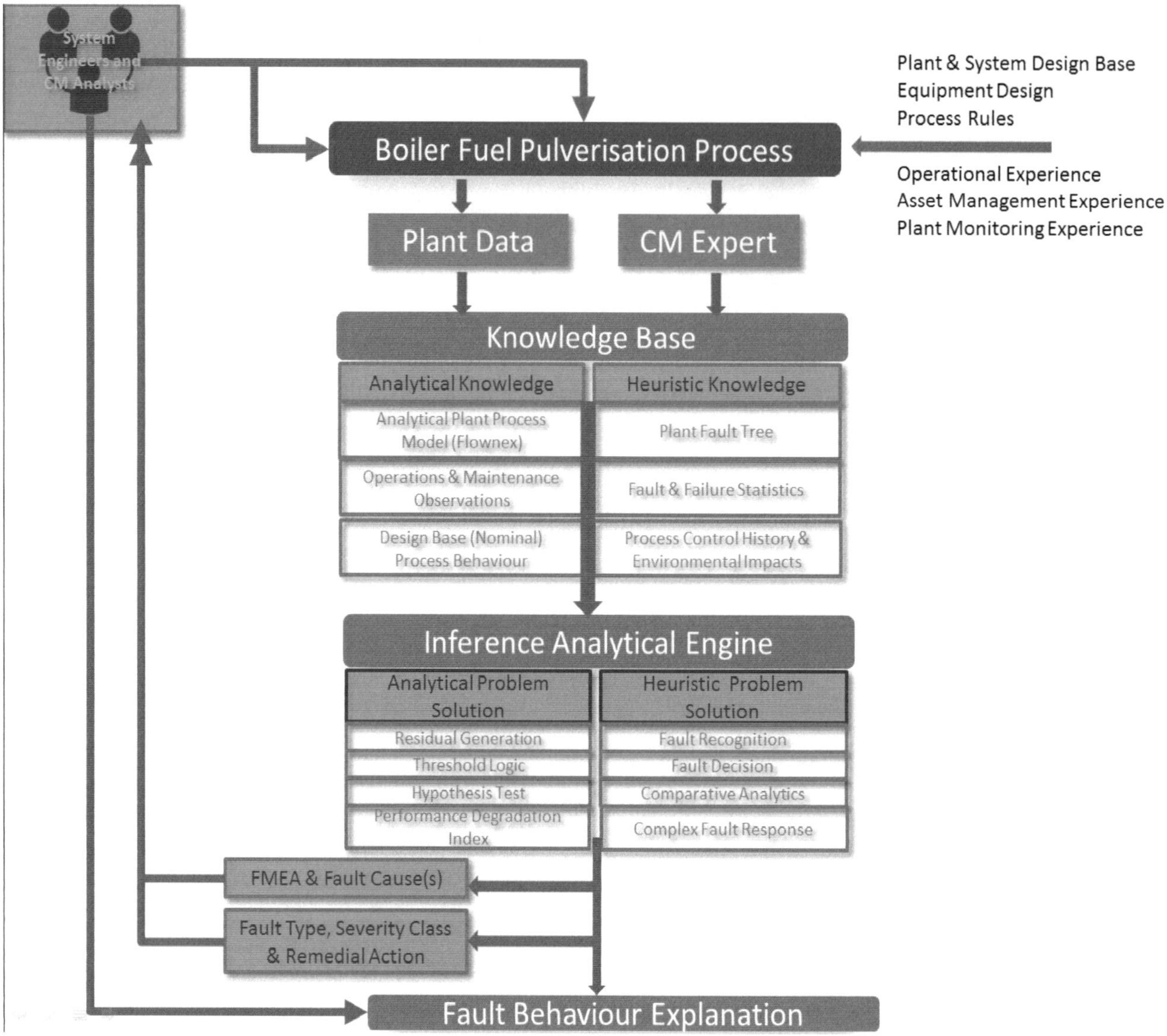

Figure 3: The CBRS Analytical Framework

2.2. The Heuristic Fault Model Framework and Implementation Methodology

The CBRS Analytical Framework is supported by a standardised implementation methodology, as well as a successfully verified Fault Model development and implementation methodology (which was validated using research study use cases).

The various elements making up the Analytical Framework were built in a phased approach, indicated in Figure 4. The colours used in Figure 4 shows alignment with the relevant MIMOSA OSA-EAI™ Specification elements of Data Acquisition (Green), Data Manipulation (Aqua), State Detection (Red), Health Assessment (Grey), Prognostic Assessment (Orange), Advisory Generation (Maroon) and linking to External Systems (Purple).

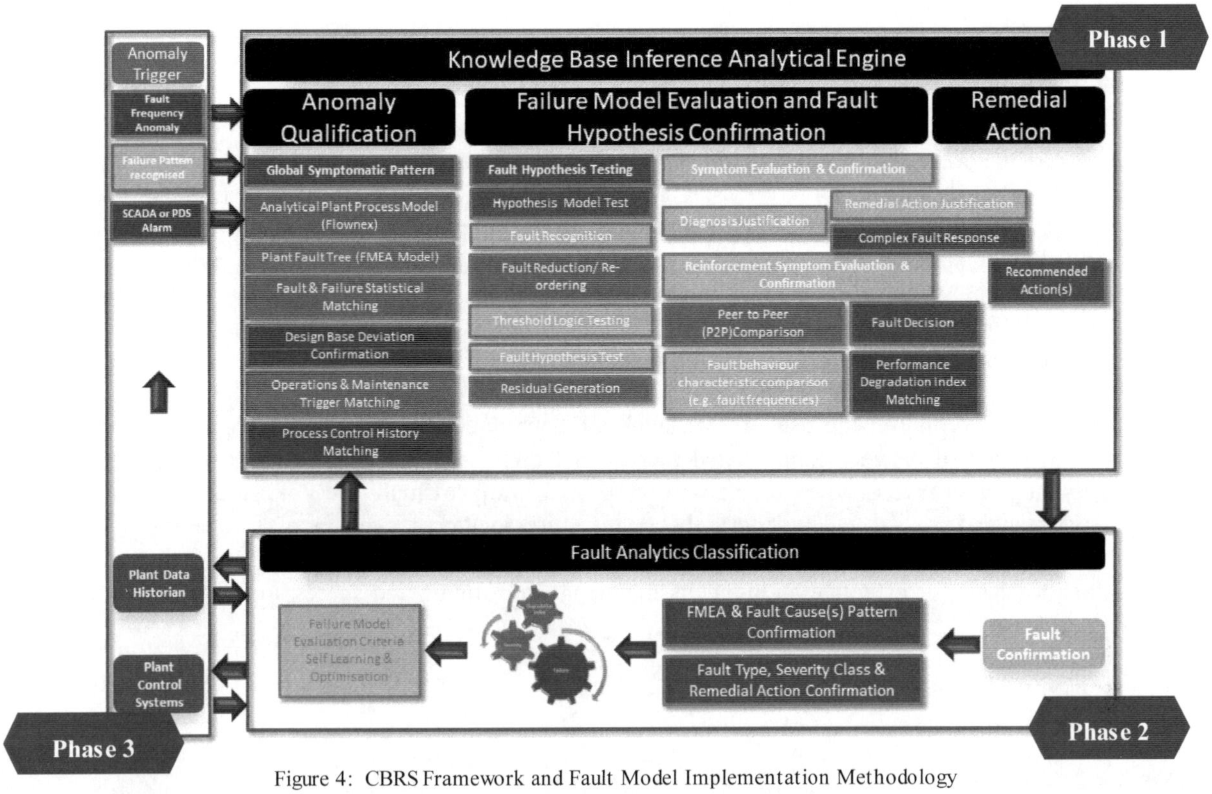

Figure 4: CBRS Framework and Fault Model Implementation Methodology

The need to integrate advanced analytics in a well-managed framework is confirmed by Emmannouilidis in their research [10]. In their research, Lei confirm the vital importance of understanding fault behaviour and characteristics in line with equipment Design Basis. This information is crucial to develop fault trees, determine failure mechanisms and failure progression patterns, and they strongly advise against using and adopting "common models" or assumptions that all equipment types will exhibit the save behaviour and fault characteristics [11].

The extent of the MIMOSA specification adoption in the CBRS Analytical Framework makes this a unique contribution to the field of specifically condition monitoring and vibration analysis, as most research in these two technology areas would focus only on elements of the specification, and none of the literature consulted in the research survey provided a holistic framework that fully encompass the MIMOSA specification in its entirety.

Researchers tend to focus their energy on only the advanced analytics algorithm aspects and very few, if any, would describe a standardised methodology with which to achieve a consistent outcome across a wide variety of equipment types (their research tend to focus on specific equipment types only). At best, they would describe *some* of the elements depicted in Figure 4 [12, 13, 14, 15, 16].

2.3. The CBRS Analytical Framework Fault Model

A significant challenge with advanced vibration analytics is the significant amount of data captured with each data point captured in the plant. This is in line with the findings of Rozados that states that machine-generated data far exceed conventional ERP and CRM system data in volume, velocity and variety of data generated [17]. Biehn shares this view and states that as little as 5% of "big data" gathered results in 95% of the value contribution of the data [18].

Marwala's extensive research suggests that modal properties, frequency response functions (FRF) and wavelet transform (WT) data is required for meaningful and correct advanced vibration data analytics and the use of AI algorithms, but points out that the use of FRF and WT data is significantly more challenging than modal property analysis due to the volumes of data collected during the data capturing process [19].

The CBRS research study developed a detailed Fault Model methodology and validated it by means of research use cases. A unique mechanism (ARMA Statistical modelling) was implemented in the methodology to address the "big data" challenges described [17, 18, 19]. Using ARMA it was possible to identify Specific Fault Frequencies of Interest (SFFoI's) in the time waveform data and these SFFoI amplitudes and trend patterns were then used to determine whether a specified fault condition exist, as well as the severity of the fault condition. The outcome from the Fault Model prediction was validated by means of detailed vibration spectrum and time waveform analysis (*internal validation*), as well as physical inspections on equipment identified to present with the fault conditions (*external validation*).

The CBRS Fault Model proposes the use of five levels of degradation as indicated in Figure 5. Degradation thresholds were required to prove the postulated theory that there is a point in the asset life (when a failure condition is present in the asset), where the cost of repair will approximate the cost of failure. At this cost intersection point (repair cost vs. failure cost), the opportunity for "cost avoidance" will be lost. Ideally the equipment should be removed in sufficient time *BEFORE* reaching this point to justify the condition monitoring effort; as well as to allow for effective and timeous maintenance planning and spares sourcing; and to minimise production loss costs.

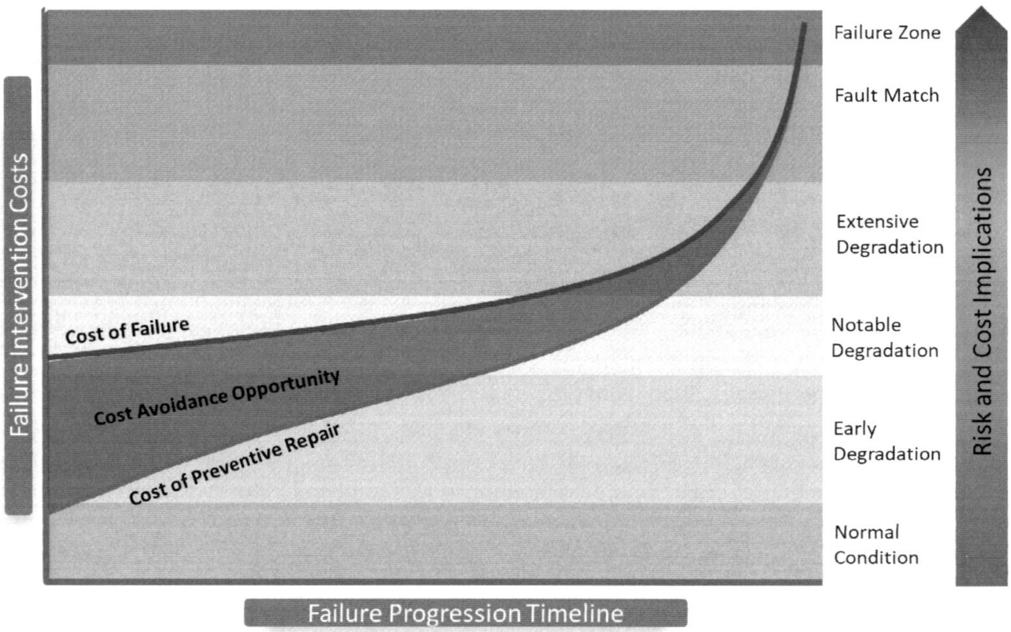

Figure 5: CBRS Fault Model Degradation Levels

And this is the primary intent of the fault models developed as part of the CBRS Analytical Framework – identify the optimal point for removing equipment from service for repairs or replacement. It was concluded that equipment should be scheduled for inspection or remedial action when "Notable Degradation" is identified, and removed for repair when Extensive Degradation is noted. Equipment can be kept in service for a limited period of time (depending on the failure degradation curve behaviour) when there is "Fault Match", but the repair cost will approximate failure cost with the resultant loss in "cost avoidance" opportunity.

3. PROVING THE ANALYTICAL FRAMEWORK AND FAULT MODELS

Three use cases were used to build and prove the required heuristic fault models. They covered mechanical, electrical and complex fault conditions. A complex fault condition in the research study is defined as a fault imposed by an external machine train component on another piece of equipment in the machine train. This paper discusses Use Case 2 which covers an electrical fault condition.

3.1. Use Case 2: Electrical Supply Phase Imbalance

Use Case 2 covered the fault condition of an electrical supply fault condition, i.e. Supply Source Phase Unbalance. Phase imbalance is typically caused by loose phase connections, although supply breaker fault conditions can generate similar fault indications. The typical fault frequencies of interest (FFoI's) are indicated in Table 1. During initial FFoI identification, acceptable levels of vibration specified by the equipment manufacturer were used as a baseline to define degradation thresholds. ARMA statistical analysis methods were then employed to determine the most statistically significant FFoI, termed the "Significant Fault Frequency of Interest (SFFoI)", which in this case is the 2 x Line Frequency (100 Hz) FFoI.

Table 1: Electrical Phase Imbalance Fault Frequencies of Interest (FFoI's)

Fault Frequency	Frequency	Unit
Motor Running Speed	993	cpm
Electrical Line Frequency (50 Hz)	3,000	cpm
2 x Line Frequency (100 Hz)	6,000	cpm
Rotor Bar Pass	69,510	cpm
Stator Slot Pass	89,370	cpm

Once the SFFoI was identified, a baseline case was identified that most typically present the failure behaviour and degradation curve to be expected. Figure 6 depicts the performance degradation curve (in red) of the motor considered to form the baseline for this fault condition, i.e. Mill 2D Motor. This is a polynomial 2nd order curve fit based on the actual vibration behaviour data taken over the time period in question.

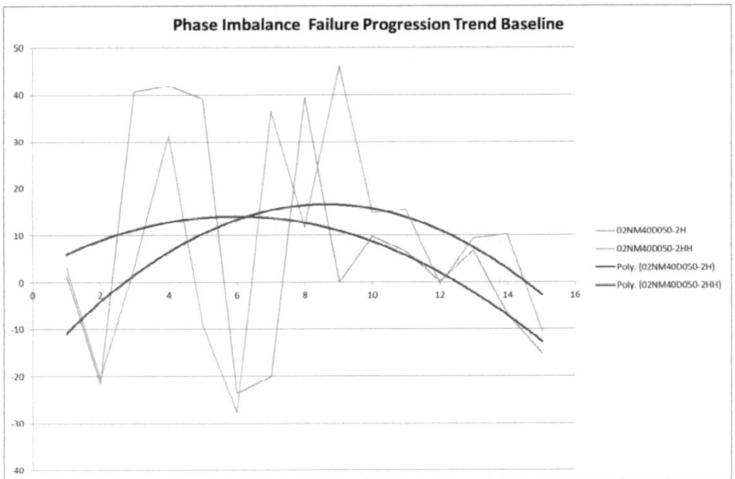

Figure 6: Phase Imbalance Baseline Degradation Curve

Once the trend curve and SFFoI for the baseline case was confirmed, additional historical cases of the fault condition were sourced as an internal validation mechanism to confirm that their SFFoI values display the same failure progression behaviour and that the identified SFFoI is indeed most indicative of the fault condition. The trend curves exhibited by these additional cases, as well as evidence from the external physical inspections performed at the time of removing the equipment from service, were used to define the failure threshold levels for the specific fault condition (in line with Figure 5).

3.2. Validating the Fault Model Use Case

Internal validation involved detailed evaluation of the vibration spectrum of Mill 2D motor (see Figure 7).

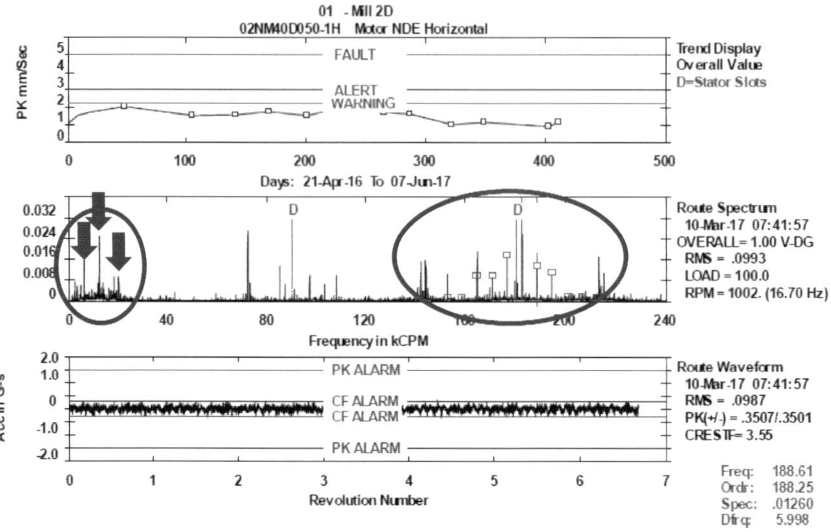

Figure 7: Motor Vibration Spectrum with electrical fault indications

Stator Slot Pass (SSP) fault frequency and its harmonics are clearly indicated in Figure 7 (marked "D"), and significant 2 x Line Frequency (100 Hz) sideband harmonic frequencies are visible around the 2nd SSP harmonic frequency. The 2nd SSP harmonic is also significantly higher in amplitude when compared to the fundamental SSP amplitude level. Amplitudes of the 2 x Line Frequency and its harmonics in the lower frequency range are also very notable (Fundamental, 2nd and 3rd harmonic).

External validation of the fault condition was achieved by evaluating the entire dataset for the 36 motors deployed in the power station's milling plant, and the SFFoI amplitude distribution is indicated in Figure 8.

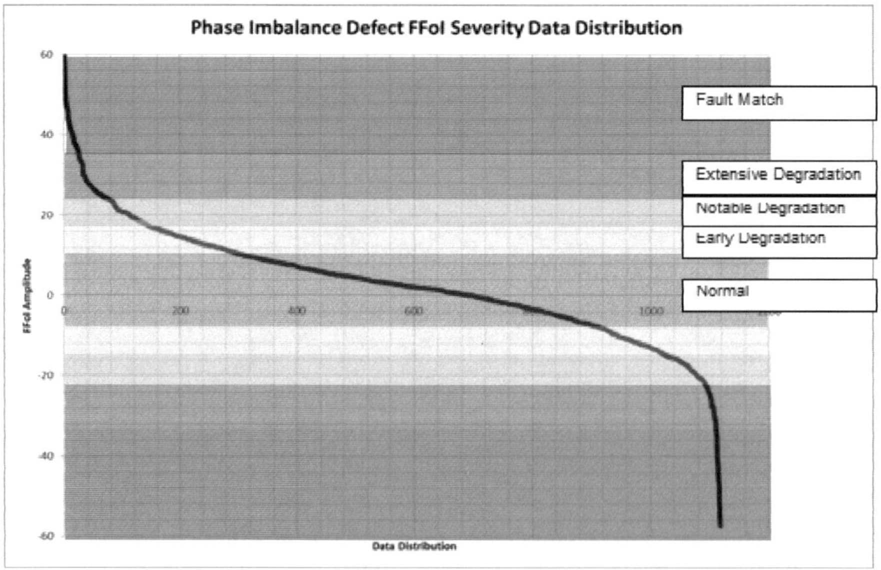

Figure 8: Phase Imbalance Baseline Degradation Curve

This distribution curve was then used to identify more motors that present with SFFoI amplitudes above the "Extensive Degradation" threshold value. As mentioned, this threshold was deemed by the research study to be the optimal point to leverage cost avoidance opportunity.

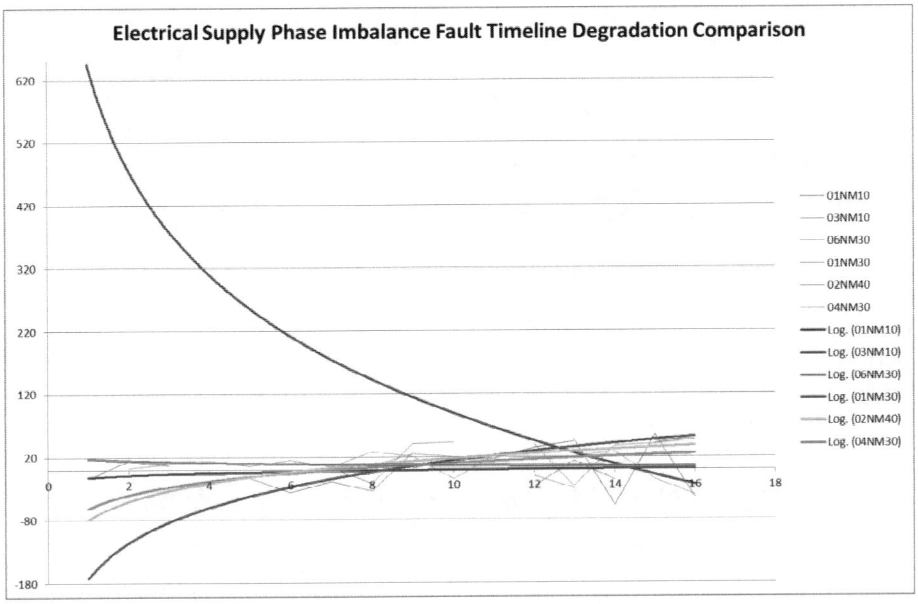

Figure 9: Phase Imbalance Baseline Degradation Curve

These motors SFFoI failure curves were mapped against the baseline case (see Figure 9) and based on severity of the failure trend curve in Figure 9; remedial action recommendations were made (summarized in Table 2).

Table 2: Phase Imbalance Fault Condition Severity Allocation

Motor Condition	Motor Candidates	Recommended Action
Fault Match	Mill 3A (03NM10)	Remove motor from service and inspect all electrical supply breakers and connections for faults and looseness, based on high SFFoI amplitudes (> 600 microns) as well as failure progression curve behaviour. **Urgent priority for remedial action.**
Extensive Degradation	Mill 1C (01NM30)	Schedule outage to inspect electrical supply components for looseness and damage. Not critical and can be part of normal scheduled outage.
Notable Degradation	Mill 6C (06NM30) Mill 2D (02NM40)	Increased monitoring and schedule electrical supply inspection to coincide with normal scheduled maintenance.
Early Degradation	Mill 2B (02NM20) Mill 5E (05NM50)	Maintain monitoring program frequency and act if degradation curve behaviour changes.
Normal	*	Normal vibration monitoring program frequency to be applied.

**Note: Due to the amount of motors falling within the "Normal Category", this was not indicated in this table.*

The relevant motors was then scheduled for a physical inspection and in all cases revealed significant issues with electrical phase connections, either at the supply junction box or at the motor phase connections. Figure 10 reflects the infrared thermography (thermal) images taken of supply cables on the Mill motors. On the left hand side is an infrared image of the normal cable temperature expected on a mill motor with healthy connections (± 36°C), and the right hand side shows increased cable temperature (45°C) on the motor with the fault condition (Mill 2D).

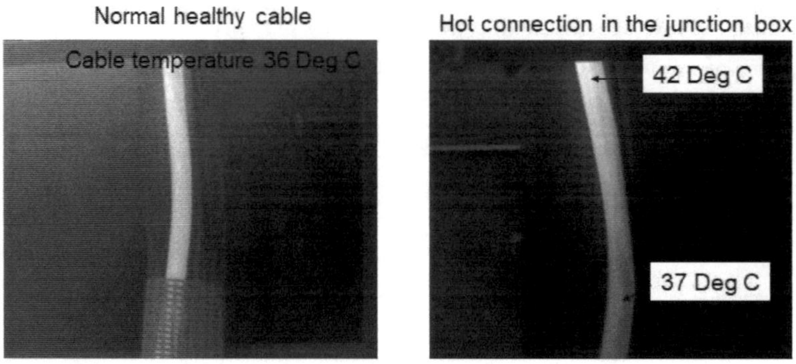

Figure 10: Infrared Thermography Image of Cable overheating due to hot connections

As part of Analytical Framework continuous improvement, more data was added to the Fault Model database for optimisation purposes. Mill 1F motor was subsequently identified as another case meeting the SFFoI fault and threshold criteria for this condition. This motor was removed from service and the phase connections inspected during a 3000hr Mill Service outage. Two of the phase connections were found to be severely damaged with one of the connections nearly destroyed by electrical arcing (Figure 11).

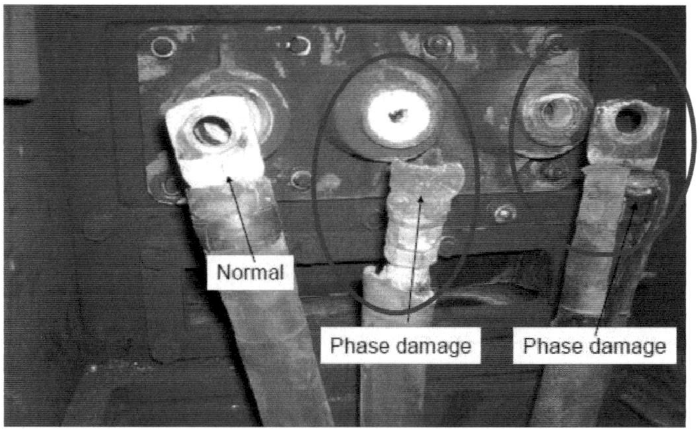

Figure 11: Severe Phase Connection Damage

4. CONCLUSION

The outcome of the research study proved that a combination of condition monitoring assessment, Design Base generated analytical fault models and understanding of the impact of operating and maintenance philosophies was vital to provide the most accurate assessment of plant health and failure prediction. The research study also successfully designed, implemented and validated the CBRS Analytical Framework and Fault Model methodology and demonstrated it with operational research and use cases as a viable tool with which failure identification and failure degradation predictions can be made using vibration data analysis on rotating machinery.

The research study outcome also demonstrated that although early warning failure detection capability is valuable, the benefit of such early detection must be balanced with the most optimal asset maintenance cost (repair versus overall failure cost). This is crucial in the drive of ensuring that the lowest total asset cost (depicted in Figure 2) can be achieved on power plant assets.

REFERENCES

[1] **Angeli, C. and Chatzinikolaou, A. (2004).** On–Line Fault Detection Techniques for Technical Systems: A Survey. Department of Mathematics and Computer Science, Greece. Technomathematics Research Foundation. International Journal of Computer Science & Applications, Volume 1, No. 1. 2004.

[2] ***Covino, L. and Hanifan, M. (2011).*** *Proactive Approach to Managing Production Assets. Bently Nevada. Orbit Magazine, Volume 31, Number 2, 2011.*

[3] **Ngubane, B. (2015).** Eskom Holdings Corporate Plan 2015/6 – 2019/20, Rev 3. Eskom Holdings (Pty) Ltd. 24 April 2015.

[4] ***ISO Standards Organisation. (1996).*** ISO 7919-2: Mechanical vibration of non-reciprocating machines - Measurements on rotating shafts and evaluation criteria - Part 2: Large land-based steam turbine generator sets.

[5] **Mobius Institute, (2010**). Vibration Training – Category I, II and III Training Modules. Developed between 2005 and 2012. Published as Revision 3.0 published in 2010.

[6] **Sondalini, M.** Don't Waste your Time and Money with Condition Monitoring. Article by Lifetime Reliability Solutions. Drawn from www.lifetime-reliability.com on 10 April 2017.

[7] **Emerson. (2003).** White Paper: Reducing Operations and Maintenance Costs. Emerson Process Management, Austin Texas. www.EmersonProcess.com. September 2003. Retrieved January 2012.

[8] **Rivera, J. and Van Der Meulen, R. (2013).** Gartner's 2013 Hype Cycle for Emerging Technologies Maps out Evolving Relationship Between Humans and Machines. Gartner, Stamford, Connecticut. 19 August 2013.

[9] **ISO Standards Organisation. (2007).** ISO 13374-2: Condition monitoring and diagnostics of machines – Data processing, communication and presentation – Part 2: Data Processing.

[10] **Potgieter, C. (2017).** Station Statistical Overview of Partial Load Loss Contributions to UCLF. Report compiled and issued on 13 March 2017. Eskom Holdings.

[11] **Emmanouilidis, C., Fumagalli, L., Jantunen, E., Pistofidis, P., Macchi, M. and Garetti, M. (2010).** Condition monitoring based on incremental learning and domain ontology for condition-based maintenance. APMS 2010 International Conference on Advances in Production Management Systems, Italy. October 2010.

[12] **Lee, JI, Ni, J, Djurdjanovic, D and Liao ,H. (2006).** Intelligent Prognostics Tools and E-Maintenance. Computers in Industry, Volume 57. August 2006.

[13] **Shaheryar, A., Yin, X.C and Ramay, W.Y. (2017).** Robust Feature Extraction on Vibration Data under Deep-Learning Framework: An Application for Fault Identification in Rotary Machines. International Journal of Computer Applications, Volume 167, No 4. June 2017.

[14] **Sheng, S. (2012).** Wind Turbine Gearbox Condition Monitoring Round Robin Study – Vibration Analysis. Technical Report by the National Renewable Energy Laboratory, US Department of Energy. July 2012.

[15] **Lei, Y., Lin, J., Zuo, M.J. and He, Z. (2014).** Condition Monitoring and fault diagnosis of planetary gearboxes: A Review. Measurement, Volume 48. February 2014.

[16] **Yang, S., Xiang, D., Bryant, A., Mawby, P., Ran, L. and Tavener, P. (2010).** Condition Monitoring for Device Reliability in Power Electronic Converters: A Review. IEEE Transactions on Power Electronics, Vol. 25, No. 11. November 2010.

[17] **Wang, W. (2003).** Modelling condition monitoring intervals: A hybrid of simulation and analytical approaches. Journal of the Operational Research Society, Volume 54, Issue 3, pp. 273-282. March 2003.

[18] **Rozados, I.V. and Tjahjono, B. (2014).** Big Data Analytics in Supply Chain Management: Trends and Related Research. Supply Chain Research Center, Cranfield University. 2014.

[19] **Biehn, N. (2013).** The Missing V's in Big Data: Viability and Value. PROS. www.wired.com/insights. May 2013.

[20] **Marwala, T. (2012).** Condition Monitoring Using Computational Intelligence Methods: Applications in Mechanical Systems. University of Johannesburg. Springer. Published in 2012.

Cloud based Real-time Condition Monitoring Model for Effective Maintenance of Machines

Ganga Dhandapani [1], Ramachandran Veilumuthu[2]

[1] Department of Electrical and Electronics Engineering, NIT Nagaland, Dimapur 797103, India
[2] Department of Information Science and Technology, CEG, Anna University, Chennai 25, India

ABSTRACT

This research aims at transforming the conventional condition monitoring model by including cloud services at the group and process control levels through virtualization of Process Control System (PCS) for effective maintenance to achieve maximum performance of machines. Application specific condition monitoring algorithms are represented as cloud services. The parameters of the machines which are operating in an industrial network are communicated to the virtualized PCS and distributed to the cloud services. An experimental setup consisting of a servomotor and an induction motor operated by S120 and V20 AC drives respectively is considered to validate the proposed cloud-based condition monitoring model in real-time. The vibration signals acquired from the servomotor and induction motor at different operating conditions have been communicated to virtual PCS. The non-stationary vibration data of the machines are processed using the cloud services to estimate the respective vibration threshold levels adaptive to the operating conditions and have been archived as contextual repository of thresholds in the virtual machine. These vibration threshold levels corresponding to specific operating conditions are received by the PCS as responses for alarm fixation, which facilitates reliable monitoring of the machine conditions together with prediction of machine behavior pattern and coordinates the processes of multiple machines in real-time. Immediate decision making, flexible analysis under real-time operating conditions and scalability are the inherent features of the proposed cloud-based condition monitoring model. The convergence of cloud applications with conventional condition monitoring models is the root cause for innovative research that provides accurate solution for Data Logging, Control and effective Monitoring in all industrial applications and leads to enhancement in coordinated machinery diagnostics and process control decisions in industrial automation.

Keywords: Condition Monitoring, Cloud, Virtual Process Control System, Vibration Alarms, Machine Maintenance

Corresponding author: D. Ganga (gangaadhan@gmail.com)

1. INTRODUCTION

Conventional technologies and their convergence with the existing technologies lead to profound changes in the field of machine condition monitoring and enable better decision making in automation which consecutively improve the performance, reduce the downtime and frequency of failures. The ever growing advancements in the technologies of smart sensors, actuators and embedded devices in association with cloud ensure efficient and upgraded method of control and monitoring of electrical machines which are located at geographically distributed industrial environments. The new paradigm of Internet of Things (IoT) leads to lot of innovation in Industrial applications due to ease in gathering of information and communication with other devices. IoT provides a common and conducive platform to sense data and communicate with heterogeneous devices, and monitor applications which are executing or functioning in a distributed environment. Automation is the process that incorporates multiple technologies in interdisciplinary manner to make any system to operate without or much less human interaction, but leading to very high performance. Enhancement in the communication technologies, advancement in the sensing and monitoring systems and in turn enrichment in gaining access to information and interaction between heterogeneous devices via Internet along with the concept of virtualization overlay a strong and innovative research direction in Industrial Automation applications. The performance of the machines which are operating under dynamically varying conditions depends on their automated maintenance through effective condition monitoring with precise data acquisition, appropriate data analysis, communication of results of the analysis to all the relevant systems for carrying out necessary fault detection, control and recovery actions. Hence, efficient condition monitoring is a highly significant component in the architecture of Industrial Automation and has been critical for mitigation of failures through prediction techniques and formulation of subsequent solutions. Accurate calibration of measurement systems, reliable data logging and exchange, real-time and collaborative analysis in platform independent manner are the various key aspects of effective condition monitoring and automation of industrial machines. In this research, the convergence of cloud and IoT technologies for analysis of real-time systems has been brought into the implementation of condition monitoring for electrical machines and enhanced with virtualized PCS to support scalable and interoperable data exchange between machines from group control level to process control level with features of flexible and collaborative analytics, fixation of adaptive alarms with contextual thresholds and control of multiple machines in real-time operating environment. The proposed cloud model for condition monitoring is implemented on an industrial prototype set up having machines and drives, IoT devices, virtualized PCS and VMs executing vibration analytics and decision making algorithms for effective maintenance. The model gives collaborative access to machine data for analysis and decision making from any geographical location.

2. RELATED WORKS

The application of cloud for industrial applications is currently being proposed and discussed by various authors. Omid Givehchi et.al [1] has presented an overview on the application of cloud computing technology for industrial automation. It has been viewed at three levels viz. Enterprise Management and Manufacturing Execution level, Process Control Level and Field level. The authors have envisaged the importance of cloud solution for lower levels of automation i.e., at control and field levels. The physical devices at the lower levels have been integrated in a single cloud by choosing service bus as information model. The integration is proposed by encapsulation of services and functions inside delivery standards of cloud. At control level, migration of soft controllers as services has been proposed. Cloud computing is seen as a solution to provide platform for integration of growing information technologies such as Internet of Things, Service Oriented Architectures and mobile computing. Similarly, the automation architecture for the optimal control of devices in process plants through Internet has been elucidated by R. Kirubashankar et.al [2]. The web based architecture proposed for the control of devices remotely consists of Programmable Logic Controller (PLC) connected to Gateway, Supervisory Control and Data Acquisition (SCADA) and Virtual Private Network (VPN). Effective point-to-point network communication architecture has been said to decrease the problems associated with Internet latency and system stability in the design of web-based remote supervisory control and information system. Larry combs [3] has stated that the functionality and reliability of conventional SCADA gets enhanced with the application of cloud for SCADA and has provided an insight on possible utilization of cloud. The SCADA can be from the cloud by itself or the data of the onsite SCADA shall be transferred to the cloud. This facilitates data availability to multiple resources. The cloud can be of public, private or hybrid nature based on the industrial requirements. Omid Givehchi et.al [4] cited about the development of Virtual PLCs on the Virtual Machines of the private cloud created using VMware's vCloud suite. The hypervisor host is connected with more network adapters to interface with multiple physical devices. The Virtual switch manages the network policies for VMs to communicate with clients. The VM communication Interface with network adapter driver for VMs supports communication of Physical systems (here sensors and actuators) using industrial protocols such as PROFINET. The practical implementation of simple virtual PLC illustrates the possibility of having an industrial application as a Service. Sandy Dunn [5] has reviewed the reasons that drive the industries to make changes in the condition monitoring techniques and the implications they will create. The condition monitoring system adopted by industries is expected to reduce Mean Time Between Failures (MTBF). The industries consider asset effectiveness as significant because it leads to higher profit from minimum investment and the author has substantiated that condition monitoring has a significant role in bringing asset effectiveness. Condition based maintenance is centered on the process of condition monitoring which diagnoses the abnormalities by evaluation and comparison of the current condition with a fixed failure limit. The reliability of the decision that has been made is associated with the predefined failure limits of the parameters monitored and the data interpretation made through analysis. The issues pertaining to the definition of failure limit and appropriate monitoring parameters to indicate the system deterioration have been pointed out as the challenges for future research. Willetts R et al. [6] insisted on the need for adaptive alarms in condition monitoring to avoid identification of failures due to false alarms. The techniques of classical statistics, multi-variant statistics and intelligent systems namely neural networks, classifiers and genetic algorithms have been reviewed for determination of adaptive alarms through prediction methods using historical data. Condition monitoring governs the diagnostics and prognostics adopted by the industries in bringing out efficient process and asset maintenance. Rosmaini Ahmad et al. [7] has given an overview of asset maintenance techniques, which are being performed in the industries. The practice of preventive maintenance initially carried out by OEMs could not meet the practical requirements of the industries as the machines function in different environmental and operating conditions, which demanded the incorporation of application specific scientific approaches for real-time data analysis for condition assessment. Mallikarjun Kande et al. [8] have extensively reviewed and discussed about existing machine condition monitoring techniques and industrial automation for plant-wide condition monitoring of rotating electrical machines. They pointed out the importance for on-equipment, on-premise and on-cloud integration of condition monitoring and the Distributed Control System to provide continuous monitoring of the equipment with high update rates from the sensors, to collect and send sensor data to diagnostics running as part of plant operations and to offer the elasticity required for the data and computational resources. While discussing about on-equipment and on-premise integration methods, the need for on-cloud monitoring using IoT gateway has been substantiated to meet the requirements of advanced diagnostics and data platforms for enhanced computation. The integration of remote services gives the benefit of making intensive data analysis even with the application basic data acquisition devices for condition monitoring. Steve Lacey [9] has stated that condition monitoring carried out with the incorporation of cloud facilitates comparative analysis of the conditions of similar machines or related machines. The adoption of cloud allows data sharing and enables implementation of new analysis techniques when unknown signal patterns are observed at the user end. The cloud environment provides an added value of being able to share and compare the local machine condition data with other similar machines across the plant, or with other machines at multiple plants wherever they are located. The cloud based condition monitoring system can infer the data from the distributed databases for effective decision making in vibration analysis. The vibration data is further processed in the cloud with the combination of data of one machine and other similar machines' data with extensive analysis options. This increases the reliability of the diagnosis information for appropriate decision making.

3. CLOUD-BASED CONDITION MONITORING MODEL - IMPLEMENTATION

The condition monitoring set up shown in Figure 1 encompasses a networked hardware architecture composed of motors, drives, cloud based Process Control System 7 (PCS), Remote Input Output module and IoT device connected over Industrial Ethernet. In this architecture, the data from all the networked components are collectively stored in cloud and control actions are initiated after implementing the data analysis with the condition monitoring and predictive maintenance services deployed in separate Virtual Machines. The SIMATIC PCS 7 software hosted on one of the virtual machines is capable of receiving the data acquired by different hardware components from multiple machines. In the proposed set up shown in Figure 1, S120 AC drive operating the servo motor reads the physical data of position and speed measured by proximity sensor and encoder. These data variables in S120 AC drive are then communicated to PCS where the actual values of the motor variables are compared with commanded or processed values. The control variables generated by the PCS are sent as operational commands to S120 AC drive which modulates the power input to the Servo motor as per the control variable. Alike to this, the vibration and speed data of three phase induction motor are acquired by IoT 2040 gateway through piezo electric and proximity sensors respectively when operated at different operating conditions of starting to no load speed and loading. The acquired values of vibration and speed are further communicated by IoT2040 gateway to the PCS over Industrial Ethernet through S7 communication. The vibration measurement made with piezo electric sensor using SPI protocol is unsupported by both S120 and V20 AC drives. This incompatibility of piezo electric sensor interface to either S120 or V20 drive has been overcome by S7 communication feature of IoT2040 gateway which communicates the vibration values to PCS without the said limitation.

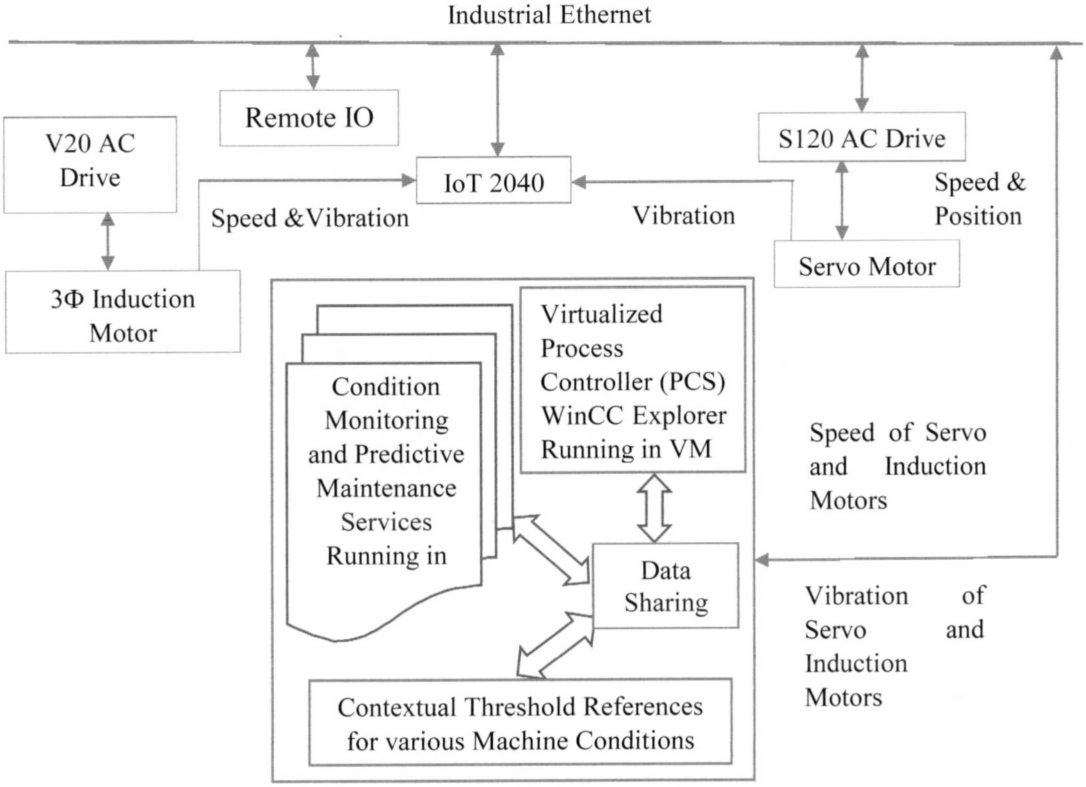

Figure 1. Cloud based condition monitoring model

The motor data acquired by IoT 2040 gateway and S120 AC drive are transferred to the data tags of Continuous Function Chart executed in PCS 7 software. The tagged data of position, speed and vibration are archived in the WinCC Explorer of PCS. The virtual machine settings of the windows VMware station are configured to share the archived data to the other VMs for the execution of vibration analysis. The bridged connection setting of the network adapter has been configured so as to enable the VMs running in the host to network among each other and also to the physical network. In order to share the data files between the VMs and host of the bridged network, the option of shared folder has always been enabled and mapped as a network drive in Windows guests. The data of induction motor and servo motor that need to be archived for any required time interval are selected from the settings of 'value column' and 'time column' of WinCC online table properties. The vibration data of the induction motor at different operating conditions namely starting to no load speed with and without disturbance and loading at standalone condition are logged and archived from WinCC online table control. Such archived vibration data specific to the operating conditions are fed as input to the Condition Monitoring and Predictive Maintenance services for the determination of the vibration threshold values. The implementation of the condition monitoring service on the shaft vibration data of the induction motor has yielded the values of upper and lower threshold class clusters for different operating conditions as furnished in Table 1. This analysis reveals that the vibration threshold class clusters identified for standalone condition cannot be maintained as alarm for monitoring of machines at all conditions in machine maintenance. The integrated vibration threshold analysis carried out with two different motors namely induction motor and servo motor, operating in the close vicinity reveals the impact of the motors running in the same environment on each other's shaft vibration pattern. The effects of external disturbance and loading on a machine's vibration are also realized to be equivalent from the closely matching values of upper and lower threshold class clusters. The fact that the neighboring machine running at constant speed of 1500 rpm creates a vibration effect equivalent to that of loading in other machines substantiate the fixation of the operational limits for mechanical and electrical loading, dynamic load duty cycle, speed and torque capacity etc., during machine maintenance.

Table 1: Vibration Alarm Fixation for Three Phase Induction Motor at Dynamic Operating Conditions

	Upper Threshold Class Cluster in g			**Lower Threshold Class Cluster in g**		
	Starting to No Load Speed Condition (Standalone Mode)	Starting to No Load Speed Condition (Disturbance From Servo Motor)	Loaded Condition	Starting to No Load Speed Condition (Standalone Mode)	Starting to No Load Speed Condition (Disturbance From Servo Motor)	Loaded Condition
	0.054	-0.007	-0.015	-0.585	-0.634	-0.630
	0.267	0.203	0.190	-0.372	-0.425	-0.425
	0.481	0.412	0.395	-0.159	-0.216	-0.220
	0.694	0.621	0.600	0.000	0.000	0.000
	0.907	0.831	0.805	0.000	0.000	0.000
Cluster Mean	0.481	0.412	0.395	-0.223	-0.255	-0.255

The values of thresholds determined are stored and used as references for the fixation of vibration alarms adaptive to operating conditions of induction motor and servo motor. The vibration alarms fixed are in turn written to the tags of PCS through SCADA layout of the WinCC (Figure 2) for carrying out precise operational decisions for motors and thereby to enhance the quality of machine maintenance. Thus, the threshold values computed from instantaneous values of the induction motor's vibration are continuously compared with that of the alarms corresponding to the operational conditions.

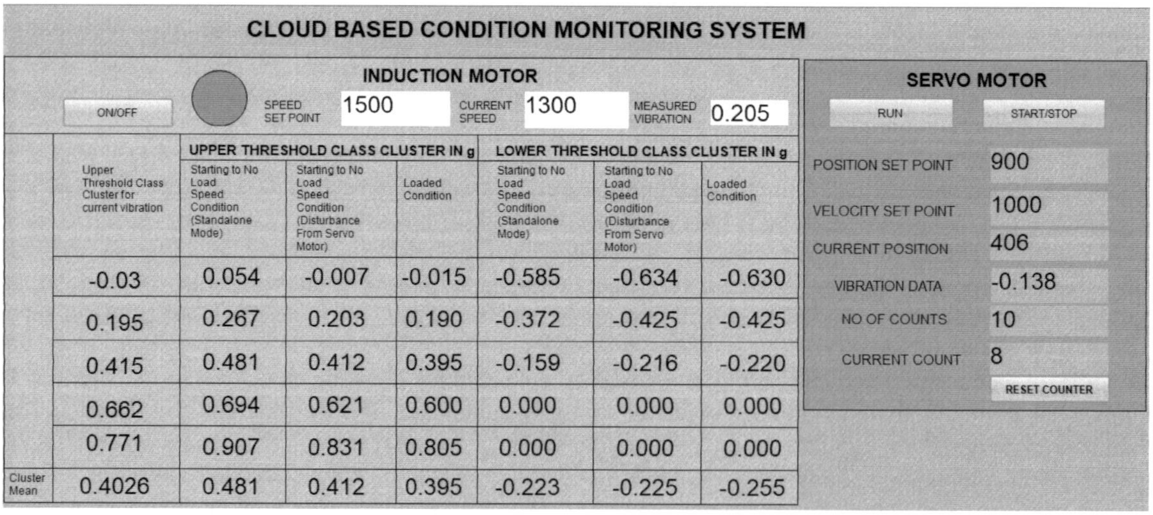

Figure 2. SCADA layout for machine condition monitoring

The operational decisions made for the induction motor based on the measured values of vibration during starting to no load speed of 1500 rpm when the servo motor is also in operation has been illustrated in Figure 2. The actual vibration thresholds of the induction motor running in the presence of servo motor are compared with the threshold class clusters meant for the operating condition of 'Starting to No Load speed with disturbance'. The analysis reveals that the motor while reaching the speed of 1300 rpm at no load possess the value of 0.4026g as upper threshold class cluster mean. This on comparison with the threshold of 0.412g fixed as vibration alarm for the said condition, brings out an insight that the value of the speed set point should be fixed between 1300 rpm and 1400 rpm during this condition instead of maintaining 1500 rpm as constant speed set point. This operational decision avoids induction motor overloading under high speeds during disturbance condition and hence will improve the durability of the motor. The new speed set point has been fed as input to the tag of Continuous Function Chart of the PCS. The PCS compares the new speed set point with actual speed of the induction motor and generates control output in the range of (0-10) V. The remote IO module ET200SP connected to PCS generates corresponding analog voltage output of (0-10) V which acts as speed reference for V20 AC drive as shown in Figure 3.

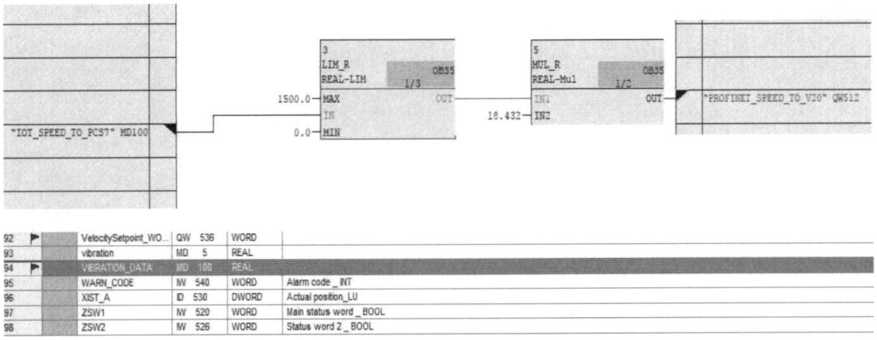

Figure 3. Data Transfer from Machines to PCS

The analog voltage generated by ET200SP modulates the power output of V20AC drive which has been fed as input to the induction motor. In this work, the operational decision of motor speed adjustment of induction motor has been done by comparing the instantaneous vibration pattern and the thresholds corresponding to the operating condition.

4. CONCLUSION

The condition monitoring of electrical machines in Industries are carried out widely with vibration signals. These signals are of non-stationary nature under dynamic conditions and hence numerous mathematical techniques are applied to extract information. Thus for effective machine maintenance, precise processing of vibration signals has been most sought and progressed. In this work, the need for a versatile condition monitoring system that possess flexible and precise analysis techniques for fixation of adaptive vibration thresholds as adaptive alarms has been elucidated. Such analysis brings outs the benefits that appropriate operational variations could bring for improved performance and life time of the machines in the working environment. The partially automated maintenance of the motors implemented by making maintenance decisions with the help of virtualized PCS, cloud services running in VMs and remote IO aid the system designers and enterprise resource managers to access the machine data and services in a collaborative and secured manner that lead to better maintenance strategies with ease. The IoT 2040 gateway also possess the feature of data exchange with the public cloud which opens up varied analysis domains for multiple and concurrent information extraction and condition monitoring decisions. Although the proposed cloud based architecture requires high end servers for mitigation of implementation overheads, the advantages of scalability, redundancy, manageability, interoperability and cost effectiveness of cloud contributes for improved performance of industrial applications. Overall, the proposed condition monitoring architecture based on cloud will address the challenges of latency in data processing, data storage and requirement of wide analytic features for predictive maintenance of machines in industries.

REFERENCES

[1] Omid Givehchi, Henning Trsek, Juergen Jasperneite (2013) Cloud Computing for Industrial Automation Systems - A Comprehensive Overview.18th IEEE Conference on Emerging Technology and Factory Automation, pp. 1- 4.

[2] Kirubashankar, R., Krishnamurthy, K., Indra, J., Vignesh, B. (2011) Design and Implementation of Web Based Remote Supervisory Control and Information System. International Journal of Soft Computing and Engineering (IJSCE), Vol.1, No.4, ISSN: 2231-2307

[3] Larry Combs (2011) Cloud Computing for SCADA.Control Engineering Magazine.

[4] Omid Givehchi, Jahanzaib Imtiaz, Henning Trsek, Juergen Jasperneite (2014) Control-as-a-service from the cloud: A case study for using virtualized PLCs.10th IEEE Workshop on Factory Communication Systems (WFCS), pp. 1- 4.

[5] Sandy Dunn (2015) Condition Monitoring in the 21st Century. Plant Maintenance Resource Center.

[6] Willetts, R., Starr, AG., Doyle, A., Barnes, J. (2005) Generating adaptive alarms for condition monitoring data. International Journal of COMADEM Vol. 8, No.3, pp. 26-36.

[7] Rosmaini Ahmad, Shahrul Kamaruddin (2012) An overview of time-based and condition-based maintenance in industrial application. Computers & Industrial Engineering Vol. 63, No.1, pp.135-149.

[8] Mallikarjun Kande, Alf J. Isaksson, Rajeev Thottappillil, Nathaniel Taylor (2017). Rotating Electrical Machine Condition Monitoring Automation—A Review. Machines Vol.5, No.4. [Online]. Available: http://dx.doi.org/10.3390/machines5040024.

[9] Dr. Steve Lacey (2017). From local to cloud-based condition monitoring. [Online]. Available: www.engineerlive.com/content/local-cloud-based-condition-monitoring.

Arc Tangent Failure Rate Distribution Method

Yerong Hu, Peng Yang, Yong Zhang, Jing Qiu, Guanjun Liu, Hukang Li
Science and Technology on Integrated Logistics Support Laboratory
National University of Defense Technology, NUDT
Changsha, China
yeronghu@foxmail.com, nudtyp7894@163.com, 95363020@qq.com, qiujing16@sina.com, gjliu342@qq.com, lihuakang856@163.com

ABSTRACT

Aiming at solving the problem that the distributed index may exceed 1 or is too small in the current fault rate distribution method, we propose an arc tangent fault rate distribution method based on arc tangent function. The distribution function model is determined by the parameter that is calculated out based on the specific object under distribution. The steps of the method are simple and easy to operate. Finally, a case is given to analyze the contrast between the conventional fault rate distribution method and arc tangent fault rate distribution method. The analysis results show that the distribution method can overcome the problem that the distributed index in the conventional fault rate distribution method is too large or too small, making the allocated indexes more accordant with the engineering practice.

Keywords: testability; fault rate; index distribution; integrated maintenance management; conceptual design

Corresponding author: Yerong Hu (yeronghu@foxmail.com)

1. INTRODUCTION

Testability distribution refers to assigning the quantitative requirement of equipment to the components of the equipment from top to bottom level according to the given principles and methods in the demonstration stage and preliminary design stage. Testability distribution is an important part of testability design stage. It has guiding significance for determining the testability index of system [1]. It lays a solid foundation for the integrated maintenance management. Currently, the classical testability index distribution methods include experience distribution method, fault rate distribution method, weight distribution method, old-product-based distribution [2], etc. In recent years, new research has been carried out on the test distribution technology in the domestic academia. Shen Qinmu has proposed testability distribution method based on block diagram [3], Wang Baolong has put forward testability optimization distribution method based on the genetic algorithm [4], Shen Qinmu [5] and Li Jinlong [6] respectively have studied the distribution method based on analytic hierarchy process (AHP).In particular, the fault rate, which is objective and reliable to better reflect the distribution needs of the actual equipment, is the data source of the fault rate distribution method. However there are some problems existing in the classical fault rate distribution method right now.

The basic principle of the fault rate distribution method is to allocate the test indexes according to the size of the unit fault rate. This method is applicable in the system that the fault rate of the component is known. We only consider the factor of fault rate in the fault rate distribution method. The distribution function of the classical fault rate method is linear. So the distributed index may varies greatly when the unit fault rate varies greatly. The distributed index of a low-fault-rate unit is low even close to 0, while the index of high-fault-rate unit is high even more than 1. This phenomenon is obviously not in line with the actual situation.

2. PRINCIPLE

The derivative of a fault rate distribution function can be used to indicate the difficulty degree of index promotion. When the derivative is large, index upgrading is relatively easy and when the derivative of curve is relatively small, it is difficult to improve the index. In a real production environment, even if the fault rate of a product's component is low, we still hope that the component can have a suitable index when the promotion of the index will not cause great technical difficulties. Though the fault rate is relatively small, the derivative of the fault rate distribution function should be larger. It's difficult to improve the index when it has reached a high situation. So when the fault rate of a component is high, the derivative of the function should be small. Therefore, the fault rate distribution function that satisfies the actual situation is not linear, and different fault rates should be corresponding to different derivatives. So the function should be a convex function.

In summary, we can assume that the distribution function of the fault rate is a convex function. Given two constants ζ_1 and ζ_2 ($\zeta_1 > \zeta_2$), the distribution $f(\lambda)$ should satisfy the following conditions:

① $\exists \varepsilon_1, f'(\lambda) > \zeta_1 (\lambda < \varepsilon_1)$,

② $\exists \varepsilon_2, f'(\lambda) < \zeta_2 (\lambda > \varepsilon_2)$,

③ $\lim_{\lambda \to \infty} f'(\lambda) = 0$,

④ $\lim_{\lambda \to \infty} f(\lambda) = 1$,

where λ represents the fault rate. In allusion to the condition①,②and③ mentioned above, we construct a function

$$f'(\lambda) = \frac{a}{b + c\lambda^{\beta}} \tag{1}$$

where a, b and c are constants greater than 0. β represents the power of λ .Given $\beta=2$,then

$$f(\lambda) = \int f'(\lambda) d\lambda = \int \frac{a}{b + c\lambda^2} d\lambda = \frac{a}{b} \arctan(\sqrt{\frac{c}{b}}\lambda) \tag{2}$$

From condition④, we can calculate out $\frac{a}{b} = \frac{2}{\pi}$. Given $\alpha = \sqrt{\frac{c}{b}}$, the function can be written as

$$f(\lambda) = \frac{2}{\pi} \arctan(\alpha\lambda) \tag{3}$$

2.1. Distribution of fault detection rate

Based on (3), we can adopt the fault detection rate distribution function

$$\gamma_{FDk} = \frac{2}{\pi} \arctan(\alpha_{FD}\lambda_k), (1 \le k \le N) \tag{4}$$

where N represents the number of system's component, γ_{FDk} represents the distributed fault detection rate of component k, λ_k denotes the fault rate of component k, and α_{FD} stands for the coefficient of fault detection rate distribution. The coefficient is determined by the following steps:

① In order to make the fault detection rates assigned to each component meet the requirement of the system index, the following equation should be satisfied

$$\sum_{i=1}^{N} \frac{2}{\pi} \arctan(\alpha_{FD}\lambda_i) \square \lambda_i = \gamma_{FDsr} \square \sum_{k=1}^{N} \lambda_k \tag{5}$$

where γ_{FDsr} represents the required index of system. The solution $\alpha_{FD} = \alpha_{FD1}$ is the minimum value that α_{FD} can be obtained to satisfy the requirement of the system index.

② If the system needs to meet the specific requirement that the fault detection rate should reach γ_{FDq} in the case that the failure rate of a unit is λ_{FDq}, the following equation should be satisfied:

$$\gamma_{FDq} = \frac{2}{\pi}\arctan(\alpha_{FD}\lambda_{FDq}). \tag{6}$$

The solution α_{FD}=α_{FD2} is the minimum value that α_{FD} can be obtained to satisfy the specific requirement.

③ In the case $\alpha_{FD2} \ge \alpha_{FD1}$, we have α_{FD}=α_{FD2}; in the case $\alpha_{FD2} \le \alpha_{FD1}$, we have α_{FD}=α_{FD1}; if the system does not give the specific requirement, we directly have α_{FD}=α_{FD1}.

2.2. Distribution of fault isolation rate

Based on (3), we can adopt the fault isolation rate distribution function

$$\gamma_{FIk} = \frac{2}{\pi}\arctan(\alpha_{FI}\lambda_{FDk}), (1 \le k \le N) \tag{7}$$

where γ_{FDk} represents the distributed fault isolation rate of component k, λ_{FDk} represents the failure rate of unit k which can be detected and calculated by λ_{FDk}=λ_k□γ_{FDk}, α_{FD} represents the coefficient of fault isolation rate distribution. The coefficient is determined by the following steps.

① In order to make the fault isolation rates assigned to each component meet the requirement of the system index, the following equation should be satisfied:

$$\sum_{i=1}^{N}\frac{2}{\pi}\arctan(\alpha_{FI}\lambda_{FDi})\square\lambda_{FDi} = \gamma_{FIsr}\square\sum_{k=1}^{N}\lambda_{FDk} \tag{8}$$

where γ_{FIsr} represents the required index of the system. The solution α_{FI}=α_{FI1} is the minimum value that α_{FI} can be obtained to satisfy the requirement of the system index.

② If the system needs to meet the specific requirement that the fault isolation rate should reach to γ_{FIq} in the case that the failure rate of a unit is λ_{FIq}, the following equation should be satisfied:

$$\gamma_{FIq} = \frac{2}{\pi}\arctan(\alpha_{FI}\lambda_{FIq}) \tag{9}$$

The solution α_{FI}=α_{FI2} is the minimum value that α_{FI} can be obtained to satisfy the specific requirement.

③ In the case $\alpha_{FI2} \ge \alpha_{FI1}$, we have α_{FI}=α_{FI2};in the case $\alpha_{FI2} \le \alpha_{FI1}$,we have α_{FI}=α_{FI1};if the system does not give the specific requirement we directly have α_{FI}=α_{FI1}.

2.3. ALGORITHM

Based on the principle mentioned above, a generalized algorithm for the proposed arc tangent fault rate distribution method is given as follows.

Step.1 Collect relevant data, including:
①The required system index to be allocated, including the fault detection rate γ_{FDsr} and the isolation rate γ_{FIsr};
② The number N of units that can be found from the product structure tree, the number of redundant modules of each unit $n_1,\ldots,n_N$, and the failure rate of each unit $\lambda_1,\cdots,\lambda_N$;

③ If it is required by the designer that the fault detection rate should reach to γ_{FDq} in the case of λ_{FDq} ,we have $a=1$; otherwise, we have $a=0$;

④ If it is required by the designer that the fault isolation rate should reach to γ_{FDq} in the case of λ_{FDq} ,we have $b=1$; otherwise, we have $b=0$;

Step.2 Calculate α_{FD} in equation (4).
① Obtain $\alpha_{FD}=\alpha_{FD1}$ through equation (5);
② In case $a=1$,obtain $\alpha_{FD}=\alpha_{FD2}$ though equation (6), if $\alpha_{FD2} \geq \alpha_{FD1}$, we have $\alpha_{FD}=\alpha_{FD2}$, if not, we have $\alpha_{FD}=\alpha_{FD1}$;in case $a=0$,we directly have $\alpha_{FD}=\alpha_{FD1}$;

Step.3 Allocate the fault detection rate through equation (4).

Step.4 Calculate out each unit's fault rate that can be detected through equation $\lambda_{FDk}=\lambda_k \cdot \gamma_{FDk} (1 \leq k \leq N)$.

Step.5 Calculate α_{FI} in equation (7).
① Obtain $\alpha_{FI}=\alpha_{FI1}$ through equation (8);
② In case $b=1$, obtain $\alpha_{FI}=\alpha_{FI2}$ though equation (6), if $\alpha_{FI2} \geq \alpha_{FI1}$, we have $\alpha_{FI}=\alpha_{FI2}$, if not, we have $\alpha_{FI2} \leq \alpha_{FI1}$; in case $b=0$, we directly have $\alpha_{FI}=\alpha_{FI1}$;

Step.6 Allocate the fault isolation rate through equation (7).

3. CASE AND CONFRONTATION

3.1. Case date and the result of the classical method

The case in figure 1 is used to allocate the system index. The system consists of 5 LRUs with requirement $\gamma_{FDsr}=0.95$ and $\gamma_{FIsr}=0.90$. It is appointed that the fault detection rate should reach 0.85 when the fault rate reaches 30($\times10^{-6}$), the fault detection rate is 0.85.It doesn't have specific requirement for the failure isolation rate.

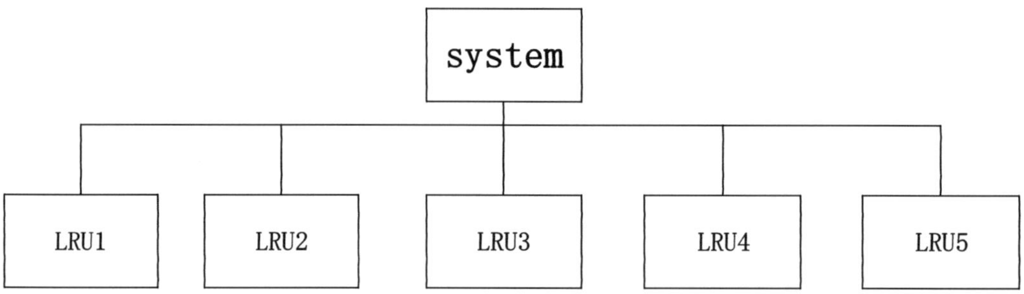

Figure 1 Allocation case

The related data of the system is filled in the distribution work sheet (see table 1), and allocation is carried on by using the classical failure rate distribution method given in the *GJB1770.2-1993 Testability and Maintainability Distribution and Prediction Method for Air Intelligence Radar* [7]. The distribution result and correction values are showed in table 1.

Table 1 Distribution work sheet of classical distribution method

name	number	λ_i $\times 10^{-6}$	γ_{FDi}		γ_{FIi}	
			result	**correction**	**result**	**correction**
LRU1	1	30	0.2788	0.90	0.2595	0.90
LRU2	1	30	0.2788	0.90	0.2595	0.90
LRU3	1	100	0.9293	0.98	0.8650	0.94
LRU4	1	150	1.3940	0.99	1.2976	0.95
LRU5	1	50	0.4647	0.90	0.4325	0.90
total	5	360	0.9677	0.9597	0.8999	0.9328

3.2. The result of arc tangent fault rate distribution method

Combined with the case in figure 1, we can calculate the failure detection rate allocation coefficient and the fault isolation rate allocation coefficient of arc tangent fault rate distribution method through the steps in the second section. The result is as follow:

$$\alpha_{FD} = 0.8075, \tag{10}$$

$$\alpha_{FI} = 0.0936. \tag{11}$$

Substitute the result (10) into equation (1), the distribution function of the fault detection rate can be obtained as follows:

$$\gamma_{FDk} = \frac{2}{\pi}\arctan(0.8075\lambda_k), (1 \le k \le N) \tag{12}$$

where the unit of λ_k is (×10-6).

Substitute the result (11) into equation (4), the distribution function of the fault detection rate can be obtained as follows:

$$\gamma_{FIk} = \frac{2}{\pi}\arctan(0.0936\lambda_{FDk}), (1 \le k \le N). \tag{13}$$

Substitute the fault rate data of the case into the equation (12) and (13) to obtain the distribution result showed in the table 2

Table 2 Distribution work sheet of arc tangent fault rate distribution method

name	**number**	$\lambda_i \times 10^{-6}$	γ_{FDi}	$\lambda_{FD} \times 10^{-6}$	γ_{FIi}
LRU1	1	30	0.8807	26.421	0.7554
LRU2	1	30	0.8807	26.421	0.7554
LRU3	1	100	0.9638	96.38	0.9297
LRU4	1	150	0.9579	143.64	0.9527
LRU5	1	50	0.9279	46.395	0.8559
total	5	360	0.95	342	0.9

4. ANALYSIS

It shows in table 2 that there will be instances when the distributed index is as low as 0.2778, or the instances when the distribute index exceed 1,even up to 1.3094.in the classical distribution method, such as the LRU1 and the LRU4. It is not consistent with the actual situation and doesn't meet the mathematics category. The result of the classical fault rate distribution method usually needs to be corrected by people with experience, which is random and unreliable. After the correction, the allocated indexes need to be checked whether they can satisfy the system requirement. If the indexes can't satisfy the total requirement of the system, the correction should be carried on again. It is hard to converge to an appropriate solution by using the classical method.

As is showed in table 2, arc tangent fault rate distribution method uses the characteristic of the tangent function to solve the problem that the allocated index is too low or too high after determining the appropriate distribution coefficient. And after completing the distribution, it's no need to have the checking of the unit index.

Based on the above case, the distribution function of arc tangent fault rate distribution method is as follows:

$$\gamma_{FDk} = \frac{2}{\pi}\arctan(0.8075\lambda_k)\ . \tag{14}$$

The distribution function of classical fault rate distribution method is as follows:

$$\gamma_{FDk} = 0.0092935\lambda_k\ . \tag{15}$$

The curves of equation (14) and (15) are shown in figure 2. In actual engineering project, the index may have the characteristic of improving easily when it's low and improving difficultly when it's high. The curve of the classical failure rate distribution method is a linear function, which obviously does not satisfy this characteristic. However, the arc fault tangent failure rate distribution method, with a large derivative at low index and a small derivative at high index in convex function curve, satisfies this characteristic well.

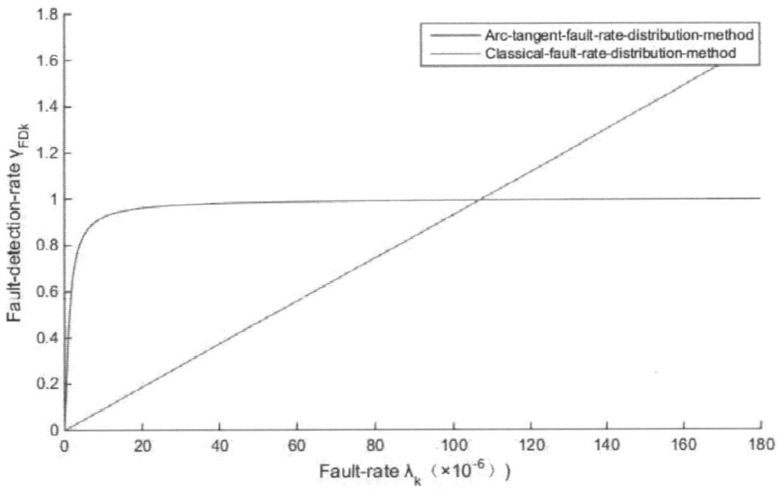

Figure 2 Comparison of two methods

5. CONCLUSION

In order to solve the problems in the method of classical failure rate assignment, we propose a method named the arc tangent failure rate assignment, which solves the problem that the distribution index exceeds 1 and the distribution index is too small, meanwhile, the distribution index meets the actual changing trend in the project. This method improves the guiding significance of the test index, and it is more in line with the engineering practice.

REFERENCES

[1] Department of Defense, MIL-STD-2165A, Military Standard Testability Program for Systems and Equipments[S]. 1993.

[2] Shi Junyou. System Testing Design Analysis and Verification [M]. Beijing: National Defense Industry Press, 2010.

[3] Shen Qinmu, Qiu Jing, Liu Guanjun, Yang Peng. Test study of estability allocation method based on block diagram of testing technology [J], 21, 2007.6 (S).

[4] WANG Baolong, HUANG Quli, SU Lin, WEI Zhonglin. Test-based Optimal Test Allocation for Complex Electronic Equipment Based on Genetic Algorithm [J] .Computer Measurement and Control, 15 (7), 2007.

[5] Shen Qin Mu. Equipment-level test-based distribution of technology research and application [M]. Changsha: National University of Defense Technology, 2004.7.

[6] Li Jinlong, Tao Fenghe, Jia Changzhi, Liang Guangjian.A Test Assignment Method Based on AHP [J]. China Testing, 36 (2), 2010.3.

[7] Zheng Xiaofu, Yang Tianliang, Fu Quanhua et al. GJB1770.2-93 Airsoft radar maintenance of the sexuality of the distribution and prediction [S] .1993: 27-28,40-41.

Deep online analysis of dielectric parameters for lubricants and insulation oils with an innovative oil sensor system: Identification of critical operation conditions of industrial gearboxes and high voltage transformers for reduction of failure rates and live time enhancement

Prof. Dr. Manfred R. Mauntz[1], Dr. Jörn Peuser[2]

[1]cmc Instruments GmbH, Rudolf Diesel Strasse 12 A, D-65760 Eschborn, Germany, CEO, Head of R&D
mrm@cmc-instruments.de
[2]cmc Instruments GmbH, Rudolf Diesel Strasse 12 A, D-65760 Eschborn, Germany, Applications
jp@cmc-instruments.de

ABSTRACT

The requirements in the renewable energy for high voltage transformers and large industrial gearboxes as installed in wind turbines on-/offshore rise. Ever more flexibility at a maximum operational reliability and a long life time are required of them at the same time, so the requirements for the oil and the oil condition monitoring grow correspondingly. This presentation provides information about a novel online oil condition monitoring system to give a solution to the mentioned priorities for both energy sectors. The focus is set to industrial gearboxes but the possibilities in monitoring applications of insulation oils are also addressed. The online oil sensor system measures the components conductivity kappa, the relative permittivity epsilon r and the temperature T independently from each other. Based on a very sensitive measurement method with high accuracy even small changes in the conductivity and dielectric constant of the oil composition can be detected reliably. The new sensor system effectively controls the proper operation conditions of industrial gearboxes, high voltage transformers and in test rigs for electrical vehicles, where lubrication and isolation of the oil has a double function. The system enables damage prevention of the gearbox by an advanced warning time of critical operation conditions and an enhanced oil exchange interval realized by a precise measurement of the electrical conductivity, the relative permittivity and the oil temperature. A new parameter, the WearSens® Index (WSi) is introduced. The mathematical model of the WSi combines all measured values and its gradients in one single parameter for a comprehensive monitoring to prevent wind turbines from damage. Furthermore, the WSi enables a long-term prognosis on the next oil change by 24/7 server data logging. Corrective procedures and/or maintenance can be carried out before actual damage occurs. Raw data and WSi results of wind turbine installations with different lubrication oils are shown. Short-term and long-term analysis of the data show significant trends and events, which are discussed more in detail. Once the oil condition monitoring sensor systems are installed on the high voltage transformer or a wind turbine's gearbox, the measured data can be displayed and evaluated elsewhere in sense of a full online condition monitoring system. 24/7 monitoring of the system (gearbox or HV transformer) during operation enables specific preventive and condition based maintenance independent of rigid inspection intervals.

Keywords: Oil sensor system, oil aging, bearing wear, online condition monitoring, oil additive degradation.

1. INTRODUCTION

In general, the field of maintenance can be divided into three sectors: preventive (time-based), intelligent (condition-based) and reactive maintenance (run to failure), which show different dependencies between costs and number of failures. Figure 1 shows the costs associated with the different strategies [1]. From this graph, the optimal point in terms of costs and number of failures can be identified within the center of the intelligent maintenance sector; intelligent maintenance can be realized with an online condition monitoring solution. Different kinds of online monitoring systems have been established over the past years: temperature [2, 3, 3], vibration [4] and particle counting [1, 5]. A vibration monitoring system analyses changes in the frequency spectrum of the observed bearing/gearbox component by Fast Fourier Transformation by converting a time-domain signal into a frequency-domain signal [6].

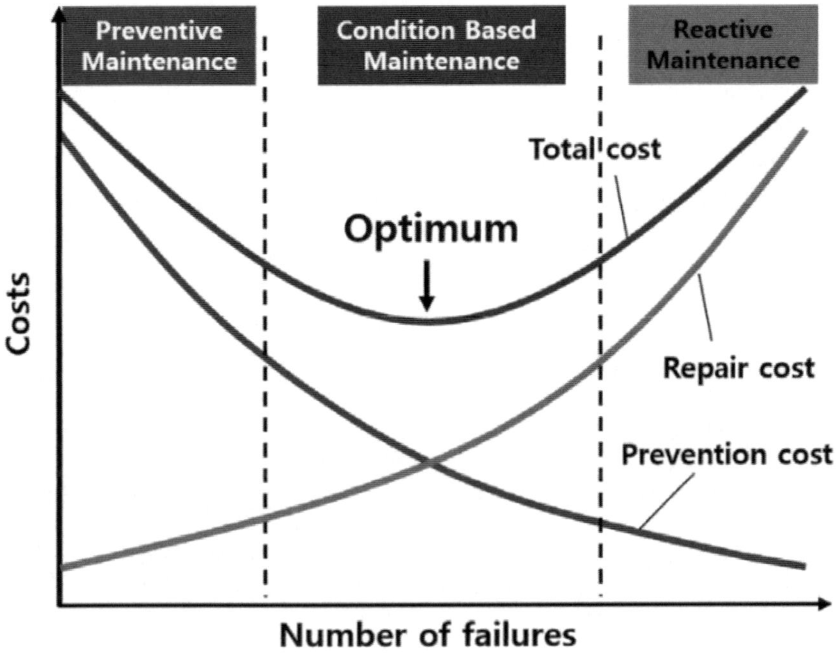

Figure 1. Costs associated with traditional maintenance strategies [1].

An optical particle-counter can detect particles larger than 4μm due to the optical resolution of the laser light source. All of these systems need a significant change in the topological contact sur-faces, which means damaged surfaces, particles and pitting.

The presented oil sensor system in contrast is already sensible at the very beginning of the damage formation stage, when the tribological layer gets depleted due to overload conditions. At this early stage, an increase of electrical charge carriers can be identified by the oil sensor system WearSens®: the electrical conductivity, relative permittivity and temperature are measured with a high precision and low noise over a broad range to enable the detection of small changes in the oil induced by variations in the tribology of the device under test [9]. Inorganic compounds occur at contact surfaces from the wear of parts, broken oil molecules, acids or oil soaps. These all lead to an increase in the electrical conductivity, which correlates directly with the wear. In oils containing additives, changes in dielectric constant infer the chemical breakdown of additives. A reduction in the lubricating ability of the oils, the determination of impurities, the continuous evaluation of the wear of bearings and gears and the oil aging all together follow the holistic approach of real-time monitoring of changes in the oil-machine system [10, 11]. Abrasive (metallic) wear, ions, broken oil molecules, acids, oil soaps, etc., cause an increase of the oil electrical conductivity kappa. It rises with increasing ion concentration and mobility. The electrical conductivity of almost all impurities is high compared to the extremely low corresponding property of original pure oils. Oils are principally electrical non-conductors. The electrical residual conductivity of pure oils lies in the range below one pS/m. A direct connection between the degree of contamination of oils and the electrical conductivity is found.

An increase of the electrical conductivity of the oil in operation can thus be interpreted as increasing wear or contamination of the lubricant. The aging of the oil is also evident in the degradation of additives, which are reflected in the relative permittivity epsilon r [12, 13]. To measure the electrical conductivity and the dielectric constant the oil is passed through an electrode array, which determines the electrical resistance and the capacitance of the sensor assembly using the base oil as a resistive material and dielectric. Figure 2 shows a detail picture of the sensor electrode array with the triple plate de-sign and the schematic electronic circuit. By the high sensitivity of a time measurement method the sensor system detects critical operation conditions much earlier than existing technologies such as vibration measurement or particle counting.

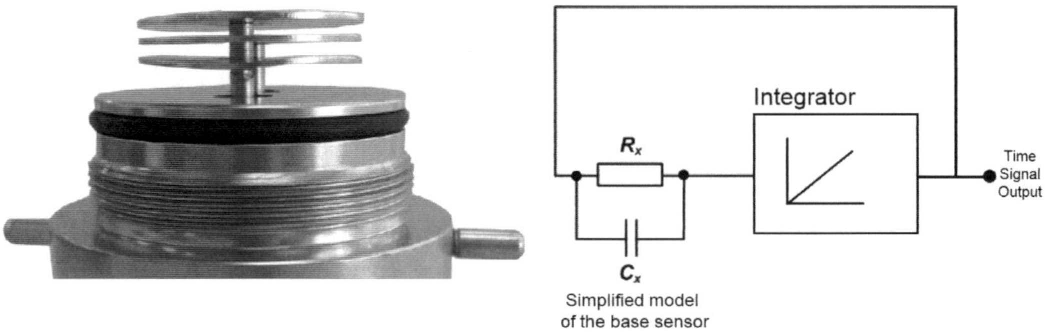

Figure 2. Detail of the triple plate design of the WearSens® base sensor and the corresponding simplified drawing of the electronic circuit.

To determine the conductivity and permittivity with a direct measurement of AC observables in a RC circuit is quite inaccurate, because oil has a very high resistance R (several GOhm) and a very low capacity C, which leads to high uncertainties, errors and a low resolution, which is necessary to follow the effects in oil. In the presented system, the electrical conductivity kappa and relative permittivity epsilon r are determined by a precise time measurement with a very high accuracy and repeatability based on an integrating measurement technique with a high time / bandwidth product: the measurement range for the conductivity starts from 0.1 pS/m up to 1,000,000 pS/m with a resolution of 0.01 pS/m; the relative permittivity is measured between 1 and 5 with a resolution of 1*10-6.

2. TEMPERATURE COMPENSATION

Ion mobility and thus, electrical conductivity kappa are dependent on the internal friction of the oil and therefore, also on its temperature. The conductivity epsilon r of the oil increases with temperature. The type of contamination and its temperature dependence cannot be assumed to be known. To improve the comparability of measurements, a self-learning adaptive temperature compensation algorithm is necessary. A change of the oil quality can then be assessed by the temperature compensated conductivity value, even though the specific contamination is not determinable [13]. Calculating the electrical conductivity and the dielectric constant at the reference temperature of 40° Celsius is realized by approximating the polynomial form of the temperature dependence.

$$\kappa_{T0} = \kappa_{T0a} + \left(a\Delta T_i + b\Delta \mathrm{T}_i^2 + c\Delta \mathrm{T}_i^3\right) \cdot \kappa_m$$

κ_{T0} is the approximate electrical conductivity of the oil at the reference temperature T0, κ_{T0a} is the previously calculated (old) electrical conductivity at the reference temperature T0, κ_m is the non-temperature compensated measured value of the electrical conductivity, a, b and c are the coefficients of the approximating polynomial to be adaptively determined during the runtime of the sensor system.

$$\Delta T_i = T_0 - T_i$$

is the temperature difference.

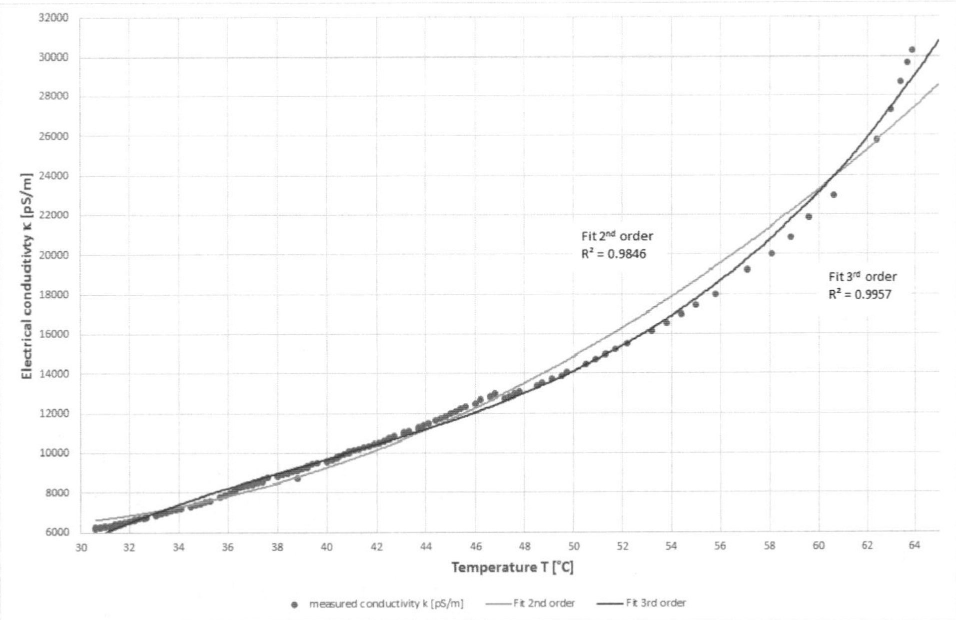

Figure 3. Electrical conductivity versus temperature with second and third order fit curves.

Figure 3 shows the dependency of the electrical conductivity with the temperature of a gearbox oil. Two different trend lines with second order fit and third order fit are plot-ted on the measured data. The approximation by a polynomial of third degree guarantees a good approximation at a reasonably low computational effort for the used microcomputer with an optimal coefficient of determination R^2. For the adaptive determination of the coefficients a, b and c of the polynomial a risk function is defined on the basis of the Gaussian method of least squares from the N measured values pairs and the approximating polynomial, whose minimization enables determination of the desired coefficients. Figure 4 shows the effect of the adaptive temperature compensation of the electrical conductivity using the example of field data from on onshore wind turbine: the wind fluctuations are changing the oil temperature and with it the electrical conductivity permanently.

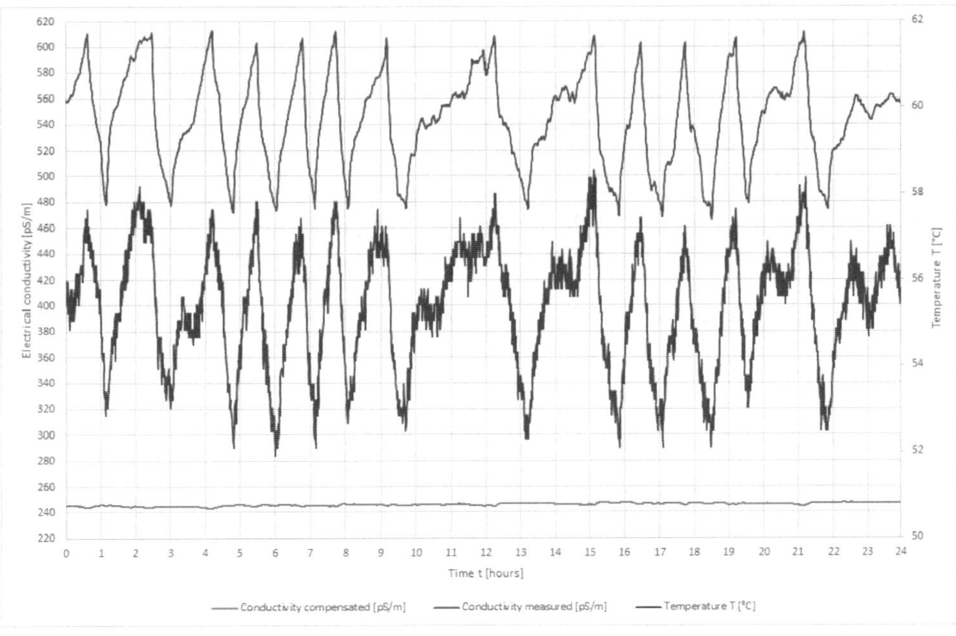

Figure 4. Graph of the measured and compensated electrical conductivity from an onshore wind turbine at a reference temperature of 40 °C.

While the measured conductivity κ changes significantly with temperature, the temperature compensated conductivity κ_{40} stays nearly constant. The implemented adaptive algorithm is working in the background of the measurement procedure autonomously; it has to be reset only after an oil exchange to adapt to the new lubricant. Without the adaptive temperature compensation, it is not possible to identify a critical operation condition in the monitored system due to the high influence of the temperature on the conductivity and the relative permittivity [17].

3. MATHEMATICAL MODEL OF THE WEARSENS® INDEX – WSI

The WearSens® Index (WS_i) has originally been developed for the lubricant analysis of a wind turbine gearbox; however, it can be adapted to any other lubricated system and different oil types with individual modifications. The following description is based on the wind turbine application.

The WS_i model considers short, mid and long-term changes in the lubricant by continuous monitoring of the conductivity, relative permittivity and temperature over a time period of several years with a high time resolution of < 45 seconds. Because of the measurement sensitivity and the high time resolution critical operation conditions can be identified much earlier and a damage can be evaded in short term analysis. The stress of the lubricant and the turbine itself is based on the actual wind condition, wind fluctuation and wind turbine settings (e.g. pitch control, torque control) resulting in instantaneous changes of the conductivity and relative permittivity and their gradients. Critical operation conditions result in an increased charge carrier generation and will change the conductivity and its gradient significantly. A big change in a short time period in the measured values leads to a high WS_i Signal; for example, a significant increase in the electrical conductivity in a short time period is an indication of an abrupt high load or depending on the increase in the conductivity a critical operation condition.

Frequent critical operation conditions lead to faster degradation of the oil additive complex. The hypothesis is that the consumption of the additives is directly correlated with the reduction of the relative permittivity of the oil: the relative permittivity ε_r is directly affected by the presence of polar elements; there is a high content of polar additives in gearbox oils [7]. The polar additives combine together with other polar elements (e.g. wear products, water contamination), so from this point of view the consumed additives are not polar anymore, which results in the reduction of the relative permittivity.

The gradient, i.e. the time derivative, of the conductivity or the dielectric constant progression respectively represents a measure of the additive degradation and consumption. After identifying initial base κ_{i40}, ε_{ir40}, $\Delta\varepsilon_{ir40}$, $\Delta\kappa_{i40}$, T_i can be feed into the simplified WS_i model below:

$$WS_i = \int_{t1}^{t2} \Big[f(\kappa_{40}, \kappa_{i40}) + f(\varepsilon_{r40}, \varepsilon_{ir40}) + f\left(\frac{\Delta\varepsilon_{r40}}{\Delta t}, \Delta\varepsilon_{ir40}\right) + f\left(\frac{\Delta\kappa_{40}}{\Delta t}, \Delta\kappa_{i40}\right) + f(T, T_i) \Big]\, dt$$

The next pages show results from an offshore installation of the WearSens® sensor system.

4. SENSOR INSTALLATION OFFSHORE WIND TURBINE

This section demonstrates first results of the 24/7 oil condition monitoring of an offshore wind turbine installation with WearSens® in the cooling bypass of a Siemens SWT 2.3MW. Figure 5a below shows the offshore wind turbine, the wind turbine gearbox in the nacelle and the installed base sensor in 5b. The communication unit in 5c transfers the data highly encrypted to the onshore located data server.

Figure 5. a) picture of the offshore wind turbine, b) WearSens® base sensor installation in the existing cooling bypass and c) sensor communication unit.

The WS_i of two identically constructed offshore wind turbines with the same oil running time are compared in figure 6. From this graph a clear difference between wind turbine WTG #01 and WTG #02 can be identified: this increased WS_i has been reported to the wind farm operator; together was other CMS data a forming damage at the fast drive shaft was identified early enough to prevent the wind turbine for a long downtime and maintenance period due to a total breakdown.

The high dynamics and the high peaks of the WS_i due to the varying load conditions (normal for WTG #01 and overload for WTG #02) and the forming damage in TGW #02 also resulted in the faster degradation of the oil and consumption of the additives, this effect is also visible in the comparative data of the temperature dependency of the electrical conductivity shown in figure 7.
The oil has been changed at the two offshore wind turbines: the data before (red colour) and after the oil exchange (green colour) show a completely different temperature dependency. From this information the active additive level and the oil quality can be determined together with the trending of the WearSens® index WS_i to estimate the next oil exchange condition based by the continuous measurement with the presented oil sensor system.
In figure 7 you see the significant difference between the wind turbine #01 with a normal behaviour and the wind turbine #02 with a forming damage in the grey curve. The additive components are getting consumed much faster due to higher stress and the increased generation of wear products, which results in a decrease of the dependency between the electrical conductivity κ and temperature T.

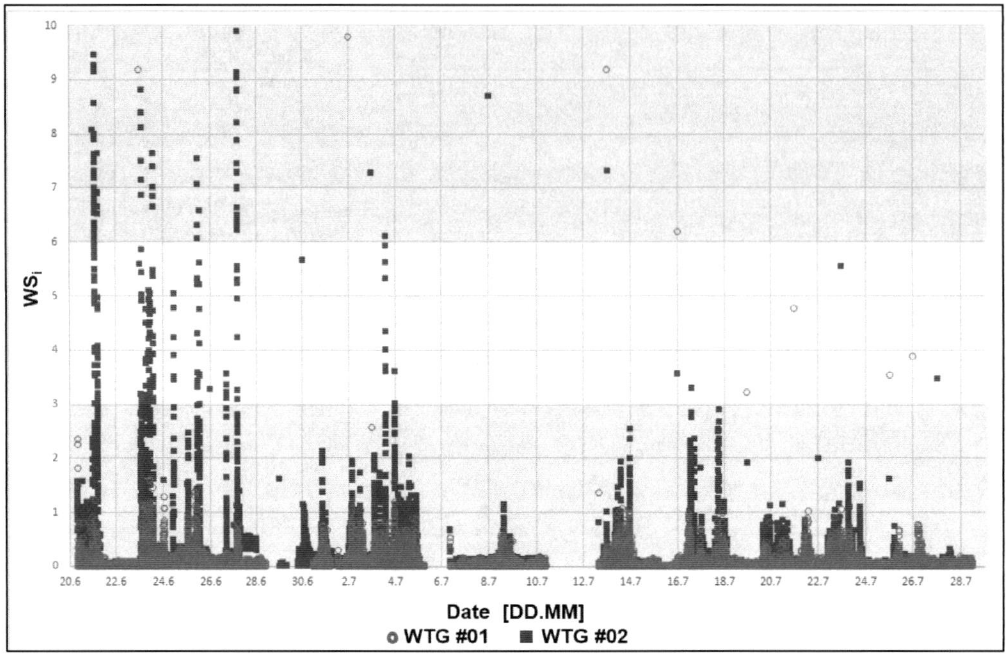

Figure 6. WearSens® Index WS$_i$ profile from two offshore wind turbines with semaphore loading indication.

Based on an empirical approach the hypothesis is, that fluctuations from normal to critical operation conditions can be identified earlier with the presented online oil sensor system by the precise measurement of the electrical conductivity and relative permittivity and the computation of the WearSens® index WS$_i$, because changes in the tribological layer, an increased charge carrier generation occur at the very beginning of an forming damage at a gearbox components due to material fatigue. Therefore, it is possible to react much faster on events of critical conditions to prevent the gearbox from damage to enhance the overall life time. By the long-term analysis over several month and years, it is possible to perform condition-based oil change on demand to preserve the environment, to protect the oil resources and to reduce costs. From this point of view, the benefits of an online oil condition monitoring are clearly eminent: the short-term analysis can avoid critical operation conditions and prevent the wind turbine from damage; a long term analysis and trending can be used to estimate the time for the next oil change – condition based.

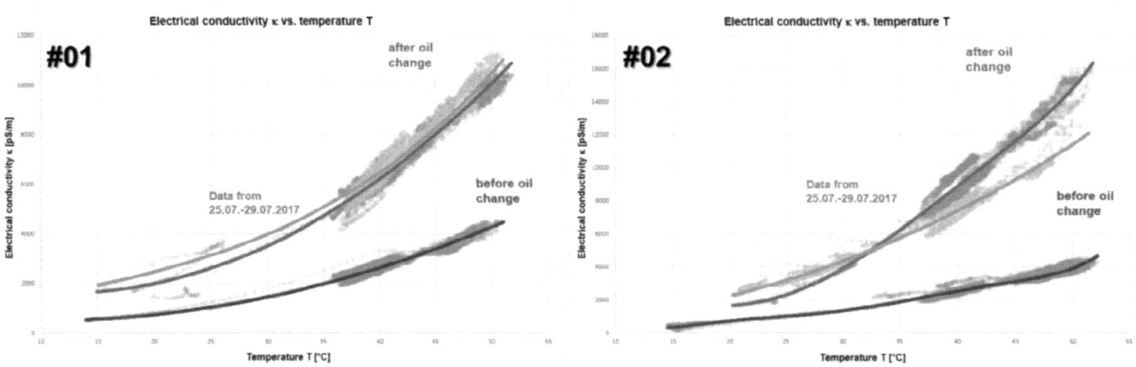

Figure 7. Time course (24 hours) of WS$_i$ with low time resolution (green), WS$_i$ high time resolution (blue) and the vibration signal (red).

5. CONCLUSION

The online diagnostics system measures components of the specific complex impedance of oils. The indication of forming stage of damage and wear is measured as an integral factor of, e.g., the degree of pollution, oil aging and acidification, water content and the decomposition state of additives or abrasion of the bearings, which is correlated to the changes in the electrical conductivity and relative permittivity. By the adaptive temperature compensation of the measured values, it is possible to identify even small variation in the actual charge carrier generation and additive consumption. For an efficient machine utilization and targeted damage prevention, the WearSens® online condition monitoring system and the WearSens® index offers the prospect to carry out timely preventative maintenance on demand rather than in rigid inspection intervals. The benefits of an extended oil change interval are reduced costs, preservation of the environment and resource protection. The oil sensor system has been installed into an offshore wind turbine performing short and mid-term analysis of the lubricant quality. The direct advantages of this online oil condition monitoring are detection of critical operation conditions and damage prevention to increase the lifetime of gearbox. The high time resolution, fast response time and accuracy of WearSens® allows earlier intervention control / optimization of the current operation in comparison to sole vibration analysis.

REFERENCES

[1] [1] P. Tchakoua, R. Wamkeue, M. Ouhrouche, F.S. Hasnaoui, T.A. Tameghe, G. Ekemb: Wind Turbine Condition Monitoring: State-of-the-Art Review, New Trends, and Future Challenges, Energies 2014,**7**, pp. 2595-2630, (2014)

[2] [2] C.J. Hellier: Handbook of Non-destructive Evaluation, McGraw-Hill Professional Publishing, New York, USA, (2003)

[3] [3] J.-Y. Park, J.-K. Lee, K.-Y. Oh, J.-S. Lee, B.-J. Kim: Design for 3MW Wind Turbine and its Condition Monitoring System, Proceedings of the International Multi Conference of Engineers and Computer Scientists, IMECS 2010, Kowloon, Hong Kong, 17–19 March 2010, Volume II, pp. 930–933,(2010)

[4] [4] B.N Madsen: Condition Monitoring of Wind Turbines by Electric Signature Analysis, Master's Thesis, Technical University of Denmark, Copenhagen, Denmark, October 2011, (2011)

[5] [5] A.C. Goncalves, J.B. Campos: Predictive maintenance of a reducer with contaminated oil under an excentrical load through vibration and oil analysis, J. Braz. Soc. Mech. Sci. Eng. 2011, 33, 1–7, (2011)

[6] [6] M.M. Khan, M.T. Iqbal, F. Khan: Reliability and Condition Monitoring of a Wind Turbine, Proceedings of the 2005 Canadian Conference on Electrical and Computer Engineering, Saskatoon, SK, Canada, 1–4 May 2005, pp. 1978–1981, (2005)

[7] [7] Noria Corporation: The critical role of additives in lubrication, Machinery Lubrication June 2012, (2012)

[8] [8] A. Jablonsky, T. Barszcz, M. Bielecka: Automatic validation of vibration signals in wind farm distributed monitoring systems, Measurement 2011, 44, 1954–1967, (2011)

[9] [9] M. Mauntz and U. Kuipers: Ölsensorsystem- Sensorsystem zur Messung von Komponenten der komplexen Impedanz elektrisch gering leitender und nichtleitender Fluide, dessen Realisierung und Anwendung, patent application no. 10 2008 047 366.9, German Patent Office, Munich, (2008)

[10] [10] M. Mauntz and U. Kuipers: Verfahren, Schaltungsanordnung, Sensor zur Messung physikalischer Größen in Fluiden sowie deren Verwendung, European patent application no. EP 09000244, European Patent Office, Munich, (2009)

[11] [11] M. Mauntz, J. Gegner and U. Kuipers: Ölsensorsystem zur Echtzeit-Zustandsüberwachung von technischen Anlagen und Maschinen, Technisches Messen 77, pp. 283-292, (2010)

[12] [12] M. Mauntz, U. Kuipers and J. Gegner: New Electric Online Oil Condition Monitoring Sensor – an Innovation in Early Failure Detection of Industrial Gears, The 4th International Multi-Conference on Engineering and Technological Innovation July 19th – July 22nd, 2011, Orlando, Florida, USA 2011, Proceedings Volume I, International Institute of Informatics and Systemics, Winter Garten, FL, USA, 2011, pp. 238-242, (2011)

[13] [13] M. Mauntz, U. Kuipers and J. Gegner: High-precision online sensor condition monitoring of industrial oils in service for the early detection of contamination and chemical aging, Sensor + Test Conf., 7.-9.6.2011, Nürnberg, AMA Service GmbH, Wunstorf, pp. 702-709, (2011)

[14] [14] M.Mauntz, J. Gegner, S. Klingauf and U. Kuipers: Continu-ous Wear Measurement in Tribological Systems to Control Operational Wear Damage with a new Online Oil Sensor Sys-tem, TAE Technische Akademie Esslingen, 19th International Colloquium Tribology, Esslingen, January 21-23, 2014, (2014)

[15] [15] A. Saeed: Online Condition Monitoring System for Wind Tur-bine, Master's Thesis, Blekinge Institute of Technology, Karlskrona, Sweden, (2008)

[16] [16] M. Mauntz, U. Kuipers and J. Peuser: Continuous, online detection of critical operation conditions and wear damage with a new oil condition monitoring system, WearSens®, 14th In-ternational Conference on Tribology - SERIATRIB '15 Proceedings, Belgrad, Serbian Tribology Society Kragujevac, University of Belgrade, Faculty of Mechanical Engineering, Belgrade, ISBN: 978-86-7083-857-4, S. 283-288, (2015)

[17] [17] M. Mauntz, U. Kuipers and J. Peuser: New oil condition monitoring system, WearSens® enables continuous, online detection of critical operating conditions and wear damage, Malaysian International Tribology Conference 2015 - MITC2015, Penang, Malaysia on November 16-17, 2015, Conference Proceedings, ISBN: 978-967-13625-0-1, S. 179-180, (2015)

Real-Time Monitoring of Temperature Distribution in a Lithium-Ion Battery Pack

A.M. Aucamp[1], A. Janse van Rensburg[1]
[1] School of Electrical, Electronic and Computer Engineering, North-West University, South Africa

ABSTRACT

Intelligent battery management systems aim to continuously diagnose the state-of-health (SOH) of cells in a battery pack and predict performance, reliability and safety issues. Cells with a higher internal resistance component will generate more heat and can be identified by the battery management system (BMS) from this thermal behavior. Placing a temperature sensor on each cell in a battery pack containing thousands of individual units might not be feasible due to practical limitations or budgetary constraints. As such, this paper describes an approach to real-time monitoring of a pack's temperature distribution using a small, configurable number of temperature sensors with the aid of interpolation and extrapolation algorithms. Of the five algorithms initially considered, only the nearest neighbor and inverse distance weighting algorithms were implemented and compared in terms of accuracy and aesthetic practicality. The temperature distributions are displayed using a 3D visualization framework inside an option-rich user interface. The effect the number of temperature sensors and sample resolution had on data accuracy and latency were also investigated. With accurate and timeous data describing the behavior of the cells in a LIB pack, the BMS can better achieve its purpose in terms of monitoring, diagnosis and control.

Keywords: Practical low-cost condition monitoring; Energy storage systems; Lithium-ion battery pack; Temperature distribution

Corresponding author: Dr. Angelique Janse van Rensburg (angeliquejvr86@gmail.com)

1. INTRODUCTION

Lithium-ion battery (LIB) technology is used in a wide variety of applications ranging from consumer electronics to electric vehicles and, more recently, for large-scale stationary energy storage [1, 2]. The majority of efforts aimed at improving the safety of LIB technology focus on effective thermal management strategies and the conditions that lead to thermal runaway [1, 3]. Internal exothermic reactions generate heat and an elevated ambient temperature also accelerates these reactions. A lithium-ion cell's performance is improved by these accelerated reactions because the energy conversion rate increases. At the same time, lifetime might be sacrificed because detrimental material changes can occur at elevated temperatures and result in capacity fade. In rare cases, thermal runaway can propagate throughout a battery pack and cause cells to ignite and/or explode [3].

Lithium-ion battery packs consist of many cells that all have slightly different individual characteristics due to manufacturing imperfections and impurities in the materials used. These characteristics will cause cells to generate differing amounts of heat. This leads to hotspots forming inside the battery pack. These hotspots can lead to thermal runaway and will damage not only the affected cell, but also all the cells surrounding it. Hotspots can therefore lead to catastrophic failures if they are not identified and mitigated [1, 3].

This paper describes an approach to real-time monitoring of a pack's temperature distribution using a small, configurable number of temperature sensors with the aid of interpolation and extrapolation algorithms. The distributions generated by the algorithms are then used to create a 3D visualization of the battery pack's temperature distribution. The low-cost approach is in response to the traditionally more expensive proprietary data acquisition systems and their accompanying software that may not comply with the needs of the user. The data gathered can then be used by a BMS system to calculate certain SOH statistics and identify and mitigate anomalies in the temperature distribution. The visualizations generated can also be recorded, reviewed and used to identify anomalies.

The small number of configurable sensors used in conjunction with a custom data acquisition system (DAQ) means that the solution can be more cost effective than placing a sensor on every cell and using proprietary data gathering hardware and software. The software also creates an interactive visualization that is subjectively more informative than a static set of numbers or 2D graphs. The temperature distribution of single cell batteries and battery packs have been well studied and simulated [4]–[6]. These simulations, however, are not real-time displays of a multi-dimensional battery pack in operation.

Maxim Integrated™ have designed a system to "Measure Multiple Temperatures in Battery- Management Systems, and Save Power Too" [4]. Current data-acquisition integrated circuits for battery packs are capable of measuring multiple cell voltages but they only scan and measure two temperatures at most [4]. This arrangement is inadequate since an accurate representation of the temperature distribution within a battery pack cannot be made with only two measurements. Maxim's solution also does not make the information available to the user since a battery management system is supposed to run in the background and manage the battery temperature and voltages according to its preconfigured set-up. Johns Hopkins Applied Physics Laboratory have developed a system they call the Battery Internal Temperature Sensor or BITS [7]. The system is reportedly capable of accurately measuring the instantaneous internal temperature of a lithium-ion battery in a non-intrusive manner. The technology is not yet commercially available. They have not tested it on all battery form factors. The 18650 is one of the form factors they have not tested yet at the time of writing. The technology is, however, largely untested and there is little documentation.

In a recent work, temperature data of a single 18650 cell in operation are collected using 4 thermocouples for the purposes of model validation. Post-processing is applied and the data are presented offline in simple 2D graphs with no option of intervening during operation [8]. An attempt to reconstruct the 3D temperature distribution of a series-connected LIB pack in real time using Kalman filtering has also been reported. The results seem to indicate visibility of only the outside surface temperature and not the distribution inside the pack itself [9]. Onboard monitoring of a cell's internal temperature distribution in 2D using impedance-based temperature detection has also yielded promising results because the method was experimentally validated [10].

The following section starts with a brief discussion on the interpolation algorithms used to obtain the resulting temperature distributions. Section 2 continues with an explanation of the process behind the hardware and software design as well as the approach to data verification. The resulting visualizations of the temperature distributions and the investigation into the number of sensors and sample resolution are presented in Section 3. The paper concludes with the final section summarizing the key findings and providing suggestions on how to improve the current work.

2. METHOD

2.1. Interpolation algorithms

Interpolating the temperature distribution requires that there be a set of known coordinates with known values to begin with. These coordinates and values correspond with the sensor's coordinates and temperature values. It is reasonable to assume that coordinates located closer together have a higher probability of having temperature values that are more similar than those of coordinates that are farther from each other. This estimation intuitively follows from the general heat equation [11] shown in Equation (1), which is based on the diffusion equation [12].

$$\frac{\partial u}{\partial t} - \alpha \nabla^2 u = 0, \qquad (1)$$

However, this formula assumes conservation of energy as well as a single heat source within a uniform material. The general principal of a divergent vector field is useful in that it allows for comparisons with approximation algorithms that will produce similar results with less computational complexity and cost. Using Big-O notation, the performance of the interpolation algorithms can be can be estimated [13]–[15]. We take n as the set of coordinates that needs interpolation and m as the number of sensors. The time needed to interpolate the values would therefore be as Table 1 describes. Since the data need to be interpolated in real-time, the faster methods will deliver more timeous results. It is for this reason that the nearest neighbor (NN) and inverse distance weighted average (IDWA) interpolation algorithms were identified as relatively simple and computationally inexpensive.

Table 1: Computational complexity of interpolation algorithms

Interpolation Algorithm	Big-O
Nearest Neighbor	$O(nm)$
Inverse Distance Weighted Average	$O(n4m)$
Linear	$O(n(6m+9))$
Spline	$O(nm^2)$

Nearest neighbor interpolation is the simplest implemented algorithm. The temperature of the discrete and evenly placed coordinates in the array used by the software are calculated by determining the closest sensor to each individual coordinate and setting that coordinate's temperature equal to that of the sensor [13]. Equations (2) and (3) show this relationship where d is the minimum distance calculated and $u(x)$, the temperature being calculated is set equal to x_i, which is the temperature of the nearest sensor.

$$d = \min\{\mathrm{d}(\mathrm{x}, x_i)\}, \quad (2)$$

$$u(x) = x_i, \quad (3)$$

The IDWA of each discrete coordinate is calculated by first calculating the distance of the discrete coordinate to each of the sensors. The inverse of each distance is calculated and the sum thereof is normalized. Each normalized inverted distance is then multiplied by its sensor's temperature and these temperatures are summed again to calculate the temperature value at the specific coordinate [14]. Equation (4) represents the mathematical formula that is used. If $\mathrm{W_i}$ is taken as the inverse distance a value u at a given coordinate x can be estimated based on the samples $u = u(x_i)$ for $i = 1, 2, \ldots n$ by applying the following function:

$$u(x) = \frac{\sum_{\mathrm{i=1}}^{\mathrm{n}} \mathrm{W_i(x)u_i}}{\sum_{\mathrm{i=1}}^{\mathrm{n}} \mathrm{W_i(x)}}, \quad (4)$$

2.2. Design process

A systems engineering approach was followed to achieve the final design that was implemented [16]–[18]. The system consists of hardware and software that work in tandem to affect the desired outcome. This section details the different components of the system as depicted in Figure 1.

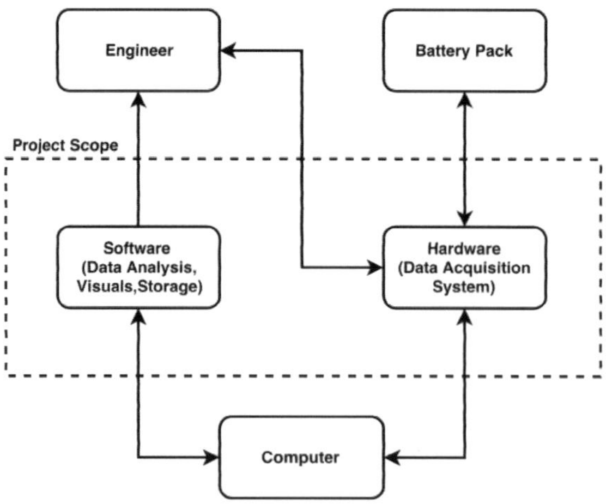

Figure 1. Proposed system architecture of the temperature distribution monitoring system

2.2.1. Hardware

The hardware system developed for the proposed system consists of a variable number of DS18B20 sensors connected to an STMF303 development-board via auxiliary cables. The auxiliary cables connect power, ground and data pins of the sensors to the STMF303. The STMF303 in turn is connected to a PC with a USB connection that carries serial data and power. The DS18B20 temperature sensors come in a TO-92 package. The sensors have a programmable resolution (accurate from 0.5-0.0625°C), onboard analogue to digital (A/D) converters and each sensor is etched with a 64-bit serial number to differentiate them when they are all connected via the same data pin. The sensors are soldered to stereo auxiliary cables to enable easy configurability. The DAQ consists of the STMF303 dev-board housed within a 3D printed casing that also encloses the debugging LED circuits as well as the stereo jacks. These components are shown in Figure 2.

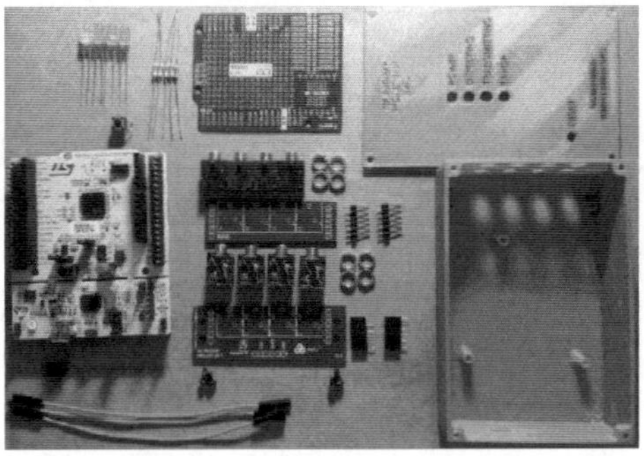

Figure 2. Components required to construct the custom DAQ

2.2.2. Software and firmware

The STMF303 is programmed using the online Mbed IDE with C. The firmware loop is represented in Figure 3. The main objective of the firmware is to detect the sensors and send the temperature data from the sensors to the PC in a serialized string each time a temperature reading is taken.

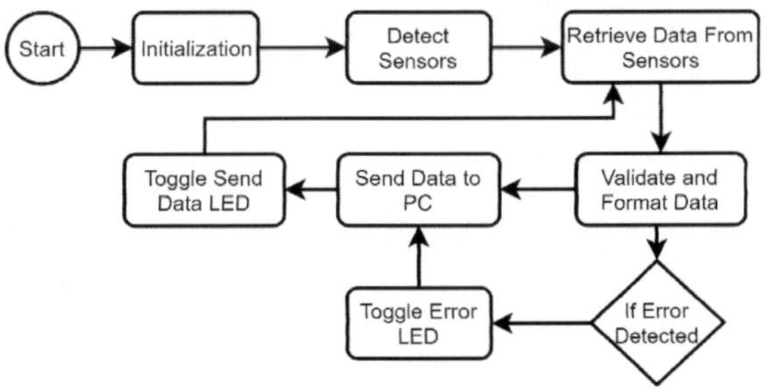

Figure 3. DAQ firmware loop

Unity 3D, a game engine based on the cross platform Mono framework, is used to generate the interactive real-time graphics. The framework itself is based on a subset of the .NET framework and allows for the use of .NET's multithreading and serial communication libraries. The software can be compiled to run on both Windows and Linux whereas the aim of achieving a low-cost design excluded a macOS compatible version.

The programming is done in C#. The main thread of the software is responsible for handling the user input and generating the graphics. Separate threads are spawned that handle the serial data communication and the interpolation of the data. The interpolation algorithms generate a set of temperature values at discrete intervals using the data from the temperature sensors and the sensors locations. This generated array is then used to generate the interactive real-time 3D graphics.

2.2.3. Interfaces

The system has two important physical interfaces. These are the interface between the sensors and the STMF303, and the interface between the DAQ and the PC. The proposed system also has an important software user interface on the PC. The sensors are connected to the STMF303 via auxiliary cables and utilize the MAXIM 1-wire interface. This enables all the sensors to connect to the DAQ while utilizing only a single digital pin for data transfers. This feature also allows the sensors to be configurable and dynamic without having to rewrite the firmware of the DAQ. The STMF303 is connected to the PC via USB and acts as a serial connection. This makes it possible to use serial communication libraries and protocols in both the DAQ and the visualization software to send and receive data. The USB connection also enables the PC to act as a convenient power supply to the DAQ and, by proxy, the sensors as well.

2.3. Verification

Each individual temperature sensor is tested and corroborated with 2 other temperature sensors to ensure accuracy. The temperatures are also verified within the visualization software to ensure that the temperature at the discrete location of the sensors correlate with the discrete locations within the visualization. The visualization was tested by deliberately heating up and cooling down certain areas of the battery pack, by using hot and cold air, and checking if the visualization emulated the temperature changes on the specific locations.

3. RESULTS & DISCUSSION

The figures and graphs in this section show that the graphics generated can be changed and interacted with to a degree by the user. This means that a better and more intuitive understanding of the temperature distribution of the battery pack can be reached in contrast to the standard 2D line graph displaying a single temperature average of the whole battery pack. The custom DAQ with sensors that was constructed is shown in Figure 4.

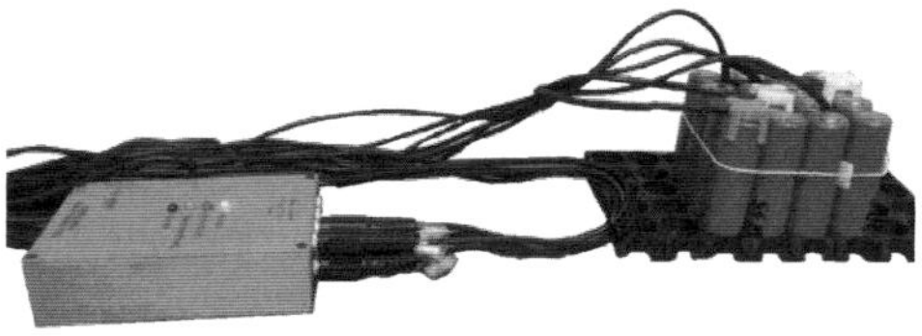

Figure 4. Battery pack setup with sensors and DAQ

The data gathered is displayed to the user instead of propagating silently in the background. It is saved as a text file with a custom extension but can be easily converted to a csv file by removing the header. The software has a built-in function for this purpose. This also means that, by proxy, any other analytic software or procedure that the user wishes to subject the data to can be done with relative ease. The data can further be analyzed with Matlab, Python, R etc. by analysts wishing to better understand the thermal behavior of battery packs. Figure 5 depicts the 2D graph on the bottom panel of the data review mode where the user reviews data that were previously recorded. It shows the maximum, minimum and average temperatures throughout the session and the locations of these maximum and minimum temperatures.

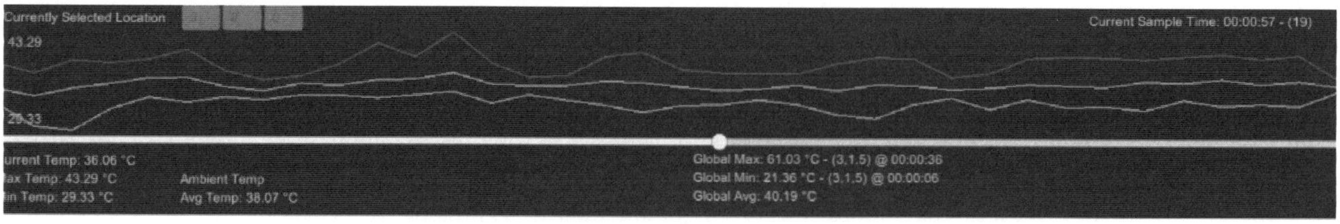

Figure 5. Timeline graph in 2D for data review with statistics

In contrast with Figure 5, Figure 6 displays the software operating in real time with the middle layer isolated to enable a better view of the inside of the array. The software is also capable of recording and reviewing data. The control panel on the right-hand side is responsible for configuring the graphics to the user's satisfaction, including setting the sensitivity of the colors, the maximum and minimum temperatures and isolating layers or rows of the battery pack to more clearly view the "inside". The bottom panel shows a graph of a single discrete location's (cell) recent temperature. Right clicking on a location or entering the location's coordinates in the text boxes changes the graph shown and the relevant max/min statistics under the graph to the location that has been selected. The software also has a warning feature that flashes either yellow or red in the right-hand panel to warn the user if the temperatures are approaching or exceeding the preset limits.

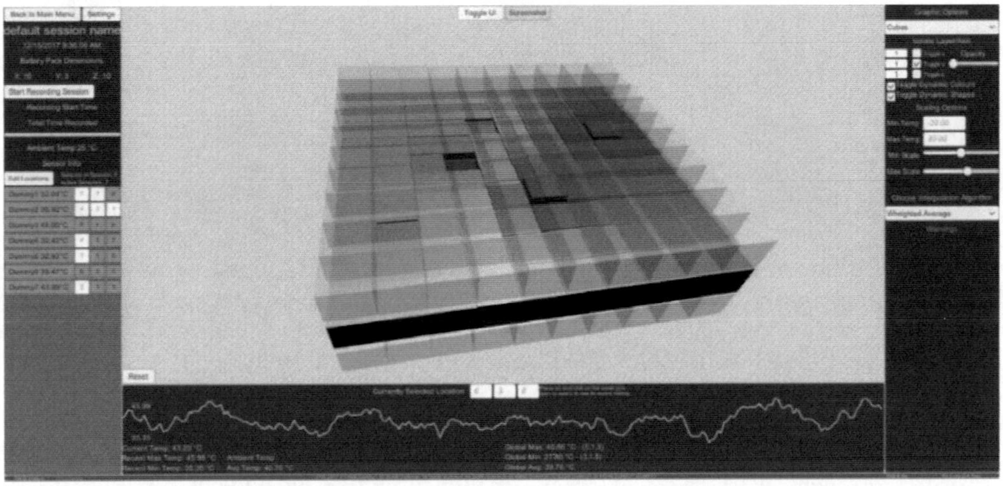

Figure 6. Software operating in real-time operation with middle layer isolated

The software lets the user pick a 3D model that they want to use to depict the temperature distribution. The choices are depicted in Figure 7. These models will scale and change color depending on the calculated temperature at the discrete coordinates.

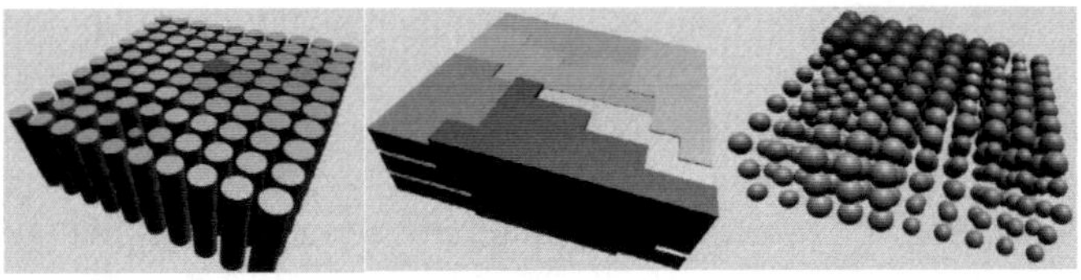

Figure 7. Visualization options (cylinders, cubes, spheres)

The system gathers data in real time with the DAQ, depicted in Figure 4, but the A/D conversion and the amount of sensor data that need to be formatted result in a delay being present when the final interactive graphics are being rendered. This delay, illustrated in Figure 8 sows that the resolution of the A/D conversion has a much larger initial impact than the number of sensors present in the system. This latency that the system experiences between each data sample proved to be the largest delay and, since the serial communication was running in a separate thread, it would not be increased by any other part of the software. This did mean that to facilitate a coherent user experience the interpolation algorithms needed to complete faster than the delay created by the DAQ.

Since the interpolation algorithms were also running in their own thread, their execution time was not affected by other parts of the software. The execution time of the interpolation algorithms was mainly affected by the battery pack size and the number of sensors. The larger the battery pack, the longer it takes to interpolate the data. The same effect is witnessed with the number of sensors connected.

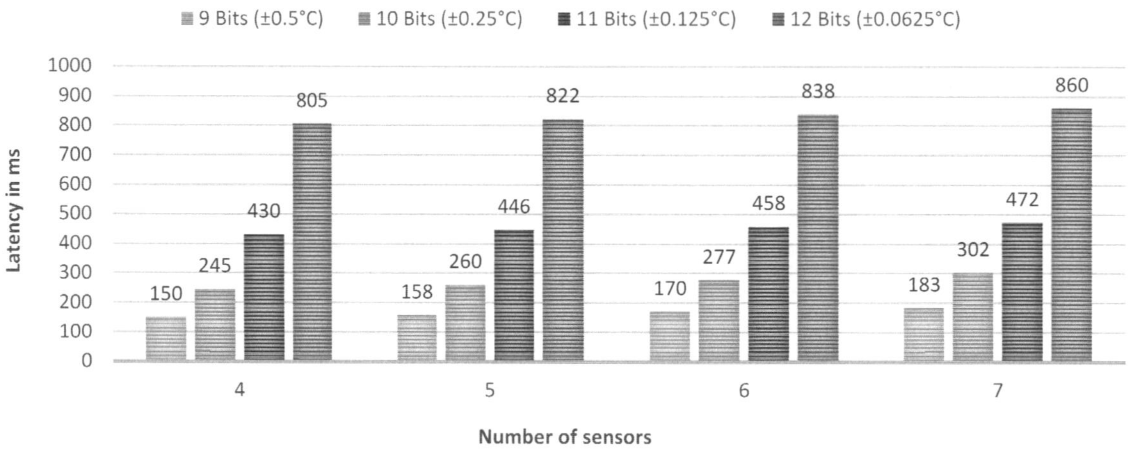

Figure 8. Average data stream latency with regards to the number of sensors and the sensor resolution

The software limits the creation of battery packs to sizes smaller than 50x50x50. Battery packs of this size and larger tended to cause a significant drop in frame rates of the software, from 60 fps to <12 fps, which is unacceptable from a usability and aesthetics perspective. This was mainly because of the graphics framework not being able to update all the dynamic objects fast enough due to the implementation inefficiencies.

There is thus a tradeoff that the final user of the system can make when using the system. Increasing the resolution and limiting the number of sensors or increasing the number of sensors (which generally result in a more accurate 3D temperature distribution generated). They can also have a high resolution and number of sensors if a longer update delay is acceptable to them. Figure 8 shows the graph generated of the average data stream latency with regards to the number of sensors and the sensor resolution. Even though only 7 sensors were tested, connecting more is possible but more expensive.

The inverse distance weighted average algorithm generated a smooth gradient in the resulting graphics, while the nearest neighbor algorithm resulted in a certain set of regions that all seemed to have the same temperature. The IDWA therefore produced more natural and accurate results, while the NN produced clearly sectioned patterns.

The system as it was developed needs to be physically connected via wires to the battery pack and the PC. This might induce difficulties in certain scenarios where the battery pack's temperature data needs to be viewed while in operation, for example, a running electric vehicle. Developing a wireless DAQ could solve this issue as well as make the system more usable.

The system (hardware and software) cost less than R 1,600 to implement. This is more cost effective than using traditional DAQ hardware and the proprietary software that they may need to function. The sensors themselves are the most expensive components (except for the STMF303), and since they are configurable, the cost of the system can be kept very low should a budget be limited.

4. CONCLUSIONS

The low-cost system for monitoring of the temperature distribution throughout a battery pack proved to be effective and can relay information in real-time with delays of less than a second. The information gathered and made available can be used by a BMS to improve the thermal management of a battery pack. By knowing the temperature distribution of the battery pack, hotspots can be identified and mitigated before they cause any harmful side effects. The BMS can apply measures to dissipate the excess heat as quickly as possible preventing thermal runaway. Large temperature variations throughout the pack can also be minimized to ensure balanced cells.

The small number of configurable sensors allows the solution to be scalable and applicable in many scenarios. The interactive nature of the software also allows for a more intuitive and user-friendly experience when manually determining the SOH of a battery pack with regards to the temperature distribution. While the system is functional and operating correctly for the purposes of its original design, it can be improved to be of much greater use and functionality. At the time of writing, it is still only a generalized real-time temperature distribution monitor of 3D structures.

The system can further be improved upon by incorporating it into other battery pack projects. This may enhance its functionality and applicability. As part of a BMS, it could potentially increase the lifetime and safety of battery packs as well as allowing for a better understanding of thermal behavior within battery packs and battery pack design. In terms of the hardware, a wireless DAQ solution can be developed. It can also be given a smaller physical footprint so that it can be used in scenarios where space is an issue. The software can be improved in terms of user friendliness and functionality. However, as new technology is being developed, the software can be rewritten with another framework that is has a more permissive license and may run certain algorithms faster and more efficiently. It can also be improved by adding more variants of interpolation algorithms that may increase efficiency and accuracy. To further improve the accuracy of the temperature distribution, the generated graphics can be enhanced and compared with an infrared camera view of the battery pack. This might increase costs but it will serve as an extra validation method that closely resembles the generated graphics.

5. ACKNOWLEDGEMENTS

This work is based on the research supported, wholly or in part, by the National Research Foundation (NRF) of South Africa (Grant numbers 91093, 103392). Opinions expressed and conclusions arrived at are those of the authors and are not necessarily to be attributed to the NRF.

REFERENCES

[1] Abada, S. (2016) Safety focused modeling of lithium-ion batteries: A review. Journal of Power Sources. vol. 306. pp. 178–192.

[2] Rezvanizaniani S. M. , Liu Z., Chen Y., L. J. (2014) Review and recent advances in battery health monitoring and prognostics technologies for electric vehicle (EV) safety and mobility. Journal of Power Sources. vol. 256. pp. 110–124.

[3] Saw, L. H., Ye, Y., and Tay, A. A. O. (2016) Integration issues of lithium-ion battery into electric vehicles battery pack. Journal of Cleaner Production. vol. 113. pp. 1032–1045.

[4] Racherla, K. Measure Multiple Temperatures in Battery Management Systems, and Save Power Too. [Online]. Available: https://www.maximintegrated.com/en/app-notes/index.mvp/id/5070.

[5] Robinson, J. B., Darr, J. A., Eastwood, D. S., Hinds, G., *et al.* (2014) Non-uniform temperature distribution in Li-ion batteries during discharge – A combined thermal imaging, X-ray micro-tomography and electrochemical impedance approach. Journal of Power Sources. vol. 252. pp. 51–57.

[6] Saw, L. H., Ye, Y., and Tay, A. A. O. (2013) Electrochemical–thermal analysis of 18650 Lithium Iron Phosphate cell. Energy Conversion and Management. vol. 75. pp. 162–174.

[7] Srinivasan, R., Carkhuff, B. G., Butler, M. H., and Baisden, A. C. (Feb. 2011) Instantaneous measurement of the internal temperature in lithium-ion rechargeable cells. Electrochimica Acta. vol. 56. no. 17. pp. 6198–6204.

[8] Panchal, S., Mathew, M., Fraser, R., and Fowler, M. (2018) Electrochemical thermal modeling and experimental measurements of 18650 cylindrical lithium-ion battery during discharge cycle for an EV. Applied Thermal Engineering. vol. 135. pp. 123–132.

[9] Tian, N., Fang, H., and Wang, Y. (2017) 3-D Temperature Field Reconstruction for a Lithium-Ion Battery Pack: A Distributed Kalman Filtering Approach. IEEE Transactions on Control Systems Technology. pp. 1–8.

[10] Richardson, R. R., Zhao, S., and Howey, D. A. (2016) On-board monitoring of 2-D spatially-resolved temperatures in cylindrical lithium-ion batteries: Part II. State estimation via impedance-based temperature sensing. Journal of Power Sources. vol. 327. pp. 726–735.

[11] Cannon, J. R. (1984) The One–Dimensional Heat Equation 1st ed. Cambridge University Press. Addison-Wesley Publishing Company.

[12] Thambynayagam, R. K. M. (2011) The Diffusion Handbook: Applied Solutions for Engineers. McGraw-Hill.

[13] Lu, G. Y. and Wong, D. W. (2008) An adaptive inverse-distance weighting spatial interpolation technique. Computers & Geosciences. vol. 34. no. 9. pp. 1044–1055.

[14] Bartier, P. M. and Keller, C. P. (1996) Multivariate interpolation to incorporate thematic surface data using inverse distance weighting (IDW). Computers & Geosciences. vol. 22. no. 7. pp. 795–799.

[15] Toraichi, K., Katagishi, K., Sekita, I., and Mori, R. (1987) Computational complexity of spline interpolation. International Journal of Systems Science. vol. 18. no. 5. pp. 945–954.

[16] Burge, S. (2011) The Systems Engineering Tool Box.

[17] US Army (2001) Systems Engineering Fundamentals. Fort Belvoir, Virginia. The Defense Acquisition University Press.

[18] Heath, S. (2002) Embedded Systems Design. Elsevier Science.

A perspective on rotating equipment technology trends and maintenance in mining

Joe Amadi-Echendu & Masiye Matheta
Department of Engineering and Technology Management, University of Pretoria, South Africa

ABSTRACT

Advances in technology suggest marked improvements in the reliability of machinery deployed in mining operations. Rotating equipment is extensively deployed in mining and as such, there is increasing expectation that such equipment should operate at higher levels of reliability and reduced total life cost. The question thus arises as to the extent to which actual maintenance practice tracks changes in the technology of the rotating equipment. This paper describes a study that compares the trends in rotating equipment technologies against the history of actual maintenance interventions on such equipment in a real mining environment. Empirical data obtained during the study provides intriguing evidence about the relationship between technology trends and actual maintenance interventions.

Keywords: Rotating Equipment Maintenance; Equipment Technology Trends; Maintenance in Mining; Integrated Maintenance Management.

1. INTRODUCTION

Rapid evolutions in technology pervade every area of human endeavor, and in particular, advances in automation, information and communication technologies, new materials, et cetera, are increasingly incorporated in machinery deployed in mining. Rotating equipment make up a large portion of machinery used in mining operations, and the maintenance of such equipment is a significant activity in minerals extraction and processing. The technology trends affect both the original equipment manufacturers (OEMs) and the mining sector businesses, and increasingly influence the relationships between OEMs, vendors, suppliers, and users of rotating equipment in mining operations.

With increasing levels of automation at the component, subsystem, equipment and machinery levels, the mining industry is advancing towards full scale mechanized automation and autonomous operations. There are claims that such full scale mechanization and automation should result in improved performance in terms of reduced costs of operations, increased production outputs, improved safety performance, increased profits, and so on. There are also claims that advanced technologies embedded in modern day rotating equipment should increase reliability and availability of such equipment during mining operations.

Granted that changes in equipment technology will continue to happen at an even greater pace, the question arises as to how actual maintenance interventions in mining operations keep track with the technology trends. This paper briefly describes a review of the technology trends in rotating equipment and examines whether actual maintenance interventions in a case study mining operation has tracked the technology trends.

Section 2 of this paper includes a summarized review of the rotating equipment technology trends. Section 3 also includes a summarized review of the history of actual maintenance interventions performed on rotating equipment deployed in a mining operation. A comparison of the technology trends against the maintenance interventions is discussed in section 4, and some conclusions are drawn based on the results.

2. TRENDS IN ROTATING EQUIPMENT TECHNOLOGY

The picture in Figure 1 provides an overview of the range of rotating machinery, equipment types and components (cf: Forsthoffer, 2005:4), classified in terms of i) driven equipment, ii) prime movers, iii) transmission devices, and iv) auxiliary systems.

I Driven Equipment

- A Compressors
  - 1 Dynamic
    - Centrifugal
    - Axial
    - Integral gear
  - 2 Positive Displacement
    - Screw
    - Rotary Lobe
    - Reciprocating
    - Diaphragm
    - Liquid Ring
- B Pumps
  - 1 Dynamic
    - Centrifugal
    - Axial
    - Slurry
    - Integral Gear
  - 2 Positive Displacement
    - Plunger
    - Diaphragm
    - Gear
    - Screw
    - Progressive cavity
- C Extruders
- D Mixers
- E Fans

II Drivers - Prime Movers

- A Steam turbines
- B Gas turbines
- C Motors
  - Induction
  - Synchronous
  - Variable Speed
- D Engines
  - Internal Combustion
  - Diesel
  - Gas turbines

III Transmission Devices

- A Gears
  - Helical
  - Double Helical
- B Clutches
- C Couplings

IV Auxiliary Equipment

- A Auxiliary Equipment
- B Buffer Gas Systems
- C Cooling Systems

Figure 1. Rotating equipment types and components (Source: Forsthoffer, 2005:4)

In terms of the scale and scope, references [1] and [2] noted that most technology changes focused on automation, embedded diagnostics and operational ergonomics. References [3], [4], [5] indicate that technology trajectory mostly involved increased modularization so as to create highly reliable components, subsystems and equipment with the aim to extend life of machinery. The approach subsumes the following hierarchy; components → subsystems → equipment → machine. In terms of life extension, the suggestion is that the components have the shortest life so as to maintain pace with technology changes.

Some authors (see, for example, [6], [7], [8], [9], [10], and [11]) surmise that the introduction of technological superior equipment complicated operation and maintenance interventions. It is not trivial to effect traditional maintenance intervention on highly reliable and automated machinery that also has embedded diagnostics capabilities. Whereas new technologies improved energy efficiency and operational ergonomics of machinery, however, in some instances, operators and users of technologically superior rotating equipment did not achieve desired availability, cost and reliability performance.

Some of the technology changes in rotating equipment deployed in the case study mine are summarized in Table 1. For example, between 1970 and 1999, the mine utilized a combination of direct-current (DC) and alternating-current (AC) drives in machinery such as trammel screens, but from 2000 to date, only AC drives are deployed in vibrating screens, albeit that the drives are obtained from various OEMs. Some of the rotating equipment deployed for mining operations are depicted in Figure 2.

Table 1: Summary of equipment technology changes

Technology changes affecting	Description of the type of technology changes observed in the mine rotating equipment	Mostly affected
Scale of operation	physical size of the equipment to increase the equipment output capacities	Dozers, loaders, locos, cranes
Equipment Life	design in terms of strength, durability and reliability to provide for increased of extended equipment life	Pumps, compressors, disintegrators, dozers
Information processing	more accurate information (in terms of the internal and external processes parameters).	Drives, clarifier plant automation
Level of control	rransfer of control from human to machinery (Automation)	Turbines, pumps
Functional range	equipment capabilities (e.g., increased power output, speed, etc.)	Vibrating screens
Scalability and modularity	physical shape of the equipment to provide for a level of scalability and modularity in the design	Drives, Screens, hydraulic pumps
Amplification of human	execution of work by shifting the energy to do work from humans to machines.	Screens
Ergonomics	reduce the physical stresses imposed on the operator by the work environment	None Recorded

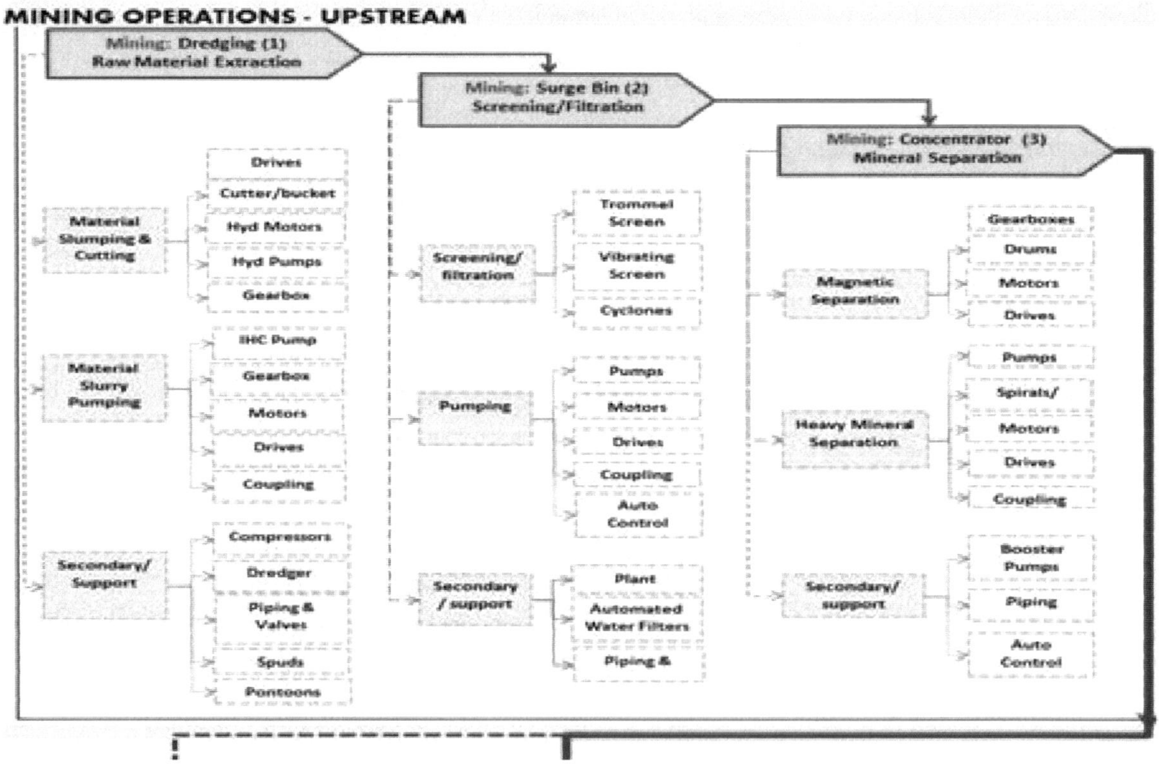

Figure 2. Some rotating equipment installed in the case study mining operation.

3. CASE STUDY OF MAINTENANCE INTERVENTIONS ON ROTATING EQUIPMENT

Data regarding maintenance interventions were obtained from the computerized maintenance management system of the case study mining operation. Remarkably, the records showed that only about 20% of the planned maintenance interventions were associated with rotating equipment, even though the mine utilizes a substantial number of such equipment. Between 2003 and 2017, it was surprising to note that out of the 83,815 maintenance interventions recorded against rotating equipment, only 827 were based on detailed lists of tasks that were performed, thus creating the impression that most maintenance interventions were adhoc. Further investigation revealed that detailed lists of tasks were only transferred from one platform to another during software upgrades, i.e., the tasks were often not updated to reflect the fact that a new version of rotating equipment had been installed to replace an earlier model.

For brevity, Table 2 provides a summarized view of maintenance interventions on some electric drives deployed in the case study mining operation between 2003 and 2017. The boxes marked with "X" represent planned maintenance interventions with detailed lists of tasks. The column of shaded boxes indicate that the exact same maintenance interventions were carried out each year, i.e., for electric drive with code A/516151/1, the detailed list of tasks performed were unchanged from 2005 to 2012, while for electric drive A516154/1, the same maintenance interventions have been carried out since 2005 unchanged. What is intriguing is that the row of contiguous shaded boxes indicates that the same maintenance interventions were carried out for electric drives in different functional locations of the mine. In 2005 for example, the same set of tasks were performed on 9 respective electric drives in different functional locations. A similar curious observation applies to other rotating equipment like pumps, trammel and vibrating screens.

Table 2. Maintenance interventions on some electric drives between 2003 and 2017

	Drive code											
Year	A/21527/1	A/516151/1	A/5161526/1	A/5161527/1	A/516154/1	A/5161541/1	A/5161542/1	A/5161556/1	A/516158/1	A/5161603/1	A/5161839/1	A/5161841/1
2003			X									X
2004	X		X									X
2005		X	X	X	X	X	X	X	X			X
2006		X	X	X	X	X	X		X		X	X
2007		X	X	X	X	X	X		X			X
2008		X		X	X	X	X		X			X
2009		X		X	X	X	X		X			X
2010		X		X	X	X	X					
2011		X		X	X	X	X		X			X
2012		X		X	X	X	X		X			
2013			X		X							X
2014			X		X							X
2015			X		X							X
2016			X		X			X				X
2017			X		X							

The pattern for the lists of tasks carried out during maintenance interventions is illustrated in Table 3. For the blank spaces, there were no recorded lists of tasks for maintenance interventions found. At least three of the columns from the sample show that the same maintenance intervention tasks have been carried out from 2003 to 2016. It seems unreasonable that the list tasks have not been changed or updated for nearly 13 years!

Table 3. List of tasks for maintenance interventions on some rotating equipment between 2003 and 2017

	Maintenance Task List code											
Year	090170818027177C	0901708180271695	09017081802717D4	09017081802716 7B	09017081802717B1	0901708180271788	09017081804241FB	09017081802716FB	090170818053F771	09017081802717A0	090170818027179D	09017081802716F3
2003	X	X	X	X	X	X	X					
2004	X	X		X	X	X	X	X	X	X		
2005	X	X			X	X	X	X	X	X	X	X
2006	X				X	X	X	X		X	X	X
2007	X				X	X	X	X		X	X	X
2008	X				X	X	X	X		X	X	X
2009					X	X	X	X		X	X	X
2010					X	X	X	X		X	X	X
2011	X	X	X	X	X	X	X	X		X	X	X
2012	X	X	X	X	X	X	X	X		X	X	X
2013	X	X	X		X	X	X	X		X	X	X
2014	X	X	X		X	X	X	X			X	X
2015	X	X	X		X	X	X	X			X	X
2016	X	X	X		X	X	X	X			X	X
2017												

A very remarkable observation is the difference in availability of trammel versus vibrating screens despite increased maintenance interventions on vibrating screens. Trammel screens were deployed between 2003 and 2009 before most were replaced with vibrating screens by 2013. The bar charts in Figure 3 indicate that, on average, the downtime for vibrating screens were higher (at 12.5 days/year) than for the trammel screens (at 2 days/year). Based on the preceding Tables 2 and 3, it is reasonable to suggest that the lack of updated list of tasks meant that the maintenance interventions were less effective such that the apparently better reliability of vibrating screens could not be achieved.

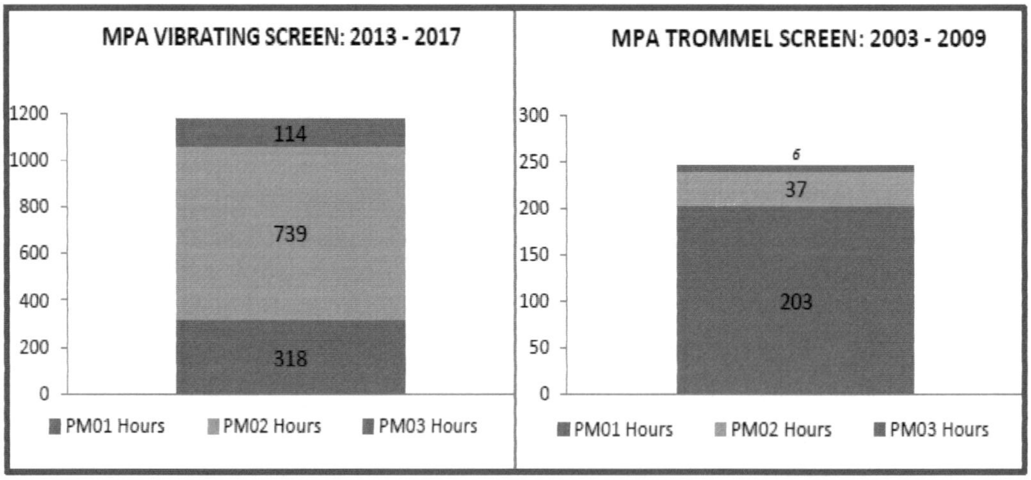

Figure 3 A comparison of downtime hours in changing from trammel to vibrating screens.

4. DISCUSSION AND CONCLUSION

The following general observations are based on a review of the recorded history of actual maintenance interventions :-

- There was little or no change in the detailed maintenance intervention tasks even after DC drives were replaced byAC drives from 2000 onwards
- Increased inherent reliability of the AC motors led to the maintenance effort being reduced and maintenance work made easier
- Maintenance effort had increased as evident in the increased number of planned and corrective maintenance work undertaken, including the additional operations support work, which was indicative of the higher level of effort required to keep the equipment running
- Even with significantly higher maintenance effort and maintenance expenditure on the vibrating screen, the availability and reliability remained much lower than in the previous case of trammel screens
- Changes in the mode of operation and equipment used introduce increased complexity. This is mostly made up of changes in speed (whilst hydraulic drive was rotational speed was low, speed on the electric motors was high), mode of operation from controlled rotation to high frequency vibrations (this inherently lead to increased cyclic loading on the structural components), power source (change from hydraulic system to electrical driven unit effectively requiring a change in the type of maintenance work performed on the equipment), nature of design (whilst defects and condition of equipment could be assessed or corrected through use of visual inspections and corrective work was easily taken with standard tools, on vibrating screens, the faults and defect are usually hidden and not easily detected by visual inspection and as such, it required a use of advanced monitoring and diagnostic tools (predictive – condition monitoring) as well as advanced repair toolkits (i.e. torque wrench instead of spanners, laser alignment on motors, highly technical skills). This scenario highlights that, with introduction of vibrating screen technology, the maintenance needed to change. The adoption of planned maintenance PMs previously used on the screen, implies that either the end user had no knowledge and technical capability to develop and execute appropriate maintenance work suitable for vibrating screen OR that the results observed on the vibrating screen was due to the inherent poor reliability and maintainability characteristics of the vibrating screen as opposed to the trammel screen.

Based on this case study, the concerning observation is that in practice, it appears that maintenance interventions are not updated with respect to changes in the technology of rotating equipment deployed in some mining operations. With increasingly automated and autonomous mining operations, there is also increased requirement to redefine the relationships between original equipment manufacturers, suppliers, vendors and the users so as to ensure that maintenance interventions track and keep pace with technology changes in rotating equipment.

Acknowledgments
The authors wish to thank the mining company for benevolently supporting the study reported here

REFERENCES

[1] Goodrum, P., Zhai, D. & Yasin, M., 2009. Relationship between changes in material technology and construction productivity. *Journal of construction engineering and management,* 135(4), pp. 278-287.

[2] Lynas, D. & Horberry, T., 2011. Human factor issues with automated mining equipment. *The Ergonomics Open Journal,* Volume 4, pp. 74-80.

[3] Bartos, P. J., 2007. Is mining a high-tech industry?:Investigation into innovation and productivity advance. *Resource Policy,* Volume 32, pp. 149-158.

[4] Brown, G. M., Ebacher, B. J. & Koellner, W. G., 2000. *Increased productivity with AC drives for mining excavators and haul trucks.* Alpharetta, IEEE Industry Applications Society, pp. 28-37.

[5] Di Febo, M. & Paganini, P., 2015. Centrifugal pump technology in oil and gas refinment. *World Pumps*, November, pp. 36-40.

[6] Swanson, L., 1997. An empirical study of the relationship between production technology and maintenance management. *International journal of production economics,* Volume 53, pp. 191-207.

[7] Sheridan, T. B. & Parasuraman, R., 2000. Human Vs Automation: An expected analysis. *Journal of human factors and ergonomics society,* 42(3), pp. 403-407.

[8] Wang, K. & Zhou, M., 2006. An analysis of technological progress contribution to economic growth in construction industry of china. *Construction and design for project,* Volume 1, pp. 77-78.

[9] Bellamy, D. & Pravica, L., 2011. Assesing the impact of driverless haul trucks in australian surface mining. *Resources Policy,* Volume 36, pp. 149-158.

[10] Ahmed, A. et al., 2012. Using design of experiment and genetic algorithm to obtain the otimum gear shifting strategy for a real driving cycle. *Applied mechanics and materials,* Volume 224, pp. 497-503.

[11] Boudreau-Trudel, B., Zaras, K., Nadeau, S. & Deschamps, I., 2014. Introduction of innovative equipment in mining: Impact on productivity. *American Journal of industrial and Business Management,* Volume 4, pp. 31-39.

A Policy Framework for Integrating Smart Asset Management within Operating Theatres in a Private Healthcare Group to Mitigate Critical System Failure

C.B.H. Nel [1], J.L. Jooste [1]
[1] Department of Industrial Engineering, Stellenbosch University, South Africa

ABSTRACT

Within the hospital environment health technologies include medical equipment, consisting of physical assets which directly affect human lives [1]. Apart from being responsible for patient well-being, the private healthcare industry is also a for-profit business, where assets are considered an investment and where utilisation, performance, downtime and maintenance need to be considered for determining the return on investment. It is therefore important, to have a well-integrated physical asset management programme for ensuring that the medical equipment is reliable, safe and available for use when it is needed.

This paper proposes a policy framework for the implementation of *smart asset management* to assist with the strategic execution of physical asset management in private healthcare businesses. The focus of the study is on operating theatres, which is regarded as critical assets, both in terms of patient safety and operational importance. A case study approach is used where data is collected with structured questionnaires and further analyses of archival records is considered. Results from the data analysis is used to develop a policy framework for implementing smart asset management at the case organisation. The results cover smart asset management applications, requirements and benefits of implementation. The integration of performance management and condition monitoring within a smart asset management approach are discussed and it is concluded that this synergetic approach can support strategic execution of physical asset management to predict and mitigate critical system failure within the private healthcare industry. Validation of the policy framework is done through face validation, which highlights the business potential of smart asset management for assisting in strategic decision-making.

Keywords: Physical Asset Management, Smart Asset Management (SAM), Integrated maintenance management, Integrated Asset Management, Industrial Asset Management, Condition Monitoring, Engineering 4.0, Internet of Things (IoT).

Corresponding author: Charles B.H. Nel (cbhnel@gmail.com)

1. INTRODUCTION

According to the World Health Organisation (WHO), health technologies are an essential basis for the correct functionality of an effective healthcare system [1]. It is within this hospital environment that medical equipment is considered assets that can directly affect human lives. However, hospitals within the private healthcare industry are also considered businesses which apart from being responsible for human lives, are also accountable to management to be profitable, where assets are also considerable investments and maintenance costs need to be factored into the return of asset investment as well as income generated from asset utilisation and performance. It is therefore important to have a well-planned and managed asset management (AM) programme that is able to keep the medical equipment in a private hospital reliable, safe and available for use when it is needed. The emphasis of this paper is towards the strategic execution of AM within an operating theatre, which is regarded by the health industry as a critical asset, both in terms of human safety and operational importance.

Smart asset management (SAM) is a concept where the term "smart", is increasingly used within the commercial environment, which relates to a perception of technological intelligence. The concept of the Internet of Things (IoT) has also become a reality, which necessitates the need for a different approach to managing physical assets

[2]. With this emergence of technological intelligence, comes the possibility of vast quantities of asset data and the analysis thereof, which has proven to add value to asset management [3]. To capitalise and expand on this value realisation, SAM was established by incorporating various proven methodologies and applying these within real-time management environments. This research aims to create industry awareness about the business potential of incorporating SAM within automated asset environments for assisting in strategic management decisions. Considering the qualitative intricacies of this the research, it is established that management at various levels need to be engaged as well as informed about the context in which SAM can be utilised to address potential medical asset-related risks. The paper presents the result of case study research about the common understanding of technical managers operating within various private hospitals across Southern Africa. Based on the results a policy framework for implementing SAM is established from the feedback and experiences from stakeholders within the industry as well as archival records withdrawn from specific private hospitals to verify results. To validate the policy framework, face validation is performed. The policy framework serves to assist senior management to steer their focus and efforts towards practises that contribute to strategic execution for risk mitigation planning associated with critical asset failure.

2. LITERATURE REVIEW

The SAM concept is associated with various keywords and concept variations as the term SAM itself is not fully explicated in current literature [2]. To define SAM within its own field of study, a suitable definition should represent both a practical implementation as well as an abstract or hypothetical optimum to strive towards. To do so the integral aspects of AM, is primarily the foundation of SAM. A brief overview of the current literature is provided.

AM is becoming a well-established field of research with a variety of definitions, academic sources and alternative variants which further branch out into separate specialist studies. AM in the true sense of the term inherently relates to the simple concept of managing an asset. The term, asset, is defined by Oxford dictionary as "a useful or valuable thing or person" [3], essentially highlighting an asset as an attribute of value contribution. Therefore, the term AM relates to a concept of managing value contribution. It is evident that the term, asset, is therefore dependent on perceived value creation and contribution. Hence, the term, asset, can therefore also be defined with respect to the value contribution from physical equipment, where AM relates to the management of these physical items to control or propagate their value contribution [4,5,6,7]. Considering this value which AM can contribute, other sources [8,9,10,11,12,13,14,15,16,18] have also highlighted the positive contribution of real-time asset information being available within an organisation to make improved AM decisions. The Aberdeen Group highlights key enablers for successful implementation of "Real Time Enterprise Asset Management" which includes, but is not limited to; real-time monitoring of asset performance, real-time data exchange between systems that reside at the plant floor, Enterprise Resource Planning (ERP), workflow automation across plant floor and enterprise applications, and open architected, interoperable applications. It is therefore evident that within these established AM practices, the requirement for a real-time asset-monitoring environment can further add value to asset-owning organisations [17]. Within this AM value contribution opportunity, the impact of SAM becomes evident as a means to establish AM principles in a "smart" technologically aware manner with respect to real time asset monitoring. By doing so, real-time information throughout the organisation is possible, where this availability of information can facilitate more informed management decision-making [5].

Quality information is another concept closely associated with AM, where [18] states: "The quality of Asset Data & Knowledge should be assessed, understood and managed in order to ensure that it provides effective support to business decision making and processes". AM literature refers to asset knowledge enablers, which consist of: asset information strategy, asset knowledge standards, asset information systems, and asset data and knowledge. These four asset knowledge enablers illustrate a flow from asset information, which leads to organisational knowledge through standards and systems. Furthermore, [19] concludes that "rational decision-making models establish a weighing mechanism between choice and value. Rational methodologies lead to the optimization of the outcomes by emphasizing the process of choosing rather than on what is chosen. A certain alternative is always selected whenever its expected value is greater than that of other potential choices". Therefore, rational decision-making is defined by the process of choosing the structures and standards which govern the choices we make. Quality information and specifically knowledge, therefore form the basis of quality decisions which may lead to quality results, when applied within a premeditated systematic, standardised operational process. From these theories, it is evident that AM standards are defined by an individual or a group who have considered which decisions may offer the best outcome, taking the probability of failure into consideration, while also acknowledging the potential risks associated with these decisions. This notion of rational decision-making illustrates how AM leaders need to make rationalised strategic decisions concerning assets. Their decision-making processes should facilitate the growth and development of structures, which focus on information becoming knowledge, in order to enable quality decisions, which allow for AM performance improvement. It is within this decision-making process that SAM can play a vital supporting role in accumulating asset information, used to enhance the quality and consistency of these strategic decisions.

The Oxford Dictionary defines smart as "having or showing a quick-witted intelligence" [3], where smart when used as an adjective to describe a device, can be defined as "programmed so as to be capable of some independent action". Smart appliances are also considered a growing consumer trend where it is a marketing-related term which is used to differentiate between ordinary household appliances with those having an ability of above average intelligence [2]. Various researchers have contributed to the term, smart object, by describing an object characterised by a degree of technological intelligence which is able to make informed decisions for itself or within a greater network of devices [15,19,20]. Smart fields, defined by de Best and van den Berg, is described as an integration of technology, processes and resources [21]. Furthermore, these researchers claim that asset managers should strive for continuous measurement to allow for improved decision-making [21]. In doing so, an association between the implementation of strategy and the risk management of assets can take place, where accurate operational feedback allows for managing the opportunity to make more informed strategic decisions. The word, smart, therefore relates to being able to interpret information into workable knowledge, which is useful to the end user.

Currently available literature about SAM does not provide an industry standard definition or structure for SAM implementation, although similar concept variations exist. Concepts such as, Internet of Things (IoT), Industry 4.0 and Big Data, is associated with creating greater interdependence between technology, with specific focus on information feedback and asset management. Within a literature review, Nel and Jooste [2] presents how these concepts are used to derive a definition for SAM, while also highlighting how SAM facilitates asset communication within organisational management structures to better facilitate the control of operational activities. SAM can therefore be considered as taking two concepts – asset management and smart equipment (physical assets) – and integrating these for technological advantage. Smart assets can communicate within a given structure to assist management and make predefined decisions, or use artificial intelligence (machine learning) to improve productivity. SAM also serve to advance AM through the use of digital capabilities to enhance the control and performance of organisational assets. A "techno-organisational" environment is created which allows technology to assist with the implementation of organisational strategy via the intercommunication between assets with organisational infrastructure [15]. This creates the impression of artificial intelligence, where assets can think for themselves and for humans.

3. METHODOLOGY

The research approach followed for this research is a case study investigation, where various sources of evidence are utilised for establishing the policy framework regarding SAM implementation. These sources are used for triangulation where the initial investigation is to establish current operational issues experienced with AM within the hospital maintenance environment.

Initial data is collected from a homogenous sample of participants involving various hospitals within the Mediclinic Southern Africa (MCSA) group. Their feedback is analysed through a thematic analytic approach, which utilises coding techniques to group feedback from participants within general themes and concepts.

Furthermore, quantitative data from the organisation's Computerised Management Maintenance System (CMMS), and hospital archival records are utilised for further data verification and motivating the concerns experienced with the current maintenance methodology. Interpretation of the results is performed by comparing the various sources of evidence to form a basis from where a policy framework can be developed from industry best practises. The validation of this policy framework is achieved by means of face validation with a heterogenous sample of participants, representing the corporate management of MCSA who governs the implementation of MCSA standards through various policy frameworks.

Within the MCSA group there are 52 hospitals each with a head of the technical department. Depending on the size of the hospital a technical supervisor or a technical manager holds this position. Assisting these hospitals on a regional level are five regional managers, four technical specialists and two senior managers in the corporate division of the organisation. The participants selection for the homogenous sample is done through purposive sampling by selecting participants based on their role and associated decision-making involvement in MCSA. For this study, the hospital technical managers were included in the homogenous sample, while for the heterogeneous sample regional or senior managers who are involved in corporate decision-making were included.

The data collection is conducted in the form of a structured questionnaire, completed online though a Google forms platform. During the process, participants are required to answer a series of predefined questions. This allows for the analysis and interpretation of the collected evidence in a standard manner where the format of the interview is the same for all participants in the sample. The questionnaire provided for both closed and open-ended answers where the analysis of this information is related to the case organisation.

4. RESULTS AND DISCUSSION

The questionnaires are analysed according to a six-phase process for conducting thematic analysis, where the researcher undergoes a familiarisation process to comprehend the data collected. This is followed by the generation of codes and theme categorisation. Association and pattern recognition is then performed, followed by interpretation and representation. Lastly, explanation and abstraction conclude the process. The outcome of the thematic analysis involves selecting vivid, compelling extract examples which demonstrate the feedback toward the questions posed in the questionnaire. The feedback is summarised in seven focus areas, namely; the life cycle approach to managing assets, asset risk management, criticality of life-support equipment, reactive notifications for asset malfunction, important asset infrastructure (or asset systems) in operating theatres, theatre unavailability and electronic data acquisition. Each of these focus areas are briefly discussed.

From a life cycle approach to managing assets the questionnaire results are indicative that effective use of the CMMS is critical for managing assets. Technical managers use the information recorded in the CMMS to make decisions concerning the procurement, maintenance and replacement interval of certain assets. Regarding the maintenance of critical assets, the CMMS is used to produce both preventive maintenance, based on a time-based maintenance strategy, as well as breakdown notifications created from end users within the hospital. Preventive maintenance, which is controlled by the CMMS, is a reliable source of information, but depends on the feedback quality collected after completion of work orders. The quality of information received from breakdown maintenance work orders is frequently questionable due to the end user creating the work request. The end user is typically from other non-technical departments, resulting in incorrectly recorded information and in many cases request with missing information.

Asset risk management commonly forms part of the life cycle approach to managing assets, where the managing of risk is covered within the asset management strategy. However, due to the nature of assets within the hospital environment, where above-average risks are involved, the importance of risk management should be highlighted. From the questionnaire results, it is evident that risks are managed from breakdown information recorded in the CMMS, which are monitored on a monthly basis. Furthermore, instantaneous alarms connected to certain critical assets, alert specific users who report faults via telephone to the technical department for corrective action.

Critical life-support equipment are assets, which are critical to the operation of the operating theatre. Within the case organisation equipment is classified in three categories with the following definitions; life support, strategic, and general. Life support is considered the most critical as it affects the patient's life directly. Strategic assets are those which can have an adverse effect on the operation of the business. Based on the questionnaire results, high priority items are those facilitating life-support equipment and include central assets systems such as the electrical reticulation, water reticulation (including emergency water), gas reticulation as well as climate control and sterilisation. These systems are central to the entire hospital where certain departments utilise more of these assets due to their support of patients' lives.

Reactive notification for asset malfunction relates to the delayed time for notifying the technical department about critical asset malfunction. The majority of respondents confirmed that these notifications are directly received from the end user. In most cases an alarm notifies the end user, where the end user then utilises telephonic contact as the fastest means of reporting system malfunction to the technical department. Although proactively, planned maintenance inspections are performed according to a time-based schedule and daily visual inspections are performed during the week, the questionnaire results are indicative that immediate notifications are lacking. An opportunity exists for proactively notifying about malfunction through the use of a real-time monitoring system.

The results from the questionnaires highlights that not all life-support assets are considered relevant to the study. This study is only concerned with assets affecting the support of infrastructure within specific departments of a hospital. The highest priority maintenance departments are indicated in Figure 1 based on questionnaire responses which highlight each departments' critical priority as well as economic revenue priority. It is indicative from the results that equipment associated with operating theatres is considered the most important to the case organisation.

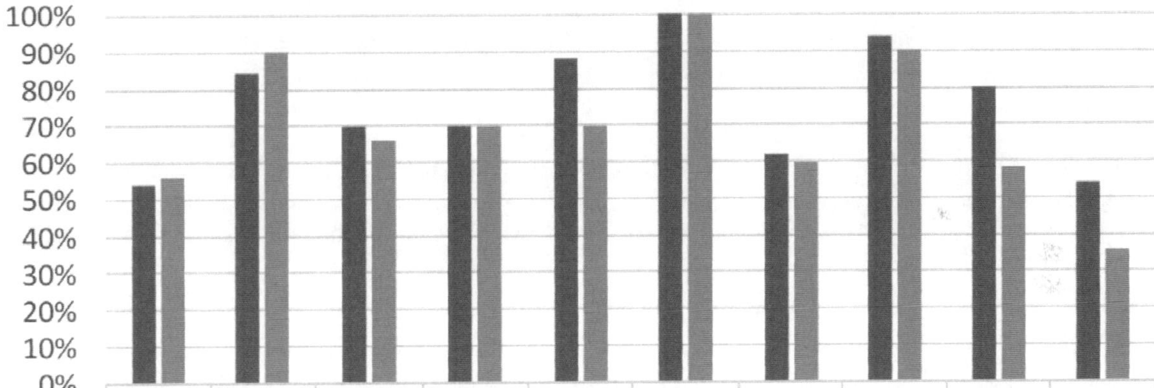

	Administration	High Care Room	Medical Wards	Cardiothoracic Ward	Emergency Centre	Operating Theatre	Pharmacy	Intensive Care Unit	Computer Server Room	Kitchen
■ Critical Nature	54%	84%	70%	70%	88%	100%	62%	94%	80%	54%
■ Economic Revenue	56%	90%	66%	70%	70%	100%	60%	90%	58%	36%

Figure 1. Interview Response for Departments with Critical Asset Infrastructures

Theatre unavailability relates to the establishment of the operating theatre as the most important department in terms of maintenance priority and strategic importance, where the infrastructure of assets within this department are considered as having the greatest priority. Further to establishing the operating theatre as the most important department, Figure 2 depicts the ranking of the most important assets within the theatre environment. If any of these assets fail, the patient's life will be in danger and the theatre would be deemed unavailable for use.

The current practise within the case organisation for data acquisition is a strategic visual monitoring inspection of assets. Strategic assets are identified by each hospital, where qualified personnel (mostly artisans) conduct these checks and information is manually recorded on a checklist. However, the possibility arises that during a visual inspection, certain aspects can be overlooked. The information, which is instantaneous when manually recorded, is also cumbersome to capture electronically. From the questionnaire results, it is evident that by utilising electronic monitoring devices, value can be derived from more timeous information to make improved decisions. This applies specifically within critical areas such as the operating theatres where centrally located asset systems support the hospital infrastructure both in terms of strategic business importance but also in terms of patient life support.

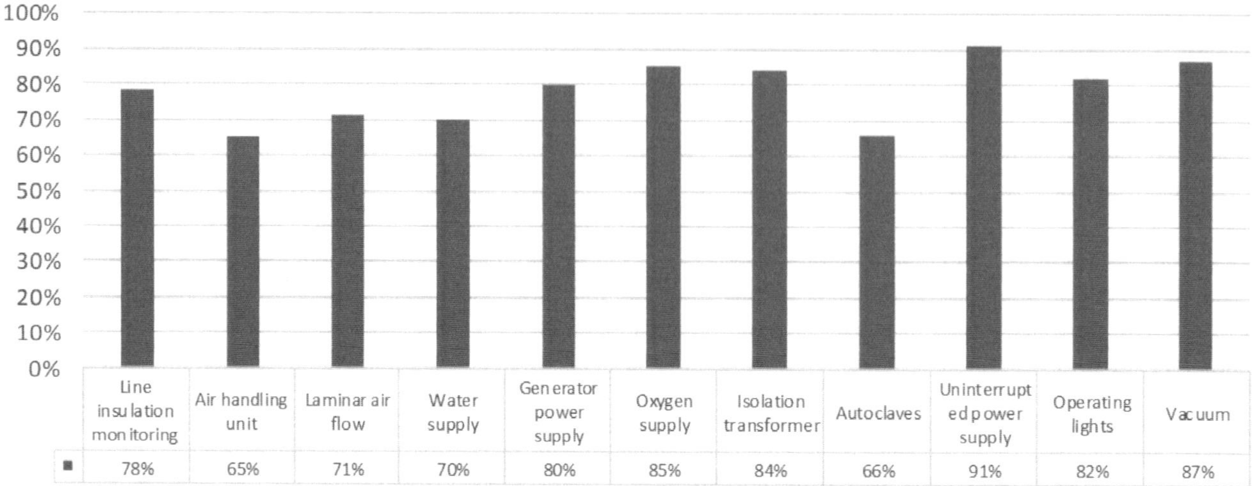

Figure 2. Interview Response for Critical Infrastructure Assets within an Operating Theatre

The analysis of archival records are used in support of the results from the questionnaires and outcomes of the thematic analysis. Information from the CMMS is extracted to highlight asset failures, whilst also looking at cases of asset failure and considering Mean Time Between Failure (MTBF). MTBF measures the average operational time between failures. It is considered an indication of reliability where historic data is used to highlight failure trends [22]. A low MTBF value represents a negative operational history; where the asset has experienced more failures over a total time, or has a low operational time representing availability. It is important to note that the time between failures is measured from the time of operation after the previous failure of the asset, to the time of the next failure. Only breakdowns are considered based on work orders created to repair damaged equipment. Table 1 presents an example of the calculations concerning MTBF data for the period June 2016 to July 2017 for a critical asset identified in Figure 2, namely an autoclave steriliser.

Table 1: MTBF calculation for Autoclave steriliser 3 at Hospital 1

Order	Order Typ	Opr. short text	Description	Description2	Barcod	Reference Date	Act.finish date	System Statu	Priority	Failure days
4000024747	PM01	PLease repiar Autoclave	CSSD 01	Autoclave, Steriliser	0647962	06/06/2016	06/06/2016	CNF TECO	2 - Urgent Work	0
4000024747	PM01	PLease repiar Autoclave	CSSD 01	Autoclave, Steriliser	0647962	06/06/2016	06/06/2016	CNF PRT TECO	2 - Urgent Work	
4000025825	PM01	Faulty autoclave (Allie)	CSSD 01	Autoclave, Steriliser	0647962	22/06/2016	28/06/2016	CNF PRT TECO	2 - Urgent Work	6
4000073123	PM01	Theatre 1-4 ccsd autoclave not working	CSSD 01	Autoclave, Steriliser	0647962	15/11/2016	17/11/2016	CNF PRT TECO	2 - Urgent Work	2
4000088057	PM01	Theatre 1-4 Auto clave faulty	CSSD 01	Autoclave, Steriliser	0647962	14/12/2016	14/12/2016	CNF PRT TECO	2 - Urgent Work	0
4000118129	PM01	theatre 1-4 auto clave rack repair	CSSD 01	Autoclave, Steriliser	0647962	06/02/2017	26/02/2017	CNF PRT TECO	2 - Urgent Work	
4000120631	PM01	Theatre 1-4 Autoclave gasket loose	CSSD 01	Autoclave, Steriliser	0647962	10/02/2017	13/02/2017	CNF PRT TECO	2 - Urgent Work	3
4000131576	PM01	Autoclave Training and Installation of p	CSSD 01	Autoclave, Steriliser	0647962	21/02/2017	21/02/2017	CNF PRT TECO	1 - Planned Work	
4000135946	PM01	Repair pump on "Allie"	CSSD 01	Autoclave, Steriliser	0647962	28/02/2017	28/02/2017	CNF PRT TECO	2 - Urgent Work	0
4000137450	PM01	Investigate faulty pump that was replace	CSSD 01	Autoclave, Steriliser	0647962	02/03/2017	02/03/2017	CNF PRT TECO	2 - Urgent Work	
4000137532	PM01	Theatre 1-4 autoclave faulty	CSSD 01	Autoclave, Steriliser	0647962	02/03/2017	03/03/2017	CNF PRT TECO	2 - Urgent Work	1
4000143760	PM01	Theatre 1-4 PLease repair autoclave	CSSD 01	Autoclave, Steriliser	0647962	08/03/2017	08/03/2017	CNF PRT TECO	2 - Urgent Work	0
4000157617	PM01	Theatre 1-4 Autoclave faulty	CSSD 01	Autoclave, Steriliser	0647962	28/03/2017	28/03/2017	CNF PRT TECO	2 - Urgent Work	0
4000284096	PM01	Theatre 1-4 ccsd autoclave faulty	CSSD 01	Autoclave, Steriliser	0647962	18/07/2017	18/07/2017	CNF PRT TECO	2 - Urgent Work	0
				Total days		175			Sum	12
									MTBF	16

It is evident that various risks highlighted due to low MTBFs support responses from the questionnaire results. From the correlation between the MTBF results and qualitative evidence from the questionnaires, it is evident that CMMS as a reporting tool is effective when good quality information is recorded in the system. However, there is proof from past cases, where critical failures occurred without data correctly monitored or recorded. Therefore, although the failure was dealt with and a solution was created, from an analysis point of view, poor quality records are on recorded in the CMMS. This data quality issue is mainly related to human intervention which is required to correctly record the information. To eliminate human intervention with respect to reporting faults, a system with characteristics identified in SAM could be utilised to correctly report faults from assets where a true indication of the number of faults and a true reflection of MTBF can be recorded.

5. POLICY FRAMEWORK FOR SAM IMPLEMENTATION

From the data collection and analysis, it is evident that the utilisation of electronic monitoring devices, such as those identified within a SAM implementation plan is required, especially within critical areas such as the operating theatres. In response, a SAM policy solution is developed to assist technical managers within the private healthcare sector to implement a SAM plan to monitor critical assets for mitigating potential asset failure. To address the concerns identified during the analysis of the questionnaire and the analysis conducted during the verification process, where a breakdown explanation and abstraction is conducted, the proposed implementation of a policy framework for SAM is considered a viable solution to electronically monitor critical infrastructure assets within case organisation.

The basic premise of the SAM implementation is characterised from an information loop as presented by Holdowsky et al. [23]. Within this research, a cyclical value-adding model is presented where conversion is established within the loop from machine intervention to real-world decisions affecting overall value contribution. Considering the external actions of the loop namely; act, create, communicate, aggregate and analyse; these actions are affected by the basic technological building blocks, which can be referred to as technological enablers, which assist in the information transfer between actions. From the basic sensory collection of raw data, converted to a digital language within a network, this information is then sorted and stored on a central platform enforcing certain criteria and standards to be implemented. Once this organised information is accessible, it is applied to preconceived knowledge criteria which are used as a reference to obtain an augmented intelligence which ultimately leads to a value-contributing decision or to augment behaviour. The steps outlined within Figure 3 are closely related to the theory presented by Holdowsky. Utilising these established procedures presents a valuable tool for AM decision-making, implementing of the collection of asset information within an SAM strategic plan.

Building from this research and further research regarding the correct implementation of policy frameworks presented by Ostrom [24], a derived concept for a policy framework is created specifically for application within the case organisation. This policy framework is presented as an implementation philosophy of SAM where the basic flow of information as well as structure of the document is captured within the Figure 3.

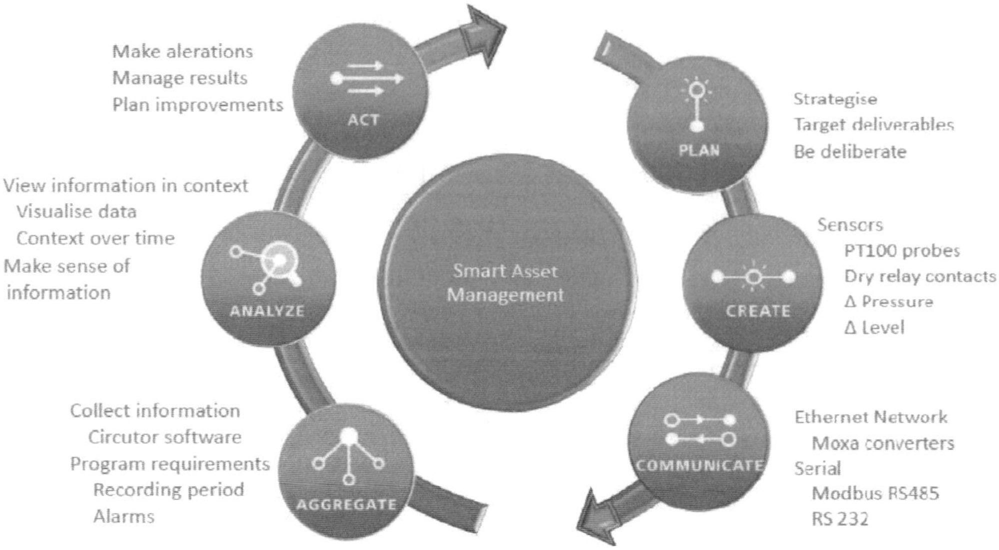

Figure 3. Proposed SAM Continuous Improvement Cycle

The premise of the policy framework resides in the amalgamation of existing fundamental MCSA standards and principles captured in established MCSA documents namely the *Maintenance Pre-amble* and *Technical Operations Management System incorporating Planned Maintenance* documents. The cyclical procedure presented in Figure 3 is further incorporated in the management of remote monitoring software applications classified within the SAM principles. Within the document fundamental principles are established outlining the intended purpose of the document, its intended implementation scope, original referenced documents as well as the responsibility of various key role players accountable for implementing the document. Key role players include the technical manager as well as the corporate technical operations managers who are required to assist with respect to implementation of current and newly established policies captured within the proposed document.

Further to execution of the document, the expected procedure for implementation briefly discusses the premise of SAM, the existing standards and quality maintenance standards and best practise applications applicable for background knowledge. The scope of the intended SAM implementation is also clearly established as means to record information to further utilise in strategic decision-making, where clear guidelines are set for specific assets critical in nature, which need to be targeted. Furthermore, the implementation principle of achieving an actionable response is also clearly established highlighting the progression of information as in Figure 4.

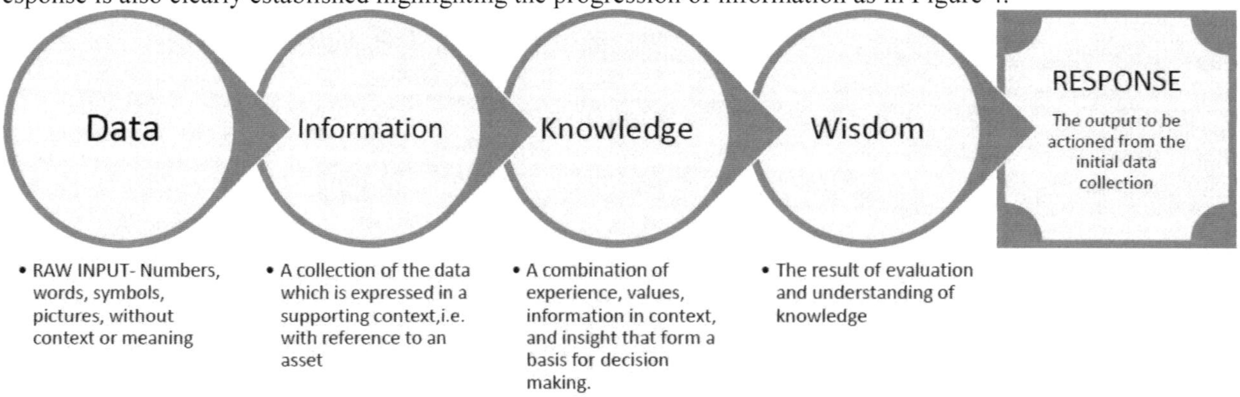

Figure 4. Progression of information flow

Building on Figure 3, each process within the cycle is further expanded, where applications within a hospital environment are established and discussed with examples for further practical implementation. Upon establishing the basis of implementation, further standards are established regarding the operation control within the case organisation, where concepts such; as training, awareness and competence, security of the system, operational checks and balances, critical equipment to be monitored, record keeping and the backup of information are aspects further deliberated highlighting the best practises for SAM implementation.

For validation of the policy framework a face validation process is followed with a heterogeneous sample representing corporate technical operation management of the case organisation who governs the implementation of standards through various policy frameworks. Their feedback was supportive toward implementation of such a policy framework, with acknowledgement that such a concept did not exist currently.

6. CONCLUSIONS

The paper proposes a policy framework for SAM implementation. It is evident from the results that current applications utilised within this specific case study are lacking with respect to monitoring critical assets as well as further mitigation of critical system failure. Therefore, the presented policy framework has potential to add value not only within the case organisation but also has potential value to industry to support the implementation of SAM. The limitations of the policy framework include the boundaries set by case study research investigation where a specific application within the private healthcare environment is considered. Furthermore, implementation of the document could not be established which leads to recommendations for future research within this field and other areas within industry.

REFERENCES

[1] WHO (2011) Medical equipment maintenance programme overview. WHO Medical device technical series, vol. 1, no. ISBN 978 92 4 150153 8, p. all.
[2] Nel, C.B.H. and Jooste, J.L. (2016). A technologically-driven asset management approach to managing physical assets - a literature review and research agenda for smart asset management. South African Journal of Industrial Engineering, vol. 27(4), pp. 50-65.
[3] Oxford (2016 May). Oxford dictionary
[4] ISO55000 (2014). Management, ISO 55000 international standard for assets. Online. Available at: http://www.assetmanagementstandards.com
[5] Schneider, J., Armin, J.G., Neumann, C., Hografer, J., Wellbow, W., Schwan, M. and Schnettler, A. (2006 March). Asset management techniques. Electrical Power and Energy Systems, vol. 28, pp. 643-654
[6] Campbell, J.D., Jardine, A.K.S. and McGlynn, J. (2011). ASSET MANAGEMENT EXCELLENCE: Optimizing Equipment Life-Cycle Decisions. Second edition ed. CRC Press Taylor and Francis Group. International Standard Book Number-13: 978-0-8493-0324-1.
[7] Hastings, N.A.J. (2009). Physical Asset Management. Springer Science & Business Media.
[8] Salonen, A. (2011 March). Strategic maintenance development in manufacturing. Mälardalen University Press Dissertations, vol. 99, no. 99. Printed by Mälardalen University, Västerås, Sweden.
[9] Berger, D. (2010 Feb). The growing value of a cmms. Available at: https://www.plantservices.com/articles/2010/02AssetManager/
[10] de Best, L. and van den Berg, F. (2006). Smart fields: Making the most of our assets. In: SPE Russian Oil and Gas Technical Conference and Exhibition, 3-6 October, Moscow, Russia. Society of Petroleum Engineers, SPE Russian Oil and Gas Technical Conference and Exhibition (3-6 October). Available at: https://www.onepetro.org/conference-paper/SPE-103575-RU
[11] Bouleau, C., Gutierrez, F., Gehin, H., Landgren, K. and Miller, G. (2007/2008). Integrated asset managment. Oil eld Review, pp. 34-48.
[12] Halgamuge, M., Liyanage, J.P. and Mendis, P. (2010). Wireless technologies for integrated e-operations in offshore environments. EJSE Special Issue: Wireless Sensor Networks and Practical Applications, vol. Special Issue, pp. 100-112.
[13] Dioguardi, R.M. and Smith, P. (2010). Use of r d asset management systems for monitoring analytical instrumentation. PerkinElmer Inc., vol. 22, no. 9, pp. 66-70.
[14] Glova, J., Sabol, T. and Vajda, V. (2014). Business models for the internet of things environment. Elsevier B.V., vol. 15, pp. 1122-1129.
[15] Liyanage, J.P. and Langeland, T. (2009). Smart assets through digital capabilities. IGI Global, pp. 3480 3485.
[16] Lopez, T.S., Ranasinghe, D.C., Patkai, B. and McFarlane, D. (2011). Taxonomy, technology and applications of smart objects. Inf Syst Front, vol. 13, p. 281 to 300.
[17] Aberdeen (2008 June) Enterprise asset management: Maximizing return on assets and emerging trends. Aberdeen group Inc.
[18] IAM, T. (2015). Asset management maturity scale and guidance. In: The Institute of Asset Management Annual Conference 2015, vol. 1
[19] Oliveira, A. (2007). A discussion of rational and psychological decision-making theories and models. Electronic Journal of Business Ethics Vaishnavi, V.K., Buchanan, G.C. and Jr., W.L.K. (1997). A data/knowledge paradigm for the modelling. IEEE

TRANSACTIONS ON KNOWLEDGE AND DATA ENGINEERING, vol. 9, no. 2, pp. 275-291.
[20] Gangyan, X., George, H.Q. and Fang, J. (2015). Cloud asset for urban flood control. Advanced Engineering Informatics, vol. 29, p. 355 to 365. and Organization Studies, vol. 12, no. 2, pp. 12 17.
[21] de Best, L. and van den Berg, F. (2006). Smart elds: Making the most of our assets. In: SPE Russian Oil and Gas Technical Conference and Exhibition, 3-6 October, Moscow, Russia. Society of Petroleum Engineers, SPE Russian Oil and Gas Technical Conference and Exhibition (3-6 October). Available at: https://www.onepetro.org/conference-paper/SPE-103575-RU
[22] Campbell, J.D., Jardine, A.K.S. and McGlynn, J. (2011). ASSET MANAGEMENT EXCELLENCE: Optimizing Equipment Life-Cycle Decisions. Second edition ed. CRC Press Taylor and Francis Group. International Standard Book Number-13:978-0-8493-0324-1.
[23] Holdowsky, J., Mahto, M., Raynor, M.E. and Cotteleer, M. (2015). Inside the internet of things (IoT). Deloitte University Press | DUPress.com.
[24] Ostrom, E. (2011). Background on the institutional analysis and development framework. The Policy Studies Journal, vol. Vol. 39, No. 1, pp. 11 - 27.

Identifying current challenges of data-based maintenance management: a case study

S. Marttonen-Arola[1], D. Baglee[1]
[1] Faculty of Engineering and Advanced Manufacturing, University of Sunderland, UK

ABSTRACT

Exabytes of data from various sources are available for maintenance decision makers. The yearly increase in data is exponential due to technological developments such as the rapid increase in the amount of interconnected systems and assets, which utilize smart sensors, cloud-based computing and eMaintenance. All of these are supported by the rapid developments in the Internet. The data provide vast possibilities for smart, autonomous assets and predictive maintenance. However, in practice, there are technical, managerial, and organizational challenges, which impede the maintenance decision makers from exploiting the information retrieved from the data analyses. The existing literature has discussed the data required in different maintenance decision making situations extensively, although there is a limited number of academic publications which explore general-level frameworks or tools to support the management of maintenance data. This paper builds upon a review of the current literature on the value of maintenance data management. The data needed to support a number of different maintenance management situations are discussed, and an approach to analyze and increase the value and resource efficiency of the maintenance data management process is suggested. The paper presents a case study example conducted in collaboration with a UK manufacturing industry. The objective of the paper is to map the current state of maintenance data exploitation paths. This makes the different value-based development needs in the data management process visible. The results of this paper will contribute to future empirical research including modelling and optimizing the use of data in maintenance decision making through adopting lean management principles. The majority of previous lean management research has focused on the optimal management of production processes and the maintenance processes. In this research, the principles of lean management will be taken to the level of optimizing the maintenance data management process.

Keywords: Maintenance data; Decision making; Case study; Value; Industrial asset management.

Corresponding author: Salla Marttonen-Arola (salla.marttonen-arola@sunderland.ac.uk)

1. INTRODUCTION

The recent technological developments regarding for instance eMaintenance, sensors, and the Internet of Things have exponentially increased the amount of data available to maintenance decision makers [1-2], providing vast possibilities for novel approaches in smart, predictive maintenance [3-4]. However, in practice many companies are struggling with the process of gathering, integrating, analyzing, and exploiting their maintenance data [5], and so far they have not been able to harvest the value of the data. There is no universally agreed definition for maintenance data. In this paper it is defined to include asset data (technical, operational, financial, environmental, and performance of the assets), data on the planned and implemented maintenance actions, and supportive data (on e.g. strategy, processes, risks, and contracts) [6-10]. The optimal amount of maintenance data to be gathered and analyzed depends on the size of the company, the type of business, the complexity of processes, the competence of people, and the complexity of assets [11].

The value of data is an ambiguous term, and systematical approaches to assessing or even defining it are still scarce both in industry and in academia. The objective of this paper is to introduce a value-based approach to studying maintenance data management processes. The current state of maintenance data exploitation paths are then mapped through a case study. The paper contributes to the understanding of the value of data in maintenance management, and helps to build the foundation for further research including value-based modelling of maintenance data and methods for lean maintenance data management.

In section two of the paper, the research design and methods are discussed. Section three addresses previous literature on maintenance data and its value. Section four presents the results and observations from the case study, and the paper finishes with conclusions in section five.

2. RESEARCH DESIGN

In this research a literature-based approach to evaluating the value of maintenance data is created and evaluated using a case study example. The selected case company operates in the food and drink industry in the UK. The company is interested in increasing the value of their maintenance through better exploitation of their maintenance data. However currently, the state of maintenance data management in the company is limited; they are operating with manual data collection, the maintenance engineers are still learning how and why they should document their work, and data is mostly used to monitor corrective maintenance tasks which constitute a majority of the company's maintenance. The case company studied which maintenance tasks take up the most of their maintenance engineers' time, and based on the results they wanted to analyze the data management processes of three specific maintenance tasks in more detail:

1) Changing a retort probe when it has been damaged,
2) Installing or removing a specific conveyor belt used when making cannelloni, and
3) Cleaning the printer head of videojets after the quality of the print has been deemed suboptimal.

To analyse the data management processes in the selected maintenance tasks, the case company's maintenance work requests from July 2017 until mid-January 2018 were studied. Two maintenance managers from the company were also interviewed to gain understanding on the current processes and practices in the company. The maintenance work requests from the studied period included 284 retort probe changes, 92 cannelloni belt installations/removals, and 117 videojet head cleanings. In total, these 493 work requests equaled 12% of all the maintenance work requests from the assembly department during the studied period, and 5.3% of all the maintenance work requests in the case company during the period. All three studied maintenance tasks are simple and quite straightforward to conduct: the time used in completing them equals only 5% of the time used in completing all the work requests from the assembly department, and 1.8% of the time used to complete all the work requests in the case company.

3. LITERATURE REVIEW

The recent technological developments have taken maintenance decision making towards real-time decisions. The time available for analyzing the data to support maintenance decisions is decreasing, and so is the time required for data collection [5]. Sun et al. [12] have recognized four different time scales in asset management decisions:

1) Strategic decisions ranging from one to five years (e.g. developing strategic asset management plans),
2) Technical decisions ranging from a month to a year (e.g. creating preventive maintenance plans for major assets),
3) Implementation decisions ranging from days to a month (e.g. scheduling maintenance actions), and
4) Reactive decisions ranging from minutes to a day (e.g. selecting maintenance actions in case of unexpected asset failure).

To form a more comprehensive view of the data needed to support maintenance decisions, the authors reviewed the literature on various decision-making situations in maintenance management. In figure 1 below, various decisions are presented according to the categorization of the four different time scales in decision making. In the empirical part of this paper, the focus is on reactive decisions related to designing and implementing corrective maintenance actions. The main elements of a corrective maintenance plan can be seen as detecting failures, and deciding on whether to repair or replace the asset, immediately or later. To optimize these decisions maintenance managers would need data related to e.g. asset condition and life cycle, costs, safety and production. [13-16]

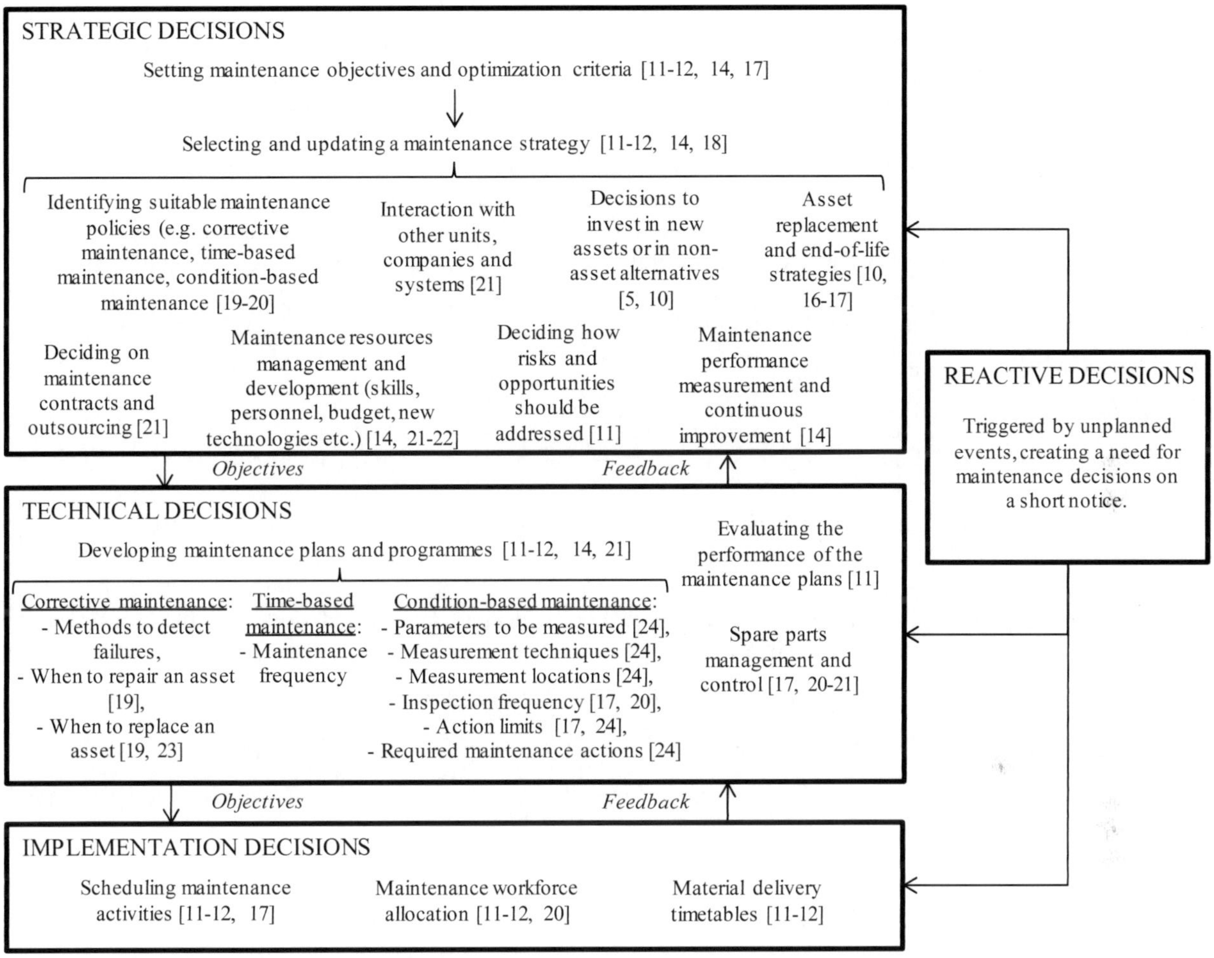

Figure 1. Classification of maintenance decision making situations.

When collecting and analyzing maintenance data it should be noted that, the objective is to create more value for the maintenance decision makers. The value of data or information still has not received much attention in scientific literature, and practical applications on the topic are mostly missing. The value of asset management data in fleet contexts has previously been discussed by e.g. [25], who consider data management to be an investment to be paid off by the benefits achieved through improved decision-making. The costs of data management can be categorized into hardware, software, and work hour based costs. The benefits, on the other hand, include saving in maintenance costs, savings in quality costs, as well as other savings and benefits. [25]

To define what constitutes value in this study, the view on value proposition in information services presented in [26] is adopted. According to this view, the value of information depends on not only the information items provided, but also on the amount, quality, format, time, place, and price or cost of the information as follows:

Value of information = Right information, in the right level of detail, in the right condition, at the right time, at the right place, and for an appropriate price [26].

The optimal extent of maintenance data to be gathered and analyzed to support decision making is highly case specific. Data collection and analysis must not be too complex or strenuous, however collecting data only on the level of e.g. large systems might invalidate the whole process if the data do not provide root causes for failures and thus do not enable improvements of maintenance plans [27]. Thus the optimal data depend on the maturity of the organization in question. Data can be collected from both internal and external sources. When considering which data to collect, [8] suggests all organizations to at least record the details of their corrective maintenance actions. The standards recommend coding data using mutually exclusive codes, and using additional free text to provide further details [8].

It is necessary to acknowledge the potential problems in data quality to evaluate and manage the impact on maintenance decisions. According to the categorization presented by [28], data uncertainty can be caused by:

1) Input data (e.g. material properties, sampling rates),
2) Measurements (e.g. sensor noise),
3) Operating environment (e.g. high variability in conditions), and
4) Modelling errors (e.g. lack of understanding about the process to be modelled).

In practice, there are often a large number of quality problems in maintenance data. According to the categorization of levels of data completeness in asset record databases presented by [29]:

- Approximately 10-20% of the records can be considered "perfect data" which do not require extrapolation or expert judgement before exploitation,
- 35-40% of the records are "imperfect data", requiring additional efforts to ensure the quality,
- 25% of the records are "verbal/inspection data", which require e.g. expert judgement for verification, and
- 10-15% of the records are "soft data" relying on human perceptions and/or memory, which require a lot of time to summarize and can be considered inconsistent quality-wise.

The concept of value of data offers an approach to analyzing data management processes and identifying their current value-destroying features. The case study in the next section presents an example of mapping the current maintenance data flows in an industrial context through analyzing the various dimensions of value of data.

4. CASE STUDY

4.1. The current maintenance process

Figure 2 depicts the current maintenance data management process in the case company. The corrective maintenance process is triggered by an asset failure, after which the asset operators use a specific manual form to file a maintenance work request. All the three studied maintenance tasks are production critical (the failures cause the production line to stop), so the maintenance engineer(s) react to the work requests as quickly as possible, executing the requested maintenance task and restoring the asset into normal operating condition. The engineers report any possible needs for follow-up maintenance work, and they use the manual forms to document the details of the conducted maintenance work. The forms are then manually transferred to the maintenance management team. The maintenance managers insert the data gathered with the manual work request forms into electronic spreadsheets once a week. Every week the maintenance managers use the spreadsheet to construct a report to describe the performance of the maintenance department to the company management. In addition to performance reporting, the maintenance management team regularly uses the spreadsheet data to identify bottlenecks and development needs in their maintenance processes.

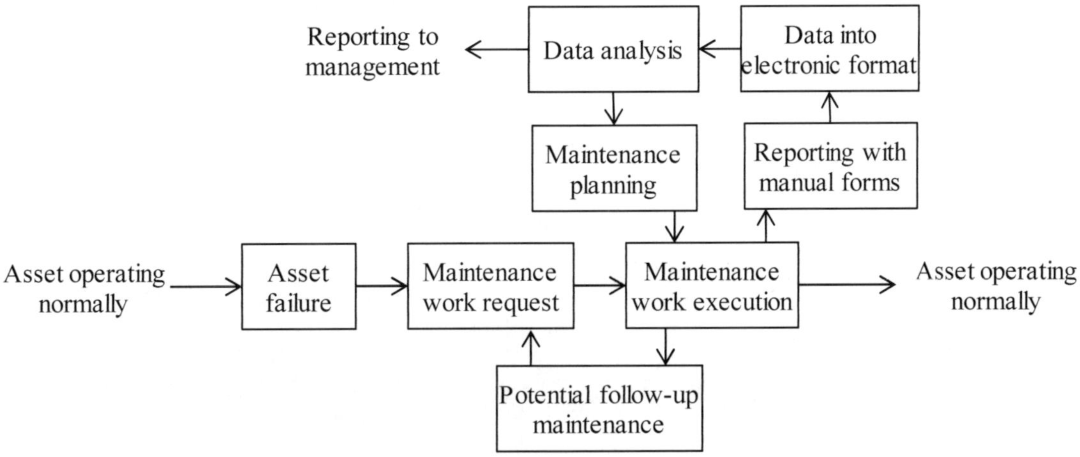

Figure 2. The current corrective maintenance data management process in the case company.

4.2. Identifying value-destroying challenges in the process

The six dimensions of value of information presented in section two have been used as the basis of the analysis addressed here. Table 1 below summarizes the identified value-destroying challenges in the current maintenance data management process of the case company. Eight challenges were identified, out of which three were deemed more significant (at the moment): poor quality data, manual data, and unnecessary data processing.

The poor quality of maintenance data is a significant challenge because it can void the entire data management process if maintenance decisions are based on incorrect data. In the future, the case company will contribute to this challenge by promoting instructions and awareness on how and why maintenance data is collected. The condition the data is processed in (documented in manual work request forms) also causes quality problems and makes data analysis slower. The case company has considered implementing technological solutions to support transferring the manual data into electronic format, or adopting a computer-based maintenance management system to make the whole data management process electronic and to increase the quality of data. However, so far the managers of the company have averted from investing in software and/or hardware before the basic process and motives of maintenance data management have been adopted and accepted by the personnel. This decision could be costly, as mentioned in table 1 the amount of time currently used in unnecessary data processing is significant and must be a cause for lots of wasted personnel costs. It would be important for the case company to conduct a quantitative investment appraisal on comparing the costs of the current process, including working hours, with the costs of implementing a computerized system.

Table 1. Identified value-destroying challenges in the current maintenance data management process of the case company.

DIMENSION OF VALUE	IDENTIFIED CHALLENGE	IMPACT ON MAINTENANCE
Right information	***1) Missing information:*** Details of the used spare parts are not documented consistently. For retort probe changes, only 11% of the work request forms documented the used spare parts at all, and most of them were missing details such as part numbers or specific location.	The costs of spare parts cannot be allocated properly, the performance of spare part management cannot be evaluated, and proactive spare part management cannot be supported by data analysis.
	2) Unnecessary information: The engineers use time documenting some information items which are constant for each of the three tasks (e.g. impact and nature of work).	The unnecessary data get carried through the process and cause both the maintenance engineers and maintenance managers to use additional time in processing it.
Right level of detail	***3) Poor quality data:*** 12% of the studied work requests had visible quality problems that made them unreliable regarding data analysis (e.g. several tasks bundled in the same work request, several work requests created on the same task, or maintenance start time later than the reported finish).	The maintenance engineers and managers use their time in inserting the poor quality data into the manual forms and spreadsheets. In addition, the incorrect data might lead to incorrect conclusions in maintenance decision making.
	4) Confusion over key terminology: There are no explicit instructions on how to document the data and what kind of terminology should be used, so the engineers use a number of different terms when writing about the same issues/assets/tasks.	The reliability of the data can be questioned because the maintenance management team has to interpret the data before inserting it into the spreadsheet. Data analysis takes more manual work.
Right condition	***5) Manual data:*** The manual forms are strenuous to fill and to interpret. The poor legibility of the forms causes additional work for the maintenance managers.	Sometimes the maintenance management team struggles to read the hand-written forms, which slows the data processing and again reduces the reliability of the data.
At the right time	***6) Slow process:*** The manual forms cause severe delays to the data management process. It can occasionally take weeks before the data of a maintenance task is inserted into the spreadsheet.	The data is not timely enough to support any kind of proactive maintenance decision making. Currently this is not a very significant problem since the data is mostly used to monitor corrective maintenance on a weekly or monthly basis, but in the future it should be taken into account.
At the right place	***7) Poor traceability:*** The traceability of the manual forms is not good and occasionally they get lost in the process.	The data management process is slower, and using data to support proactive decision making in maintenance is not possible.
For an appropriate price	***8) Unnecessary data processing:*** The arduous process causes additional work, for instance the maintenance managers examine the forms again when inserting the data into the spreadsheet.	If inserting the data from one form takes on average 5 minutes, there is 62 hours/year of unnecessary work related to the three studied maintenance tasks, and 1165 hours/year considering all work requests.

5. CONCLUSIONS

This paper has suggested an approach to analysing and increasing the value of the maintenance data management process. An industrial case study example has been presented to demonstrate how the current state and main challenges of the data management process can be mapped. The paper contributes to theory through showing how the data management process can be analysed and improved by adapting the concept of value of information, which has been presented in previous literature. Industry and academia agree on the importance of data in maintenance management, but in practice many organizations are struggling with developing their data management processes. The main managerial contribution of this paper is related to focusing on value-centred thinking, which enables achieving business feasibility in data management. It is critical to remember that data do not create value unless exploited in decision making. It is easy to forget this and implement various technological solutions to create more and more data without stopping to think whether it is actually needed.

Due to the limited extent of this paper, developing the data management process systematically to maximise the value, as well as measuring the value of information quantitatively were not included. These topics will be included in the later phases of the research project. The value-based analysis presented in this paper will be integrated into a method of adopting lean principles in the maintenance data management process to maximise the value while minimizing the waste. The majority of existing lean management research has focused on optimizing production processes, with a few exceptions of studying the maintenance process. In this research project the principles of lean management will be adapted to the data management process in maintenance. Further research will also include creating a value-based cost model for measuring and improving maintenance data management to better communicate the value of information to company managers.

ACKNOWLEDGEMENTS

This project has received funding from the European Union's Horizon 2020 research and innovation programme under the Marie Skłodowska-Curie grant agreement No 751622.

REFERENCES

[1] Baglee, D., Marttonen, S., Galar, D. (2015) The need for big data collection and analyses to support the development of an advanced maintenance strategy. *Proceedings of the 11th International Conference on Data Mining*. Las Vegas. July 27-30. pp. 3-9.

[2] Candell, O., Karim, R., Söderholm, P. (2009) eMaintenance – Information logistics for maintenance support. *Robotics and Computer-Integrated Manufacturing*. 25 (6). pp. 937–944.

[3] Bumblauskas, D., Gemmill, D., Igou, A., Anzengruber, J. (2017) Smart Maintenance Decision Support Systems (SMDSS) based on corporate big data analytics. *Expert Systems With Applications*. 90 (C). pp. 303–317.

[4] Crespo Márquez, A. (2007) *The maintenance management framework. Models and methods for complex systems maintenance.* Springer series in reliability engineering. ISBN 978-1-84628-820-3.

[5] Kinnunen, S.K., Marttonen-Arola, S., Ylä-Kujala, A., Kärri, T., Ahonen, T., Valkokari, P., Baglee, D. (2016) Decision making situations define data requirements in fleet asset management. *Proceedings of the 10th World Congress in Engineering Asset Management (WCEAM 2015)*. Lecture Notes in Mechanical Engineering (2195-4356). Springer. pp. 357-364. ISBN 978-3-319-27062-3.

[6] Marttonen-Arola, S., Baglee, D., Kinnunen, S.-K., Holgado, M. (2018) Introducing lean into maintenance data management: a decision making approach. *Submitted to the 13th World Congress on Engineering Asset Management (WCEAM)*. Stavanger, Norway. September 24-26.

[7] BS EN 13306 Std. (2010) *Maintenance. Maintenance terminology*. BSI Standards Ltd. ISBN 978-0-580-64184-8.

[8] BS EN ISO 14224 Std. (2006) *Petroleum, petrochemical and natural gas industries – Collection and exchange of reliability and maintenance data for equipment*. BSI Standards Ltd. ISBN 978-0-580-50138-8.

[9] Wang, L., Qian, Y., Li, Y., Liu, Y. (2017) Research on CBM information system architecture based on multi-dimensional operation and maintenance data. *IEEE International Conference on Prognostics and Health Management*. Allen, TX, USA. June 19-21. ISBN 978-1-5090-0382-2.

[10] BS ISO 55002 Std. (2014) *Asset management. Management systems – Guidelines for the application of ISO 55001*. BSI Standards Ltd. ISBN 978-0-580-86468-1.

[11] BS ISO 55001 Std. (2014) *Asset management. Management systems – Requirements*. BSI Standards Ltd. ISBN 978-0-580-75128-8.

[12] Sun, Y., Fidge, C., Ma, L. (2008) A generic split process model for asset management decision-making. *Proceedings of the 3rd World Congress on Engineering Asset Management and Intelligent Maintenance Systems*. Beijing, China.

[13] Ahonen, T. (2005) Eri tietolähteiden käyttö kunnossapidon tukena. In: Helle, A. (Ed.) *Kunnossapito ja prognostiikka. Prognos-vuosiseminaari 2005.* VTT Symposium 239. Otamedia Oy. Espoo. ISBN 951-38-6302-6.
[14] BS EN 16646 Std. (2014) *Maintenance – Maintenance within physical asset management.* BSI Standards Ltd. ISBN 978-0-580-83321-2.
[15] Karim, R., Westerberg, J., Galar, D., Kumar, U. (2016) Maintenance analytics – the new know in maintenance. *IFAC-PapersOnLine.* 49 (28). pp. 214–219.
[16] Shafiee, M., Animah, I. (2017) Life extension decision making of safety critical systems: an overview. *Journal of Loss Prevention in the Process Industries.* 47. pp. 174–188.
[17] Jardine, A.K.S., Montgomery, N. (2009) Optimal maintenance decisions for asset managers. *Industrial Engineer.* 41 (6). pp. 44–49.
[18] Dong, Y.-L., Gu, Y.-J., Yang, K. (2004) Research on the condition based maintenance decision of equipment in power plant. *Proceedings of the Third International Conference on Machine Learning and Cybernetics.* Shanghai. 26-29 August. IEEE. pp. 3468–3473. ISBN 0-7803-8403-2.
[19] Kareem, B., Jewo, A.O. (2015) Development of a model for failure prediction on critical equipment in the petrochemical industry. *Engineering Failure Analysis.* 56. pp. 338–347.
[20] Sharma, A., Yadava, G.S., Deshmukh, S.G. (2011) A literature review and future perspectives on maintenance optimization. *Journal of Quality in Maintenance Engineering.* 17 (1). pp. 5–25.
[21] Marttonen-Arola, S., Ali-Marttila, M., Ylä-Kujala, A., Saunila, M., Pekkola, S., Sinkkonen, T., Kärri, T., Pekkarinen, O., Rantala, T., Ukko, J. (2016) Towards comprehensive value management in inter-organizational industrial maintenance. *International Journal of Condition Monitoring and Diagnostic Engineering Management.* 19 (3). pp. 27–32.
[22] BS ISO 55000 Std. (2014) *Asset management. Overview, principles and terminology.* BSI Standards Ltd. ISBN 978-0-580-86467-4.
[23] Ahmad, R., Kamaruddin, S. (2011) Maintenance decision making method for repairable system by using output-based maintenance technique: a case study at pulp manufacturing industry. *IEEE.* pp. 15–20. ISBN 978-1-61284-486-2.
[24] BS ISO 17359 Std. (2011) Condition monitoring and diagnostics of machines – General guidelines. BSI Standards Ltd. ISBN 978-0-580-67132-6.
[25] Kinnunen, S.-K., Marttonen-Arola, S., Kärri, T. (2016) Value of fleet information in asset management. In: Galar, D., Seneviratne, D. (Eds.) (2016) *Proceedings of the 6th International Conference on Maintenance Performance Measurement and Management (MPMM 2016).* Luleå, Sweden. November 28. pp. 76–80. ISBN 978-91-7583-841-0.
[26] Bucherer, E., Uckelmann, D. (2011) Business models for the Internet of Things. In: Uckelmann, D., Harrison, M., Michahelles, F. (Eds.) *Architecting the Internet of Things.* Springer. 352 p. e-ISBN 978-3-642-19157-2.
[27] Kunttu, S., Kortelainen, H. (2004) Supporting maintenance decisions with expert and event data. *Proceedings of the annual Reliability and Maintainability Symposium.* Los Angeles, CA, USA. January 26-29. IEEE. pp. 593–599.
[28] Javed, K., Gouriveau, R., Zerhouni, N. (2017) State of the art and taxonomy of prognostics approaches, trends of prognostics applications and open issues towards maturity at different technology readiness levels. *Mechanical Systems and Signal Processing.* 94. pp. 214–236.
[29] Hale, P.S. Jr., Arno, R.G. (2009) Operational and maintenance data collection for determining site reliability or availability. *IEEE Industrial Applications Magazine.* Sept/Oct. pp. 21–24.

Investigations on Augmented Reality based maintenance practices within SMEs

M. Müller[1,2], D. Stegelmeyer[2], R. Mishra[1]
[1] School of Computing and Engineering, University of Huddersfield, United Kingdom
[2] Institut für Interdisziplinäre Technik, Frankfurt University of Applied Sciences, Germany

ABSTRACT

Maintenance services provided by Original Equipment Manufacturers (OEMs) play a fundamental role in industrial asset management. The vast majority of OEMs provide maintenance services to their customers, as most customers now outsource complex maintenance tasks. For many OEMs the economic value of operations has even shifted from manufacturing traditional products, to providing services required to operate and maintain their installed base. In general downtime is expensive and maintenance activities are often conducted under time pressure. However, maintenance processes are knowledge intensive and field service technicians have to cope with increasingly complex and diverse systems. To stand out from competitors and to satisfy customer needs, productivity improvement in maintenance processes is urgently required. Augmented Reality (AR) technologies, e.g. head mounted displays (HMD) can assist field service technicians in improving efficiency of maintenance operations. Systems for mobile collaborative AR supported maintenance are commercially available, and the benefits of AR supported maintenance work are generally evident. Still, on a wider scale in the industrial maintenance service delivery, the technology is far from established. From the perspective of small and medium-sized enterprises (SME), with a limited number of field service technicians, remote support for customers and service partners might be an opportunity to enhance resources availability. This study investigates ten business cases of mobile collaborative AR systems using HMDs for services delivery. The research strategy follows an exploratory approach. Based on focus group discussions with representatives of SMEs and semi-structured interviews with industry experts, enablers and barriers for the use of mobile collaborative AR systems are determined.

Keywords: augmented reality, head mounted displays, industrial asset management, maintenance, servitization

Corresponding author: Maike Müller (maike.mueller@hud.ac.uk)

1. INTRODUCTION

In recent decades, traditional manufacturing companies of different kinds of capital goods have moved up the value chain and become service or solution based businesses in order to distinguish themselves from their competitors, better meet customer's needs, and thus survive in increasingly competitive product markets [1–5]. For many OEMs, regardless of their position in the supply chain, the economic value has shifted from manufacturing traditional products, to providing services required to operate and maintain their installed base [3][6]. The vast majority of capital goods manufacturers provides at least basic services, such as spare parts supply or repair and maintenance-on-demand [6]. In general downtime is expensive and maintenance activities are often conducted under time pressure. Besides, customers often expect availability commitments for specialised service technicians for recurring maintenance activities as well as in cases of unscheduled downtime. Maintenance processes, however, are knowledge intensive and maintenance personnel have to cope with increasingly complex and diverse systems [7]. To stand out from competitors and satisfy customer needs, productivity improvement in maintenance processes is required. With respect to their limited resources and difficulties to implement service-oriented strategies, SMEs are challenged in particular [8][9]. Nevertheless, SMEs are an important part of most economies. In Germany, for instance, 85 % of machinery and equipment industries are SMEs. They account for nearly 30 % of the total turnover generated and 37 % of the people employed by the sector as a whole [10].

One concept, to address aforementioned challenges and increasingly complex maintenance operations, is the real-time collaboration between experts and on-site field service technicians via mobile collaborative AR systems using HMDs [11][12]. Generally, remote expert assistance produces quicker and more accurate troubleshooting and repairs [13] and is one of the most popular application areas for HMDs [14]. For SMEs, considering their limited resources, mobile collaborative AR might be an opportunity to increase customer satisfaction, withstand competitive pressure, and broaden their sales network [15]. But even though AR and HMDs have been a research interest for more than 50 years [16], there are very few concrete examples of implementation in industry and if implemented, most applications are on a field test level in large enterprises (LE) [17][18]. In the past, researchers developed prototypical systems for mobile collaborative AR and tested those systems in experiments or case studies [11][12][15][19][20]. Those systems usually include a wearable computing system (e.g. HMD) worn by a technician and a user interface to enable collaborative communication between remote experts and technicians in the field. The main function of the HMD is to capture a live audio and video stream [11]. The user interface displays the live stream captured by the HMD's camera and features communication and documentation functionality, e.g. video recording, screenshots, document sharing, chat, language translation and laser pointing into the live stream. HMDs used can be either see-through HMDs or video display HMDs (non-see-through). See-through HMDs are based on waveguides, semi-transparent monocular or binocular mirrors that reflect computer generated images to augment the real world (e.g. Epson BT-350, Google Glass, Microsoft HoloLens). Non-see-through HMDs capture the real world, overlay the computer-generated information and display the composited image in a small monitor placed in front of one of the user's eyes (e.g. RealWear HMT-1, Vuzix M300) [18]. Mobile collaborative AR systems, both hardware and user interfaces (e.g. web-based communication platforms), are commercially available. However, on a broad scale in industrial maintenance service delivery, the technology is far from established. Only 3.4 % of companies currently use HMDs in the manufacturing context [14]. Previous work on mobile collaborative AR has mostly been focused on the development of systems and aspects of usability of features [21][22], whereas now that the technology seems to be capable of handling complex tasks, our focus is on service business cases enabled through the technology.

In particular, we would like to explore barriers and enablers for the use of mobile collaborative AR systems supporting services provided by small and medium-sized machinery manufacturers (OEMs) on their installed base. Even though we investigate OEMs that offer divers products, the service delivery is characterised by some similarities: Those similarities are: a multi-variant technically complex installed base, customers that typically outsource complex maintenance tasks and often expect availability commitments. Especially, availability commitments for specialised service technicians are a major concern for SMEs. In this paper we solely take the perspective of the OEM providing paid services for their installed base. Therefore, the term maintenance describes service offerings, the OEM charges for. We do not take the perspective of the customer and/or their maintenance function.

2. METHODOLOGY AND DATA COLLECTION

Given the comparative lack of existing literature on the business perspective of mobile collaborative AR, especially on barriers and enablers, we aim to explore business cases, barriers and enablers for the use of the technology in service delivery. In order to address the aforementioned research aim, a qualitative research approach with a strong exploratory notion has been used. This approach was selected on conceptual grounds rather than for its representativeness [23].

Data was collected in two focus group discussions and individual semi-structured interviews with 20 representatives from 17 German companies (table 1). The companies used in our sample were classified following the definition of the Institut für Mittelstandsforschung [24]. According to this definition SMEs are the companies that employ less than 500 people. The sample consists of 14 machinery manufacturers, referred to as OEMs, of which ten are SMEs and four are LEs. Focus group participants and interviewees all held positions of strategic responsibility for the service business to ensure validity of results. In Addition to the OEM perspective, representatives of two HMD manufacturers and a system integrator, referred to as industry experts, are also part of the sample to broaden the perspective and provided additional insights. The industry experts provide insight from their experience of many AR implementation projects in capital goods industries. Additionally, 24 use cases of AR supported service delivery found in non-academic publications on the internet were analysed and compared to the findings derived from the empirical data (table 2). This aimed to fill the above mentioned lack of academic literature.

Table 1: Sample of the interview study

Position of Interviewees	Company / Industry	Company Size	Company Type	Interview Method
General Manager 1 General Manager 2	Air Cleaning Plant Technology	SME	OEM	Focus Group and Individual Interview
Head of Service Sales & Repair	Coating Equipment	LE	OEM	Focus Group
Manager After Sales 1 Manager After Sales 2	Control and Regulation Systems	LE	OEM	Focus Group
Head of Customer Support	Food Processing Plant Engineering	LE	OEM	Focus Group
CEO	HMD Hardware/Software System Integrator	SME	HMD provider / industry expert	Focus Group and Individual Interview
Business Development Manager	HMD Manufacturer	SME	HMD provider / industry expert	Focus Group Individual Interview
Business Development Director	HMD Manufacturer & Software Provider	SME	HMD provider / industry expert	Focus Group and Individual Interview
Innovation Manager	Machine Tools	LE	OEM	Focus Group
CEO	Machine Tools	SME	OEM	Focus Group
Head of Customer Support Service Engineer	Machine Tools	SME	OEM	Focus Group
Quality and Process Management Engineer	Machine Tools	SME	OEM	Focus Group
Service Manager	Machine Tools	SME	OEM	Focus Group
Director of Service and Quality Management	Machine Tools	SME	OEM	Focus Group
Head of Mechanical Engineering	Plastics Processing Machinery	SME	OEM	Focus Group
Head of Customer Support	Power Plant Engineering	SME	OEM	Focus Group
CEO	Steam Generator Technology	SME	OEM	Individual Interview
Sales Director	Wastewater Technology	SME	OEM	Focus Group

The focus group discussions took place in December 2017 and January 2018. We selected companies that intended to use HMDs for their service delivery in the near future and do not compete in the same markets. The absence of completion was important to ensure that participants share their views openly. However, the experience of focus group participants with the technology varied. Some companies tested HMDs in the past, while other participants had no experience with HMDs. This limited experience was addressed within a half-day workshop on features of the technology prior to the discussion. In addition to the focus group discussions, five semi-structured interviews with industry experts and companies that already tested HMDs within their service operation were conducted. These interviews aimed to gain a deeper understanding of business cases, enablers and barriers previously discussed in the focus groups (table 1). The interviews were conducted between February and March 2018 via telephone and lasted between 60 and 120 minutes. Data collected from focus groups and interviews were recorded and transcribed for analysis. The empirical data was analysed by applying thematic coding according to Miles et al. (2014) using NVivo 11 [25].

Table 2: Analysed internet use cases of AR systems

No.	Use Case	Main features of AR system	No.	Use Case	Main features of AR system
# 1	Bosch Automotive	Unidirectional AR assembly instruction, contextual data provision, access to knowledge data base	# 13	MA micro automation	MCAR-system
# 2	Bosch Rexroth	MCAR-system	# 14	MAN Truck & Bus	Unidirectional AR aassemly/repair instructions, 3D-tutorial training, contextual data provision
# 3	Caterpillar	MCAR-system	# 15	Mekorot	Unidirectional AR assembly instruction, MCAR-system
# 4	Daikin Netherlands	MCAR-system, access to knowledge data base and error codes	# 16	NORDEN	MCAR-system
# 5	GE Aviation	AR aassemly/repair instructions with feedback on performance of task, video tutorials, MCAR-system	# 17	Porsche	MCAR-system
# 6	Gebhardt Fördertechnik	MCAR-system	# 18	Schmidt & Heinzmann	MCAR-system, overlay of process data
# 7	Getinge	Unidirectional AR training	# 19	Schulz Systemtechnik	Unidirectional AR assembly instruction, overlay of process data
# 8	Koenig & Bauer	MCAR-system, access to knowledge data base (projected)	# 20	Systronik	AR-based measure monitoring
# 9	Kieback & Peter	MCAR-system	# 21	TGW Lifetime	MCAR-system
# 10	Kurtz	MCAR-system	# 22	Thyssenkrupp	Unidirectional AR projection of 3D-models of products (pre-sales), MCAR-system
# 11	Lee Company	MCAR-system	# 23	VMT - Tunnelbau	MCAR-system
# 12	Leybold	Unidirectional AR assembly instruction, contextual data provision	# 24	Wärtsilä	MCAR-system, overlay of process data

MCAR-system = mobile callaboration AR-system

3. RESULTS AND DISCUSSION

The following section presents the results of our investigation and compares the findings of present work to earlier works. First, ten business cases for the technology are presented. Thereafter, enablers and barriers for the use of the technology, which we derived from the two focus group discussions, the five interviews and the supporting 24 use cases, are discussed.

Our research identified ten business cases listed in table 3. Besides supporting their own service technicians with mobile collaboration AR systems, SMEs also intend to equip service partners, sub-contractors and customer's personnel. Inspection, diagnosis and repair, assembly/installation, and training are the main service operations companies propose to support with the use of HMDs. Our findings correspond with Palmarini et al. (2018) who analysed 30 academic articles on the state of the art of AR in maintenance and found that dis/assembly (33%), repair (26%), inspection/diagnosis (26%) and training (15%) were the maintenance operations described in those articles [18]. In addition, we learned that application support to improve customer production processes was identified as a valuable business case for mobile collaborative AR systems. Application issues usually occur sporadically. Thus, the remote approach enables the application engineer to support customers in the moment of occurrence.

In correspondence with Quint et al. (2017), the use cases analysed (table 2) are confirmed to be demonstrators or test level systems of predominately LEs [17]. Therefore, those use cases provided little information on organizational barriers of the implementation of mobile collaborative AR systems from the perspective of SMEs. Nonetheless, the use cases analysed did confirm the interest of companies in mobile collaborative AR systems as 18 out of 24 feature this functionality. Other functionalities found in the use cases were unidirectional AR application supporting step-by-step instructions and training scenarios that aim to reduce errors and minimize training effort.

Table 3: Business cases and value of mobile collaborative AR systems for service delivery

Business Case	No.	Description	Value for SMEs	Discussed in focus group
Inspection, diagnosis and repair	# 1	OEM's service technicians are equipped with HMDs and are remotely supported by a specialist.	Time saving for diagnosis/repairs/identification of needed spare parts, spontaneous involvement of further specialists, new technicians are faster able to work independently, automatic video/screenshot documentation in session protocols	both
	# 2	Service technicians of service partners and sub-contractors are equipped with HMD and are remotely supported by an OEM's specialist.	Reduction of traveling time and expenses, increase of partners service quality though remote OEM back up, consistent service portfolio also for distant customers	both
	# 3	Customer's maintenance personnel is equipped with HMD and is remotely supported by a OME's specialist.	Reduction of traveling time and expenses, avoidance of unnecessary technician deployment, briefer respond time to service inquiries (increase of service quality through ad hoc support if no spare parts are needed), precise spare part identification	both
Training	# 4	Customer's operators are equipped with HMDs and are remotely instructed by a OEM's trainer, e.g. in case of frequent fluctuation.	Resource saving for operator on-site trainings, better training results compared to showroom trainings (actions have been really executed by operator and not only simulated by trainer, session video can be recalled by customer operators)	both
	# 5	Service technicians of service partners and sub-contractors are equipped with HMDs and are remotely instructed by a OEM's trainer.	Increase of service quality (better skilled technicians), reduction of training cost could lead to more frequent trainings, increase of customer loyalty through international quality service, less "firefighting" through headquarter	one
	# 6	Service personnel of subsidiaries are equipped with HMDs and are remotely instructed by a OEM's trainer.	Time and cost savings for trainer travels, more frequent training through decreased time and costs, increase of international service quality, less "firefighting" through headquarter	both
System installation	# 7	OEM's service technician is equipped with HMD and is supported by OEM's or component supplier's specialist, e.g. software engineers or design engineers.	Relaxation of human resources situation (saving specialist travels), spontaneous involvement of further specialists, installation condition documented immediately (as is versus as engineered)	one
System acceptance tests	# 8	OEM's service technicians is equipped with HMD and remotely supported in the process of system acceptance.	Saving specialist travels, spontaneous involvement of further specialists, video documentation of performance approval according to sales contract	one
	# 9	Pre-delivery acceptance test of sub-supplier's components. OEMs technicians is equipped with the HMD and components are remotely accepted by the sub-supplier.	Faster delivery of system to end customer, decrease of cost for acceptance tests (sub-suppliers traveling expenses)	one
Application support	# 10	OEMs remote application specialist collaboratively optimises customer's production process, while customer's operator is equipped with HMD and executes instructions on-site.	Specialist support in the moment of occurrence (application issues often occur sporadic), saving specialist travels	one

3.1. Enablers and Opportunities

Improvement of OEM service delivery (Business cases #1, #4, #7, #8, #10, table 3)

Companies expect that hands-free collaboration and "seeing through the users' eye" will lead to productivity improvement in service delivery. If the problem can't be solved immediately, at least errors occurred and spare parts needed can be identified more quickly and more precisely. The mobile collaborative AR approach is superior compared to today's hotline support in combination with emailed photos or videos. The findings can be compared to Fussell et al. (2003), who found that technology enabled shared visual spaces are superior compared to audio only collaboration [26]. The latest generation of HMDs seem to be mature enough to provide those shared visual spaces. Equipping technicians with HMDs also enables companies to deploy them to customers with less training, as they are covered by remote specialists. *„With the glasses [mobile collaborative AR system] new field service technicians don't have to be deployed alongside an experienced technician as a ‚second man' for such a long time to become familiar with our diverse systems."* Predictably initial training is one of the most popular application areas for HMDs [14].

Reduction of service costs for OEM (Business cases #1, #2, #7, #8, #9)
Cost reduction, unsurprisingly, is an enabler of mobile collaborative AR. Cost reduction primarily derives from the reduction of traveling time and expenses for deploying field service technicians, trainers, and specialists. In SMEs the company's specialised know-how is often concentrated in few people. A general manager of an SME stated: *"[...] we can multiply our experience without going on an airplane. This is really valuable to us."* Mourtzis et al. (2017) presents a case of a robotics SME that could lower costs from 1.370 € to 150 € for a battery replacement, a problem which can also be solved by instructing the customer's workers remotely, instead of deploying an OEM's technician to the customer's production site [15]. Especially during the warranty period of products, field service technician deployment is aimed to be avoided, as cost can't be charged. If the first error detection process can be conducted remotely, the value obtained is twofold. First, unnecessary technician deployment can be diminished, which reduces costs. A statement of a focus group participant illustrates the problem: "*[...] It happens that someone needs to go to Poland [from Germany] and then he takes two wires, reconnects them and drives back home. For this purpose the glasses are perfect. [...]*". Second, service enquiries can be responded to faster, which may avoid contractual penalties. *"Particularly the immediate [response is valuable to us]. I really can't tell the customer, 'it is a warranty case, so I will have time for you only next week'."*

Ability to respond to service enquiry (Business cases #2, #4, #10)
Specialised field service technicians are hardly available on labour markets. Equipping customers with HMDs enables ad hoc problem solving and minimises the effects of the human resource shortage. Deploying technicians to customer's sites for minor problem solving leads to serious problems for SMEs. With their limited number of field service technicians, the use of mobile collaborative AR systems can ensure the ability to respond to customer errors in case of simultaneous service enquiries. As one manager puts it *"We are a relatively small company and we have a major problem with capacity. We can't recruit 200 percent technicians, because there is no one available. [...] For me it is not so much about money, but about being able to respond at all."*

International consistent quality of service (Business cases #2, #3, #4, #5, #6, #10)
During the discussion the participants stressed that generally, domestic service is of higher quality than service to distant customers. The use of mobile collaborative AR systems also enables SMEs to provide their customers with an international service network. Equipping subcontractors, service partners and customers with HMDs supports quality standards. *"On the one hand, I train the employees [of the service partner], on the other hand we enable them to provide services to the customers which they can't provide today. Because they are just not capable to provide these services. Besides, the training can be performed while he [the service partner's technician] is having the problem at the end customer's site."* The qualification of service partners is important because SMEs are generally more reliant on a broader partner network to perform their field services [27] than LEs with subsidiaries in major markets. Above that, the reduction of training costs is expected to lead to more frequent training of subsidiaries, service partners and customer's operators. This increases service and product quality through improved operation of machinery and superior service of international service networks.

Increase of customer satisfaction (Business cases #2, #3, #4, #5, #6, #10)
Training, application service, error diagnosis and the identification of spare parts are major value propositions for OEMs to their customers. More efficient and effective management of the technicians leads to an increase of customer satisfaction. With the help of mobile collaborative AR systems, previously neglected distant customers may enjoy similar service levels as domestic customers. *"[...] When customers buy new systems, they always ask how we can support them. And if I have to say, 'of course you can call the hotline support in Germany [...]', the Vietnamese certainly says 'nice offer, but you are not available during my operating hours'. Therefore, we certainly benefit form qualifying service partners. Because it increases customer loyalty, if our systems can be serviced properly."* Thus, customer satisfaction will increase.

Remarkably, we found only one OEM (use case #2, table 2) promoting lower total cost of ownership (TCO) to the customer. Neither in the focus group discussions, nor in the interviews did we observed this enabler.

3.2. Barriers

Connectivity (Business cases #1, #2, #7, #8, table 3)
Maintenance service usually takes place in the field or at the customer's production site. However, customers are often reluctant to open their Wi-Fi networks for external devices. Moreover, production sites may be situated in rural regions with poor bandwidth of wireless data connectivity. *"A problem that we definitely do have with the glasses is mobile connectivity, since our machinery is often located in some factory in the back of beyond. [...] Mobile reception is already bad outside at the car park and when you enter the building, then it becomes even worse."* Bottecchina et al. (2010) found that under experimental conditions experts preferred video resolution of at least 640x480 [12], which might not always be possible. Thus, connectivity is a major barrier to mobile collaborative AR field service.

Privacy protection (All ten business cases)
A major advantage of mobile collaborative AR systems is the possibility of storing screenshots and videos of maintenance or training sessions. However, the documentation function of the system is a barrier with respect to privacy concerns. Data security representatives and trade unions often insist that consent of all recorded participants is required. Although privacy protection is more likely to be an issue of LEs, rather than for SMEs, privacy concerns can also be a barrier from the personal perspective of technicians. The feeling of being monitored could become an issue.

Protection of intellectual property (Business cases #1, #2, #5, #6, #7, #8, #9)
Many companies within our sample serve customers that are rigorous in protecting their intellectual property. Interviewees assume, that technicians wearing HMDs might be banned from working at customer's production sites. As a participant puts it: *"We are forced to hand over our smartphone when we enter [...] and now we should convince them to walk in wearing glasses that feature a camera system with a video store function."*. However, it was also stressed that when it comes to machine breakdowns, customers usually leave all concern behind since unavailability of production facilities is a much larger concern. *"Even the automotive industry gets this idea, [...] either your production is down for four hours or a day until my technician is there, or there is someone wearing glasses and your production continues [...]."*

Resistance of technicians (All ten business cases)
As with every new technology getting people to use HMDs might be a challenge. The reason for the resistance can be diverse, ranging from fashionable appearance to privacy concerns [28]. Besides, even if the benefits of the system are apparent from the business perspective, the personal perspective of technicians might be different. *"[...] experienced technicians often diagnose problems in a hands-on process. Some just don't see the advantages [of the technology] and therefore are reluctant. These older service guys are used to their tool kit and some even have problems in using a smartphone."*

Loss of service orders (Business case #3, #4, #5, #9, #10)
Deploying service technicians is an important source of revenue for many companies [29]. Equipping customers with HMDs poses the risk of losing field service orders, by training customer's personnel with every remote maintenance session. Mourtzis et al. (2017) outline a maintenance case of a robotics SME that illustrates this risk. In their case, a battery replacement is described to be crucial for securing the robots function, and usually requires the OEM to deploy a technician to the customer's site, although the task only requires simple assembly actions and could easily be explained remotely [15]. With this procedure the OEM qualifies the personnel of the customer. This qualification could lead to decreased service revenues, when customers execute simple tasks on their own. Obviously, this barrier is contrary to the aforementioned *ensure ability to respond* enabler.

Business model innovation (All ten business cases)
With respect to innovation enabled by the industrial internet of things, SMEs are typically challenged by their lack of resources, management know-how, selection of suitable solutions and the lack of transparency of the value of technologies [30]. Against those findings, the "transparency of value" does not seem to be a barrier concerning the use of mobile collaborative AR systems. All companies within our sample formulated the value of the technology

quite clearly. Consequently, a consensus was found in one focus group discussion that companies must be prepared with a smart service business model for the after-warranty period to monetize the improvements in service quality. However, as Kowalkowski et al. (2013) indicate, SMEs have more difficulties than larger companies to charge for service at all, as their revenue models are often based on product unit sales and large customers expect services for free [9]. Thus, using mobile collaborative AR systems might be an opportunity to innovate the business model and apply a more service-orientated strategy. Managers do not seem to have worked out such business models.

4. CONCLUSION

With the description of concrete business cases, enablers and barriers for AR supported service delivery, we contribute to the scarce research of servitization from the perspective of SMEs.

Our investigation indicates that use of mobile collaborative AR systems is an opportunity for SMEs to improve their service quality and extend their service businesses. New business opportunities arise, when providing mobile collaborative AR systems to service partners, sub-contractors and customers. The reduction of time that adds value (e.g. traveling time, lengthy error diagnostics, unnecessary technician deployment) and frees capacity to perform additional service orders subsequently might increase service revenues. The technology also facilitates international consistent quality of service to distant customers, which is a major challenge for SMEs. For SMEs who are generally more dependent on external service partners to provide customer support, mobile collaborative AR systems are an opportunity to expand their service business. Notably, no interviewee took the customers' perspective and considered the decreased total cost of ownership. Decreasing total cost of ownership might lead to higher sales of new equipment.

However, we were able to identify barriers to be addressed. Connectivity problems might be removed in the near future with the introduction of 5G networks. Barriers regarding protection of individual's privacy and company's intellectual property might be solved with further technology developments and/or contractual negotiation. Reluctance of human beings might disappear once the technology becomes more widespread. As of now, the loss of service orders and with it the lack of new business models to monetize the innovation is an unsolved issue.

Unlike the study by BMWi (2015) [30] suggests, we could not confirm the lack of transparency of the value barrier for mobile collaborative AR systems.

5. LIMITATIONS AND FURTHER RESEARCH

The research strategy and data collection method are not without limitations. Especially external validity and transferability are issues when using a qualitative research approach. All results and conclusions are derived from information that relies on views from a limited number of individuals, stemming from a limited number of companies. The companies and interviewees were selected to ensure the validity of this study. However, to describe the investigated phenomena more precisely and confirm validity of the results, a larger number of companies should be included in our study. As more and more companies are implementing mobile collaborative AR systems, the outcome of this investigation should be tested in a quantitative large sample study. Furthermore, the economic value of the ten identified business cases should be assessed. A better understanding of barriers and enablers of mobile collaborative AR systems in field service is indispensable for developing robust business models for SMEs. From the perspective of servitization research it would be interesting to investigate whether or not AR is an enabler for SME servitization.

REFERENCES

[1] VanDerMerwe S, Rada JF (1988) Servitization of business: Adding value by adding services. *European Management Journal* 6(4).
[2] Martin CR, Horne DA (1992) Restructuring towards a Service Orientation: The Strategic Challenges. *International Journal of Service Industry Management* 3(1):25–38.
[3] Wise R, Baumgartner P (September/Oktober Issue 1999) Go downstream: The new profit imperative in manufacturing. *Harvard Business Review* 77(5):133–41.
[4] Mathieu V (2001) Product services: From a service supporting the product to a service supporting the client. *Journal of Business & Industrial Marketing* 16(1):39–61.
[5] Davies A, Brady T, Hobday M (2006) Charting a path toward integrated solutions. *MIT Sloan Management Review* 47(3).

[6] Lay G, Copani G, Jäger A, Biege S (2010) The relevance of service in European manufacturing industries. *Journal of Service Management* 21(5):715–26.

[7] Aromaa S, Väätänen A, Aaltonen I, Heimonen T (2015) A model for gathering and sharing knowledge in maintenance work. *Proceedings of the 33rd Annual Conference of the European Association of Cognitive Ergonomics*, pp. 1–8.

[8] Storey DJ, Greene FJ (2010) *Small business and entrepreneurship*. Financial Times Prentice Hall, Harlow.

[9] Kowalkowski C, Witell L, Gustafsson A (2013) Any way goes: Identifying value constellations for service infusion in SMEs. *Industrial Marketing Management* 42(1):18–30.

[10]The Federal Statistic Office. *Destatis Datenbank. Beschäftigte und Umsatz der Betriebe im verarbeitenden Gewerbe: Deutschland, Jahre, Beschäftigtengrößenklassen (Wirtschaftszweige WZ2008 2-4-Steller Hierarchie) [Employees and Sales of German manufacturing companies of industry sector WZ2008]*.

[11]Alem L, Tecchia F, Huang W (2011) Remote Tele-assistance System for Maintenance Operators in Mines. *11th Underground Coal Operators' Conference, University of Wollongong & the Australasian Institute of Mining and Metallurgy*:171–7.

[12]Bottecchia S, Cieutat J-M, Jessel J-P (2010) T.A.C: Augmented Reality System for Collaborative Tele-Assistance in the Field of Maintenance through Internet. *Proceedings of the 1st Augmented Human International Conference*, Article No. 14.

[13]Kraut RE, Miller MD, Siegel J (1996) Collaboration in Performance of Physical Tasks: Effects on Outcomes and Communication. *Proceedings of the 1996 ACM conference on Computer supported cooperative work*:57–66.

[14]Plutz M, Große Böckmann M, Siebenkotten P, Schmitt R (2016) *Smart Glasses in der Produktion [smart glasses in manufacturing]: Studienbericht des Fraunhofer-Instituts für Produktionstechnologie IPT [Survey report of the Fraunhofer Institute for Production Technology]*. Fraunhofer IPT, Aachen.

[15]Mourtzis D, Zogopoulos V, Vlachou E (2017) Augmented Reality Application to Support Remote Maintenance as a Service in the Robotics Industry. *Procedia CIRP* 63:46–51.

[16]Sutherland IE (1965) The Ultimate Display. *Proceedings of IFIP Congress*:506–8.

[17]Quint F, Loch F, Bertram P (2017) The Challenge of Introducing AR in Industry - Results of a Participative Process Involving Maintenance Engineers. *Procedia Manufacturing* 11:1319–23.

[18]Palmarini R, Erkoyuncu JA, Roy R, Torabmostaedi H (2018) A systematic review of augmented reality applications in maintenance. *Robotics and Computer-Integrated Manufacturing* 49:215–28.

[19]Kurata T, Sakata N, Kourogi M, Kuzuoka H, Billinghurst M (2004) Remote Collaboration using a Shoulder-Worn Active Camera/Laser. *ISWC 2004: Eighth International Symposium on Wearable Computers proceedings 31 October-3 November, 2004, Arlington, Virginia*. IEEE Computer Society. Los Alamitos, Calif, pp. 62–69.

[20]Zhong XW, Boulanger P, Georganas ND (2002) Collaborative Augmented Reality: A Prototype for Industrial Training. *21th Biennial Symposium on Communication, Canada.*

[21]Alem L, Huang W (2011) Developing Mobile Remote Collaboration Systems for Industrial Use: Some Design Challenges. in Campos P, Graham N, Jorge J, Nunes N, Palanque P, Winckler M, (Eds.). *Human-computer interaction - INTERACT 2011: 13th IFIP TC 13 international conference, Lisbon, Portugal, September 5 - 9, 2011 ; proceedings, part IV*. Springer. Berlin, pp. 442–445.

[22]Huck-Fries V, Wiegand F, Klinker K, Wiesche M, Krcmar H (2017) Datenbrillen in der Wartung [smart glasses in maintenance]. in Eibl M, Gaedke M, (Eds.). *Informatik 2017 - Bände I-III: Tagung vom 25.-29. September 2017 in Chemnitz*. Gesellschaft für Informatik. Bonn.

[23]Miles MB, Huberman AM (1994) *Qualitative data analysis: An expanded sourcebook.* 2nd ed. Sage, Thousand Oaks, Calif.

[24]Günterberg B (2012) *Unternehmensgrößenstatistik [company size statistics]: Unternehmen, Umsatz und sozialversicherungspflichtige Beschäftigte 2004 bis 2009 in Deutschland*. Ergebnisse des Unternehmensregisters (URS 95), Bonn.

[25]Miles MB, Huberman AM, Saldaña J (2014) *Qualitative data analysis: A methods sourcebook.* 3rd ed. Sage, Los Angeles, London, New Delhi, Singapore, Washington DC.

[26]Fussell, Susan R., Setlock, Leslie D., Kraut, Robert E. (2003) Effects of Head-Mounted and Scene-Oriented Video Systems on Remote Collaboration on Physical Tasks. *New horizons conference proceedings, Conference on Human Factors in Computing Systems* 5(1):513–20.

[27]Gebauer H, Paiola M, Edvardsson B (2010) Service business development in small and medium capital goods manufacturing companies. *Managing Service Quality* 20(2):123–39.

[28]van Krevelen DWF, Poelman R (2010) Survey of Augmented Reality Technologies, Applications and Limitations. *The International Journal of Virtual Reality* 9(2):1–20.

[29]VDMA (2016) *VDMA-Benchmarks Kundendienst 2016 [customer service benchmarking survey]*, Frankfurt.

[30]BMWi (2015) *Industrie 4.0 - Volks- und betriebswirtschaftliche Faktoren für den Standort Deutschland [Study on macroeconomic and microeconomic factors in the context of Industry 4.0 in Germany]: Eine Studie im Rahmen der Begleitforschung zum Technologieprogramm AUTONOMIK für Industrie 4.0*, Berlin.

Discrepancy analysis for gearbox condition monitoring: A comparison of different healthy data models

S. Schmidt[1], P.S. Heyns[1], K.C. Gryllias[2,3]
[1] Centre for Asset Integrity Management, Department of Mechanical and Aeronautical Engineering, University of Pretoria, South Africa
[2] Department of Mechanical Engineering, PMA division, KU Leuven, Belgium
[3] Dynamics of Mechanical and Mechatronic Systems, Flanders Make, Belgium

ABSTRACT

Discrepancy analysis is a novelty detection technique that uses diagnostic information, extracted from vibration signals that were acquired from a rotating machine, and a model of the healthy historical machine data to infer the condition of the gearbox under varying operating conditions. In discrepancy analysis, various healthy data models such as Gaussian mixture models can potentially be used to model the healthy features, however, the performance of these models have not been compared in a systematic way for discrepancy analysis. In this work, a principal component reconstruction-based model is considered for discrepancy analysis. Moreover the ability of Gaussian models, Gaussian mixture models, hidden Markov models, and principal component-based reconstruction models to detect bearing damage in gearboxes operating under varying speed conditions is evaluated and compared. The gearbox data are generated using a phenomenological numerical gearbox model and the performance of the different models is compared by investigating the sensitivity of the models to detect changes in the magnitude of impulses induced by bearing damage and to compare the models' robustness to changes in operating conditions. The results provide further insight into the properties of the different models and can assist with model selection when investigating new datasets with discrepancy analysis.

Keywords: Condition monitoring; Diagnosis; Discrepancy analysis; Fault detection and localisation

Corresponding author: Stephan Schmidt (stephanschmidt@zoho.com)

1. INTRODUCTION

Condition monitoring is performed on critical assets such as gearboxes to ensure that they operate as intended. Under varying operating conditions, it is difficult to discern between changes in machine condition and changes in operating condition [1] and therefore diagnostic techniques, which allow the condition of the machine to be inferred under varying operating conditions, are actively investigated.

Discrepancy analysis is a novelty detection approach that uses the measured condition monitoring data and a model of the historical condition monitoring data of a healthy machine to generate a discrepancy signal. This discrepancy signal quantifies the deviation of the machine's response from the response of a healthy machine for each time step and can be processed to infer the condition of gears and bearings of gearboxes. The healthy data model forms a central part of discrepancy analysis, with different models such as Gaussian [2], Gaussian mixture [3,4] and hidden Markov models [5] that can be used. A comparison of the aforementioned models are required to assist with model selection in future investigations.

In this paper, different healthy data models are investigated and compared on numerical gearbox data simulating bearing damage under varying speed conditions, with principal component analysis also investigated to obtain a healthy data model for discrepancy analysis.

2. DISCREPANCY ANALYSIS

Discrepancy analysis consist of (1) data acquisition and processing, (2) feature extraction, (3) feature modelling and discrepancy signal generation, and (4) discrepancy signal processing. A very brief overview of each step is given in this section, with a more detailed description can be found in the paper by Schmidt et al. [2].

2.1. Data acquisition and processing

Vibration measurements are used as condition monitoring data, because the data are rich with diagnostic information and can easily be acquired. The damage components are generated from physical rotating components which result in the signals to be angle periodic [6], hence the discrepancy signal is analysed in the angle domain by extracting features from the angle domain. Therefore, the rotational speed or phase information of a shaft has to be measured or estimated from the vibration data as well.

2.2. Feature extraction

The purpose of the feature extraction process is to extract diagnostic information from a signal that is rich with information pertaining to the condition, the operating conditions and the operating environment of the machine. The bearing features used in this investigation are based on the features used by Schmidt et al. [2]. The wavelet packet transform is used with a Daubechies db1 wavelet function to decompose the time domain vibration signal into a set of wavelet coefficients. The Daubechies db1 basis function is used because it is able to detect singularities induced by bearing damage. A four level wavelet packet decomposition is performed which results in 16 wavelet coefficients, associated with different frequency bands, to be obtained. This allows the frequency bands to have a relatively fine resolution to detect damage manifesting in specific frequency bands and the features to have an acceptable dimensionality to avoid potential problems associated with the curse of dimensionality.

The wavelet coefficients of each frequency band are separately order tracked, whereafter the localised RMS is extracted from a moving average window with a length of $2\pi/50$ radians and a 50% overlap between adjacent windows. The short window length makes it possible to detect localised changes in the vibration signal, however, a method to select the optimal window length for bearing diagnostics still needs to be determined. Hence, instead of obtaining a single 16 dimensional feature for a single vibration measurement i.e. by calculating the RMS of the wavelet coefficients for all shaft angles, many localised 16 dimensional features are extracted from a single vibration measurement by extracting features from the moving windows. A detailed description of the windowing and feature extraction procedures can be found in Ref. [2]. Order tracking is performed on the wavelet packet coefficients instead of the vibration signal so that the angle-time cyclostationary properties of the bearing damage impulses are preserved.

2.3. Feature modelling and discrepancy signal generation

The features of a healthy gearbox, extracted from the previous section, is subsequently modelled by the models investigated in this section, where new features extracted from a single window, denoted by $\mathbf{b}$, are evaluated with the discrepancy function $S(\mathbf{b})$ to obtain a discrepancy measure. The discrepancy measures that correspond to the features extracted from each window are combined to form a discrepancy signal which is subsequently analysed for damage [2]. The dimensionality of the feature space is 16, which means that the dimensionality of the features $\mathbf{b}$ are 16 x 1. A brief overview of each model and its associated discrepancy measure are given in this section.

2.3.1. Gaussian Model (GM)

A multivariate Gaussian Model (GM) is one of the simplest models that can be used for discrepancy analysis and has been successful for bearing diagnostics under varying speed conditions [2]. The Mahalanobis distance for the features $\mathbf{b}$

$$S(\mathbf{b}) = \sqrt{(\mathbf{b}-\boldsymbol{\mu})^T \boldsymbol{\Sigma}^{-1}(\mathbf{b}-\boldsymbol{\mu})}\,, \quad (1)$$

uses the mean $\boldsymbol{\mu}$ and covariance $\boldsymbol{\Sigma}$, estimated from the healthy data, and is used as a discrepancy measure. If the features have a dimensionality of D, then the mean $\boldsymbol{\mu}$ is a $D \times 1$ vector while the covariance $\boldsymbol{\Sigma}$ is a $D \times D$ matrix. The features investigated in this work result in $D=16$ and the mean and covariances have the same size throughout this work, unless stated otherwise. The GM has the benefit that it is easy (i.e. it has closed form maximum likelihood solutions) and computationally efficient to implement, however, it is incapable of describing multi-modal and non-Gaussian distributions.

2.3.2. Gaussian Mixture Model (GMM)

The Gaussian Mixture Model (GMM) is a generalisation of the GM, where a weighted sum of *N*-Gaussian densities are used to approximate the density of the features under investigation. The GMM has the benefit that it can approximate any density to an arbitrary accuracy if a sufficient number of mixture components is used [7]. The GMM has been successfully used for discrepancy analysis-based gear diagnostics [3,4], with the NLL of the discrepancy signal

$$S(\mathbf{b}) = -\log\left(\sum_{i=1}^{N} \pi_i G(\mathbf{b} \mid \boldsymbol{\mu}_i, \boldsymbol{\Sigma}_i)\right), \tag{2}$$

used as a discrepancy measure for the GMM. In Equation (2), π_i is the weight of the *i*th mixture component and $G(\mathbf{b} \mid \boldsymbol{\mu}_i, \boldsymbol{\Sigma}_i)$ denotes the Gaussian probability density function associated with the *i*th mixture component. Even though the GMM is very flexible, it requires the estimation of many parameters (i.e. *N*-means, *N*-covariances and *N*-weight parameters) with an optimisation procedure, and the appropriate number of mixture components needs to be estimated from the data as well. GMMs are also susceptible to overfitting and care needs to be taken when optimising the models [7]. In this investigation, the Expectation-Maximisation (EM) algorithm is used to obtain the model parameters using a maximum likelihood framework.

2.3.3. Hidden Markov Model (HMM)

The aforementioned models are optimised based on the assumption that each data point is independent and identically distributed (i.i.d.), which may be limiting. HMMs overcome the i.i.d. limitation by assuming that the features are generated by a latent state that follows a Markov process and the features are related to the latent states by an observation density. HMMs have been successful in the discrepancy analysis field [5] and in this investigation, a first-order Markov process is used with a Gaussian observation distribution for each latent state. The discrepancy measure for the HMM is calculated with

$$S(\mathbf{b}_i) = -\log\left(p(\mathbf{b}_i \mid \mathbf{b}_{i-1}, \ldots \mathbf{b}_1)\right), \tag{3}$$

for the feature at window *i*. The HMM is even more computationally expensive than the GMM, because the transition probability matrix of the Markov process needs to be estimated as well. The additional temporal dimension gives the HMM additional discriminatory power for detecting novelties over the GMM and GM. The mean and covariance of the Gaussian observation density of each state, the initial state, and the transition matrix are obtained using the Baum-Welch algorithm using a maximum likelihood framework in this investigation.

2.3.4. Principal Component Analysis-based Reconstruction (PCA-R)

Principal Component Analysis (PCA) is a popular dimensionality reduction technique, where the eigenvectors $\mathbf{V}$ of the covariance matrix of the features are used to transform the features to a lower subspace by removing redundant information. The dimensionality reduction is performed with

$$\mathbf{z} = \mathbf{V}_{1:d}^{T}(\mathbf{b} - \boldsymbol{\mu}), \tag{4}$$

where $\mathbf{V}_{1:d}$ represents the eigenvectors associated with the d-largest eigenvalues and has a dimensionality of $D \times d$, $\boldsymbol{\mu}$ denotes the mean of the healthy features and $\mathbf{z}$ denotes the principal components with a dimensionality of $d \times 1$. It is possible to obtain a one dimensional signal from the original signal with PCA (i.e. $d = 1$), however, this is not a discrepancy signal and very poor results are obtained when using it in a discrepancy analysis context.

In this paper, a new discrepancy analysis approach based on PCA is investigated. The mean and the eigenvectors of the healthy features are used to map the lower dimensional principal component features with a dimensionality of $d \times 1$ back to the original feature space with

$$\mathbf{b}_{rec} = \mathbf{V}_{1:d}\mathbf{V}_{1:d}^{T}(\mathbf{b} - \boldsymbol{\mu}) + \boldsymbol{\mu}, \tag{5}$$

where the reconstructed features are denoted by $\mathbf{b}_{rec}$ which is a $D \times 1$ vector. The root-mean-square of the reconstruction error $\mathbf{b} - \mathbf{b}_{rec}$ is used as a discrepancy measure

$$S(\mathbf{b}) = \sqrt{\frac{1}{D}\left(\mathbf{V}_{1:d}\mathbf{V}_{1:d}^{T}(\mathbf{b} - \boldsymbol{\mu}) + \boldsymbol{\mu} - \mathbf{b}\right)^{T}\left(\mathbf{V}_{1:d}\mathbf{V}_{1:d}^{T}(\mathbf{b} - \boldsymbol{\mu}) + \boldsymbol{\mu} - \mathbf{b}\right)}. \tag{6}$$

If $d < D$, then the reconstruction error will be non-zero and if the machine condition changes, the reconstruction error will increase. The dimensionality d is estimated by selecting the minimum PCA dimension that has an accumulative contribution rate of 95%.

2.4. Discrepancy signal post-processing

It is possible to calculate a statistic of the discrepancy signal and use it to infer changes in the condition of the machine, however, further processing is required to infer the condition of a specific machine component. The spectrum of the discrepancy signal can be used with the characteristic frequencies of the rotational components (e.g. ball-pass frequency) to determine which components generate the damage. As a result, similarly to Ref.[2], the spectrum of the discrepancy signal is used to infer the condition of the bearings.

3. NUMERICAL VALIDATION

A phenomenological gearbox model proposed by Abboud et al. [8] is used to compare the different healthy data models. The motivation for the phenomenological gearbox model is that it is able to simulate many of the complex phenomena seen in actual gearboxes (e.g. dependence of the signal's amplitude on speed) and the exact operating conditions and the exact condition of the gearbox are known.

3.1. Overview of model

The casing vibration signal of the gearbox comprises of the contributions of the gear mesh components x_g, the damaged bearing x_b, and broadband noise x_n in the following form

$$x_c(t) = x_g(t) + FS \cdot x_b(t) + x_n(t), \tag{7}$$

where the fault severity FS is a factor that linearly increases the bearing components' magnitude. The different components can be decomposed in terms of the signal at the source, convoluted with the impulse response function of the system

$$x_g(t) = h_g(t) \otimes \left(M(\omega(t)) \sum_{k=1}^{N_{dg}} A_{dg}^{k} \cos\left(kN_{teeth} \int_0^t \omega(\tau)d\tau + \phi_{dg}^{k} \right) \right), \tag{8}$$

$$x_b(t) = h_b(t) \otimes \left(M(\omega(t)) \sum_{i=1}^{N_{imp}} \delta(t - T_i) \right), \tag{9}$$

$$x_n(t) = \varepsilon(t) M(\omega(t)), \tag{10}$$

where $h_i(t)$ denotes the impulse response function of characteristic *i*, the magnitude and the phase of the *k*th harmonic of the gear component are denoted by A_g^k and ϕ_g^k. The rotational speed of the shaft is denoted by ω, with the magnitude of the vibration component increasing with rotational speed and is modelled with $M(\omega) = a\omega + b$. During the measurement, N_{imp} bearing impulses occur, with the *i*th bearing impulse occurring at time step T_i. The raw noise is sampled from a zero mean Gaussian distribution and is denoted by $\varepsilon(t)$ at time *t*. Outer race bearing damage is simulated with the model, with the characteristic frequency of the damage being 4.12 shaft orders. The number of teeth on the gear that is connected to the shaft under consideration is $N_{teeth} = 20$.

Four different rotational speed profiles are investigated for the phenomenological gearbox model and presented in Figure 1. The four operating conditions help to compare the robustness of the discrepancy signals, obtained from the different models, to varying speed conditions.

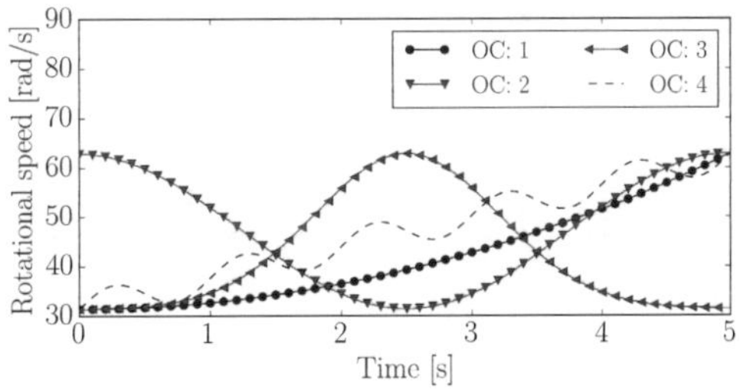

Figure 1: The Operating Condition (OC) profiles investigated for the phenomenological gearbox model.

Five different FS, namely 0.0, 0.02, 0.04, 0.06 and 0.08 are investigated here as well. The FS merely quantifies the relative magnitude of the bearing impulses of the different cases e.g. a FS of 0.02 contains impulses that are on average four times smaller than a FS of 0.08, while FS = 0.0 simulates a healthy bearing.

The three vibration components and the casing vibration signal are presented in Figure 2 for a FS = 0.02 with operating condition profile 2 (i.e. OC: 2) being used during the simulation.

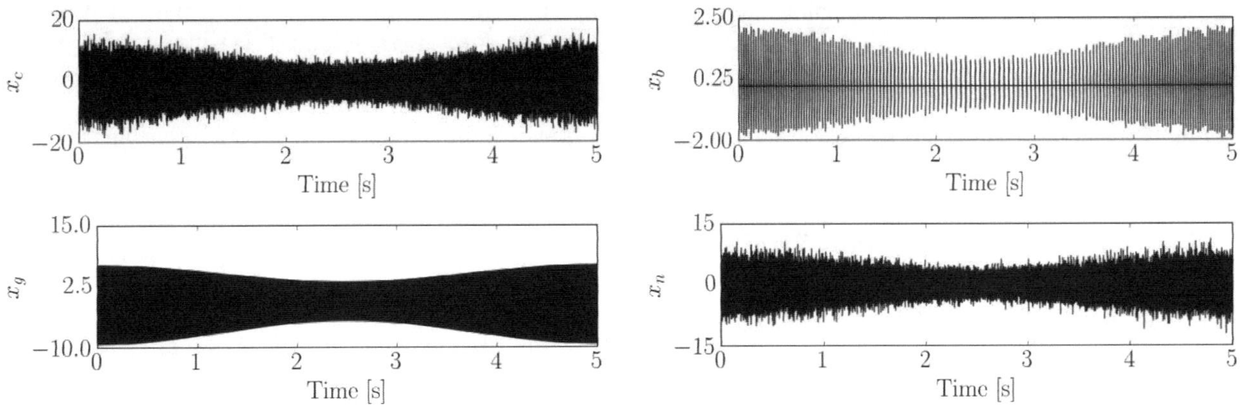

Figure 2: The casing vibration signal of operating condition 2 and its components for the case where the bearing damage FS is 0.02.

The casing vibration signal and the corresponding rotational speed are used in the discrepancy analysis process investigated in the next section.

3.2. Results

The casing vibration signal and the rotational speed of the healthy vibration data are used with the wavelet packet transform to extract 16-dimensional features as described in Section 2.3. These features are modelled separately with the models under consideration. The different discrepancy measures, denoted by S(**b**), are subsequently used with the features extracted from the windowed wavelet coefficients of the new data to generate a discrepancy signal associated with each vibration measurement. In this section, three distinct investigations are performed (1) the spectra of the discrepancy signals are investigated for diagnostic information, (2) the sensitivity of the discrepancy signal to damage is investigated, and (3) the sensitivity of the discrepancy signal to changes in operating conditions is investigated as well.

3.2.1. Spectrum of the discrepancy signals

The frequency content of the discrepancy signals, where the bearing has a Fault Severity (FS) of 0.02, is shown in Figure 3 for the four investigated models. This is obtained by calculating the spectrum of the discrepancy signal of each measurement in the dataset. The model complexity (e.g. number of mixture components) with the best cross-validation results are presented in the figure where appropriate.

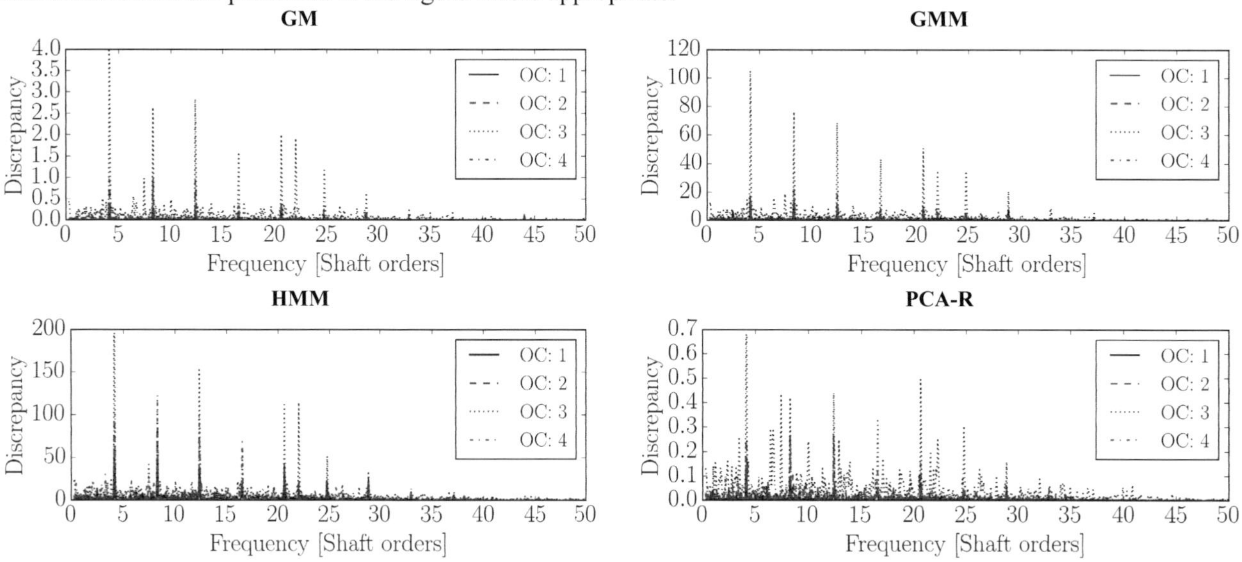

Figure 3: The spectrum of the discrepancy signals obtained from the different models for a bearing with a FS of 0.02 are shown.

The damage has a fundamental component at 4.12 shaft orders, with the fundamental component and a few harmonics being seen in the spectra of all the models. However, the spectra are definitely sensitive to changes in operating conditions as seen by the variation of the amplitudes between different operating condition profiles. The spectra of the discrepancy signals generated with the GM, GMM and the HMM contain mostly diagnostic information, whereas the PCA-R spectrum contains a lot of noise which impedes the condition inference process.

It is difficult to quantify the robustness of the different models to varying operating conditions and the sensitivity of the models to damage from the spectra and therefore two separate investigations are performed in the next sections to quantify the performance of the models.

3.2.2. Sensitivity to damage

The sensitivity of the discrepancy signals obtained from the different models to damage is investigated in this section. The amplitude of the fundamental component of the characteristic frequency (i.e. 4.12 shaft orders) in the spectrum is used to quantify the magnitude of the damage for the different FS values and the different models in this section.

The amplitude of the fundamental components in Figure 4(a) are averaged over the four operating condition profiles, and presented over bearing fault severity. The HMM has the largest discrepancy values with respect to damage, where in contrast the PCA-R and the GM have the smallest discrepancy values. However, it is not desired to quantify the absolute values, but instead to quantify the relative increase in the values as the damage progresses. Therefore, the amplitudes in Figure 4(a) are normalised by the amplitude of the component from a healthy machine, with the results presented in Figure 4(b).

It can be observed that the GMM is the most sensitive to damage, while the GM is second most sensitive to damage, performing even better than the HMM. The PCA-R model does not perform very well relative to the other models, however, it successfully detects changes in the bearing's condition. The GM is very simple relative to the GMM and HMM and may not represent the exact density of the healthy features, however, it is clearly capable of detecting small changes in the spectrum of the machine and is therefore well suited for discrepancy analysis.

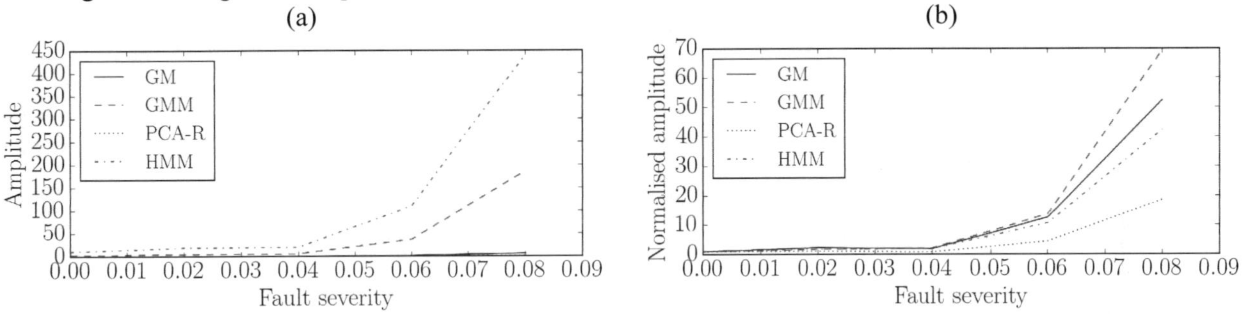

Figure 4: The sensitivity of the models to bearing damage are compared. The amplitude (a) and normalised amplitude (b) of the fundamental ball-pass outer race component are shown over fault severity.

3.2.3. Sensitivity to operating conditions

The discrepancy signal of a healthy gearbox is presented over rotational speed in Figure 5 for the different models. The conditional mean and variance of the discrepancy signal, given the rotational speed, varies over rotational speed. The GM, GMM and the PCA-R models contain very similar characteristics, however, the HMM's statistics vary significantly over rotational speed.

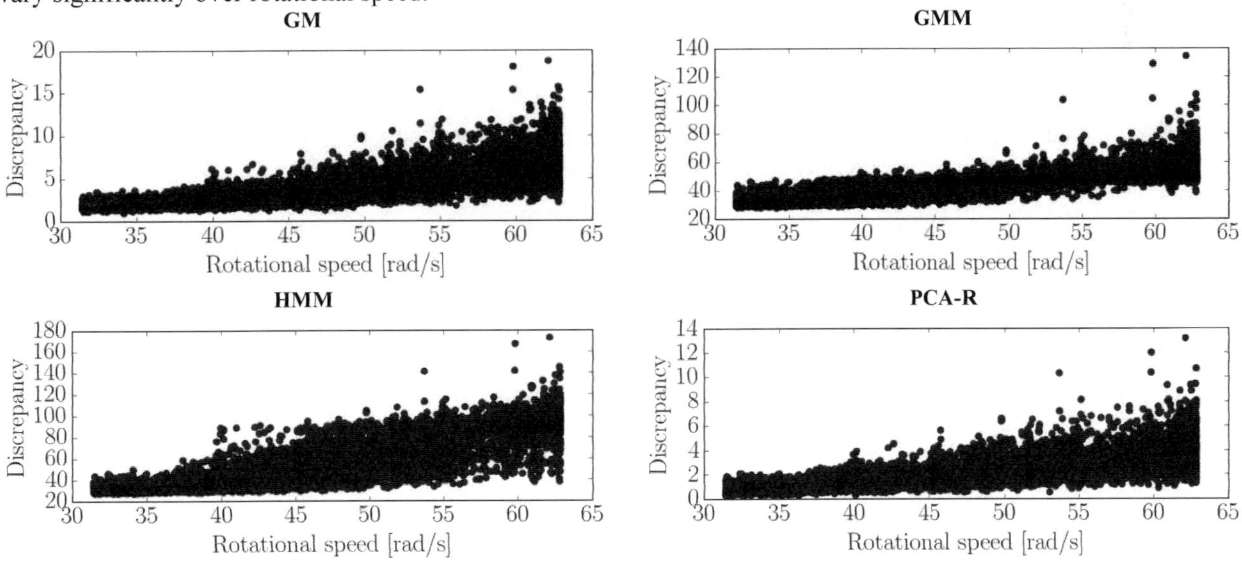

Figure 5: The discrepancy signal of a healthy gearbox versus the rotational speed is presented for the different models.

It is desired to quantify the sensitivity of the discrepancy signal to rotational speed and therefore the conditional mean of the discrepancy signal given the rotational speed is calculated with the procedure used by Ref. [2]. The raw conditional mean results are presented in Figure 6(a). It can be observed that the HMM contains the largest changes in discrepancy value, while the GM and the PCA-R models contain the smallest changes.

Similar to the investigation in Section 3.2.2., the normalised discrepancy, obtained from dividing the discrepancy value by the discrepancy value that correspond to the lowest rotational speed, is investigated as well. The results in Figure 6(b) indicate the PCA-R and the GM are the most sensitive to changes in rotational speed, while the GMM is the most robust. Even though the GM results are sensitive to rotational speed variations, it can easily be compensated for by the method proposed in Ref. [2].

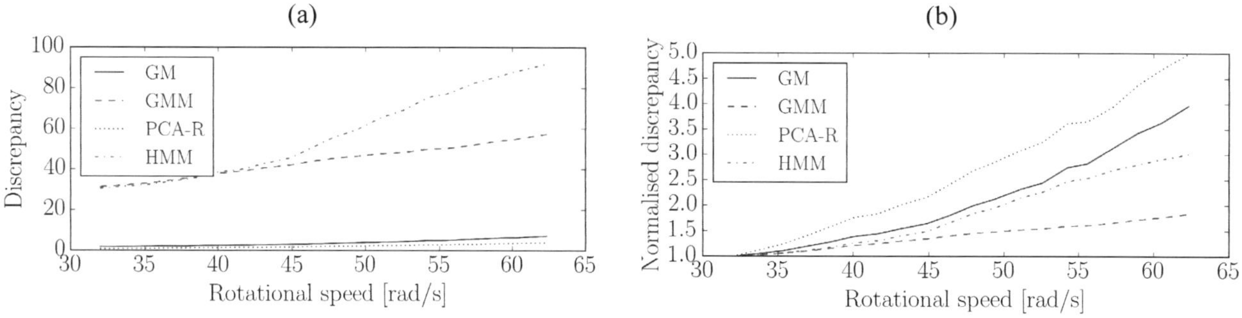

Figure 6: The sensitivity of the discrepancy signal to rotational speed is compared for the different models by comparing the conditional mean (a) and the normalised conditional mean (b) of the results in Figure 5.

4. CONCLUSION

In this investigation, different models for discrepancy analysis were investigated and compared. The Gaussian mixture model is the most sensitive to damage and the most robust to varying speed conditions and is therefore best suited for discrepancy analysis according to this investigation. The hidden Markov model is quite sensitive to operating conditions which may be ascribed to the Markov process of the model which struggles to distinguish between changes in machine conditions and operating conditions. The Gaussian model performed very well with the bearing diagnostics task which has the advantage that it can be easily implemented without considering convergence and hyperparameter optimisation issues. Lastly, even though the principal component reconstruction-based model did not perform as well as the other models, the investigation indicated how the reconstruction model framework can be used for discrepancy analysis. More sophisticated reconstruction-based models can be investigated in future work. Future investigations can also compare the different models on experimental data and this investigation can be performed for gear diagnostics as well.

REFERENCES

[1] Stander, C.J., Heyns, P.S. and Schoombie, W.: Using vibration monitoring for local fault detection on gears operating under fluctuating load conditions. Mechanical Systems and Signal Processing, 16(6), pp.1005-1024 (2002).

[2] Schmidt, S., Heyns, P.S. and Gryllias, K.C.: A discrepancy analysis methodology for rolling element bearing diagnostics under variable speed conditions. Submitted to Mechanical Systems and Signal Processing (2017).

[3] Heyns, T., Heyns, P.S. and De Villiers, J.P.: Combining synchronous averaging with a Gaussian mixture model novelty detection scheme for vibration-based condition monitoring of a gearbox. Mechanical Systems and Signal Processing, 32, pp.200-215 (2012).

[4] Heyns, P.S., Vinson, R. and Heyns, T.: Rotating machine diagnosis using smart feature selection under non-stationary operating conditions. Insight-Non-Destructive Testing and Condition Monitoring, 58(8), pp.417-422 (2016).

[5] Schmidt, S., Heyns, P.S. and De Villiers, J.P.: A novelty detection diagnostic methodology for gearboxes operating under fluctuating operating conditions using probabilistic techniques. Mechanical Systems and Signal Processing, 100, pp.152-166 (2018).

[6] Antoni, J., Bonnardot, F., Raad, A. and El Badaoui, M.: Cyclostationary modelling of rotating machine vibration signals. Mechanical systems and signal processing, 18(6), pp.1285-1314 (2004).

[7] Bishop, C.M., 2006. Pattern recognition and machine learning. Springer (2006).

[8] Abboud, D., Antoni, J., Sieg-Zieba, S. and Eltabach, M.: Envelope analysis of rotating machine vibrations in variable speed conditions: As comprehensive treatment. Mechanical Systems and Signal Processing, 84, pp.200-226 (2017).

A Generalized Synchroextracting Transform for Fast and Strong Frequency Modulated Signal Analysis

Peng Chen[1], Kesheng Wang [1], Ming J. Zuo[1,2]
[1] School of Mechanical and Electrical Engineering, University of Electronic Science and Technology of China, China
[2] Department of Mechanical Engineering, University of Alberta, Canada

ABSTRACT

High-quality time-frequency representation (TFR) is vital for signal analysis and condition monitoring. The blurred time frequency (TF) energy distribution in TF plane may lead to uncertainty in TFR and inevitably result in ambiguous results. Synchroextracting, with an adaptive and reconstructive natured transform, is used to improve the readability of STFT-based TFR. However, the standard synchroextracting method is based on the assumption that the analyzed signal is purely harmonic, as a result, it is more applicable to deal with signals with small amplitude and slow frequency modulations, whereas, signals with fast and strong nonlinear and non-stationary nature are of not suitable for the current synchroextracting method. To tackle this problem, a generalized synchroextracted transform (GSET) is proposed to cope with fast and strong nonlinear and nonstationary signals. In the proposed method, the generalized Fourier transform is first used to map the original frequency time-varied signal to a constant frequency signal. The constant frequency signal is further analyzed through synchroextracting operation so that the energy concentrated TF ridge can be obtained for a fast and strong nonlinear and nonstationary signal. Numerical simulated case study is provided for the demonstration of the efficiency of the proposed GSET method.

Keywords: Time frequency representation (TFR); Instantaneous frequency; Generalized Fourier transform; synchroextracting; short time Fourier transform

Corresponding author: Kesheng Wang (keshengwang@uestc.edu.cn)

1. INTRODUCTION

Rotating machinery vibration signals [1–3] in industry are commonly analyzed through time frequency analysis (TFA), and such signals may suffer from unstable operational conditions. In this case, these operating conditions may give rise to frequency and amplitude modulated vibrations or nonlinear and nonstationary vibrations, and these vibration signals are often featured with fast and strong frequency time-varying nature. Recently, many TFA methods have been introduced to extend traditional Fourier analysis to deal with nonlinear and nonstationary signals, thus resulting in a body of literature that is referred to [4–8]. The bottleneck of traditional time frequency analysis methods such as short time Fourier transform (STFT), continuous wavelet transforms (CWT) and Wigner-Ville distribution (WVD), to deal with fast and strong frequency modulation natured signal, is to obtain a high-quality time and frequency representation. In this regard, recently, several advanced time frequency analysis methods such as reassignment method (RM) [10], synchrosqueezing transform (SST) [6,11] and synchroextracting transform (SET) [8] have been introduced. RM can provide TFA result with high readability, but it needs to reassign the time frequency spectrogram into an IF ridge along two-dimensional TF directions. It, therefore, loses ability to reconstruct the component of interest, it is a non-reconstruction method. To cope with this shortcoming, Daubechies et al. [6] introduced the SST method, it squeezes the TF coefficients into the IF ridge only in frequency direction. Consequently, SST not only can provide a TFR with a better readability but also enable signal reconstruction. By comparing the traditional methods such as STFT,WT and advanced SST, Iatsenko et al. [12] pointed out that SST also has its drawbacks for signal representation and reconstruction, i.e., it squeezes all the TF coefficients no matter the concentrated TF coefficients are related to signal time-varying features or not. In other words, TF coefficients due to the noise are possible to be squeezed. Based on this work, Yu et al. provide a novel post-processing strategy SET [8]. This SEST technique aims to retain the closet TF information which is related to the intrinsic time-varying characteristics of the analyzed signal. It provides a high resolution TFR and a better robustness to noise pollution with signal reconstruction ability. Though SET features so many advantages, it should also be noted that the standard synchroextracting transform method assumes that the analyzed signal is purely harmonic. Therefore, it is more applicable to deal with signals with small amplitude and slow frequency modulations, whereas, signals with fast and strong nonlinear and non-stationary nature may not be handled properly by the current synchroextracting transform method. To overcome this problem, a generalized sysnchroextracting transform (GSET) is proposed in this paper to cope with "strong" nonlinear and nonstationary signals. In the proposed method, the generalized Fourier transform is used to map the original time-varying frequency signal to a constant frequency signal. The constant frequency signal is further analyzed through synchroextracting operation so that the energy concentrated TF ridge can be obtained for a fast and strong nonlinear and nonstationary signal. Numerical simulated case study is provided for the demonstration of the efficiency of the proposed GSET method.

2. THE THEORY OF GENERALIZED SYNCHROEXTRACTING TRANSFORM

An enhanced generalized synchroextracting transform (GSET) method is illustrated in this section in detail. In the method, the generalized Fourier transform (GFT) is first used to map the original time-varying frequency signal to an analytic signal with a constant frequency. It can be used to remove smeared TFR by the fast and strong time-varying effect. The constant frequency signal is further analyzed through synchroextracting procedure so that the energy concentrated TF ridge can be obtained for a fast and strong nonlinear and nonstationary signal. As a result, it may result in an energy concentrated TFR, and avoid the ambiguous time frequency analysis results.

2.1. Revisit the generalized Fourier Transform

The generalized Fourier transform (GFT) proposed was used in [13]. GFT can be employed to map a signal with time-varying instantaneous frequency to an analytic signal with a constant frequency.
For a signal $x(t)$, the GFT is defined as follows,

$$\mathrm{x}_G(f) = F_G(x(t)) = \int_{-\infty}^{+\infty} x(t) e^{-i2\pi(ft+s_0(t))} dt \tag{1}$$

where $s_0(t)$ is a real-valued function depending on time, and it satisfies the evolutionary phase behavior of original signal $x(t)$. It is noted that equation (1) can be viewed as the standard Fourier transform to $x(t)e^{-i2\pi s_0(t)}$. About the inverse GFT, it can be computed from the inverse Fourier transform to $\mathrm{x}_G(f)$, followed by multiplication of the result by $e^{i2\pi s_0(t)}$ i.e.,

$$\begin{aligned} x(t) = F_G^{-1}(\mathrm{x}_G(f)) &= \int_{-\infty}^{+\infty} \mathrm{x}_G(f) e^{i2\pi(ft+s_0(t))} df \\ &= e^{i2\pi s_0(t)} \cdot \int_{-\infty}^{+\infty} \mathrm{x}_G(f) e^{i2\pi ft} df \end{aligned} \tag{2}$$

If $\mathrm{x}_G(f) \equiv \delta(f - f_0)$, then $x(t) = e^{i2\pi(f_0 t+s_0(t))}$. That means that signal $x(t)$ with the instantaneous frequency (IF) $f(t)$ will be projected to the desired horizontal GFT frequency ridge $f_0 = f(t) - s_0'(t)$, where $s_0'(t) = d(s_0(t))/dt$. The above equations (1) and (2) show that the signal $x(t)$ with time-varying frequency IF $f(t)$ can be transformed to a constant frequency signal $d(t) = y(t) \cdot e^{-i2\pi s_0(t)}$, where $y(t) = x(t) + H(x(t))$, via mapping function $e^{-i2\pi s_0(t)}$. Because the signal $d(t)$ is not an analytic signal and in order to remove the effect of negative frequency components during the frequency mapping, Hilbert transform is again exploited to generate a new analytic signal $z(t)=d(t)+i\cdot H(d(t))$. Then the maximal overlap discrete wavelet packet transform (MODWPT) is applied to decompose $z(t)$ into $z(t)=\sum_{i=1}^{J} c_i(t)$. Hence the interested signal component $c_i(t)$ can be obtained by the GFT method.

2.2. Generalized synchroextracting transform (GSET)

In this section, the current synchroextracting transform [8] will be reviewed. Assuming that the signal is a purely harmonic signal $x(t)=A_0 e^{i2\pi\phi(t)}$ i.e., a constant frequency $\phi'(t)=f_0$ and with invariant amplitude $A(t)=A_0$,

$$x(t)=A_0 e^{i2\pi f_0 t} \tag{3}$$

The Fourier transform (FT) of $x(t)$ can be calculated as,

$$\hat{x}(f)=A\cdot\delta(f-f_0) \tag{4}$$

To obtain the result of STFT, $S_x^g(t,f)$ designates STFT of the original signal $x(t)$ using a Gaussian analyzed window g, and it can be calculated as [8],

$$\begin{aligned} S_x^g(t,f) &= \int_0^{+\infty} x(u)g(u-t)e^{-i2\pi f(u-t)}du \\ &= A\cdot\hat{g}(f-f_0)e^{2i\pi f_0 t} \\ &= x(t)\cdot\hat{g}(f-f_0) \end{aligned} \tag{5}$$

where $\hat{g}$ is the FT of g.
To obtain the estimation of local instantaneous frequency, based on equation (5) and it can be measured by the derivative of $S_x^g(t,\omega)$ with respect to t as,

$$\begin{aligned} \omega_{ins}(t,f) &= \phi'(t)=\partial_t\phi_x^g(t,f) \\ &= \frac{1}{2\pi}\partial_t \arg S_x^g(t,f) \\ &= \frac{1}{2i\pi}\frac{\partial_t S_x^g(t,f)}{S_x^g(t,f)} \end{aligned} \tag{6}$$

where $\omega_{ins}(t,f)$ is the estimate of instantaneous frequency of $x(t)$.

To reassign the complex time frequency coefficients based on STFT result, the details of mapping $(t,f)\rightarrow(t,\omega_{ins}(t,f))$ may be referred to [8].

$$T_e^x(t,f)=S_x^g(t,f)\cdot\delta(\omega-\omega_{ins}(t,f)) \tag{7}$$

where $T_e^x(t,f)$ is SET result of $x(t)$.
The original time domain signal $x(t)$ can be reconstructed by,

$$\begin{aligned} x(t) &= \frac{1}{\hat{g}(\omega-\omega_0)\cdot\delta(\omega-\omega_{ins}(t,f))}\Bigg|_{\omega=\omega_0,\,\omega=\omega_{ins}(t,f)} \cdot T_e^x(t,f) \\ &= \frac{1}{\hat{g}(0)}\cdot T_e^x(t,f) \end{aligned} \tag{8}$$

For SET method, the local instantaneous frequency (IF) is estimated by equation (6). This equation clearly interprets that IF is calculated by the results of STFT and one order derivation of $S_x^g(t,\omega)$ regarding to time t. As a result, an analyzed signal with slow frequency modulated characteristics can be distinguished by SET method. It also means that, in case of strong frequency modulated signal, these characteristics cannot be identified by the current SET method. For adapting to this strong frequency modulated signal and motivated by the advantage of GFT which can map a time-varying signal to an analytic signal with a constant frequency, the approach of GSET is proposed. To realize GSET, three necessary steps are implemented. Firstly, the GFT is applied to extract the interested signal component and it can be projected into a constant frequency ridge. Secondly, the constant frequency component, then, can be used for SET extraction. The extraction results can be calculated via equations (5) and (6). Lastly, the interested component $c_i(t)$ can be reconstructed by inverse GFT. The proposed GSET procedures are illustrated in Figure 1.

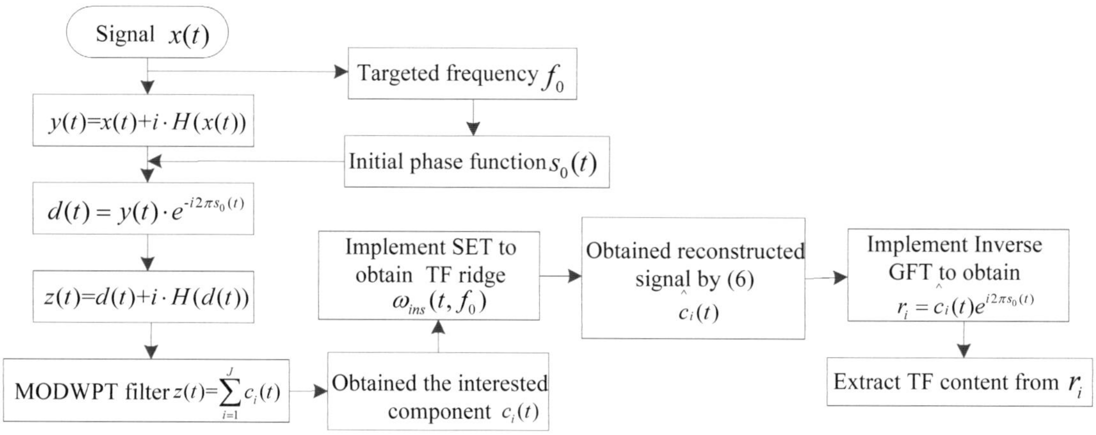

Figure 1. Flowchart of the proposed GSET method for signal TFA

3. NUMERICAL SIMULATED CASE STUDY

To validate the application procedures and effectiveness of GSET method for the fast and strong frequency modulated signal analysis, an indicator to measure the rate of change of the frequency (RCF) per unit time is defined as follows.

For a general multi-component signal (MCS) reads

$$y(t)=\sum_{i=1}^{N} y_i(t)=\sum_{i=1}^{N} A_i(t)e^{2i\pi\phi_i(t)} \tag{9}$$

where the mono-component $y_i(t)$ satisfy $\phi'_N(t)>...>\phi'_i(t)>...\phi'_1(t)>0$. For all $t>0$ and a MCS from the equation (9), instantaneous amplitude $A_i(t)$ and the instantaneous frequency $\phi'_i(t)$ satisfy $A_i(t)>0$, $\phi'_i(t)>0$, and the RCF can be calculated as,

$$RCF=\max\left(\left|\frac{d^2\phi_i(t)}{dt}\right|\right)=\max\left(\left|\phi''_i(t)\right|\right) \tag{10}$$

In this section, a numerical simulation signal with two components of RCF=118.43 and 37.89 are used for analysis. The signal is defined as $y(t)=y_1(t)+y_2(t)$

$$y(t)=\begin{cases} y_1(t)=\sin(2\pi(30t+1.5\cos(2\pi\cdot 0.8\cdot t))) \\ y_2(t)=\sin(2\pi(65t+3.0\cos(2\pi\cdot t))) \end{cases} \tag{11}$$

In this study, the sampling frequency and time are set as 4000Hz and 6 seconds, respectively. The time wave and Fourier spectrum are presented in Figure 2 (a) and (b). The ideal TF trajectories are also plotted in Figure 2 (c).

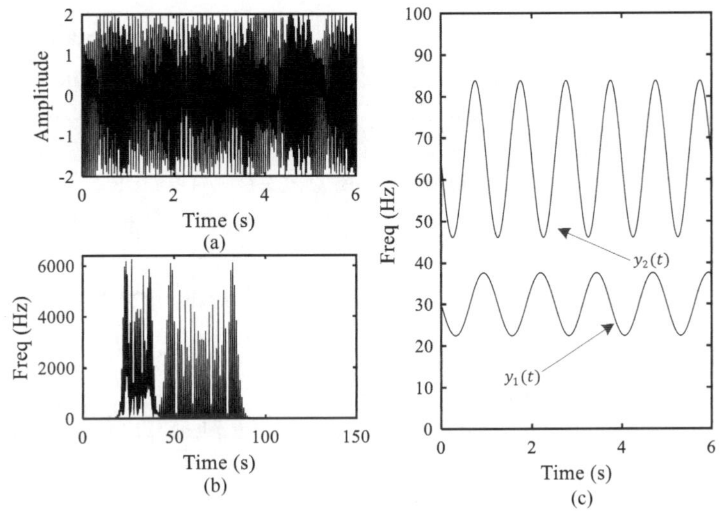

Figure 2. (a) Time waveform of multicomponent signal (b) Fourier spectrum (c) Ideal time frequency trajectories

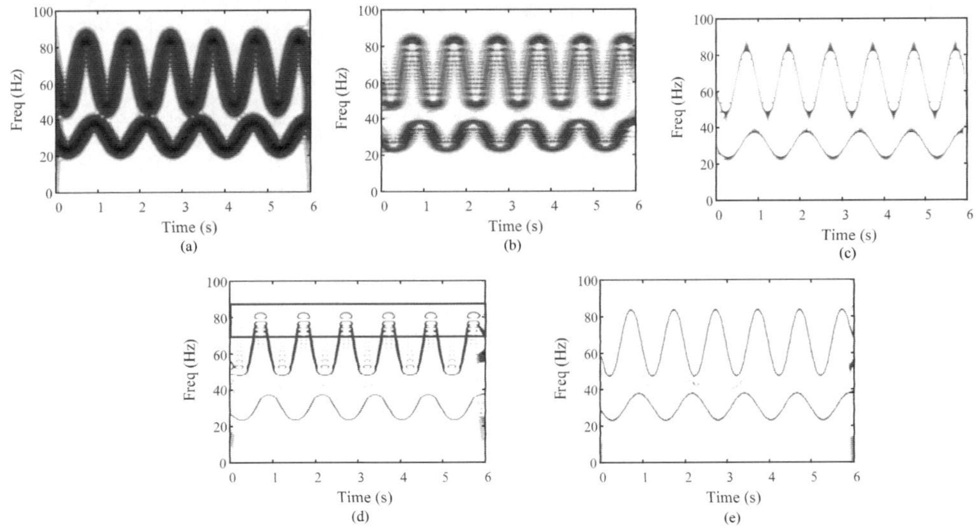

Figure 3. Time frequency representations by (a) STFT (b) SST (c) RM (d) SET (e) GSET

From the time frequency analysis results of Figure 3, it can be found that the TF nature of component $y_1(t)$ can be precisely described by RM and SET. However, due to the severely smeared TFRs in Figure 3 (a) and (b), TF information of $y_1(t)$ and $y_2(t)$ cannot be characterized properly. A careful inspection of the component $y_2(t)$ with a fast and strong frequency modulation, these four TFAs such as STFT, SST, RM and SET cannot accurately distinguish this TF signature. In addition, because the RCF of component $y_2(t)$ at zoned version circled with squared in Figure 3 (d) to the maximum, thus the four aforementioned methods can not deal with this signal component.

To overcome these limitations and problems, GSET is proposed. According to the proposed algorithm flowchart illustrated in Figure 1, the first step is to implement the GFT to the interested component and, in this study, it is $y_2(t)$. The GFT result for the interested component $y_2(t)$ is shown in Figure 4 (b) in which the frequency modulated $y_2(t)$ is transferred into a constant frequency component. Next, because the mapped signal component from $y_2(t)$ and $y_1(t)$ are distributed in different frequency region, thus, the interested signal component $y_2(t)$ can easily filtered by MODWPT and the result can be seen in Figure 4 (c). Then, the component $y_2(t)$ can be restored by inverse GFT (see in Figure 4 (d)). Consequently, the constant frequency signal in Figure 4 (c) is further analyzed through synchroextracting operation so that the energy concentrated TF ridge can be obtained, and then the synechroextracted result is used for the procedure of inverse GFT. At last, the interested component signal $y_2(t)$ can be restored by completed algorithm, and the corresponding result is displayed in Figure 5 (a).

To evaluate the accuracy of GSET, the results of extracted IF trajectory and phase function $s_0(t)$ (referring to flow chat of Figure 1) are superimposed in Figure 5 (a) and Figure 5 (b) for comparisons. It can be found that the extracted TF trajectory is almost the same with the ideal TF ridge which is shown in Figure 2 (c). Utilizing the IF ridge measured by GSET, the extracted IF ridges of three components in equation (11) are depicted in Figure 3 (e). And it is observed that the results of GSET almost perfectly match the ideal TF features shown in Figure 2 (c). Compared with STFT, SST, RM and SET (see in Figure 3 (a-d)), the results of GSET (Figure 3 (e)) feature advantages in TFR with a satisfactory result for characterizing fast and strong frequency modulated signals.

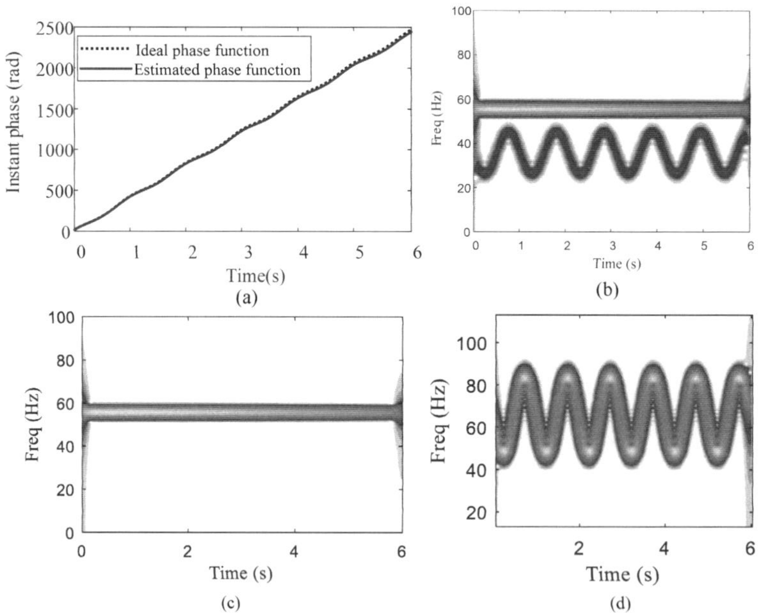

Figure 4 (a) Estimation of instantaneous phase function (b) Extracted the y2(t) component by generalized transform (c) The y2(t) component is separated by MODWPT (d) Restoring the y2(t) component by inverse GFT

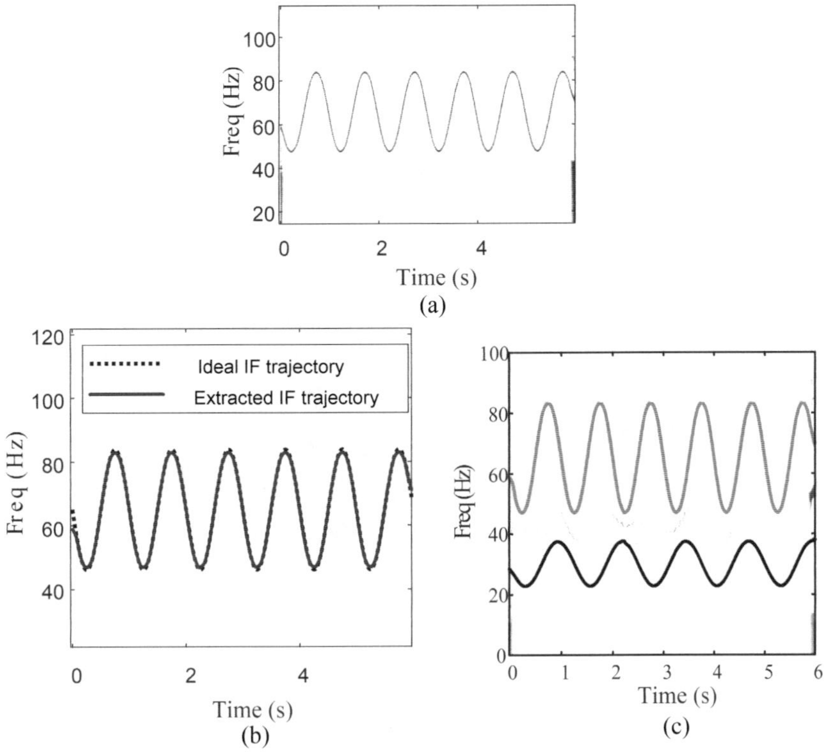

Figure 5. (a) Restoring the $y_2(t)$ component by GSET (b) Instantaneous frequency estimation result (d) Extracted TF trajectories by GSET

4. CONCLUSIONS

In this paper, the method of GSET is introduced. GSET is an extension of the recently proposed synchroextracting transform. The shortcoming of SET in dealing with fast and strong frequency modulated signal is overcome by introducing GFT in the proposed GSET. The simulation studies demonstrate that the proposed GSET exhibits advantages in TFR over traditional methods, STFT, SST, RM and SET. Future experimental studies and practical applications will be further investigated.

ACKNOWLEDGEMENTS

This work is supported by the National Key Research and Development Program of China (Project No. 2017YFC0108401), Fundamental Research Funds for the Centered University (ZYGX2016J111), National Natural Science Foundation of China (51305067 and 51537010), and the Natural Sciences and Engineering Research Council of Canada (Grant #RGPIN-2015-04897).

REFERENCES

[1] K.S. Wang, P.S. Heyns, Application of computed order tracking, Vold–Kalman filtering and EMD in rotating machine vibration, Mechanical Systems and Signal Processing. 25 (2011) 416–430.

[2] K.S. Wang, P.S. Heyns, The combined use of order tracking techniques for enhanced Fourier analysis of order components, Mechanical Systems and Signal Processing. 25 (2011) 803–811.

[3] P. Chen, K. Wang, K. Feng, Application of order-tracking holospectrum to cracked rotor fault diagnostics under nonstationary conditions, in: Prognostics and System Health Management Conference, 2017: pp. 1–6.

[4] F. Auger, P. Flandrin, Improving the readability of time-frequency and time-scale representations by the reassignment method, IEEE Transactions on Signal Processing. 43 (1995) 1068–1089.

[5] F. Hlawatsch, G.F. Boudreauxbartels, Linear and quadratic time-frequency signal representations, IEEE Signal Processing Magazine. 9 (1992) 21–67.
[6] I. Daubechies, J. Lu, H.-T. Wu, Synchrosqueezed wavelet transforms: An empirical mode decomposition-like tool, Applied and Computational Harmonic Analysis. 30 (2011) 243–261.
[7] I. Daubechies, Y. (Grace) Wang, H. Wu, ConceFT: concentration of frequency and time via a multitapered synchrosqueezed transform, Philosophical Transactions of the Royal Society A: Mathematical, Physical and Engineering Sciences. 374 (2016) 20150193.
[8] G. Yu, M. Yu, C. Xu, Synchroextracting Transform, IEEE Transactions on Industrial Electronics. 64 (2017) 8042–8054.
[9] F. Auger, P. Flandrin, Y.-T. Lin, S. McLaughlin, S. Meignen, T. Oberlin, H.-T. Wu, Time-Frequency Reassignment and Synchrosqueezing: An Overview, IEEE Signal Processing Magazine. 30 (2013) 32–41.
[10] K. Kodera, R. Gendrin, C. Villedary, Analysis of time-varying signals with small BT values, IEEE Transactions on Acoustics Speech & Signal Processing. 26 (2003) 64–76.
[11] G. Thakur, E. Brevdo, N.S. Fučkar, H.T. Wu, The Synchrosqueezing algorithm for time-varying spectral analysis: Robustness properties and new paleoclimate applications, Signal Processing. 93 (2013) 1079–1094.
[12] D. Iatsenko, P.V.E. McClintock, A. Stefanovska, Nonlinear mode decomposition: A noise-robust, adaptive decomposition method, Physical Review E. 92 (2015).
[13] S. Olhede, A.T. Walden, A generalized demodulation approach to time-frequency projections for multicomponent signals, Proceedings Mathematical Physical & Engineering Sciences. 461 (2005) 2159–2179.

A Deep Statistical Feature Learning Method Based on Stacked Auto-Encoder for Intelligent Diagnosis of Rolling Bearing

Te Han[1,2], Quan Long[2], Chao Liu[1,3], Dongxiang Jiang[1,2]
[1] State Key Laboratory of Power System and Generation Equipment, Tsinghua University, Beijing
[2] Department of Energy and Power Engineering, Tsinghua University, Beijing
[3] Key Laboratory for Thermal Science and Power Engineering of Ministry of Education, Tsinghua University, Beijing

ABSTRACT

To avoid redundancy and interferential information, feature fusion and dimension reduction is a critical step to improve the recognition accuracy in the fault diagnosis of mechanical system. The paper presents a novel feature fusion method based on one kind of deep neural networks (DNN), namely stacked auto-encoder (SAE). First, 29 popular statistical features in time domain and frequency domain are extracted from the raw vibration signals to form a statistical feature set. Then, the SAE is utilized for deep statistical feature learning to reduce redundancy. With the aid of supervised fine-tuning, the feature learning ability can be further promoted, and the diagnostic results can be given by the top-layer classifier of SAE. An experiment analysis using rolling bearing fault dataset indicates that SAE-based method possesses superior feature learning ability in comparison with conventional dimension reduction algorithm. Meanwhile, contrast results demonstrate that the proposed method can achieve higher recognition accuracy.

Keywords: Deep Learning, Stacked Auto-Encoder, Feature Learning, Intelligent Fault diagnosis, Rolling Bearing.

1. INTRODUCTION

Rolling bearing is a critical component in rotating machinery and its health conditions have a significant influence on the performance and vibrating noise of mechanical equipment. Due to the harsh working condition, rolling bearing is prone to failure, which may lead to serious accidents. Given this, developing the effective fault diagnosis techniques is beneficial to enhance the security and reliability of rotating machinery (Yuan and Lu, 2017; Han and Jiang, 2016; Han et al., 2016).

In the intelligent diagnosis approaches, the procedure mainly include the three key stages: (1) signal pre-processing; (2) feature extraction and (3) fault classification (Han et al., 2017). In the stage of feature extraction, most of the extraction algorithms have been explored in the time domain, frequency domain, and time-frequency domain. Different feature may have different sensibility to bearing faults. Moreover, the extracted features inevitably contain useless and redundant information, which brings difficulties to fault recognition and classification. Therefore, the feature selection based optimization has a great significance (Zhang et al., 2016). The most popular feature selection methods can be mainly classified into: filter approach, wrapper approach, hybrid scheme and dimension reduction strategy (Liu et al., 2014). Shen et al. (2012) established a distance evaluation technique (DET)-based filter approach method to select salient features from 116 time- and frequency domain features. Liu et al. (2016) combined the filter and wrapper approaches to search optimal feature space for diagnosing the engine operating states. Qin et al. (2016) applied the principal component analysis (PCA) to reduce the dimension of 16-dimensional primitive inputs in the diagnosis of aileron actuators. Yang and Jiang (2017) utilized the nonlinear local linear embedding (LLE) for dimension reduction from the feature parameters in time- and frequency domain. All these techniques achieve several successful cases, while they all suffer from their own defects. For instance, the evaluation metrics in filter approach is difficult to be determined; the wrapper approach will cause high computational burden; nonlinear dimension reduction algorithm always need extra optimization of structural parameters.

In recent years, deep neural networks (DNN) has received considerable attention in the felid of diagnostics (Lu et al., 2017; Lu et al., 2017). Deep convolutional neural networks (DCNN) was utilized for multi-sensor data fusion in the fault diagnosis of planetary gearbox (Jing et al., 2017). An optimized DCNN was applied for extracting features from raw vibration signal of bearing faults (Zhang et al., 2017). A stacked auto-encoder (SAE)-based deep fault recognizer model was established to automatically mine the essential feature from frequency spectra of fault signal (Guo et al., 2016). In essence, DNN can be treated as one kind of nonlinear transformation, converting high-dimensional feature to low dimension (Guo et al., 2016; Shao et al., 2015). And further, the signal pattern can be legibly reflected by the learned feature in low dimension. In this paper, we will introduce the novel SAE to deep statistical feature learning from a 29 statistical features in time domain and frequency domain, and diagnose the bearing faults with the fused features. The remaining part is organized as follows. In section 2, the principles of SAE and the proposed framework are introduced. In section 3, a case study using bearing fault datasets is conducted. A comparative analysis will be given to demonstrate the effectiveness and superiority of proposed method. Finally, the conclusions can be drawn in section 4.

2. INTELLIGENT DIAGNOSIS METHOD USING PROPOSED METHOD

2.1. Brief Introductions of stacked auto-encoder

Proposed in 1986, the auto-encoder is a special neural network, whose structure is shown as the left part in figure 1. The auto-encoder contains 3 layers: input layer, hidden layer and output layer. It should be noted that the neuron number of output layer is same as that of input layer. The training process of auto-encoder is similar to a guiding neural network learning process, with the purpose of getting the transformation function y = h_wb (x) using the data as a guide. The training process enables that output is equal to the original input signal, i.e., y=x. The detailed procedure can be generalized as following steps: (1) at the beginning of the training, use the input x and the initialized weights of encoding W^e to get the output value of the hidden layer, namely L2 after encoding process from L1; (2) use the output value of the hidden layer and the initialized weights of decoding W^d to get the estimated value y for the input x, namely L3 after decoding process from L2. (3) according to the deviation of estimated value y and actual value x, train the W^e and W^d by back-propagation (BP). The training algorithm is the classical gradient descent algorithm. Substantially, the auto-encoder is .a kind of neural network, whose output is approximately equal to the input. It is intuitional to find that the output of encoder may not make much sense. On the contrary, the output of hidden layer neurons is valuable, since it can be regarded as another kind of expression of the data. A series of fused features can be learned from original input with the coding process. Furthermore, on the basis of multilayer restricted boltzmann network, that is deep boltzmann machine (DBM), the stacked auto-encoder (SAE) was also proposed in 2006, whose structure is presented in the right part of figure 1. In each layer, only the encoder is retained. At the end of the network, a BP classifier is add for pattern recognition. After the unsupervised pretraining, the fine-tuning is performed as a supervised learning method to promote the feature learning ability.

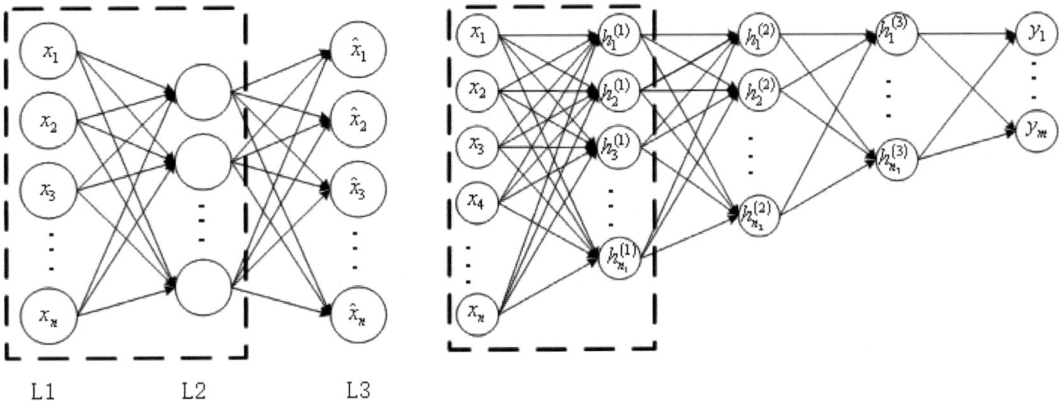

Figure 1. Illustration of the auto-encoder network and the stacked auto-encoder.

2.2. Proposed framework

Table 1. The statistical characteristic parameters in time-domain

Number	Feature expression	Number	Feature expression	Number	Feature expression
tp1	$\frac{\sum_{n=1}^{N} x(n)}{N}$	*tp7*	$max(x(n))$	*tp13*	$\frac{tp_7}{tp_4}$
tp2	$\sqrt{\frac{1}{N}\sum_{n=1}^{N} x_n^2}$	*tp8*	$min(x(n))$	*tp14*	$\frac{tp_7}{tp_3}$
tp3	$\left(\frac{1}{N}\sum_{n=1}^{N} \sqrt{\lvert x_n \rvert}\right)^2$	*tp9*	$tp_7 - tp_8$	*tp15*	$\frac{tp_5}{(tp_2)^3}$
tp4	$\frac{\sum_{n=1}^{N} \lvert x(n) \rvert}{N}$	*tp10*	$\frac{\sum_{n=1}^{N}(x(n)-tp_1)^2}{N-1}$	*tp16*	$\frac{tp_6}{(tp_2)^4}$
tp5	$\frac{\sum_{n=1}^{N}(x(n)-tp_1)^3}{N}$	*tp11*	$\frac{tp_2}{tp_4}$		
tp6	$\frac{\sum_{n=1}^{N}(x(n)-tp_1)^4}{N}$	*tp12*	$\frac{tp_7}{tp_2}$		

Where $x(n)$ is a sampling series of raw signal with N points.

Table 2. The statistical characteristic parameters in frequency-domain

Number	Feature expression	Number	Feature expression	Number	Feature expression
fp1	$\frac{\sum_{k=1}^{K} y(k)}{K}$	*fp6*	$\frac{\sum_{k=1}^{K} f_k y(k)}{\sum_{k=1}^{K} y(k)}$	*fp11*	$\frac{\sum_{k=1}^{K}(f_k - fp_5)^3 y(k)}{K(fp_6)^3}$
fp2	$\frac{\sum_{k=1}^{K}(y(k)-fp_1)^2}{K-1}$	*fp7*	$\sqrt{\frac{\sum_{k=1}^{K} f_k^2 y(k)}{\sum_{k=1}^{K} y(k)}}$	*fp12*	$\frac{\sum_{k=1}^{K}(f_k - fp_5)^4 y(k)}{K(fp_6)^4}$
fp3	$\frac{\sum_{k=1}^{K}(y(k)-fp_1)^3}{K(\sqrt{fp_2})^3}$	*fp8*	$\sqrt{\frac{\sum_{k=1}^{K} f_k^4 y(k)}{\sum_{k=1}^{K} f_k^2 y(k)}}$	*fp13*	$\frac{\sum_{k=1}^{K}(f_k - fp_5)^{1/2} y(k)}{K\sqrt{fp_6}}$
fp4	$\frac{\sum_{k=1}^{K}(y(k)-fp_1)^4}{K(fp_2)^2}$	*fp9*	$\frac{\sum_{k=1}^{K} f_k^2 y(k)}{\sqrt{\sum_{k=1}^{K} y(k)\sum_{k=1}^{K} f_k^4 y(k)}}$		
fp5	$\sqrt{\frac{\sum_{k=1}^{K}(f_k - fp_5)^2 y(k)}{K}}$	*fp10*	$\frac{fp_6}{fp_5}$		

Where $y(k)$ is the frequency spectrum of the discrete signal and k means the sequence number of spectrum line; f_k represents the frequency value of the kth spectrum line.

In the intelligent diagnosis of rolling bearing, the first step is signal acquisition and pre-processing. In this work, vibration signal is adopted as analysis object because it contains rich information about the bearing condition (Nembhard et al., 2015; Yunusa-Kaltungo et al., 2015). A variety of vibration samples can be collected by the acceleration sensor. Then, the sensitive features need to be extracted from the signal samples so as to diagnose the bearing condition. As listed in Table 1 and 2, 29 popular statistical features are employed, in which 16 features are from time-domain and the other 13 features are from frequency-domain (Lei et al., 2007). These features can describe healthy degree of mechanical components from different aspects. For example, the root mean square (tp2) can reflect the amplitude and energy of vibration signal in time-domain, the kurtosis feature (tp6) is sensitive to the impact characteristics in the signal and the parameter fp 6 may reflect the position change of main frequencies.
However, the dimensionality of the feature space is high and these features inevitably contain useless and redundant information, making the signal pattern unrecognizably. Consequently, the SAE-based deep neural network (DNN) is applied to statistical feature learning. The proposed method is a nonlinear dimensionality reduction and feature fusion method. With the learned features, the decision about the bearing condition can be made by the classifier in the end of DNN. The execution steps of proposed method is presented in figure 2. The comparative analysis and the visualization of the feature distribution are conducted in the further case study.

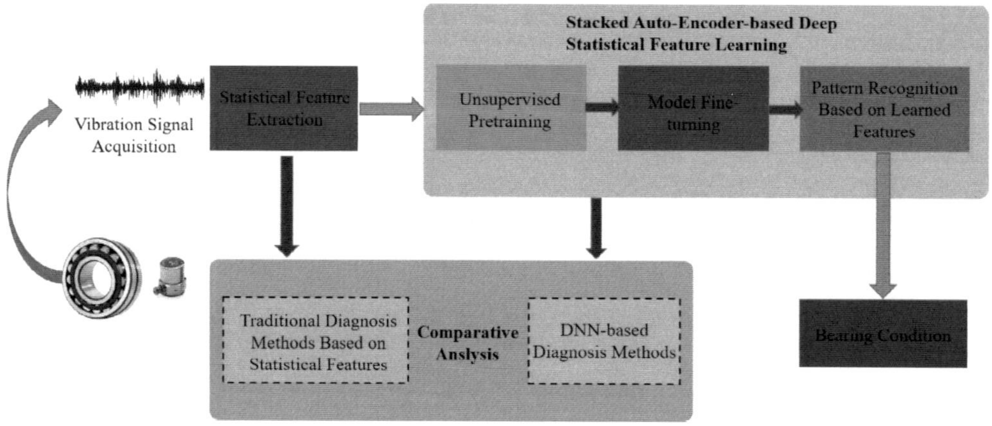

Figure 2. The architecture of this work for intelligent diagnosis of rolling bearing.

3. EXPERIMENT AND DISCUSSION

3.1. Experimental setup

Table 3. Description of bearing fault dataset

Bearing condition	Fault diameters (mil)	Motor Speed (RPM)	Label of classification	Sample size
Health	\	1797, 1772, 1750, 1730	1	2048
Inner race fault	7, 14, 21	1797, 1772, 1750, 1730	2	
Outer race fault	7, 14, 21	1797, 1772, 1750, 1730	3	
Ball fault	7, 14, 21	1797, 1772, 1750, 1730	4	

To demonstrate the effectiveness of the DNN-based diagnosis method, a case study using the bearing fault dataset from Case Western Reserve University bearing data center is performed. The test stand consists of a motor, a torque transducer, a dynamometer. The analysis focus on the drive end bearings, which support the motor shaft. Single point faults were introduced to the test bearings using electro-discharge machining with three fault diameters of 7 mils, 14 mils, 21 mils. Four bearing conditions, i.e., health, inner race fault, outer race fault and ball fault, are classified with the corresponding label. In each bearing status, the experiments are carried out under four load conditions, namely 1797rpm, 1772rpm, 1750rpm and 1730rpm. Totally, 40 operating conditions are taken into consideration and 100 samples with the length of 2048 points are collected in each operating condition. In real practical applications, it is difficult to acquire all the labelled training sets under different load conditions. To conform this condition, the samples in 1797 rpm and 1730rpm are used for training, and the others in 1772 rpm and 1750 rpm are used for testing in this study. The waveforms of one sample in the four bearing conditions are presented in figure 3.

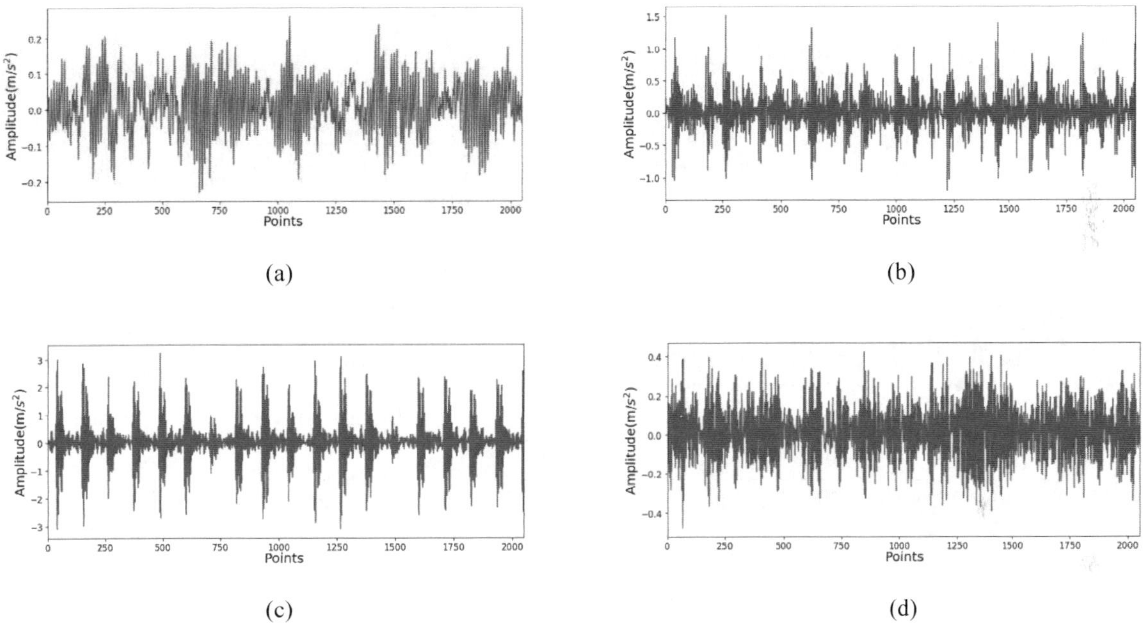

Figure 3. Vibration waveform of bearing with four states: (a) health; (b) inner race fault; (c) outer race fault; (d) ball fault.

3.2.Results and discussion of rolling bearing fault diagnosis

The architecture of the SAE is 29-30-25-20-10-6, that is, the input of the first layer is the 29 statistical features and the hidden neurons in the next few layers are 30, 25, 20, 10 and 6, respectively. In this work, to explore the inner operation conditions, we present the networks visualization for each layer in figure 4. The simple and effective PCA is adopted to generate the principal components of the output features in each layer for training samples and the first two principal components are plotted. Clearly, there are still some overlap areas in the raw 2-D principal components space for the 29 statistical features, although PCA can reduce dimensions according to the sensitive directions of maximal variance. By using SAE-based deep statistical feature learning, it is obvious that the separability of different fault patterns becomes better and better with the increase of layer number. The visualization results demonstrate that DNN not only can efficiently mine the essential features through the nonlinear transformation layer-by-layer, but can avoid the effect of speed fluctuation and crack depth change, presenting a valuable prospect for practical applications.

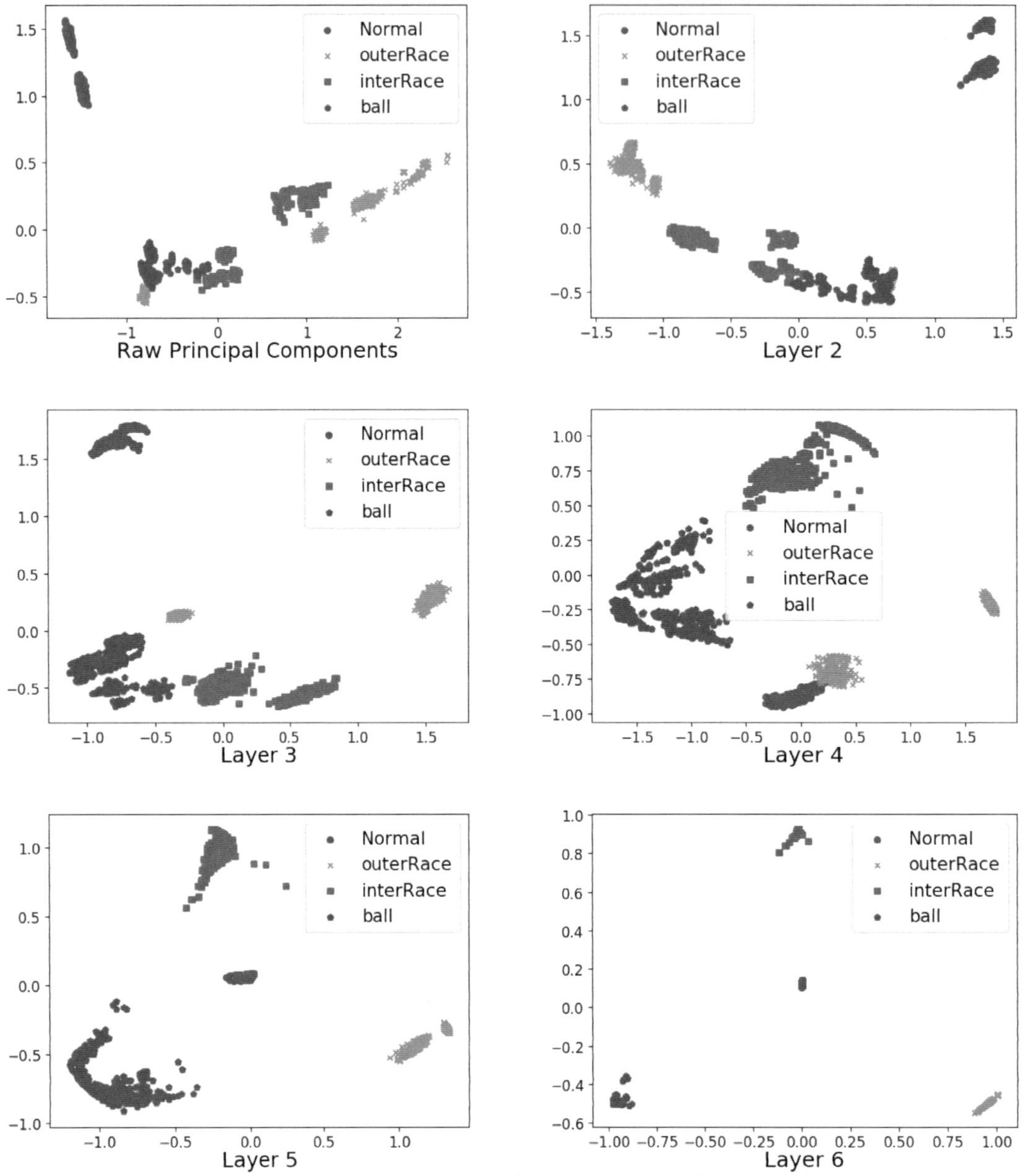

Figure 4. Visualization of the first two principal components of training samples from each layer.

Furthermore, the Table 4 gives the diagnosis results of testing samples using the proposed deep statistical feature learning method. For comparison, two traditional methods are executed, whose results are shown in Table 5 and Table 6. The principal components of raw statistical features by PCA are served as input features and the support vector machine (SVM) is utilized for classification and diagnosis. The contribution rate of adopted principal components is above 90%. Another method uses the raw statistical features as the input features for SVM directly, without prior feature fusion and dimension reduction. The results verify the learned features can remarkably reflect the nature of information hidden in raw statistical features, and further improve the diagnosis results.

Table 4. Diagnosis accuracy using deep statistical feature learning method

Test classes	Classification results(%)				Accuracy (%)	Average (%)
	Normal	OR fault	IR fault	B fault		
Normal	100	0	0	0	100	98.07
OR fault	0	96.17	3.83	0	96.17	
IR fault	0	0	100	0	100	
Ball fault	0	2.33	1.83	95.84	95.84	

Table 5. Diagnosis accuracy using statistical features, PCA and SVM

Test classes	Classification results(%)				Accuracy (%)	Average (%)
	Normal	OR fault	IR fault	B fault		
Normal	100	0	0	0	100	96.37
OR fault	0	87.17	12.83	0	87.17	
IR fault	0	0	100	0	100	
Ball fault	0	2.33	0.67	97	97	

Table 6. Diagnosis accuracy using statistical features and SVM

Test classes	Classification results(%)				Accuracy (%)	Average (%)
	Normal	OR fault	IR fault	B fault		
Normal	100	0	0	0	100	95.75
OR fault	0	84.33	15.67	0	84.33	
IR fault	0	0	100	0	100	
Ball fault	0	1.50	2.33	96.17	96.17	

4. CONCLUSIONS

With the purpose of avoiding redundancy and interferential information, a novel deep neural network (DNN) whose structure is stacked by five hidden layers of auto-encoder is proposed for feature learning from 29 raw statistical features. The proposed method is a nonlinear dimensionality reduction and feature fusion method. With the aid of the fine-tuning after the unsupervised pretraining, this method is also performed as a supervised learning process. A case study using a bearing fault dataset is conducted to demonstrate the effectiveness of this method. The networks visualization in each layer illustrates the excellent ability for capturing essential features. To make a quantitative comparative analysis, the diagnosis result of deep statistical feature learning method and another two conventional diagnosis methods are presented. The diagnosis accuracy further verify the superior fault diagnosis ability of proposed method.

REFERENCES

[1] Yuan H and Lu C. (2017) Rolling bearing fault diagnosis under fluctuant conditions based on compressed sensing. Structural Control and Health Monitoring 24(5): e1918.

[2] Han T and Jiang D. (2016) Rolling Bearing Fault Diagnostic Method Based on VMD-AR Model and Random Forest Classifier. Shock and Vibration 2016: 1-11.

[3] Han T, Jiang D and Wang N. (2016) The Fault Feature Extraction of Rolling Bearing Based on EMD and Difference Spectrum of Singular Value. Shock and Vibration 2016: 1-14.

[4] Han T, Jiang D, Zhang X, et al. (2017) Intelligent Diagnosis Method for Rotating Machinery Using Dictionary Learning and Singular Value Decomposition. Sensors 17(4).
[5] Liu C, Jiang D and Yang W. (2014) Global geometric similarity scheme for feature selection in fault diagnosis. Expert Systems with Applications 41(8): 3585-3595.
[6] Zhang X, Jiang D, Han T, et al. (2016) Feature Dimension Reduction Method of Rolling Bearing Based on Quantum Genetic Algorithm. In: Zuo MJ, Xing L, Li Z, et al. 2016 Prognostics and System Health Management Conference.
[7] Shen Z, Chen X, Zhang X, et al. (2012) A novel intelligent gear fault diagnosis model based on EMD and multi-class TSVM. Measurement 45(1): 30-40.
[8] Liu P-y, Li B, Han C-e, et al. (2016) Feature Extraction and Selection Scheme for Intelligent Engine Fault Diagnosis Based on 2DNMF, Mutual Information, and NSGA-II. Shock and Vibration 2016: 1-13.
[9] Qin W-L, Zhang W-J and Lu C. (2016) A Method for Aileron Actuator Fault Diagnosis Based on PCA and PGC-SVM. Shock and Vibration 2016: 1-12.
[10] Yang Y and Jiang D. (2017) Casing Vibration Fault Diagnosis Based on Variational Mode Decomposition, Local Linear Embedding, and Support Vector Machine. Shock and Vibration 2017: 1-14.
[11] Lu C, Wang Z-Y, Qin W-L, et al. (2017) Fault diagnosis of rotary machinery components using a stacked denoising autoencoder-based health state identification. Signal Processing 130: 377-388.
[12] Lu C, Wang Z and Zhou B. (2017) Intelligent fault diagnosis of rolling bearing using hierarchical convolutional network based health state classification. Advanced Engineering Informatics 32: 139-151.
[13] Jing L, Wang T, Zhao M, et al. (2017) An Adaptive Multi-Sensor Data Fusion Method Based on Deep Convolutional Neural Networks for Fault Diagnosis of Planetary Gearbox. Sensors 17(2).
[14] Zhang W, Peng G, Li C, et al. (2017) A New Deep Learning Model for Fault Diagnosis with Good Anti-Noise and Domain Adaptation Ability on Raw Vibration Signals. Sensors 17(2).
[15] Guo X, Shen C and Chen L. (2016) Deep Fault Recognizer: An Integrated Model to Denoise and Extract Features for Fault Diagnosis in Rotating Machinery. Applied Sciences 7(1): 41.
[16] Guo L, Gao H, Huang H, et al. (2016) Multifeatures Fusion and Nonlinear Dimension Reduction for Intelligent Bearing Condition Monitoring. Shock and Vibration 2016: 1-10.
[17] Shao H, Jiang H, Zhang X, et al. (2015) Rolling bearing fault diagnosis using an optimization deep belief network. Measurement Science and Technology 26(11): 115002.
[18] Nembhard AD, Sinha JK and Yunusa-Kaltungo A. (2015) Development of a generic rotating machinery fault diagnosis approach insensitive to machine speed and support type. Journal of Sound and Vibration 337: 321-341.
[19] Yunusa-Kaltungo A, Sinha JK and Nembhard AD. (2015) A novel fault diagnosis technique for enhancing maintenance and reliability of rotating machines. Structural Health Monitoring 14(6): 604-621.
[20] Lei Y, He Z, Zi Y, et al. (2007) Fault diagnosis of rotating machinery based on multiple ANFIS combination with GAs. Mechanical Systems and Signal Processing 21(5): 2280-2294.

Research and Application of Weak Fault Diagnosis Method Based on Asymmetric Potential Stochastic Resonance in Strong Noise Background

Zhixing Li [1], Jianguo Wang [2] , Songjiu Han [3] , Jian Kang [4], Zedong Li [5], Furui Shi [6], Wenjing Liu [7]
[1] School of Mechanical Engineering , Inner Mongolia University of Science and Technology
[2] School of Mechanical Engineering , Inner Mongolia University of Science and Technology
[3] School of Mechanical Engineering , Inner Mongolia University of Science and Technology
[4] School of Mechanical Engineering , Inner Mongolia University of Science and Technology
[5] School of Mechanical Engineering , Inner Mongolia University of Science and Technology
[6] China University of Petroleum
[7] School of Mechanical Engineering , Inner Mongolia University of Science and Technology

ABSTRACT

When machinery is used for fault diagnosis and status detection, mechanical equipment is often in a strong noise environment, The fault signal is often weak and can easily be submerged in other noise. Therefore, For this reason, this paper studies a method of stochastic resonance based on noise enhancing weak fault signals to detect weak fault signals. In this paper, a stochastic resonance phenomenon based on an asymmetrical potential system is studied. For the symmetry of the traditional stochastic resonance potential model, The shape of the potential model cannot achieve structural optimization problems. Study the effect of asymmetric potential stochastic resonance model on weak signal extraction, The effect of the asymmetric potential stochastic resonance model on weak signal extraction was studied. A non-symmetrical stochastic resonance model is proposed to achieve optimal structure. Comparing this model with traditional classical stochastic resonance. Finally, combined with experimental data of bearing weak fault diagnosis and computer simulations. Numerical simulation is consistent with the approximation theory.

Keywords: Strong noise, Stochastic resonance, Weak fault diagnosis, Asymmetric potential function

1. INTRODUCTION

With the development of modern industry, the fault diagnosis technology of mechanical equipment has received more and more attention, and it has become an important means and key technology to ensure the stable operation of production systems and improve product quality. Once a device fails to continue to work, it will often affect the normal operation of production, and serious failures may even lead to machine crashes [1-4].

In an industrial system that is more complicated and multiple devices are operating together, the faint fault signals of mechanical equipment will be submerged in a strong noise background and difficult to diagnose [5-7]. Scholars have developed many detection methods and techniques for weak signal research. such as, Wavelet analysis [8-10], singular value decomposition [11-13], empirical mode decomposition [14-16], local mean decomposition [17-19]. These methods can achieve the detection of weak signals, but there are also problems with the noise signal is filtered out together, resulting in weak signal detection result is not ideal. In 1981, Benzi [20] and other scholars first discovered and proposed the phenomenon of stochastic resonance in the interpretation of the periodic recursion of the ice age. The phenomenon of stochastic resonance shows: In a nonlinear system, noise and periodic signals are added at the same time, noise is enhanced within a certain range, and the system output signal-to-noise ratio is not reduced but will be enhanced. This has great significance for the detection of weak signals. In the early days, studies on stochastic resonances were based on common bistable studies. The classic SR requires very small parameters, and the actual fault caused a frequency greater than 1 Hz. Therefore, early stochastic resonances are not suitable for fault diagnosis[21,22]. To solve this technical problem. T Wang et al. proposed a scale transformation stochastic resonance method. The SR phenomenon can be achieved under large-parameter conditions[23-25]. In recent years, scholars have further studied stochastic resonance in the field of fault diagnosis. In terms of over-damped SR, Lu et al. [26] proposed an overdamped monostable SR method with Woods-Saxon potential to diagnose early bearing failures. Qiao et al. [27] discussed the influence of multiplicative and additive noises on the potential asymmetry of overdamped SRs. The results show, the asymmetric structure of the potential contributes to enhancing the enhancement of SR in weak feature extraction. Qiao et al. [28] established a new segmented overdamped SR model. To overcome the output saturation of the traditional bistable SR, and by using the proposed method to diagnose further planetary gearbox faults. From the above, it can be seen that over-damped SR is widely used in the field of fault diagnosis. In addition, Lu et al. [29] improved the enhanced performance of the OBSR through the full-wave signal construction and further achieved bearing fault diagnosis. Shi et al. [30] proposed a method based on multi-scale noise-based stochastic resonance tuned weak signal detection for multi-frequency signal detection in colored noise background. In addition, some scholars combine traditional detection methods with stochastic resonance to better achieve the purpose of detection. Such as, Shi et al. [31] proposed a novel weak signal detection method based on wavelet transform and parameter compensation bandpass multi-stability stochastic resonance for the detection of multi-frequency signals under heavy background noise.

In summary, Most scholars' researches are based on the classical symmetry potential bistable model, or other symmetric models such as multi-steady state and tri-stable state. However, there are few studies on the characteristics of the asymmetric potential function stochastic resonance. For this reason, this paper proposes a weak fault diagnosis method based on ASR. Compared with existing methods, ASR not only has the property of stochastic resonance with symmetrical potential, but also can independently adjust the position of the barrier, the slope of the potential wall and the potential well, and can obtain a more abundant shape of the potential structure. Therefore, this method solves the problem of local adjustment of the potential function of the existing method from the asymmetry of the potential structure, the optimization of the potential structure is achieved so as to better extract weak fault features in a strong noise background.

Section II of this paper model potential characteristics were analyzed to explore the potential effects of various parameters on the model potential model; In the third section, the method of extracting the weak signal of asymmetry potential stochastic resonance is studied. Section IV simulation verification; Section V were bearing experimental verification; Section VI concludes summarized.

2. ASYMMETRIC POTENTIAL STOCHASTIC RESONANCE MODEL

The basic principle of stochastic resonance is the synergistic effect of input signals and noise signals in nonlinear systems. As mentioned earlier, The process of detecting the weak signal by stochastic resonance can be described by the movement of particles in the potential well. The main factor that hinders the particle's transition is the shape of the potential. Adjusting parameters of traditional stochastic resonance a , b can achieve the purpose of changing the potential shape. When a is fixed, adjust the size of b , The bigger b is, the smaller the barrier height is, and the smaller the width of the potential wall is. When b is fixed, adjust the size of a , The larger a is, the larger the barrier height is, and the width of the potential wall is correspondingly increased. That is, adjusting the size of a and b , the barrier height changes, and the corresponding potential wall also changes.

Similarly, the potential function expression for asymmetric bistable stochastic resonance is as follows:

$$V(x)=V_0\left(\frac{\{1-\exp[-a(x+1)]\}\{1-\exp[b(x-1)]\}}{\{1-\exp[-a(x_0+1)]\}\{1-\exp[b(x_0-1)]\}}\right)^2 \qquad (1)$$

The smallest position of potential is at $x=\pm1$. For the sake of convenience, parameters x_0 and V_0 have been introduced to illustrate the position and height of the potential, respectively. In addition to x_0, the tunable parameters a and b control the slope of the bottom of the well. In order to compare the similarities and differences between the bistable system and the asymmetrical bistable system, the curve of the adjustment function of the different parameters is shown in Figure 1.

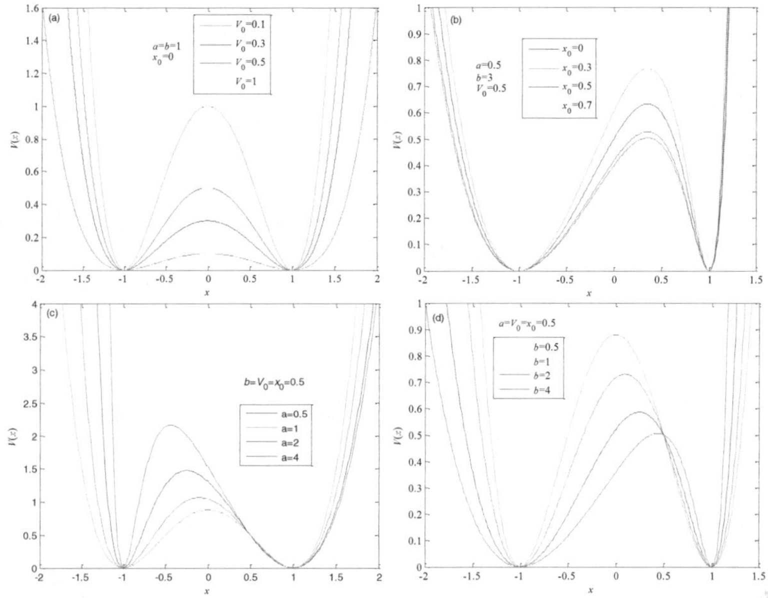

Fig. 1 Curves of different parameters to adjust the potential function: (a) The potential model when $a=b=1, x_0=0$, V_0 changes; (b) When $a=5, b=3$, $V_0=0.5$, the potential model for x_0 changes; (c) When $b=V_0=x_0=0.5$, the potential model for a changes; (d) When $a=V_0=x_0=0.5$, the potential model for b changes;

As mentioned above, the asymmetric potential stochastic resonance system also has two potential wells, a potential barrier, fixed a, b and x_0, adjusting the size of the parameter V_0, which can also achieve the adjustment of the barrier height, with the characteristics of classical bistable stochastic resonance, the difference is that the adjustment parameters can also realize the asymmetry of the potential shape. When a, b and V_0 are fixed, and the size of x_0 is adjusted individually, the barrier height and inclination change, that is, the position of the barrier changes, and the slope of the right potential wall remains basically the same; Fixed b, x_0 and V_0, adjust the size of a alone, the barrier height changes, not only the slope of the barrier changes, and the slope of the barrier side remains unchanged. In addition, when a, $V_0$and x_0 are fixed, and the size of b is individually adjusted, not only the barrier height is changed, but also the potential barrier is tilted to the right side and the gradient of the barrier is changed. From the above analysis, it can be seen that the classical bistable stochastic resonance can only adjust the height of the barrier and the width of the potential wall. The asymmetric stochastic resonance can not only regulate the height and position of the barrier. It can also ensure that the slope of one side of the potential barrier and the potential wall remains unchanged, with a richer potential shape and more adjustment characteristics.

3. ASYMMETRY POTENTIAL RANDOM WEAK SIGNAL EXTRACTION METHOD

The essence of stochastic resonance is that the particle oscillates periodically between two potential wells under the influence of noise, periodic signal and potential model, so as to achieve the purpose of improving the output of the system. In the actual project, the periodic signal and noise are fixed, and only the shape of the adjustment potential can make the random resonance effect the best. In general, when the barrier height is too high and the distance between the two sides of the wall is too far, the particles consume too much energy and cannot return to their original position from one potential well and can only oscillate within a potential well. "Under resonance"; When the inclinations of the two sides of the potential wall are too large and the barrier height is too low, the particles get too much energy to cause excessive acceleration, which causes the particles to jump back and forth between two potential wells to form an "over resonance". Therefore, only the best potential shape can enable the particles to periodically oscillate between the two potential wells with the help of weak signal and noise, thus forming the best random resonance effect. As we all know, classical bistable stochastic resonance achieves the best random resonance effect by adjusting system parameters a and b. When the height of the classical stochastic resonance barrier changes, the width of the potential wall also changes. Assuming that the width of the potential wall has reached its optimal state, the height of the barrier may not be the best. Therefore, the final state may not be the best potential shape. However, asymmetric stochastic resonance can not only independently adjust the height of the barrier, guarantee the same characteristics as the traditional stochastic resonance, but also can adjust the slope and the position of the barrier, the slope of the potential wall and the barrier. For example, when the noise is too small, in order to achieve a smooth transition of particles, classical stochastic resonance achieves its goal by reducing the height of the barrier. The asymmetry stochastic resonance can reduce the height of the barrier and at the same time, adjust the slope of the potential barrier and the slope of the potential wall, so that the asymmetric potential resonance has a better effect. Therefore, in order to extract the weak fault feature frequency in the strong noise background, The asymmetric stochastic resonance optimizes the system parameters a, b, x_0, and V_0 using the chaotic ant colony algorithm to obtain the best potential shape, and the output signal-to-noise ratio is used to evaluate the output effect. The specific steps are as follows:

(1)Signal preprocessing. The original vibration signal simulated or acquired is frequency shifted and scaled to meet the requirements of small parameters. Calculate the time domain waveform, spectrum, and envelope spectrum of the original signal.

(2)Parameter optimization. Envelope signal is input to asymmetry potential stochastic resonance system. The chaotic ant colony algorithm is used to optimize the system parameters a, b, x_0, and V_0. The parameter optimization range is set to [0,10].

(3)Calculate the signal-to-noise ratio. All the optimization parameter combination sequences are input to an asymmetric potential random resonance system and the output signal-to-noise ratio is calculated.

(4)The parameters are determined. Finding the maximum output signal-to-noise ratio results in the best combination of parameters a, b, x_0, and V_0.

(5)Signal processing. The optimal combination parameters are input to the non-symmetric potential stochastic resonance system, and finally the time domain waveform and spectrum of the asymmetric potential stochastic resonance are obtained.

4. SIMULATION VERIFICATION

In order to illustrate the effectiveness of the above ASR method, Firstly, by simulating the bearing signal to verify and construct a simulation signal, the time waveform obtained is shown in Figure 2(a). The frequency spectrum obtained by making FFT changes is shown in Figure 2(b). Due to strong noise interference, it is difficult to obtain characteristic frequencies through spectrum analysis. Then, as shown in Figure 2(c), through envelope analysis, the fault feature frequency cannot be found in the figure. In order to further extract weak fault characteristics, an asymmetric model is established. Using chaos ant colony algorithm to optimize the parameters were $a = 3.7997, b = 1.3019, x_0 = 2.623$, $V_0 = 4.7259$, Then the signal is subjected to a stochastic resonance process through an asymmetric potential model. Figure 3(a) shows the output signal after being processed by random resonance. The final Fourier transform is shown in Figure 3(b). It can be clearly seen from the figure that the movement of the particles between the two potential wells reaches resonance. The Fourier-transformed frequency domain signal is more easily identified at the characteristic frequency, and the characteristic frequency amplitude is 0.04377 higher than the surrounding noise. Therefore, the proposed asymptotic potential stochastic resonance model is effective and feasible. It can enhance the eigenfrequency and make the feature frequency easier to identify. So it can be used for weak feature fault detection. Comparing the classical symmetry stochastic resonance weak feature frequency extraction method, the same simulation signal is processed by the classical symmetric stochastic resonance method. Chaos ant colony algorithm optimizes the potential system parameters a and b , the optimization result is $a = 0.1177, b = 4.5016$, The time domain signal and frequency spectrum after the stochastic resonance processing are shown in Figure 3(c) and (d). As can be seen in Figure 3(d), the characteristic frequency is also higher than the ambient noise, but the amplitude of the characteristic frequency is only 0.01517 higher than the ambient noise. Therefore, it can be determined that the classical symmetry potential stochastic resonance restricts the change of the potential shape due to the symmetry of the potential and forms the local optimum of the potential shape. Comparing Fig. 3(b) and Fig. 3(d), it is found that the peak value of the characteristic frequency output by the proposed method is significantly enhanced compared with the conventional method, indicating that the phenomenon of stochastic resonance is more obvious. Therefore, the proposed asymmetry potential stochastic resonance method is better than traditional methods for detecting weak features.

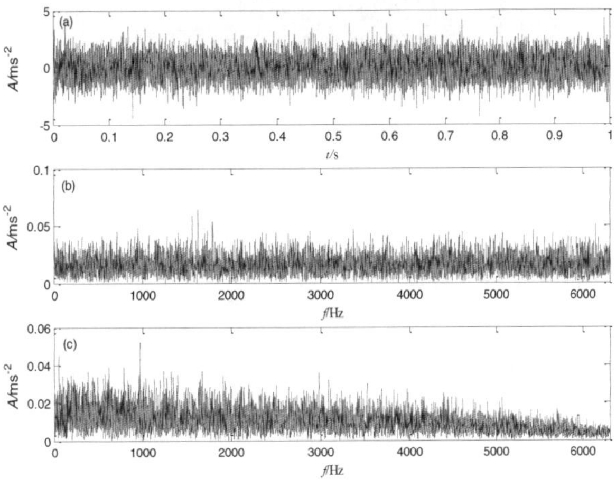

Fig. 2 Simulation signal: (a) Time domain waveform (b) Spectrum (c) Envelope spectrum

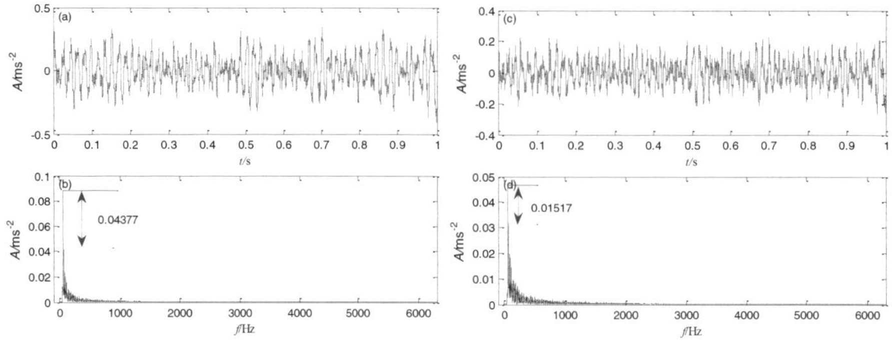

Fig. 3 Symmetrical potential stochastic resonance method for simulated signals: (a) Time domain waveform (b) Spectrum; Asymmetry potential stochastic resonance method for simulated signal: (c) Time domain waveform (d) Spectrum

5. BEARING TEST VERIFICATION

In order to demonstrate the effectiveness of ASR, bearing tests have been used. First of all, In order to meet the requirements of small parameters, the use of frequency-shifting scales to preprocess bearing fault data, the envelope signal is input to the asymmetrical system. The parameters of the asymmetric potential stochastic resonance system obtained by using the chaotic ant colony algorithm are $a = 1.3004, b = 1.4739, x_0 = 3.795, V_0 = 3.0895$, The time domain plot and frequency domain plot of the output signal are shown in Figure 4. From the spectrum diagram in Figure 4(b), it can be seen that the frequency of the fault signature is clearly demonstrated, being 0.03857 higher than the ambient noise. Therefore, It can be concluded that the proposed method can be applied to fault diagnosis of bearings. and achieved remarkable results.

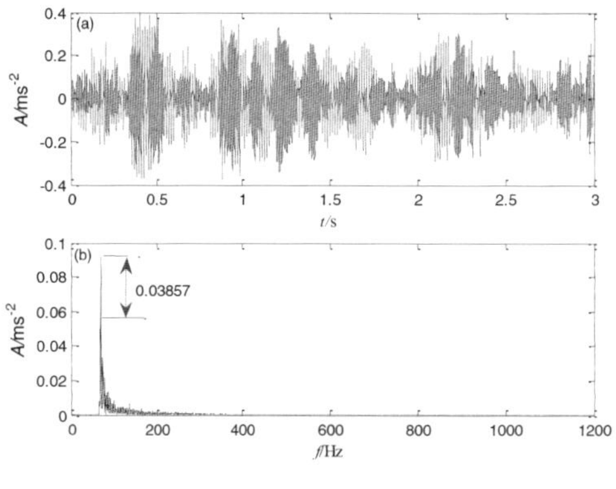

Figure 4 Asymmetric potential resonance method for bearing signals: (a) Time domain waveform (b) Spectrum

The non-symmetric potential stochastic resonance proposed in this paper not only has the symmetry potential shape adjustment characteristics, but also through multi-parameter optimization. The inclination of the barrier, the multi-angle adjustment of the slope of the potential wall and the barrier can be realized, and a more abundant potential shape can be obtained, and the movement of the particles between the potential wells on both sides can be realized more easily. The proposed method for adaptive non-symmetry stochastic resonance obtains better results in the diagnosis of weak faults in rolling bearings.

6. CONCLUSION

Aiming at the symmetry of the classical classical stochastic resonance potential shape, this paper presents a new method of asymmetry potential stochastic resonance weak fault diagnosis. And applied to weak feature signal extraction under strong noise background, draw the following conclusions: (1) The structural properties of the potential function play a key role in the effect of stochastic resonance. The multi-parameter optimization of asymmetric potential stochastic resonance can make the shape of the stochastic resonance system diversify. Not only can the potential height and potential width be adjusted individually, but also the slope of the barrier, the slope of the potential wall and the barrier, can be independently adjusted. Therefore, the best potential shape can be obtained. (2) Simulation and bearing pitting corrosion experiments show that the proposed method is effective in weak fault diagnosis.

REFERENCES

[1] [1] N. Sawalhi, R.B. Randall, Vibration response of spalled rolling element bearings: Observations, simulations and signal processing techniques to track the spall size, [J] . Mechanical Systems and Signal Processing, 2011 , 25 (3) :846-870

[2] [2] J.J. Zakrajsek , R.F. Handschuh , H.J. Decker, Application of fault detection techniques to spiral bevel gear fatigue data,[J]Advanced Materials and Process Technology for Mechanical Failure Prevention , 1994 , 14 :93-104

[3] [3] J. Zheng , H. Pan , J. Cheng, Rolling bearing fault detection and diagnosis based on composite multiscale fuzzy entropy and ensemble support vector machines,[J]Mechanical Systems & Signal Processing , 2017 , 85 :746-759

[4] [4] J. Lin , Q .Chen, Fault diagnosis of rolling bearings based on multifractal detrended fluctuation analysis and Mahalanobis distance criterion,[J]Mechanical Systems and Signal Processing , 2013 , 38 (2) :515-533

[5] [5]Z. Li , B. Shi, An adaptive stochastic resonance method for weak fault characteristic extraction in planetary gearbox,[J]Journal of Vibroengineering , 2017 , 19 (3) :1782-1792

[6] [6]S. Lu , Q. He , H. Zhang , F. Kong, Rotating machine fault diagnosis through enhanced stochastic resonance by full-wave signal construction,[J]Mechanical Systems and Signal Processing, 2017 , 85 :82-97

[7] [7]P. Zhou , S. Lu , F. Liu , Y. Liu , G. Li, Novel synthetic index-based adaptive stochastic resonance method and its application in bearing fault diagnosis,[J]Journal of Sound and Vibration , 2017 , 391 :194-210

[8] [8]F. Hemmati , W. Orfali , M.S. GadalaRoller, bearing acoustic signature extraction by wavelet packet transform, applications in fault detection and size estimation,[J]Applied Acoustics , 2016 , 104 :101-118

[9] [9]Z. Gao , J. Lin , X. Wang , X. Xu, Bearing Fault Detection Based on Empirical Wavelet Transform and Correlated Kurtosis by Acoustic Emission,[J]Materials , 2017 , 10 (6) :571

[10] [10]KCD. Kompella , MVG. Rao , R.S. Rao, Bearing fault detection in a 3 phase induction motor using stator current frequency spectral subtraction with various wavelet decomposition techniques,[J]Ain Shams Engineering Journal , 2017

[11] [11]Z. Li , B. Shi, Extracting weak fault characteristics with adaptive singular value decomposition and stochastic resonance,[J]Transactions of the Chinese Society of Agricultural Engineering , 2017 , 33 (11) :60-67

[12] [12]Q. Lu , T. Wang , Z. Li , C. Wang, Detection Method of Series Arcing Fault Based on Wavelet Transform and Singular Value Decomposition,[J]Transactions of China Electrotechnical Society , 2017 , 32 (17) :208-217

[13] [13]R. Golafshan , K.Y. Sanliturk, SVD and Hankel matrix based de-noising approach for ball bearing fault detection and its assessment using artificial faults,[J]Mechanical Systems and Signal Processing , 2016 , s 70 – 71 :36-50

[14] [14]Z. Li , B. Shi, Research of Fault Diagnosis Based on Sensitive Intrinsic Mode Function Selection of EEMD and Adaptive Stochastic Resonance,[J]Shock and Vibration,2016,(2016-12-4) , 2016 , 2016 (11) :1-12

[15] [15]V.T. Do , C.N. Le, Adaptive Empirical Mode Decomposition for Bearing Fault Detection,[J]Journal of Mechanical Engineering 62(2016)5, 281-290

[16] [16] J. Xiang , Y. Zhong, A new fault detection strategy using the enhancement ensemble empirical mode decomposition,[J]Journal of Physics Conference Series , 2017 , 842 (1) :012002

[17] [17]C. Zhang , Z. Li , C. Hu , S. Chen , J. Wang ,An optimized ensemble local mean decomposition method for fault detection of mechanical components,[J]Measurement Science & Technology , 2017 , 28 (3) :035102

[18] [18]Z. Wu , S.P. Yang , J.C. Zhang, Bearing fault feature extraction method based on LMD adaptive multiscale morphology and energy operator demodulating,[J]Journal of Vibration and Shock , 2016

[19] [19]G. Cheng , H. Li , X. Hu , X. Chen , H. Liu, Fault diagnosis of gearbox based on local mean decomposition and discrete hidden Markov models,[J] ARCHIVE Proceedings of the Institution of Mechanical Engineers,2016 , 231 (14)

[20] [20] R. Benzi , G. Parisi , A. Sutera , A. Vulpiani , A Theory of Stochastic Resonance in Climatic Change,[J]. Siam Journal on Applied Mathematics , 1983 , 43 (3) :565-578

[21] [21] Q. He , J. Wang , Y. Liu , D. Dai , F. Kong , Bearing defect diagnosis by stochastic resonance with parameter tuning,[J]Prognostics and System Health Management Conference , 2011 :1-5

[22] [22]J. Hao , D.U. Taihang , C. Jiang , S. Sun , F.U. Chao, Application of parameter-tuning stochastic resonance for detecting weak signal with ultrahigh frequency,[J]Journal of Computer Application》 , 2016

[23] [23] F. Shengbo , T. Wang , Y. Leng , Wang. Wenjin, Detection of Weak Periodic Impact Signals Based on Scale Transformation Stochastic Resonance,[J]China Mechanical Engineering , 2006 , 17 (4) :387-390
[24] [24]Y.G. Leng , T.Y. Wang , L.I. Rui-Xin , Y.S. Peng , X.X. Deng, SCALE TRANSFORMATION STOCHASTIC RESONANCE FOR THE MONITORING AND DIAGNOSIS OF ELECTROMOTOR
[25] FAULTS,[J]Proceedings of the Csee , 2003
[26] [25]Y .Zhou , H. Wang , Y. Zuo , M. Yao, Bearing fault diagnosis based on scale transformation stochastic resonance,[J]Journal of Beijing Information Science & Technology Unversity, 2015 , 8916 (1) :36
[27] [26]S. L. Lu, Q.B. He, F.R. Kong. Stochastic resonance with Woods-Saxon potential for rolling element bearing fault diagnosis[J]. Mechanical Systems and Signal Processing, 2014 , 45 (2) :488-503.
[28] [27]Z. J. Qiao, Y. G. Lei, J. Lin, et al. Stochastic resonance subject to multiplicative and additive noise: the influence of potential asymmetries[J]. Physical Review E, 2016 , 94 (5-1) :052214.
[29] [28]Z. J. Qiao, Y. G. Lei, J. Lin, et al. An adaptive unsaturated bistable stochastic resonance method and its application in mechanical fault diagnosis[J]. Mechanical Systems and Signal Processing, 2017 , 84 :731-746.
[30] [29]S. L. Lu, Q. B. He, H. B. Zhang, et al. Rotating machine fault diagnosis through enhanced stochastic resonance by full-wave signal construction[J]. Mechanical Systems and Signal Processing, 2017 , 85 :82-97.
[31] [30]P.Shi , X. Ding , D. Han, Study on multi-frequency weak signal detection method based on stochastic resonance tuning by multi-scale noise,[J] Measurement , 2014 , 47 (1) :540-546
[32] [31]D .Han , P. Li , S .An , P. Shi, Multi-frequency weak signal detection based on wavelet transform and parameter compensation band-pass multi-stable stochastic resonance,[J] Mechanical Systems and Signal Processing , 2016 , s 70 – 71 :995-1010

Fault diagnosis method of rolling element bearing based on relative wavelet packet energy and Hilbert envelope analysis

Junchao Guo[1], Haiyang Li[2], Dong Zhen[1,*], Hao Zhang[1], Zhanqun Shi[1], Fengshou Gu[2]
[1]. School of Mechanical Engineering, Hebei University of Technology, Tianjin, 300140, P.R.China
[2]. Centre for Efficiency and Performance Engineering, University of Huddersfield, HD1 3DH, UK

ABSTRACT

Rolling element bearings are widely used in rotating machinery, faults occurring in bearings may lead to fatal breakdowns in rotating machinery and such failure can be catastrophic, resulting in costly downtime. Therefore, it is significant to accurately diagnose the existence of faults at an early stage. Vibration signals collected from bearings contain rich information on machine health condition . Here, it is possible to obtain vital characteristic information from vibration signals through the use of advanced signal processing techniques due to their intrinsic advantage of revealing bearing failure. This paper proposes a new study to take advantage of the relative wavelet packet energy (RWPE) and Hilbert envelope analysis. Relative wavelet packet energy (RWPE) can reflect the energy distribution information of the signal in each wavelet packet transform frequency band. Then apply the Hilbert envelope analysis separate the modulation components to extract fault frequency. The results show that the outer ring fault features of the rolling bearing can be clearly identified through simulation and experimental analysis. It verify the proposed method is effective and feasible in the condition monitoring and fault diagnosis for rolling bearing.

Keywords: relative wavelet packet energy; Hilbert envelope analysis; rolling element bearings; fault frequency; health condition

Corresponding author: d.zhen@hebut.edu.cn

1. INTRODUCTION

Rolling element bearings are one of the most important and common components in rotary machines, and their failures can cause both personal damage and economic loss, if the fault cannot be detected and diagnosed well in advance. Therefore they have received much attention in the field of vibration analysis as they represent an area where much can be gained from the early detection of faults. According to the statistics, about 90% of the faults occur in the surface of the inner race, outer race or rolling elements of the rolling element bearing, in particular most of them are local faults [1]. Therefore, an effective signal processing method is desired to provide more evident information for fault diagnosis of rolling element bearings.

Signal processing has gone through the development of time domain, frequency domain and time-frequency analysis. Time and frequency domain analysis only can be applied to stationary and linear signals. To deal with non-stationary signals, the time-frequency analysis method that can extract local features in both time and frequency domains simultaneously has been used widely for machinery fault diagnosis. The regular time-frequency analysis methods include short time Fourier transform (STFT), Wigner-Vill distribution (WVD), wavelet pack transform (WPT), empirical mode decomposition (EMD), etc. WPT can be used for multi-scale analysis of a signal through dilation and translation, so it can extract signal features from both time domain and frequency domain effectively. Therefore, WPT has been successfully applied for the condition monitoring and fault diagnosis of electromechanical equipment [2-3]. In recent years, WPT has been used as a popular method in the field of condition monitoring and fault diagnosis [4–7].

Relative wavelet packet energy (RWPE) can reflect the energy distribution information of the signal in each wavelet packet transform frequency band. Therefore, the RWPE of the rolling bearing signal can reflect the bearing in the event of a fault, the energy distribution of the vibration signal can be as a bearing fault diagnosis eigenvector. In this paper, the proposed method integrate the benefits of RWPE with the strengths of the Hilbert envelope analysis to extraction the fault characteristic frequency. Firstly, the wavelet packet is used to decompose the vibration signal of rolling element bearing. Then, calculated the RWPE of each frequency band, select the frequency band with high RWPE for signal reconstruction. Finally, the fault feature frequency is extracted by Hilbert envelope analysis. The diagnosis result show that the proposed method can extract the fault characteristic frequency. Which is shown that the efficiency of the proposed method in fault diagnosis for rolling element bearings.

1. RELARIVE WAVELET PACKET ENERGY

The WPT has well-known properties of being orthogonal, complete, and local. The basic operation of WPT is to filter the signal being analyzed through a series of low-pass and high-pass filters recursively. A signal can be decomposed into a set of WP nodes with the form of a full binary tree by the WPT. Let $V_{0,0}$ be a vector space of R^n, corresponding to the node 0 of the parent tree. Then at each level the vector space is split into two mutually orthogonal subspaces given by the following equation:

$$V_{j,k} = V_{j+1,2k} \oplus U_{j+1,2k+1} \tag{1}$$

where j indicates the level of the tree, and k represents the node index in level j, given by $k = 0,\ldots,2^j - 1$. This process is repeated until level J, giving rise to2J mutually orthogonal subspaces.

The principle of WPT can be mathematically described as follows. First, the WP function $W^n{}_{j,k}(t)$ can be mathematically expressed as below [8]:

$$W^n_{j,k}(t) = 2^{j/2} W^n(2^j t - k) \tag{2}$$

where the integers j and k are the scale and translation parameters, respectively; $n = 0,1,\ldots$ is the oscillation parameter. The first two wavelet packet functions with $j = k = 0$ are the scaling function $\Phi(t)$ and mother wavelet function $\Psi(t)$ as below:

$$W^0_{0,0}(t) = \Phi(t) \tag{3}$$

$$W^1_{0,0}(t) = \Psi(t) \tag{4}$$

The remaining WP functions for $n = 2,3,\ldots$ are defined through the following recursive relationships:

$$W^{2n}(t) = \sqrt{2}\sum_k h(k) W^n_{1,k}(2t-k) \tag{5}$$

$$W^{2n+1}(t) = \sqrt{2}\sum_k g(k) W^n_{1,k}(2t-k) \tag{6}$$

Where $h(k) = 1/\sqrt{2}\langle \varphi(t), \varphi(2t-k)\rangle$ and $g(k) = 1/\sqrt{2}\langle \psi(t), \psi(2t-k)\rangle$ are the filter coefficients of low-pass and high-pass filters respectively, and they are orthogonal with the relationship $g(k) = (-1)^k h(1-k)$. Here, $\langle \cdot,\cdot \rangle$ represents the inner product operator. The wavelet packet coefficients $Q^n_{j,k}$ are obtained by the inner product between the signal $x(t)$ and the wavelet packet functions $W^n_{j,k}$, as below:

$$Q^n_{j,k} = \langle x, W^n_{j,k}(t)\rangle = \int_{-\infty}^{+\infty} x(t) W^n_{j,k}(t)\,dt \tag{7}$$

where $Q^n_{j,k}$ denotes the nth set of WP coefficients at the jth scale parameter and k is the translation parameter.

Mathematically, for a discrete signal $x(t)$ with N data points, assume $N = 2^{n0}$, the wavelet coefficients of node (j,n) can be denoted by $\{Q^n_j(k), k = 1,2,\ldots,2^{n0-j}\}$.The signal energy contained in the WP node (j,n) is calcualted as:

$$E^n_j = \sum_{k=1}^{2^{n0-j}} \left[Q^n_j(k)\right]^2 \tag{8}$$

Following the energy, define the average power at each WP node as:

$$AE^n_j = \frac{1}{2^{n0-j}} \sum_{k=1}^{2^{n0-j}} \left[Q^n_j(k)\right]^2 \tag{9}$$

With this assumation, the signal energy vector for the j th level can be written as below :

$$AE_j = \left[\overbrace{\left[AE^1_j\right]_{2^{n0-j}\times 1} \cdots \left[AE^n_j\right]_{2^{n0-j}\times 1} \cdots \left[AE^{2j}_j\right]_{2^{n0-j}\times 1}}^{n=1,\ldots,2^j} \right] \in R^{1\times 2^{n0}} \tag{10}$$

where $\left[AE^n_j\right]_{2^{n0-j}\times 1}$ is a $2^{n0-j} \times 1$ constant matrix with the value of AE^n_j .

For diffenrent layer, the total energy, that is the sum of average power in vector AE^n_j , will be the same. Put the average power vectors at layer 0 to J together, then deliver the average power matrix as follows:

$$AE = \left[AE_0{}^T, AE_1{}^T, \cdots, AE_J{}^T\right]^T \in R^{(1+J)\times 2^{n0}} \tag{11}$$

As there will be redundant average power values inside a same WP node whithin the average power vector AE_j , the size of the matrix AE, $(J+1)\times 2^{n0}$, can be reduced to be $(J+1)\times 2^J$ by reorganizing the average power vector AE_j as

$$RAE_j = 2^{n0-J} \cdot \left[\overbrace{\left[AE^1_j\right]_{2^{J-j}\times 1} \cdots \left[AE^n_j\right]_{2^{J-j}\times 1} \cdots \left[AE^{2j}_j\right]_{2^{J-j}\times 1}}^{n=1,\ldots,2^j} \right] \in R^{1\times 2^J} \tag{12}$$

The above reorganization is just like changing the container area of node (j,n) from 2^{n0-j} to 2^{J-j} ,so the average power value is increased. At last, the reorganized average power matrix, also called WPE matrix here, becomes a new one with the size of $(J+1)\times 2^J$ as follows:

$$WPE = \left[RAE_0{}^T, RAE_1{}^T, \cdots, RAE_J{}^T\right]^T \in R^{(1+J)\times 2^J} \tag{13}$$

where specifically, the final row of the matrix, RAE_j , is called the WPE vector that corresponds to the WP nodes energy at the final level J of the WPT.

2. HILBERT ENVELOPE ANALYSIS

The vibration signals of interest have repetitive high frequency impulses as a consequence of interaction between rolling element components and defects. These burst sounds generated by the bearing defects are modulated in amplitude by the sequence of repetitive impacts and by the damping effect. The direct FFT analysis of the signals does not provide much information, since noise and high frequency components caused by movement of rolling elements against each other will be mixed with the characteristic frequencies of bearing faults. These repeating fault frequencies are, however, easily measured in the signal envelope. The envelope detection method provides an important and effective approximation to analyze fault signals in high frequency vibrations. The signal envelope can be calculated by the Hilbert transform [9] which acts as a linear operator on the signal to produce a function with the same domain. Given a signal x(t) in the time domain, the Hilbert transform is the convolution of x(t) with the signal $\frac{1}{\pi t}$ which produces a new signal in the time domain. Hilbert transform can be defined as:

$$\hat{x}(t) = \frac{1}{\pi t} * x(t) = \frac{1}{\pi}\int_{-\infty}^{+\infty} \frac{x(\tau)}{t-\tau} d\tau \tag{14}$$

By coupling the x(t) and $\hat{x}(t)$, analytical signal h(t) can be calculated as,

$$h(t) = x(t) + j\hat{x}(t) = a(t)e^{j\varphi(t)} \tag{15}$$

where a(t) is the envelope of the signal

$$a(t) = \sqrt{x^2(t) + \hat{x}^2(t)}\ , \tag{16}$$

3. SIMULATION ANALYSIS

The vibration signature of a rolling element bearing with local defects can be typified by an amplitude modulation process. Taking into account the existence of random sliding phenomenon, rolling element bearing vibration signal can be expressed by the following formula:

$$\begin{cases} x_1(t) = \sum_{m=1}^{M} A h_1(t - \frac{i}{f_i} - \tau_i) \\ h_1(t) = \exp(-C_1 t)\cos(2 \times 8000\pi t) \end{cases} \tag{17}$$

Where A is the amplitude of signal, equal to 1, C_i is attenuation coefficient equal to 900, f_i is the fault characteristic frequency (equal to 100 Hz), τ_i is used to simulate the randomness caused by slippage, which is subject to a discrete uniform distribution . The simulated data was acquired with a sample rate of 20 kHz and the data length is 8192. The time domain waveform of the bearing vibration signal in figure 1. It is difficult to identify the fault of bearing from the time waveform , indicating that a further analysis is necessary.

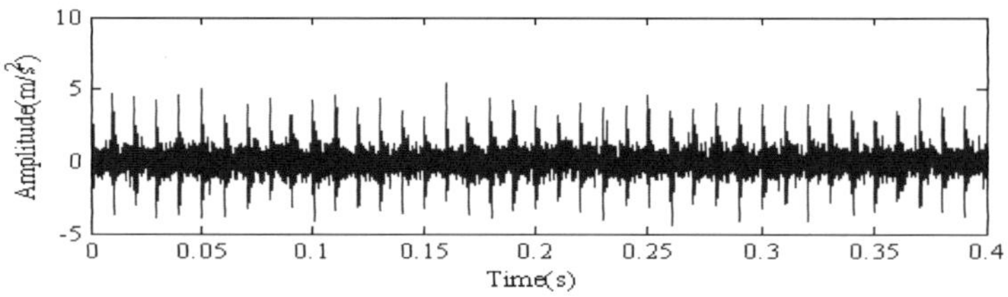

Figure 1 The waveform of the outer ring fault signal

The RWPE algorithm is used for the characteristic extraction from the separated fault signals. The wavelet function db 3 is applied, decomposition level is 3, 8 decomposition bands are obtained. The RWPE are obtained, and proportional relationships of the various energy bands are made into a series of histograms. The sum of the height of the histogram is 1, therefore, the height of each histogram represents the ratio of the frequency band energy to the total energy. figure 2 show the energy distribution proportional histogram of the outer race fault signals.

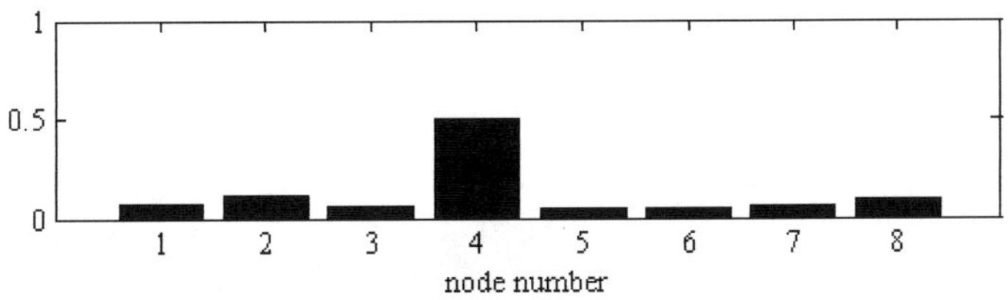

Figure 2 Frequency energy of the outer ring fault signal

As is shown in figure 2, the changes in the ratios of energy distribution in the node [3,3]. (for which the frequency band is 7500–10000 Hz). The energy distribution ratio coefficients of the one band is therefore used as the eigenvalue of the signals of the outer-race fault. The reconstruction signal as shown in figure 3. To the nonlinear modulation components in the vibration signal for the rolling element bearing, it is proposed that the Hilbert envelope analysis to the decompose modulation components. figure 4 illustrate the results of the proposed method. Four dominant frequency spectrum lines located at the harmonics of the outer race bearing rotation frequency (i.e. f_o, $2f_o$, $3f_o$, $4f_o$) can be recognized in figure 4. This strongly demonstrates that the fault occurs in the outer race bearing rolling. Therefore, it is reasonable to conclude that the proposed method is more effective in bearing fault diagnosis.

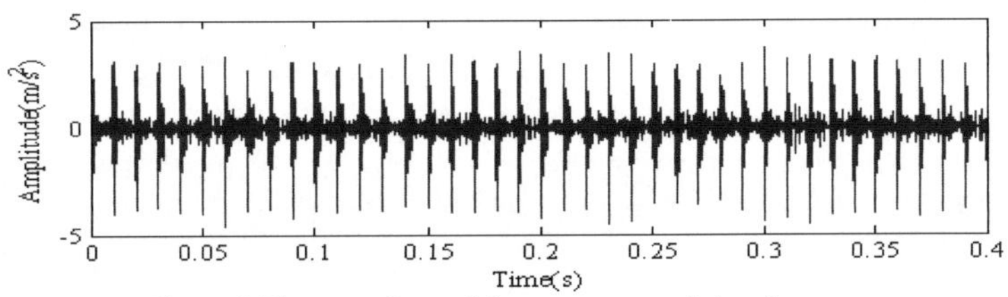

Figure 3 The waveform of the reconstructed signal

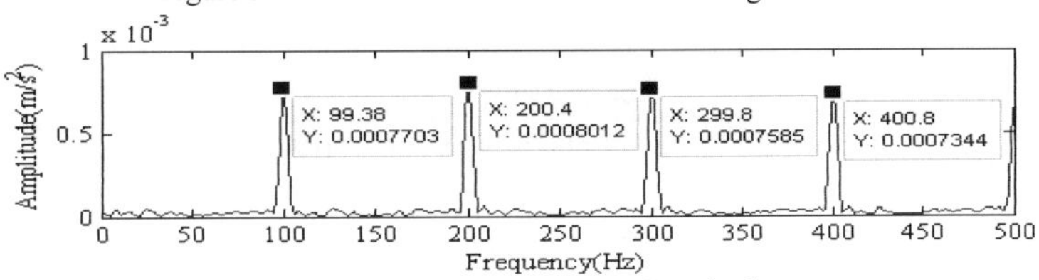

Figure 4 The result of the proposed method

4. EXPERIMENTS VALIDATION

To validate the effectiveness of the proposed method in terms of rotating machinery fault diagnosis, rolling element bearings are analyzed. The vibration signal was collected from the bearing test rig is shown in figure 5. It is comprised of a motor, coupling, intermediate shaft, supporting bearings and electrical brake. The vibration signals were acquired by piezoelectric accelerometer, which was mounted in the vertical direction on the motor drive end bearing housing and the sensitivity was 1.04 mV/ms^2. Data were sample at 20 kHz. The structural parameters and kinematical parameters of the experiment bearings are listed in Tab.3.

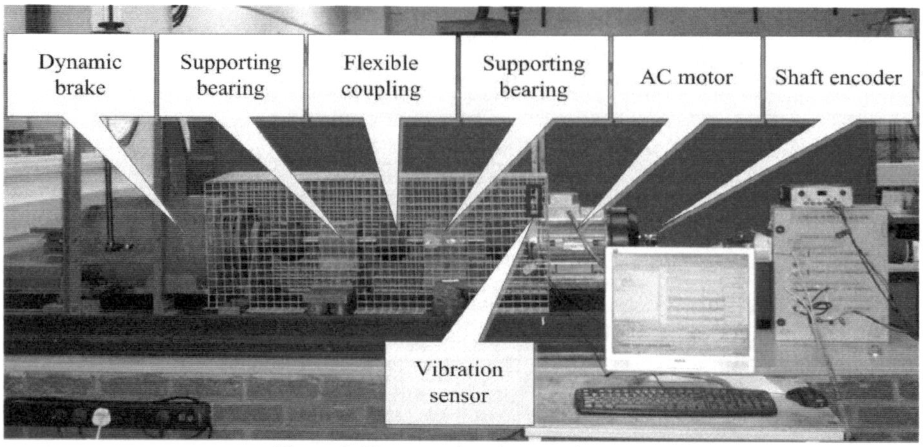

Figure 5. The motor bearing test rig

Table1: Structural parameters and kinematical parameters of the experiment bearings

Bearing designation	Ball numbers d	Pitch Diameter D_m	Ball Number z	Contact Angle β
	9	46.4	9.53	0°
6206ZZ	f_o	f_i	f_b	f_c
	89.33	130.99	62.42	9.93

The bearing characteristic frequencies of a rolling element bearing can be calculated with Eqs. (18)-(21). The outer race fault frequency f_o, the inner race fault frequency f_i, the rolling element fault frequency f_b, the fundamental cage frequency f_c were formulated as follows respectively [10]:

$$f_o = \frac{z}{2}(1 - \frac{d}{D_m}\cos\beta)f_r \tag{18}$$

$$f_i = \frac{z}{2}(1 + \frac{d}{D_m}\cos\beta)f_r \tag{19}$$

$$f_b = \frac{D_m}{d}[1 - (\frac{d}{D_m}\cos\beta)^2]f_r \tag{20}$$

$$f_c = \frac{1}{2}(1 - \frac{d}{D_m}\cos\beta)f_r \tag{21}$$

Where z is the number of balls, d is the diameter of the rolling element, D_m is the groove section size, β is the contact angle, f_r is the shaft frequency.

Figure 6 presents the time domain waveform. It is difficult to identify the fault of gearbox bearing from figure 6. Therefore, indicating that a further analysis is necessary.

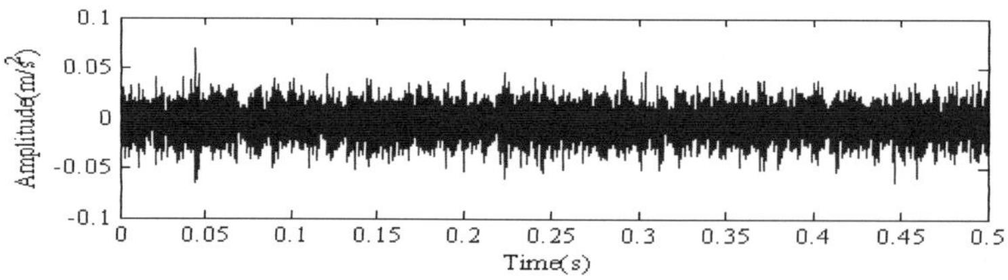

Figure 6 The waveform of the outer ring fault experimental signal

The RWPE algorithm is used for the characteristic extraction from the separated fault signals. The wavelet function db 3 is applied, decomposition level is 3, 8 decomposition bands are obtained. Figure 7 show the energy distribution proportional histogram of the outer race fault signals.

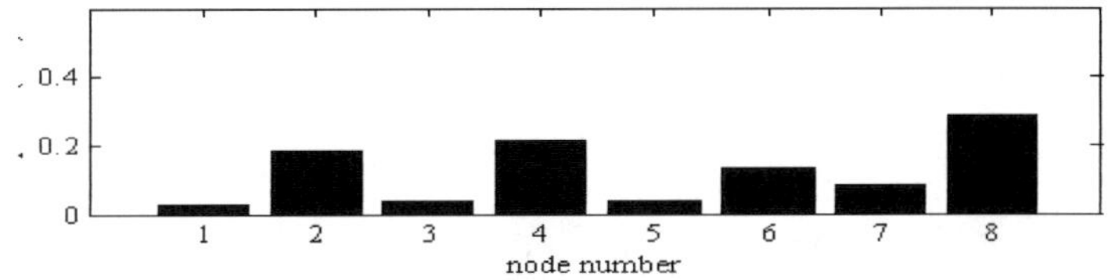

Figure 7 Frequency energy of the outer ring fault experimental signal

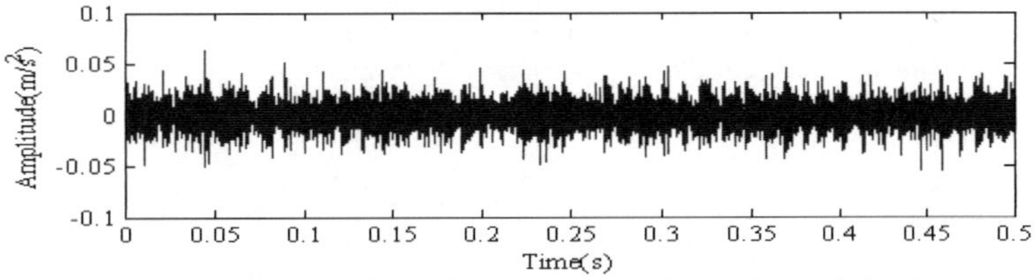

Figure 8 The waveform of the reconstructed experimental signal

As is shown in figure.9, the changes in the ratios of energy distribution in the node [3,1] (for which the frequency band is 2500–5000 Hz), node [3,3] (for which the frequency band is 7500–10000 Hz) the node [3,5] (for which the frequency band is 15000–17500 Hz) and the node [3,7] (for which the frequency band is 17500–20000 Hz). The energy distribution ratio coefficients of the four bands are therefore used as the eigenvalue of the signals of the outer-race fault. The reconstruction signal as shown in Figure. 8. To the nonlinear modulation components in the vibration signal for the rolling element bearing, it is proposed that the Hilbert envelope analysis to the decompose modulation components. Figure 9 illustrate the results of the proposed method. Five dominant frequency spectrum lines located at the harmonics of the outer race bearing rotation frequency (i.e. f_o, $2f_o$, $3f_o$, $4f_o$,$5f_o$) can be recognized in figure 9. It can be seen that there are numerous frequency peaks associated with the outer race fault and their harmonics. In conclusion, the proposed method is more effective method in rolling element bearing fault diagnosis.

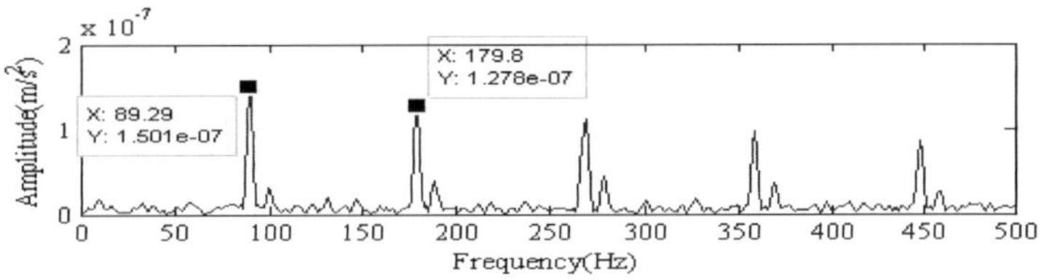

Figure 9 The result of the proposed method

6. CONCLUSION

This paper presents a new method based on relative wavelet packet energy (RWPE) and Hilbert envelope analysis method for fault diagnosis of rolling element bearing. In the simulation and experimental verification, the proposed method have the ability to extract the bearing fault signatures. It demonstrates that the proposed method can be considered as an effective method for rolling element bearing faults diagnosis.

ACKNOWLEDGMENT

This research was supported by the National Natural Science Foundation of China (Grant no. 51605133; 51705127) and Hebei Science and Technology Projects of China (Grant no. 17394303D).

REFERENCES

[1] Dong, Y., Liao, M., Zhang, X., et al. Faults diagnosis of rolling element bearings based on modified morphological method. Mechanical Systems and Signal Processing, 25(4) (2011) 1276-1286.

[2] Peng, Z.K., Chu, F.L.. Application of the wavelet transform in machine condition monitoring and fault diagnostics: a review with bibliography. Mechanical Systems and Signal Processing, 18 (2) (2004) 199-221.

[3] Xing, Y.F., Wang, Y.S., Shi, L., et al. Sound quality recognition using optimal wavelet-packet transform and artificial neural network methods. Mechanical Systems and Signal Processing, 66 (2015).

[4] Mavaddaty, S., Ahadi, S.M., Seyedin, S. Speech enhancement using sparse dictionary learning in wavelet packet transform domain. Computer Speech and Language, 2017, 44:22-47.
[5] Bianchi, D., Mayrhofer, E., Gröschl, M., et al. Wavelet packet transform for detection of single events in acoustic emission signals. Mechanical Systems and Signal Processing, 64-65 (2015) 441-451.
[6] He, Q. Vibration signal classification by wavelet packet energy flow manifold learning. Journal of Sound and Vibration, 332 (2013) 1881-1894.
[7] Han, L., Li, C.W., Guo, S.L., et al. Feature extraction method of bearing AE signal based on improved FAST-ICA and wavelet packet energy. Mechanical Systems and Signal Processing, 62–63 (2015) 91-99.
[8] Wang, Y., Xu, G., Lin. L., et al. Detection of weak transient signals based on wavelet packet transform and manifold learning for rolling element bearing fault diagnosis. Mechanical Systems and Signal Processing, 54-55 (2015) 259-276.
[9] Cizek, V. Discrete hilbert transform. Audio Electroacoust, IEEE Trans 1970;18:340 - 3.
[10] Randall, R.B., Antoni, J. Rolling element bearing diagnostics—A tutorial. Mech. Syst. Signal Process. 25 (2011) 485-520.

Application of Adaptive Variable Scale Stochastic Resonance in Bearing Fault Diagnosis

Jian-guo WANG [1],Chao ZHANG [1], Yuan-yuan HE[1, 2],and Teng-fei ZHU [1]
[1]School of Mechanical Engineering, Inner Mongolia University of Science and Technology, 014010, China
[2]Inner Mongolia Tongwei，014010，China

ABSTRACT

An algorithm of bearing fault diagnosis is proposed in this paper, which is based on human cognitive self-regulating particle swarm optimization and variable-scale stochastic resonance. In the algorithm, an objective function is first determined with the stochastic resonance structural parameters a and b considered. Then the objective function is optimized with self-regulating particle swarm optimization algorithm. Thus the optimal values of a and b corresponding to the maximum of signal-noise ratio are obtained. Bring optimal a and b into variable-size stochastic resonance system and filter out noise signal in bearing vibration signal. Finally, spectrum analysis is applied find the bearing fault characteristic frequency. Experiment results show that this method is effective in bearing fault diagnosis.

Key words: particle swarm optimization; stochastic resonance; fault diagnosis

Corresponding author: Chao ZHANG (zhanghero123@163.com)

1. INTRODUCTION

For the fault diagnosis of rotating machinery, the feature extraction of fault signals is still a hard problem in current fault diagnosis area[1]. Particularly, the effect of traditional signal processing methods in practical applications is not ideal [2].

In 1995, Particle Swarm Optimization (PSO) algorithm was proposed by Kendad et al. [3], which is a self-adaptive evolutionary computing technology based on population search. The algorithm is easy to implement, has fewer parameters and is capable of solving complex optimization tasks effectively[4]. As a new type of stochastic search algorithm, PSO algorithm suffers from a series of problems, such as falling into local extreme points, slow convergence and poor accuracy in the late evolution. In literature [5], an adaptive escape particle swarm optimization algorithm is proposed. The algorithm changes the flying speed of particles in the search space by changing velocity self-adaptively. In literature [6], the cognitive self-regulating particle swarm optimization(SRPSO) algorithm is proposed. The algorithm develops self-regulating and self-recognizing PSO algorithms on the basis of PSO algorithm.

Stochastic resonance (SR), which was proposed by Benzi et al. [7], is only applicable to small parameters signals. To expand stochastic resonance to large-parameter signals, Leng Yong-gang [8] proposed a variable-scale stochastic resonance method. By setting an appropriate compression ratio R, stochastic resonance can be used under large-parameter conditions. However, the structural parameters a and b of system are usually fixed to 1 and the system does not have self-adaptation ability.

In order to visualize the weak impact signal in the fault signal, stochastic resonance is widely used by scholars to extract weak fault feature signal. Literature [9] proposed an adaptive stochastic resonance method and used SNR as performance index to evaluate the output signal of the stochastic resonance system.

This paper presents a stochastic resonance method of SRPSO. First, signal-to-noise ratio of output signal is defined to be the objective function. Then SRPSO is used to optimize the stochastic resonance structural parameters a and b. Finally, the parameters a and b corresponding to the fitness value, which maximize the signal-to-noise ratio, are brought into a scale-scale stochastic resonance to filter noise out of input fault signal. Then the time-frequency diagram is obtained and the fault characteristic frequency of the bearing is searched in the frequency spectrum.

2. SRPSO THEORY

SRPSO was built on the basis of PSO algorithm, the speed update equation in SRPSO is as follows:

$$V_{id}^{t+1} = w_i V_{id}^{t} + c_1 r_1 p_{id}^{se}(P_{id}^{t} - X_{id}^{t}) + c_2 r_2 p_{id}^{so}(P_{gd}^{t} - X_{id}^{t}) \tag{1}$$

where V_{id} and X_{id} are the velocity and position of the *i*-th particle in the d-dimensional space, respectively. t and t_1 represent the current and next iteration, respectively. $^{t}_{id}$ is the optimal position of the *i*-th particle in d-dimensional space. $^{t}_{gt}$ represents the optimal position found by all the particles in d-dimensional space. c_1 and c_2 represent the acceleration coefficients. r_1 and r_2 represent two random numbers distributed uniformly between [0,1]. w_i denotes the inertia weight of the *i*-th particle, which is self-regulated by the optimal particle.

P_{id}^{se} is the particle's individual self-awareness, P_{id}^{so} is the particle's global perception. They are defined as follows:

$$P_{id}^{se} = \begin{cases} 0, & \text{optimal particles} \\ 1, & \text{other particles} \end{cases}; \quad P_{id}^{so} = \begin{cases} 0, & \text{optimal particles} \\ \gamma, & \text{other particles} \end{cases} \tag{2}$$

where γ is binary (i.e., 0 or 1) depending on the threshold value for defining the confidence. The self-adjusting inertia weight strategy is defined as follows:

$$w_i = \begin{cases} w_i(t) + \eta \Delta w, & \text{Optimal particles} \\ w_i(t) + \Delta w, & \text{Other particles} \end{cases} \tag{3}$$

where w_i is the current inertia weight, and

$$\Delta w = \frac{w_I - w_F}{N_{Iter}} = \frac{0.55}{N_{Iter}}$$

where N_{Iter} is the number of iterations. w_I and w_F are the initial and final inertia weights. η is a constant that controls the rate of acceleration (usually set to 1). a obeys the uniform distribution on [0,1], which is the direction of all global optimal positions. Threshold (λ) is set to 0.5 to achieve any choice of directions, the social awareness of particles as follows:

$$P_{id}^{so} = \begin{cases} 1 & \text{if} \quad a > \lambda \\ 0 & \text{otherwise} \quad i = 1,2,\ldots,m \end{cases} \tag{4}$$

Self-awareness apparently affects the search direction of the particle. But for the optimal particle, it has the best orientation and it will discard social and self-perception. Therefore, the velocity update equation of the optimal particle is as follows:

$$V_{id}^{t+1} = w_i V_{id}^{t}, \quad (i \text{ is the best position}) \tag{5}$$

Other particles:

$$V_{id}^{t+1} = w_i V_{id}^{t} + c_1 r_1 (P_{id}^{t} - X_{id}^{t}) + c_2 r_2 p_{id}^{so}(P_{gd}^{t} - X_{id}^{t}) \tag{6}$$

3. STOCHASTIC RESONANCE (SR)

In the study of stochastic resonance, the expression of potential function is:

$$U(x) = -\frac{1}{2} a x^2 + \frac{1}{4} b x^4 \tag{7}$$

where, a and b are the structural parameters of bistable system. The deterministic dynamic system equation for one-dimensional nonlinear bistable systems without considering noise is:

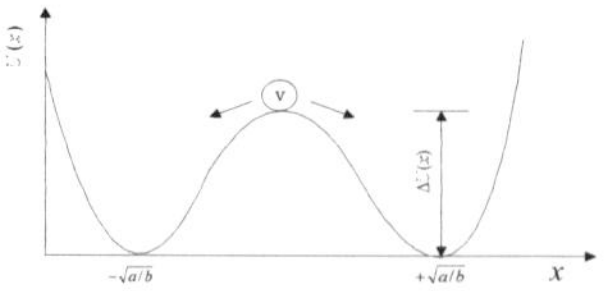

Figure1. Potential functions of nonlinear bistable system

As can be seen in Figure 1, the structural parameters a and b can both change the level of the barrier of the bistable system and also change the distance between the two barriers. If the input signal is a mixed signal of $Asin(2\pi f_0 t)$ and gaussian white noise $n(t)$, the langevin equation corresponding to the expression is:

$$\frac{dx}{dt} = ax - bx^3 + A\sin(2\pi f_0 t) + n(t) \tag{8}$$

where $n(t)$ is Gaussian white noise (mean 0, noise intensity D). If we choose a, b and D in the bistable system properly, the output response of the system will satisfy the stochastic resonance condition, that is, the jumping frequency of the particles between the two potential wells reaches the signal frequency and enhances the weak periodic signal.

For the random noise reduction effect is not obvious and can only detect the small parameters of the signal (adiabatic approximation small parameter conditions $A<< 1$, $D<< 1$, $f_0<< 1$). The basic idea is as follows:

(1) Set a frequency compression ratio R;

(2) Define a compressed sampling frequency $f_{sr}=f_s/R$;

(3) Set the Runge-Kutta method to calculate the step size $h=1/f_{sr}$.

4. ADAPTIVE STOCHASTIC RESONANCE METHOD

If traditional adaptive optimization methods are chosen to optimize the structural parameters in stochastic resonance, there are some limitations. Therefore, this paper proposes an algorithm of human cognitive self-regulating particle swarm optimization to optimize the structural parameters a and b of stochastic resonance, aiming to make the selection of a and b adaptive.

The specific steps of the algorithm proposed in this paper are as follows:

(1) Initialization: Set the size of the population and randomly select individuals to form an initial population.

(2) Set the objective function: the objective function of the genetic algorithm is defined as the output signal to noise ratio of the system, $F(a,b)=SNR_{out}(sr(a,b))$.

(3) Initialize the position and velocity of the particles in the particle group, and set the particle position as the SR parameter $[a,b]$.

(4) Stochastic resonance denoising: The parameters a and b corresponding to the SNR-maximizing fitness value are brought into variable-scale bi-stable stochastic resonance to perform denoising on the input fault signal.

(5) Calculate the frequency spectrum: Calculate the frequency spectrum of fault signal denoised by stochastic resonance.

(6) Make a decision: Calculate the fault frequency according to the bearing parameters to determine the bearing fault type.

5. SIGNAL SIMULATION

Set the simulation signal as follows: $x_1(t)=0.6sin(2\pi 30t)+n(t)$, $n(t)=\sqrt{2D}\,g(t)$, white noise intensity D is 9.1, the mean is 0, the variance is 1, the sampling frequency f_s is 2000 Hz, the number of sampling points N is 2000. And the time-frequency spectrum of the simulation signal is shown in Fig. 2.

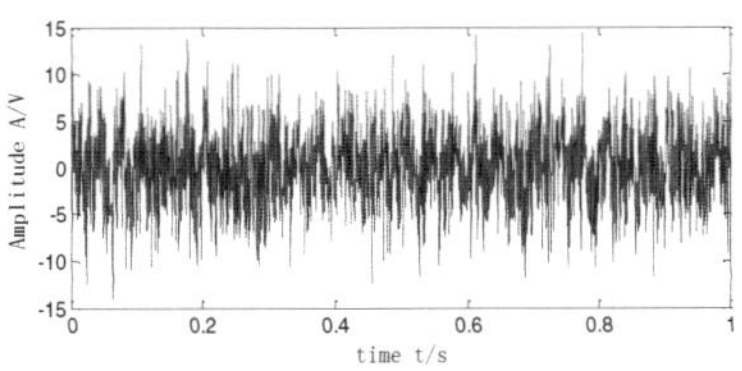

Figure 2. simulation signal time-domain diagram

In accordance with the traditional random resonance simulation signal denoising (a = 1, b = 1), the signal output results shown in Figure 3. The output signal of the SR optimized by the SRPSO is shown in Fig. 4. It can be seen from the time-frequency spectrum that the noise reduction effect is improved. The amplitude of the characteristic frequency of the simulated signal is more obvious than in Fig. 3.

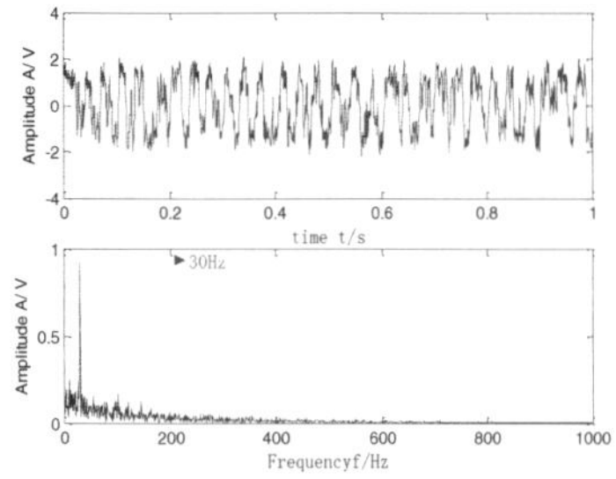

Figure 3. SR signal SR simulation results

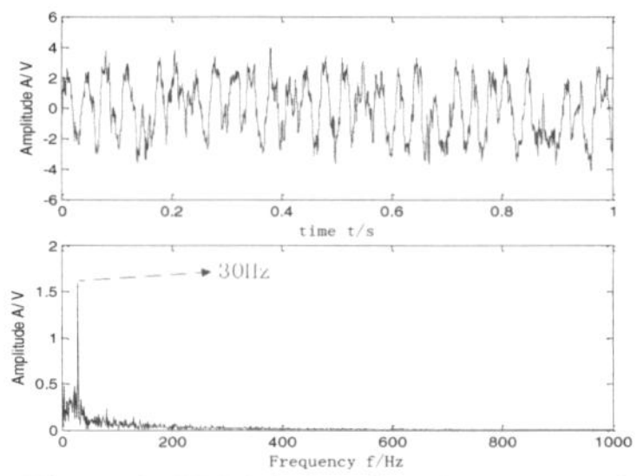

Figure 4. SRPSO SR signal output optimization results

6. EXPERIMENTAL ANALYSIS

Experimental data comes from the Rotary Machinery Diagnostics Laboratory, bearing model ER-10K deep groove ball bearings, rolling element diameter of 7.9mm, rolling body diameter of 30.3mm, the number of 8, the contact angle is 0°. The test set consists of a 1/3 horsepower 3-phase motor, a coupling, a tachometer and corresponding electrical controls. Bearing fault simulation test rig shown in Figure 5, the motor speed is set to 2400r/min. The accelerometers were mounted on the vertical and radial measuring points of the bearing housing. The vibration signals were collected by an 8-channel ZonicBook/618E data logger with sampling frequency of 2560 Hz. According to the parameters of the bearing, it is calculated that the characteristic frequency of the bearing inner ring and the rolling element fault is as shown in Table 1.

Figure 5. bearing failure simulation equipment

Table 1. MB ER-10K-type rolling element bearing the characteristic frequency

Shaft rotation frequency	Inner ring	Unit
40	197.73	*Hz*

6.1 . Bearing inner ring failure analysis

Select 5120 data points in the inner ring for analysis. Inner ring fault bearing signal time domain waveform is shown in Fig. 6.

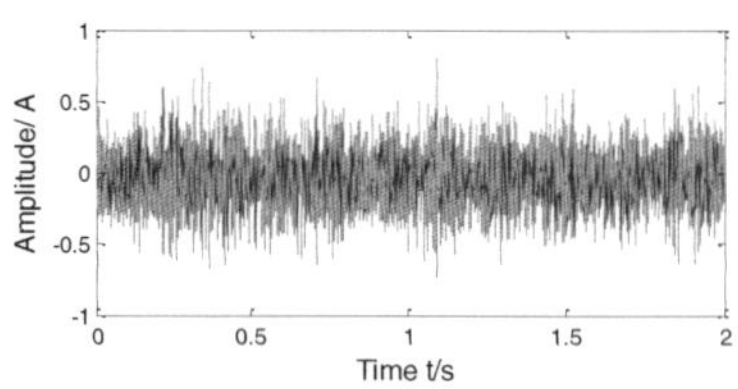

Figure 6. inner ring bearing fault signal waveform

Traditional random resonance (a=1, b=1) results of noise cancellation is shown in Fig.7,Fig. 8 shows the denoising results of the bearing signals optimized by SRPSO. The optimal parameters are a = 0.001 and b = 0.05. It can be seen from the frequency domain in the figure 8, relative to Figure 7. The frequency of bearing fault signatures is more pronounced and the amplitude is greater.

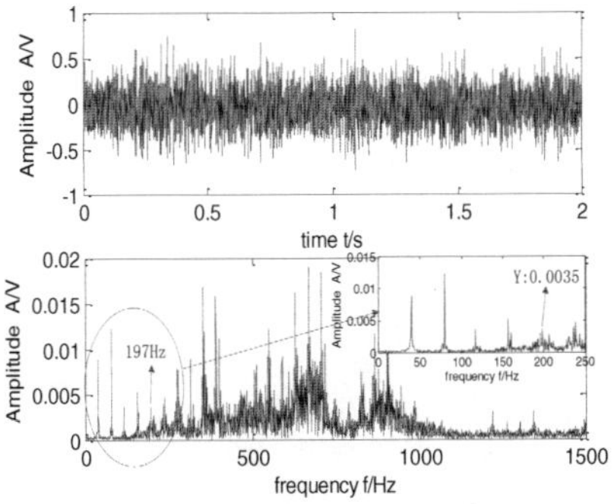

Figure 7. traditional random noise reduction results

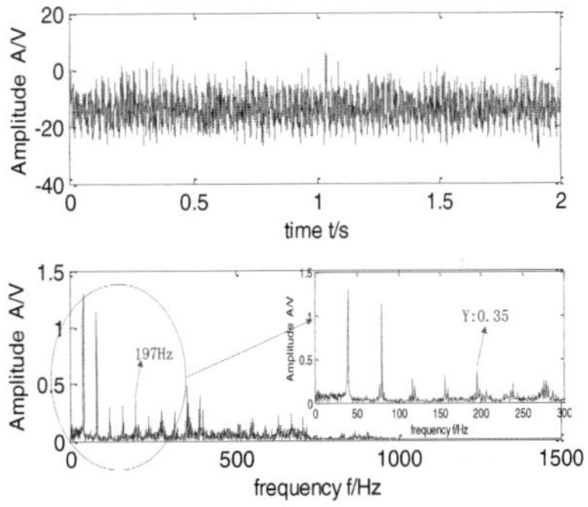

Figure 8. SRPSO-optimized stochastic resonance denoising results

7. CONCLUSION

This paper studies a fault diagnosis method based on human cognitive self-tuning particle swarm optimization and stochastic resonance algorithm. Conclusions are listed below:

(1) SRPSO overcomes artificially selected random resonance parameters *a* and *b*.

(2) Scaled bistable stochastic resonance algorithm based on SRPSO optimization compared with the traditional stochastic resonance algorithm, denoising effect is better.

REFERENCES

[1] Dai G P. Mechanical Fault Intelligent Diagnosis Based on EMD-Approximate Entropy And LS-SVM[J]. Journal of Mechanical Strength,2011,33(2):165-169.

[2] Zhang C, Chen J S. Diagnosing Faults of Bearings with LMD Approximate Entropy and SVM[J].Mechanical Science andTechnology for Aerospace Engineering, 2012,31(9):1539-1548.

[3] KENNEDY J, EBERHARTR C. Particleswarm Optimization[C]//Proc of IEEE International Conference on Neural Networks. Piscataway, NJ:IEEE Press,1995:1942-1948.

[4] PARSOPOULOS K E,VRAHATIM N. On the computation of all global minimizers through particle swarm optimization[J]. IEEE Transon Evolutionary Computation,2004, 8(3):211-224.

[5] HE Ran,WANG Yong-Ji, WANG Qing,et al. An Improved Particle Swarm Optimization Based on Self-Adaptive Escape Velocity[J].Journal of Software,2005, 16(12):2036-2044.

[6] M.R. Tanweer ,S. Suresh, N. Sundararajan. Self Regulating particle swarm optimization algorithm[J].Information Sciences 294 (2015)182-202.

[7] Benzi R, Parisi G, Vulpiani A. Theory of stochastic resonance in climatic chaner[J]. SIAM Journal on Applied Mathematics,1983, 43(3):565-578.

[8] Leng Y G , Wang T Y. Numerical research of twice sampling stochastic resonance for the detection of a weak signal submerged in a heavy Noise[J]. Acta Physica Sinica,2003,52(10): 2432-2437.

[9] ZHANG Z H,WANG D,WANG T Y. Self adaptive step-changed stochastic resonance using partical swarm optimization[J].Journal of Vibration And Shock,2013,32(19):125-130.

Characterization and modelling of a customs operation

A.J. Hoffman[1]
[1] School of Electrical, Electronic and Computer Engineering, North-West University, South Africa

ABSTRACT

Effective risk management is a prerequisite to find an acceptable balance between the objectives of a customs operation and the streamlined flow of goods. This requires the use of well-designed customs risk management models that scrutinize all cargo consignments in cyber space based on the analysis of rich data sets that can be used to accurately determine the risk represented by a cargo consignment without physically stopping it. The use of such models results in much reduced physical inspections without increasing the risk to customs of either losing income or allowing the influx of illegal contraband. It therefore represents a much more optimal compromise between the interests of customs and those of trade, reducing the economic cost to the region and making the region more attractive to global economic partners. In this paper we utilize different classification techniques to recognize patterns in electronic data transacted between customs and trade that characterize the risk attributes of cargo consignments. We subsequently extract models that can be applied in real time to minimize disruption of trade flows while reducing customs risks to below set thresholds. We quantify the impact of a variety in input factors and demonstrate how an optimal set of inputs can be selected to arrive at an effective risk management model. The diagnostic abilities of linear regression, neural network and classification tree techniques to predict both customs stops and infractions before they occur are compared.

Keywords: Diagnosis, pattern recognition, customs operations, root cause analysis, input-output modeling.

Corresponding author: Alwyn Hoffman (alwyn.hoffman@nwu.ac.za)

1. INTRODUCTION

There is very little empirical evidence that customs authorities are using well-designed statistical systems to identify possible high-risk or illegal transactions. The need therefore exists for the development of diagnostic models that can be used by customs authorities to more efficiently target declarations that should be inspected, in the process reducing the cost of unnecessary stops or inspections of compliant shipments. Komarov [1] described the use of risk management systems by Ukrainian customs to create risk profiles to enable the automated selection of high risk transactions by dividing transactions into risk categories: high risk or 'red' (representing 4.34% of transactions) are subjected to physical inspections, 'yellow' (15.60%) to documentation checks, 'light green' (6.48%) to 'information massage' while the remaining 'green' transactions (73.58%) are not subjected to further checks. Komarov however provided little details about the specific data being used and how the relationships between inputs and risk outcomes are established.

In a study applied to Senegalese customs data Laporte [2] applied regression models to calculate a risk outcome that reflects the probability of an infraction on a per transaction basis, using six input variables to predict risk: importer, freight agent, HS classification, origin, provenance and customs regime (Laporte 2011). Laporte claimed that by using this model it was possible to filter high risk cargo consignments to such a level of accuracy that more than 96% of all infractions could be found by inspecting only 20% of consignments. In a subsequent study Davaa and Namsrai [3] extracted a similar model from Mongolian customs data to predict infraction probability, using the following list of input variables: HS classification, importer, country of origin, customs terminal code, customs broker and type of transportation means. The level of risk prediction accuracy that they achieved was not quite as impressive as those obtained by Laporte; their model could classify consignments in such a way that the incidence of infractions increased from 0.05% in the lowest risk category to 0.22% in the highest risk category.

The work of Laporte and others provides evidence of the value of a non-intrusive data analytics based approach to customs risk management. No systematic method or procedure is however described of how to decide which inputs factors to consider for inclusion into a customs risk management model. The cited references furthermore provided no quantified indication of the relationships between the various input factors (e.g. customs office) and the operational customs performance in terms of time delays experienced by commercial trade.

In a previous paper we characterized the effectiveness of an existing customs operation in terms of the contribution of specific inputs factors on outcomes as well as the efficiency of current procedures used to select cargo consignments for inspections [4]. This paper will continue our previous work by utilizing input-outcome relationships for customs processes currently applied in South Africa to extract risk models using a variety of empirical modelling techniques, including linear regression, neural networks and decision trees. The specific research questions to be addressed include the quantification of the capability of each identified input factor to predict customs outcomes before these have occurred and a comparison of empirical modelling technique to capture input-output relationships the most effectively. To the best of our knowledge no previous published research have produced accurate answers to the above list of questions in the southern African context.

The rest of the paper is structured as follows: In section 2 we describe the data used in the study and in section 3 the process to select suitable explanatory variables to model the outcomes. Section 4 describes the methodology used to extract input-output models while section 5 discusses results and findings. In section 6 we conclude with recommendations for customs operations and suggest future research work.

2. DESCRIPTION OF THE DATA SET

The data that was used in this study represents EDI transactions exchanged between SARS Customs and consignors of goods imported into South Africa from September 2014 to September 2016. The available data set includes approximately 3.5 million transactions submitted to SARS Customs over the given time period. For each transaction information was obtained about the nature of the customs declaration as described in Table I below. In addition the data included the customs response codes representing the customs outcomes that we would like to be able to predict (e.g. a decision to stop and inspect a consignment, or finding an infraction), as described in Table II below. From the customs response codes we derived customs outcomes as described in Table III below. All cases that required an amendment were regarded as Infractions.

Table I: Input factors explaining nature of declaration

Input Factor	Number of Categories	Examples
Customs Office	36	Durban, Cape Town, etc.
CPC Code	31	10, 11, 12, etc.
Previous CPC Code	23	00, 14, 20, etc.
Country of Origin	237	GB, CN, GE, etc.
Country of Export	222	GB, CN, GE, etc.
Country of Import	197	ZA, ZM, ZW, etc.
Transport Code	9	Ocean, Road, Rail, etc.
Consignors	310	#0, #1, #7, etc.
HS Code	18	Animal, Chemical, etc.

Table II: Customs response codes

Customs Response Code	Description
1	Release
2	Stop for physical inspection at unpack depot or X-ray scanning
4	Refer to Other Governmental Agency (OGA)
6	Reject Declaration
13	Supporting Documents required
26	Request adjustment to declaration
31	Request additional supporting documents
33	Supporting documents received
36	Booked for physical inspection

Table III: Description of customs outcomes

Outcome	Condition
Not Stopped	no code 2, 36, 13, 31, 4
Stopped	code 2
Not Inspected	no code 36
Not Amended	no code 6 or 26
Amended	code 6 or 26
Inspected	code 36
Not Amended	no code 6 or 26
Amended	code 6 or 26
Request Additional Docs	code 13 or 33
Not Amended	no code 6 or 26
Amended	code 6 or 26
Refer Other Government Agency	code 4
Not Amended	no code 6 or 26
Amended	code 6 or 26

3. SELECTION OF EXPLANATORY VARIABLES AND OUTCOMES

In the development of empirical classifiers it is important to reduce the set of input factors, and thus the number of model degrees of freedom, as far as possible before the model is trained to help prevent overtraining and subsequent bad generalization capability [5]. We used Analysis of Variation as well as linear correlation analysis to evaluate the respective explanatory variables before we proceeded to extract empirical models. In Table IV we show the results of an ANOVA analysis applied to determine the ability of the explanatory variables in the first column to explain the customs outcomes listed in the top row. The F-statistic measures the ratio of between-class-variations and within-class-variations. A high F-value indicates on average larger differences between categories than within categories, implying that the respective category variable does have the ability to help explain the respective outcome values. In the table it can be seen that for all of the outcomes at least some of the explanatory variables have significance. Duration and Request for Additional Documents produce on average larger F-statistics, indicating that they may be easier to predict using the available explanatory variables. For all of the outcomes the F-values are sufficiently large to justify proceeding with the development of prediction models based on the available set of explanatory variables.

Table IV: F-statistics extracted through one way ANOVA

Input Factor	Duration	Req_Add_Docs	Stopped	Stopped_Inspected	Infraction
CustomsOffice	1150.8	1356.2	43.6	168.8	82.2
HSCode	892.1	8934.3	32.6	114.1	105.8
TransportCode	2105.4	4533.6	204.1	731.9	499.1
SPSource	206.7	809.5	84.2	234.2	126.5
CpcCode	423.4	352.7	2.6	10.6	93.2
PreviousCpc	118.4	834.1	4.2	22.6	75.8
CountryOfOrigin	50.1	329.2	4.9	13.3	17.8
CountryExport	59.9	352.0	4.8	18.1	17.7
CountryImport	25.4	29.7	1.8	2.0	3.7

In order to perform a direct comparison between the impacts of different input factors on the various outcomes we implemented a linear correlation analysis. As all of the input factors are categorical in nature rather than continuous, this was done in the following manner:

i. For each category of each input factor (e.g. Durban or Cape Town for category Customs Office) the accumulated average value of each outcome (e.g. fraction of infractions) was calculated as function of time as from the start of the observation period up to the end of each specific month. These averages represent the behavioural characteristics of the respective categories up to that point in time.

ii. For each new observation falling in a subsequent month its category memberships were determined (e.g. Customs Office: Durban; HS Code: textiles; Transport Mode: maritime). That observation was then allocated the accumulated averages for the categories to which it belongs as determined at the end of the previous month. Each observation thus inherits the attributes of the categories that it belongs to; as these attributes are continuous variables, it can be used as basis for a correlation analysis with the observed outcomes.

Figure 1 displays the correlations of all outcomes with respect to all the input factors. The Consignor identity is the Input Factor with strongest correlation for Requests for Additional Documents, Stop&Inspections and Infractions. The correlations with Infractions are in general however lower compared to the correlations with outcomes that represent customs decisions (Requests Additional Documents, Stop&Inspections and Refer OGA).

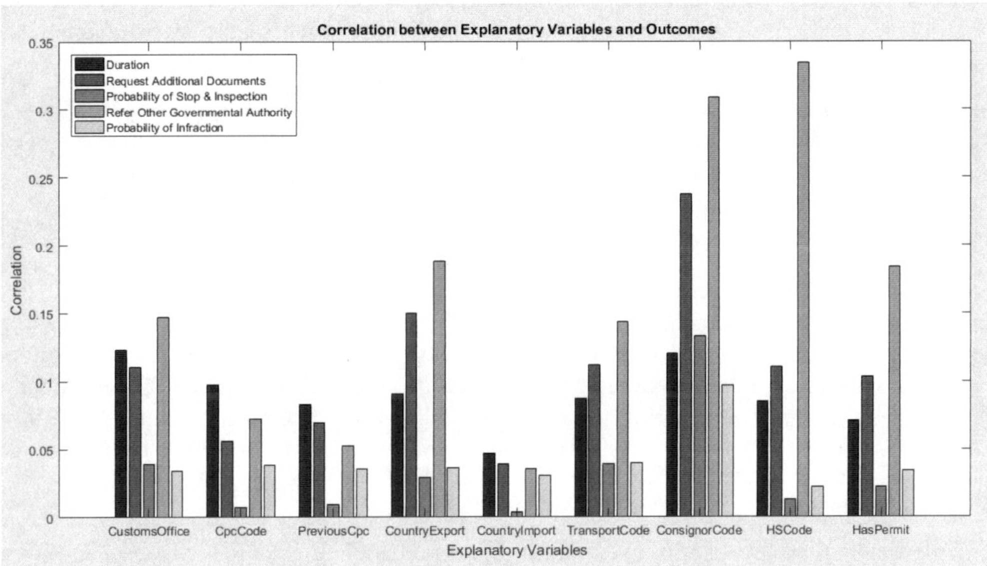

Figure 1 Correlation between explanatory variables and customs outcomes

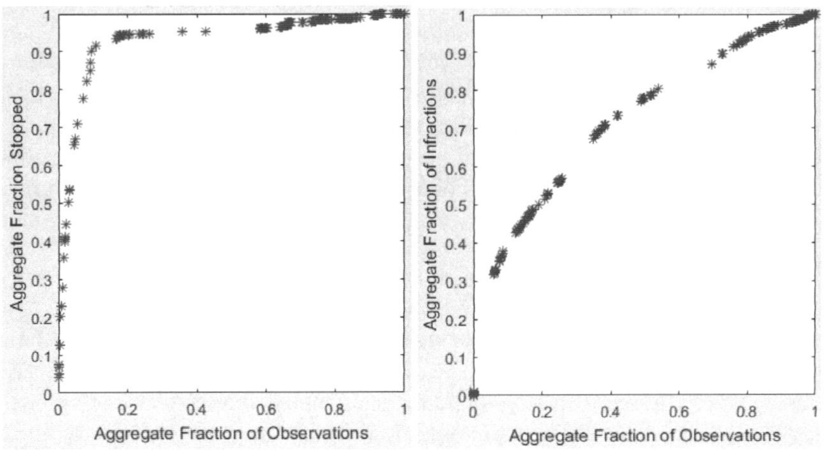

(a) (b)

Figure 2 Aggregate fraction of (a) stopped consignments and (b) infractions predicted vs aggregate fraction of observations selected based on Consignors

The ability of a single input factor to correctly classify outcomes is demonstrated in Figure 2 for the case of predicting Stops and Infractions using Consignor as input. These graphs were constructed by predicting outcomes to be true, starting with consignments from those categories that historically displayed the largest true outcome levels, and adding more categories until eventually all have been selected. It can be seen that a much higher selection accuracy for stopped consignments will be achieved when selecting a given fraction of total observations, compared to the selection accuracy for infractions. This indicates that when customs select consignments for stops, the identity of the consignor seems to play a very important role, but that the success of this strategy to actually find infractions is not quite as high.

4. EXTRACTING CLASSIFICATION MODELS

This classification problem is characterized by the fact that the target data is highly unbalanced – for our data set only about 0.8% on cases contained infractions and only a few percent were selected for some form of scrutiny. If such an unbalanced data set is directly used for training a regression or neural model, the model parameter optimization process will tend to be dominated by to the large number of samples with false outcomes. In order to overcome this problem we used weighted training sets, by duplicating each of the true outcomes to arrive at a balanced training set that still contains all of the original data.

We firstly implement both linear regression and logistic regression models, similar to the approach of Laporte [2]. We then proceed to implement a neural network based model as well as a decision tree model to verify if these more sophisticated techniques can improve upon the performance of simpler techniques.

4.1. Linear regression models

The same explanatory variables that were used for the correlation analysis in section 3 above were used as inputs for both linear and logistic regression models. We experimented with different numbers of input factors, selecting inputs based on their linear correlations with the outcome being modelled. To convert the continuous output a threshold value is set; model outputs that exceed this threshold are classified as true. Theoretically we would expect logistic regression models to outperform linear regression as the outcomes are categorical in nature. We found that model performance did not improve significantly beyond the use of three input factors.

4.2. Neural network models

Neural network models will have benefits over linear regression models should there be significant nonlinear aspects in the input-output relationships [5]. We extracted feed-forward multilayer perception networks that use the same sets of inputs as the regression models; the size of the single hidden layer is adjusted based on the number of inputs, with sigmoidal transfer functions in the hidden layer and linear transfer functions in the output layer. We used a validation set that made up 10% of the training set values; training was terminated at the point where the modelling error on the validation set started to increase. As both the input and hidden layers were restricted to not more than ten nodes each; the overall number of degrees of freedom in the models would be less than 120. Given the size of the population it was not necessary to apply specific regularization techniques as the large training set would help prevent the model from being over fitted to specific training samples.

4.3. Classification tree models

For our selected problem the available input data is inherently categorical in nature. For the previous set of modelling techniques we deliberately translated these categorical values into continuous ones as these types of model perform best when fed with continuous input data. It is however also possible to directly use the categorical input values using decision trees. A classification tree consist of many branches: at each branch point one or more categorical variables are used to make a decision regarding which way to branch. Once the probability of a specific outcome is high enough the branch terminates in a true or false outcome.

In the training of decision trees care must be taken not to over train, as a sufficiently large number of branching points, each containing an if-then-else rule, will allow any number of unique inputs observations to be correctly classified. Over training is prevented by limiting the maximum number of levels in the tree: a tree with only one level will classify all observations as either ones or zeros; with two levels at least one rule will be applied to separate the two classes; as the number of levels are increased more conditions can be defined to use as basis for more refined classification. To practically determine at which level the tree must be terminated, we used a validation set defined similarly as in the case of neural network training sets. An initial tree was extracted using no limit on tree level, and the performance of the tree was tested on the validation set. The number of tree levels was then gradually reduced until the classification performance for the validation set reached its maximum value. All of these trees were then also applied to the test set to verify if the selected tree also performed best on the test set.

5. RESULTS AND DISCUSSION

Each of modelling techniques as described above were applied to the available data to predict both the probability of specific customs decisions (e.g. to stop and inspect) and the probability of actually finding an infraction. It must be appreciated that the process that generated the available data set was not the true process that generated the incidence of infractions, but only the process as applied by the respective customs authority. It can be assumed that customs applied a set of rules and procedures to generate decisions to stop, inspect, etc.; once such a decision was implemented an infraction that was potentially present in the respective consignment may or may not have been found. Any deficiency in this process, both to correctly stop or scrutinize risky consignments and to find infractions that are present, will be directly reflected in the data. The models that were extracted mimic the customs process rather than the true incidence of infractions, and should therefore be evaluated primarily on that basis. This is supported by the results in section 3 where it could be seen that there are stronger correlations between input factors and the incidence of specific customs decisions than between the same input factors and the incidence of infractions.

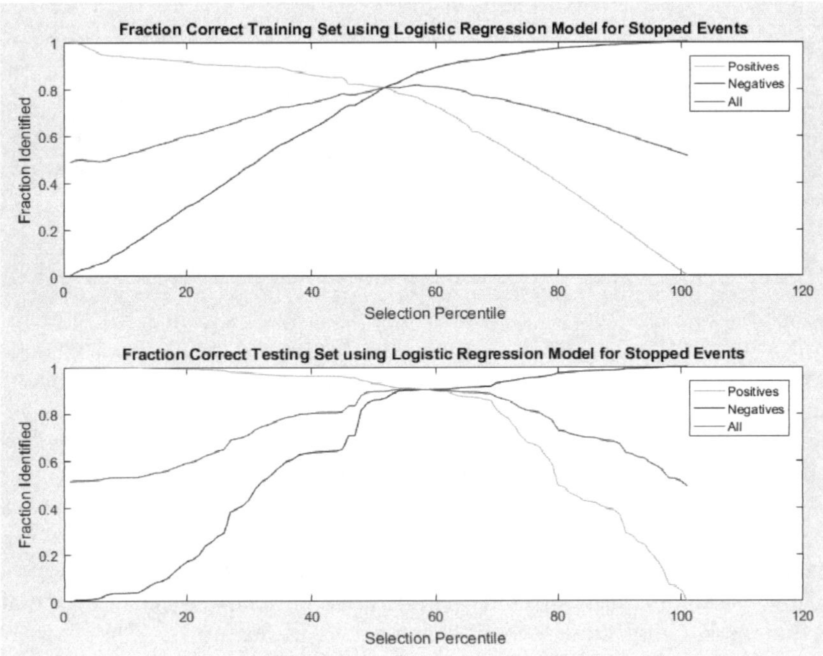

Figure 3 Fraction Stopped Consignments correctly identified by logistic regression model as function of fraction selected

Outcome accuracies are calculated for both the training and test sets and for both positive (true) and negative (false) outcomes. The impact of gradually changing the threshold level to separate positive and negative outcomes is illustrated in Figure 3; separate graphs display the results for training and test sets, and a separate graph is provide for Positive, Negative and All outcomes. Starting off with a zero threshold level all observations are classified as Positive, with resulting 50% accuracy: all positive target values are classified correctly and all negative target values are classified incorrectly. As this threshold is increased more observations will fall into the Negative category; if the model possesses any ability to correctly classify observations then the classification accuracy for All should increase, reaching a maximum value before decreasing again as the threshold value approaches 1.

The results based on regression and neural network models are shown in Table V below. The coefficients of determination (or R^2) are somewhat lower that the figures reported by Laporte; this indicates that for our data set the input-output relationships were somewhat weaker. This coefficient is also lower for the test set compared to the training set, indicating that the models extracted from the training data do not fit the test data quite as well, also in line with expectations. It can be seen that the neural networks perform slightly superior compared to the regression models, indicating that their nonlinear modelling capability is of some benefit in capturing the input-output relationships. In Table VI we show similar results to predict the incidence of customs stops. It can be clearly seen that all techniques perform much better to predict customs stops compared to the case of predicting infractions. This confirms that the modelling techniques do have the ability to capture with reasonable accuracy the underlying behaviour present in the customs decision making process.

Table V Classification results for predicting infraction using different model types

	Linear Regression		Logistic Regression		Neural Network	
Model Type	Train	Test	Train	Test	Train	Test
R^2	0.18	0.15	0.20	0.16	0.20	0.16
Fraction Infractions Found	Fraction of Observations Inspected					
50	0.32	0.48	0.32	0.48	0.32	0.46
80	0.64	0.74	0.63	0.73	0.65	0.66
90	0.81	0.85	0.81	0.83	0.83	0.75
95	0.90	0.90	0.89	0.90	0.91	0.81

Table VI Classification results for predicting stops using different model types

	Linear Regression		Logistic Regression		Neural Network	
Model						
	Train	Test	Train	Test	Train	Test
R^2	0.16	0.26	0.39	0.61	0.43	0.66
Fraction Infractions Found	Fraction of Observations Inspected					
50	0.27	0.22	0.30	0.21	0.26	0.20
80	0.49	0.34	0.65	0.30	0.49	0.29
90	0.68	0.42	0.81	0.83	0.68	0.39
95	0.74	0.65	0.99	0.53	0.77	0.57

A different approach has to be followed in the case of classification trees, as this technique does not allow a threshold to be adjusted in order to achieve a desired selection accuracy; instead the model performance can be moderated by limiting the allowed tree level. The case with tree level equal to 1 represents the trivial case where all observations are classified as true. As the tree level increases fewer observations from the training set are classified as true; at the same time the fraction of events that are correctly selected also goes up as more branches are added to the tree. The more realistic test for classification trees is the results achieved when applied to the prediction of customs decisions, rather than finding infractions, as the available data reflects the customs processes more accurately than the true presence of infractions. As shown in Table VII below the maximum tree level when extracting the model from the training set was 456. The fraction correctly selected infractions from the test set reaches a maximum value for an optimal tree level of about 20. By selecting 16-17% of consignments 93-94% of customs decisions to stop can be predicted out-of-training-sample. The technique therefore work well on data sets that contains the required level of consistency in input-output relationships. For infractions the results are less accurate: 23% of test set observations are selected to produce a selection accuracy of 58% of infractions. This is slightly superior compared to the best results for the regression and neural network models.

We can also compare our results with those of Davaa and Namsrai [3]: their risk engine increased the incidence of infractions from 0.05% in the lowest risk categories to 0.22% (i.e. by approximately a factor of 4) in the highest risk categories. Our classification tree increase the average incidence of infractions from 0.6% (when all observations are selected) to 1.7% in the test set when the optimal tree level is used, i.e. by approximately a factor of 3. We can therefore state that for our data set the performance is comparable to the results achieved by [3].

Table VII Classification results for Stops using Classification Trees

	Tree Levels	1	2	3	15	20	104	456
Training	Fraction Selected	0.00	0.25	0.23	0.17	0.16	0.10	0.08
	Fraction Correct	0.51	0.86	0.87	0.89	0.90	0.94	0.96
	Fraction Correct No Event	1.00	0.75	0.77	0.83	0.84	0.90	0.92
	Fraction Correct Event	0.00	0.97	0.97	0.96	0.96	0.99	0.99
Testing	Fraction Selected	0.00	0.17	0.16	0.14	0.13	0.09	0.06
	Fraction Correct	0.49	0.89	0.88	0.86	0.86	0.77	0.63
	Fraction Correct No Event	1.00	0.83	0.84	0.87	0.87	0.91	0.94
	Fraction Correct Event	0.00	0.94	0.93	0.86	0.86	0.63	0.34

6. CONCLUSIONS AND RECOMMENDATIONS

We summarize our conclusions and recommendation by addressing the stated research questions. The capability of each input factor to predict customs outcomes before these have occurred are reflected by the results in 3. Consignor identify was clearly the individual explanatory variable that contains the most prediction ability; it is in all likelihood used by the customs authority to select consignments for inspections. Apart from Consignor the other inputs that made a significant contribution includes Country of Origin, Transport Mode (which in the case of our data set is closely related to Customs Office as the two largest customs offices are served by primarily one transport mode each) and CPC Code.

Neural networks slightly outperformed regression models, using 3 or 4 inputs. Increasing the number of explanatory variables to a maximum of 9 did not appreciably improve prediction ability. Classification trees provided the best classification accuracies within the set of techniques that were investigated. For all of these techniques the ability to correctly predict the outcome of the customs process was much superior to the ability to predict the incidence of infractions; this was to be expected as the data set was effectively created by the customs decision making process rather than by the true incidence of infractions.

From the results in section 5 it is clear that the risk models can improve current customs processes, as they all produced outcome incidence rates in the highest risk categories that are significantly higher than the incidence rates in the lowest risk categories. The level of improvement will depend on the minimum accuracy that is required for finding infractions. Based on our results it is however suspected that the primary limitation to improve the current risk engine is not the quality of the modelling techniques but the quality of the infraction incidence data that was generated by historical customs processes. In terms of the choice of modelling technique, we conclude that if a technique is required that allows accurate control over the fraction consignments selected for infractions, then neural networks combined with setting an optimal risk score threshold value will be the most suitable; if a technique is required that allows the selection of the smallest fraction of consignments for a reasonably high selection accuracy then classification trees may be the best choice.

Future work will focus on specific vertical markets that are severely impacted by customs operations. This work will extend the set of data fields extracted and will measure time delays in the process across the entire value chain, rather than focusing on the customs process only. This should indicate in which part of the value chain the most improvements can be achieved based on the ability to predict and detect eventualities at an early stage.

REFERENCES

[1] Komarov, O. V. (2016) Risk management systems in Customs: the Ukrainian context. World Customs Journal, 10(1), pp. 35 – 44.
[2] Laporte, B. (2011) Risk management systems: Using data mining in developing countries' customs administration. *World Customs Journal,* 5(1), pp. 17-28.
[3] Davaa, T. Namsrai, (2015) B. Ways to modernise customs risk management in Mongolia. World Customs Journal, 9(2), pp. 24-37.
[4] Hoffman, A.J., Grater, S., Schaap, A., Maree, J. & Bhero, E. 2016. A simulation approach to reconciling customs and trade risk associated with cross-border freight movements. South African Journal of Industrial Engineering, 27(3), pp. 251-264.
[5] Bishop, C. (1995) Neural networks for pattern recognition.

The Improvement of Instantaneous Angular Speed Estimation using Signals from a Dual Read Head for Monitoring Planetary Gearboxes

Qiang Zeng, Ibrahim Alqatawneh,,Yuandong Xu, Yimin Shao, Fengshou Gu, Andrew Ball

ABSTRACT

The Instantaneous Angular Speed (IAS) signals obtained from the shaft encoders are influenced by inevitable manufacture and installation errors of encoder wheels. These errors can lead to low accuracy or unsatisfied estimation of IAS and fault diagnostics. To improve the accuracy of IAS estimation, this paper presents a novel IAS extraction method based on a dual read head sensing system. The analysis of common encoder errors shows that the amplitude of error changes with angular positions on the encoder whereas the IAS to be monitored is independent of the angular position. Therefore, a dual read head IAS estimation method is proposed to estimate the erroneous IAS online, and subsequently is taken from overall IAS measurement. In this way, the estimate of rotor IAS is significantly improved for more accurate monitoring and diagnostics. Experimental evaluation shows that the new method allows the small planet fault in a planetary gearbox to be detected over a wide range of operating conditions.

Keywords: Instantaneous angular speed; Planetary gearbox; Encoder error modeling; Dual read head; Diagnosis

Corresponding author: Qiang Zeng (q.zeng12357@gmail.com)

1. INTRODUCTION

Planetary gearboxes (PGs) play a vital role in power transmission systems based on their major advantages including compact structure, high transmission ratio, error tolerance etc., and they are widely used in many important industries such as aerospace vehicles, energy productions, power stations, mining equipment etc. PG failures can result in significantly economic losses and casualties. According to the previous studies in wind turbine failures [1], the PG failure leads to a second long down time. A malfunction PG with crack was founded in a UH-60A helicopter transmission system [2]. Therefore, monitoring the PG health conditions and diagnosing their faults are crucial to maintain production activities and avoid such severe tragedies.

Vibration based fault diagnosis methods are widely used in the fixed-axis gearboxes and usually have a satisfied results. However, due to the unique transmission structures of PGs, Amplitude Modulation (AM), Frequency Modulation (FM) and vibration neutralization, conventional vibration based methods presents rich of signal components, making it difficult to identify the most fault correlated one. Many approaches are needed to extract the weak and distorted fault features. A number of parameters such as Four-Order figure of merit (FM4) and NA4, which is developed by calculate the kurtosis of the residual signal, are presented [3] to indicate the PG faults. Autocorrelation based time synchronous average [4], map strategy [5] and data fusion methods [6] are presented to extract the individual vibration signal from planets or sun gears for minimising the influence of the transmission path induced modulations. The attenuation of the transmission through multi-interfaces is suppressed by calculating the ratio before and after the potential damages [7]. Although those post processing method improves the detection performances, it is still presents less satisfied results.

IAS signals are directly related to the rotation dynamics and can be obtained by a very cheap encoder. It has been long interested in the field of condition monitoring. In addition, it also has received an extensive attention in fields such as body sensing, navigation, servo control, position control and so on [8-11]. Specifically, IAS signal from a PG has no influences from the transmission paths and the nonlinear attenuation.

However, the IAS signal obtained from encoder also has considerable disadvantages. Its accuracy is highly depended to the encoder resolution, the shaft rotating speed and the sampling rate of data acquisition [12].

Especially, IAS signal can also be seriously deteriorated by the erroneous IAS components caused by the encoder imperfections or errors due to manufacturing, installation and operational deviations.

The encoder imperfections can be in the form of slot width variation at the point of encoder read-heads, and can generate jitters at the pulse and erroneous IAS signal in the estimation process. In general, erroneous IAS signals can be correlated to the low frequency errors caused by eccentricity, swash and distortion [13]; medium frequency error induced by graduation errors (pith error profile error) [11]; and high frequency errors from the measurement phase error and amplitude error [11]. A number of encoder calibration methods and error compensation methods were studied in the fields of manufacture fields. In [13, 14], a higher accurate device was used to obtain a compensation data sheet. In [11, 15, 16] a multiple read-head scheme was adopted to eliminate the low frequency errors. It is this multiple read-head scheme is of particular interesting in this study as it is can be potentially implemented online at accepted cost.

Therefore, to obtain an accurate IAS estimation, this paper focuses on suppressing the imprecisions induced by encoder errors through a dual read head (DRH) measurement method. It firstly obtains the imprecisions and then taken away from overall IAS estimation. The paper is organized into 5 sections. After this introduction, Section 2 outlines different types of encoder errors among which the periodic ones can cause more effect and should be removed. Based on general characteristics of the error, Section 3 develops the theoretical basis of DRH method and evaluates the DRH performances based on simulated signals. Section 4 verifies the effectiveness of using this IAS based for PG diagnosis. Finally, Section 5 draw main conclusion attained by this study.

2. ENCODER ERROR MODELING

For cost-effective and robust operation, IAS based monitoring usually employs a rotatory encoder which is a round disk having a number of teeth or slots uniformly distributed around the circumference of the disk. It can be off the shell products or available in a mechanical system such as engine starting ring. Whichever encoder is used there are always different types of errors: typically in the form of eccentricity and slot or tooth variations, that degrades IAS estimation.

As aforementioned, the different types of errors of an encoder can induce imprecisions in different frequency bands. The high frequency content can be suppressed by converting the pulses to a series of rectangle waves and applying a low pass filter for the estimated IAS signal. The low and medium frequency encoder errors are mainly contributed by the encoder manufacture errors and installation errors, which are inevitable especially in the in-house made encoders. These errors can lead to the slot width variation with the encoder. Consequently, these errors can be phase modulated (PM) into the encoder pulse train signal and generate an erroneous IAS signal within the shaft rotating harmonics that is often based for fault diagnosis.

2.1. Eccentric error analysis

Figure 1 shows the eccentric and swash errors between an ideal encoder wheel and an abnormal encoder wheel in a geometric manner. The blue dotted line is the ideally position of the encoder wheel, and the black line represents the encoder wheel position deviation due to errors. The geometric centre of the encoder wheel is point O, which is also the ideal rotating centre. The point R is the location of read-head, which pick up pulses when the slots pass over it. r, from O to R, is the radius of the encoder. θ is the measured angle pick up by the read-head and the θ' is the real rotated angle. $\vec{\mathbf{n}}$ is the normal vector of the ideal encoder wheel position and $\vec{\mathbf{n}}'$ is the normal vector of the swashed encoder wheel position with a swash angle α. In Figure 1 (a), due to the eccentric error, the encoder is slightly deviated from the ideal position and rotating around the centre O'. The angle $\varphi = \theta' - \theta$ is the angle error between measured angle θ and the real angle θ'. According to the geometric relation, angle θ' can be found as:

$$\tan(\theta+\phi_e)=\tan\theta'$$
$$\frac{\tan\theta+\tan\phi_e}{1-\tan\theta\tan\phi_e}=\frac{\sin\theta}{\cos\theta-\Delta r/r} \quad \text{...} (1)$$
$$\phi\approx\left.\frac{\rho\sin\theta}{1-\rho\cos\theta}\right|_{\rho=\Delta r/r,\phi<1^\circ}$$

In addition, the IAS caused by the eccentricity ω_{ef} can be given as:

$$\omega_{ef}=\frac{d\phi_e}{dt}=\frac{d\phi_e}{d\theta}\frac{d\theta}{dt}=\frac{\rho(\cos\theta-1)}{(1-\rho\cos\theta)^2}(\omega_c+\Delta\omega)\approx\left.\frac{\rho(\cos\theta-1)}{(1-\rho\cos\theta)^2}\omega_c\right|_{\Delta\omega<<\omega_c}, \quad \text{..............................} (2)$$

where ω_c is the constant angular speed, $\Delta\omega$ is the angular speed oscillation.

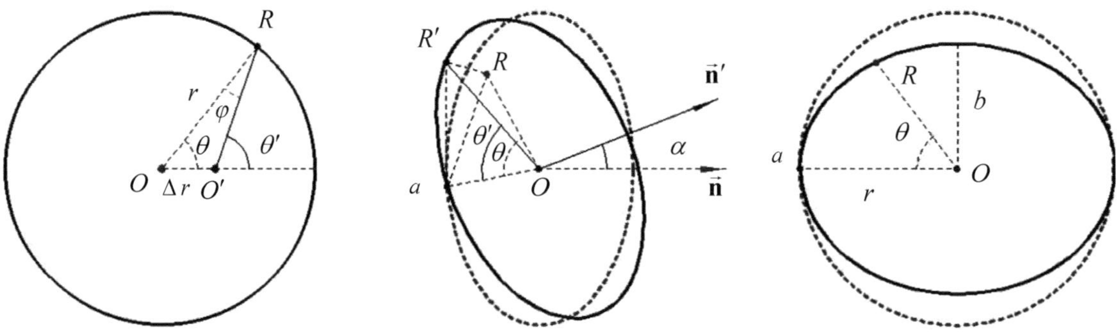

(a) (b) (c)

Figure 1. Encoder errors: (a) wheel eccentric error (b) wheel swash error and (c) swashed encoder's projection in $\vec{\mathbf{n}}$ direction

2.2. Swash error analysis

The tilt axis of the encoder wheel results in swash errors as shown in Figure 1 (b), in which the blue dotted line is the ideal position of the wheel and the black line is the swashed encoder wheel. The swash angle between the ideal wheel normal direction $\vec{\mathbf{n}}$ and swashed wheel normal direction $\vec{\mathbf{n}}'$ is α. O is the rotating centre and Oa is the intersection of the two wheels. θ and θ' are measured angle and real angle respectively. R and R' are measured encoder reading point and real reading point respectively. In Figure 1 (c), the swashed wheel was projected onto the surface of ideal position, or $\vec{\mathbf{n}}$ direction, and the projected wheel is an ellipse, where Oa is the semi-major axis and Ob is the projected semi-minor axis. In the ellipse, $Oa=r$ and $Ob=r\cos\alpha$. Therefore, for a measured θ, the $OR(\theta)$ can be given as:

$$OR(\theta)=\frac{r^2\cos\alpha}{\sqrt{(r\cos\alpha\cos\theta)^2+(r\sin\theta)^2}}, \quad \text{...} (3)$$

In $\Delta OaR'$, $OR'=Oa=r$; in $\Delta ORR'$, $OR'=r$; in $\Delta aRR'$, $RR'\perp aR$, therefore:

$$[aR(\theta')]^2=r^2-2r^2\cos\theta'+[OR(\theta)]^2, \quad \text{...} (4)$$

In ΔOaR,

$$[aR(\theta)]^2=r^2+[OR(\theta)]^2-2rOR(\theta)\cos\theta, \quad \text{..} (5)$$

For any measured $\theta \in (0, 2\pi]$, $\exists \theta' \in (0, 2\pi] \Leftrightarrow [aR(\theta')]^2 = [aR(\theta)]^2$, therefore:

$$\cos\theta' = \frac{OR(\theta)\cos\theta}{r} = \cos\theta \frac{\cos\alpha}{\sqrt{(\cos\alpha\cos\theta)^2 + (\sin\theta)^2}}, \quad (6)$$

The erroneous angular error caused by the swash can be given as:

$$\varphi_s = \theta' - \theta = \arccos(\cos\theta \frac{\cos\alpha}{\sqrt{(\cos\alpha\cos\theta)^2 + (\sin\theta)^2}}) - \theta, \quad (7)$$

In addition, the erroneous IAS ω_{sf} due to the swash error can be given as:

$$\omega_{sf} = \frac{d(\theta' - \theta)}{d\theta}\frac{d\theta}{dt} \approx \left[\frac{(1-\cos\alpha)}{2}\cos(2\theta) - \frac{(1-\cos\alpha)}{2}\right]\omega_c \qquad \theta \in (k\pi, (k+1)\pi), k = 0,1,2,3,\ldots, \quad (8)$$

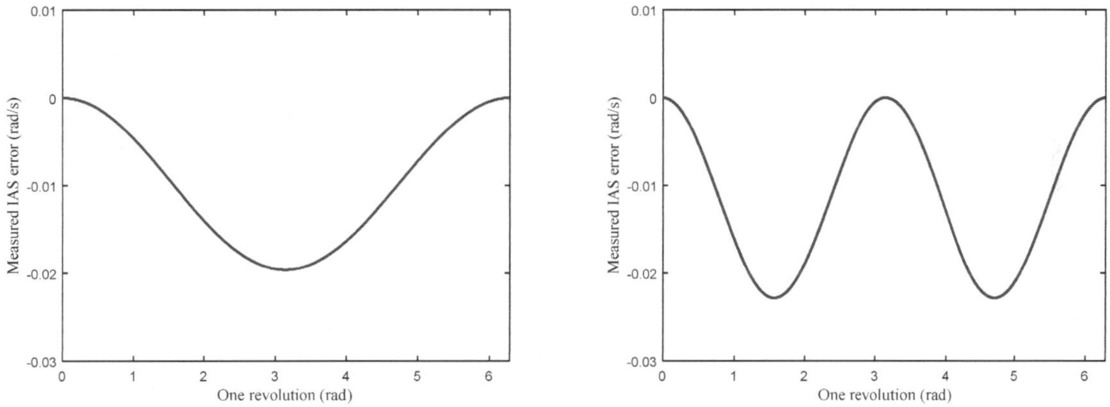

(a) (b)

Figure 2. IAS measured error caused by (a) eccentric error ($\omega_c = 150 \quad rad/s$, $\rho = 0.01$) (b) swash error ($\omega_c = 150 \quad rad/s$, $\alpha = 1°$)

The IAS measured errors are illustrated in Figure 2 (a) and (b). The measured IAS error caused by the wheel eccentric error is related to the rotating speed ω_c and the eccentric ratio ρ. The frequency of the erroneous IAS signal is the rotating frequency. The measured IAS error caused by the swash is related to rotating speed ω_c and swash angle α. The frequency of the erroneous IAS signal is at twice as much as the shaft rotating frequency. Nevertheless, both types of errors induce additional IAS components at the integer of rotational frequency, which needs to be suppressed to ensure the performance of detection when faults such as misalignment, unbalances etc. appear at such frequencies.

3. DUAL READ-HEAD BASED IAS ESTIMATION

3.1. Dual read-head IAS estimation method

To eliminate the measured IAS error caused by the encoder errors, a dual read-head error estimation method is proposed to achieve the encoder error self-estimation without any other device. The schematic diagram of the dual read-heads using optical sensors and the flowchart of the new proposed method are shown in Figure 3.

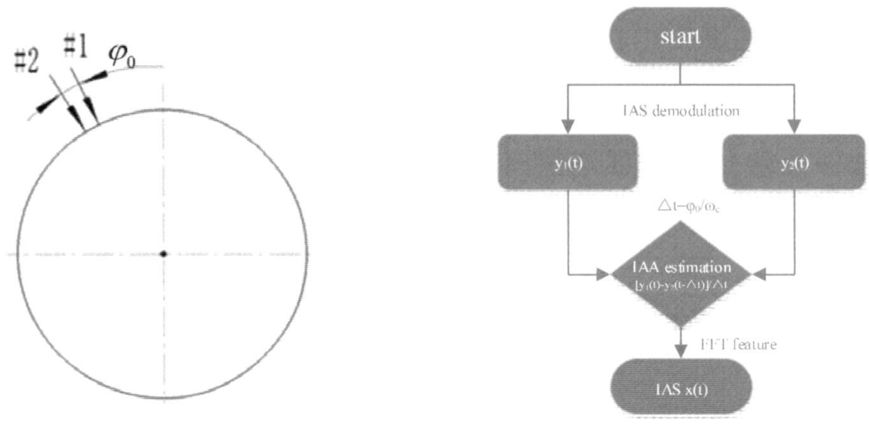

(a) (b)

Figure 3. Dual read-head IAS estimation (a) schematic diagram of the dual read-heads (b) flowchart of the IAS estimation

Read-heads #1 and #2 are mounted on the encoder with an angular interval φ_0, the smaller the φ_0 the better estimated result. To eliminate the highest order of the erroneous IAS components caused by encoder errors, the φ_0 should be smaller than $2\pi / N$, where N is the encoder resolution. The flowchart of new IAS estimation method is illustrated in Figure 3 (b). The shaft IAS $x(t)$ includes constant part ω_c and oscillation part $\Delta\omega$. Theoretically, despite the angle interval φ_0 between the two read-head, the measured shaft rotating IAS $x(t)$ from read heads are the same, and the measured erroneous IAS caused by encoder errors are $e(t)$ and $e(t-\Delta t)$,which the erroneous IAS of second read-head is delayed Δt caused by the interval φ_0. This method can also applied to the speed varying condition, where the Δt determining is more complicated. Firstly, the IAS signals $y_1(t)$ and $y_2(t)$ are demodulated from the encoder pulse train signals of the dual read-heads. The demodulated IAS signals are given as:

$$\begin{cases} y_1(t) = x(t) + e(t) \\ y_2(t) = x(t) + e(t-\Delta t) \end{cases}, \quad (9)$$

Secondly, the error IAS delay time Δt between two read-heads is determined by $\Delta t = \varphi_0 / \omega_c$. Then the instantaneous angular acceleration (IAA) $x'(t+\Delta t/2)$ is approximately expressed as:

$$x'(t+\frac{\Delta t}{2}) \approx \frac{x(t+\Delta t)-x(t)}{\Delta t} = \frac{y_2(t+\Delta t)-y_1(t)}{\Delta t}, \quad (10)$$

Finally, the IAS signal is converted from the IAA signal by using the FFT. Therefore, the IAS signal can be estimated as:

$$x(t) = F^{-1}\left(\frac{F(x'(t+\frac{\Delta t}{2}))}{j\omega e^{j\omega\frac{\Delta t}{2}}}\right), \quad (11)$$

where F is the Fourier transform, F^{-1} is the inverse Fourier transform.

3.2. IAS estimation evaluation

To evaluate the proposed method, a simple IAS signal model in Eq. (9) is used. The real shaft rotating IAS component includes the speed variation term and the fault IAS term.

$$x(t) = a\sin(2\pi f_r) + b\sin(2\pi f_f), \quad (12)$$

In addition, consider the encoder wheel eccentric and swash error, the measured error IAS signal can be given as:

3 rad/s, ...(13)

where a is the amplitude of IAS variation, b is the amplitude of fault IAS variation, c is the erroneous IAS caused by the eccentric and d is the erroneous IAS caused by swash. f_r and f_f are the rotating frequency and fault frequency respectively.

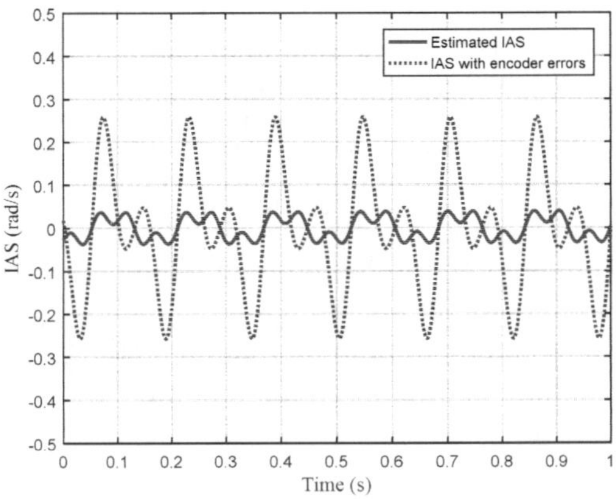

(a)

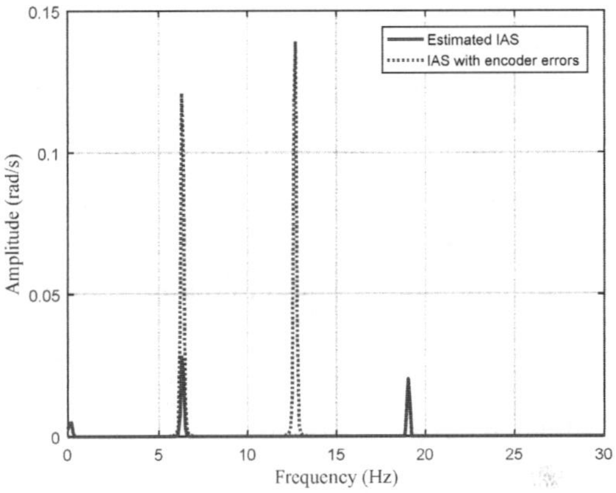

(b)

Figure 4. IAS estimation method evaluation (a) waveform (b) frequency spectra ($a = 0.03$, $b = 0.02$, $c = 0.1$, $d = 0.15$, $f_r = 6.33$, $f_f = 19$)

The waveform of the estimated IAS signal is smaller than the original IAS signal as illustrated in Figure 4 (a). Some IAS components were eliminated from the original IAS signal. The frequency spectra in Figure 4 (b) show that the estimated IAS signal retain the speed variation component and the fault component, which are influence free from the encoder errors. As illustrated in the estimated IAS spectrum, the low frequency component, <0.5 Hz, has been contaminated by the IAA integration process. The estimated IAS frequency spectrum has the same frequency components as the simulated IAS, which speed variation amplitude and fault amplitude are 0.03 and 0.02 individually. Therefore, the proposed method can also be used to identify the misalignment or run out errors.

4. EXPERIMENTAL STUDY

4.1. Experiment setup

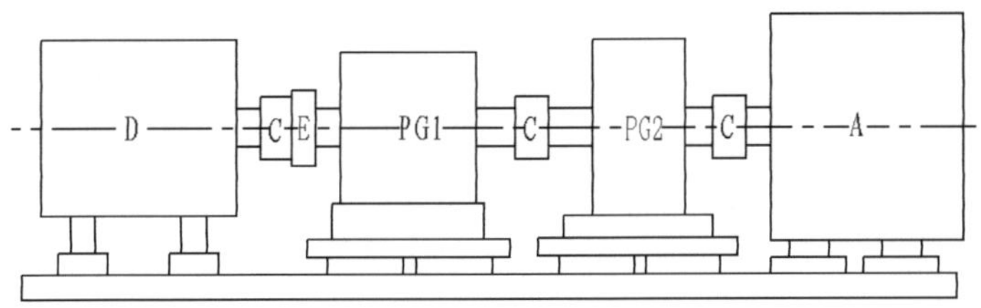

(a)

(b)

Figure 5. Test rig (a) schematic (b) planet tooth damage

A test rig was developed to evaluate the proposed IAS estimation method based on monitoring PG health conditions. A schematic diagram of the test rig is shown in Figure 5 (a). It consists of a DC generator (D), an AC motor (A), the tested PG (PG1), an accompany PG (PG2), three couplings (C) and a 60-tooth encoder (E) made from 3-D printing. The DC generator applies loads to the test rig and the maximum load is 360 Nm. PG1 with a transmission ratio of 7.2 and PG2 with a transmission ratio of 5.77 are connected to the carrier shafts by a back-to-back layout. The test rig is driven by the AC motor with a full speed up to 1500 rpm (100%) which can be adjusted by a variable speed controller for different operating conditions.

Two cases were studied in this paper. One is the planet gear spalling damage fault as illustrated in Figure 5 (b), which is denoted as PF for the brevity of discussion. The other is the healthy case considered as the baseline, denoted as BL. To verify the performance of IAS based monitoring, the tests were carried out under at 60% of AC motor full speed and four loads (25% 50% 75 % and 90% of AC motor maximum load). Two pulse trains reflecting the effect of encoder tooth passing the two optical readers were acquired at 96 kHz for IAS calculation.

The characteristic frequencies of PG1 are fault frequencies, mesh frequency and rotating frequencies. According to the tooth number in PG1: the sun gear tooth number $z_s = 10$, the planet tooth number $z_p = 26$, and the ring gear tooth number $z_r = 62$, the transmission ration is $r = 1 + z_r / z_s = 7.2$ and the characteristic frequencies are calculated as shown Table1, based on which the fault diagnostics can be implemented.

Table 1: Characteristic frequencies of PG1

Events	**Expression**	**60% speed**
Sun gear rotating frequency	f_{sr}	18.8Hz
Carrier frequency	$f_{cr} = f_{sr} / r$=0.1389 f_{sr}	2.61Hz
Mesh frequency	$f_m = z_s(f_{sr} - f_{cr}) = 8.6111 f_{sr}$	161.85Hz
Planet rotating frequency	$f_{pr} = f_m / z_p - f_{cr} = 0.1923 f_{sr}$	3.62Hz
Planet fault frequency	$f_{pf} = 2(f_{pr} + f_{cr}) = 2 \times 0.3312 f_{sr}$	2×6.22=12.44Hz
Sun fault frequency	$f_{sf} = 3(f_{sr} - f_{cr}) = 3 \times 0.8611 f_{sr}$	3×16.18=48.55Hz

4.2. IAS data analysis

4.2.1. Error reduction evaluation

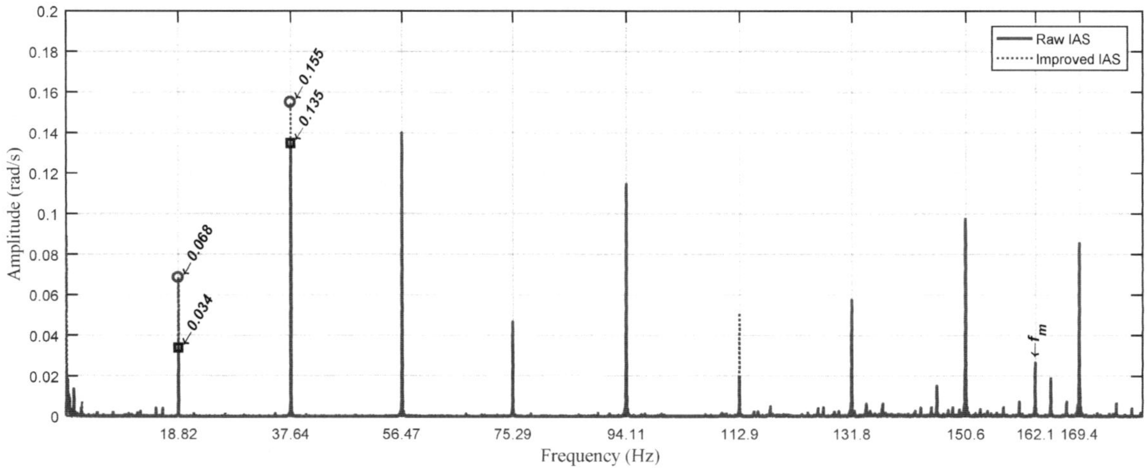

Figure 6. IAS spectrum comparison for BL case at 40% speed 90% and load

The raw IAS and improved IAS spectra obtained from BL case are illustrated in Figure 6. According to encoder error model in Section 2, the eccentric and swash error will lead to the erroneous IAS, illustrated in Figure 2, which makes the measured IAS amplitude smaller than the real shaft IAS signal. In the experiment, the proposed method compensate this IAS reduction caused by encoder errors. Moreover, the amplitudes at higher orders of rerating frequency n*18.82 Hz have been suppressed significantly. In the meantime, other non-rotating harmonics remains unaffected. Therefore, the proposed DRH IAS estimation method efficiently canceled the encoder errors, and pave the way to fault diagnosis.

4.2.2. IAS based detection and diagnosis

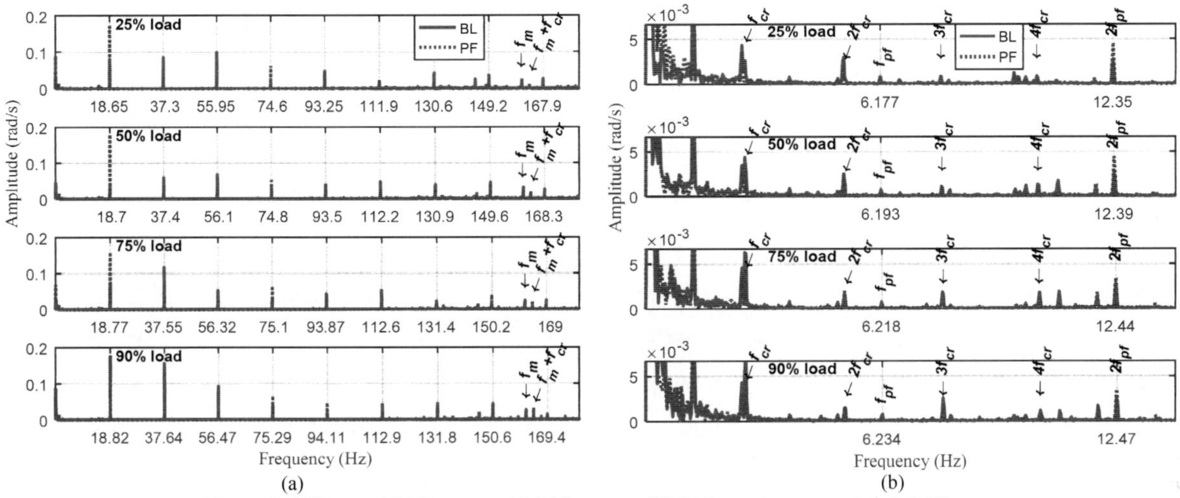

Figure 7. 60% speed IAS spectra (a) IAS spectra (b) IAS spectra zoomed in 0-14Hz

The IAS spectra at the 60% speed are illustrated in Figure 7. It can be seen that the spectra are dominant by sun gear shaft rotating harmonics nf_{sr}, mesh frequency f_m and sidebands of mesh frequency like $f_m + f_{cr}$, as shown in Figure 7 (a). In the magnified spectra of Figure 7 (b), the planet fault related frequency component $2f_{pf}$, generated by fault planet tooth mesh with sun gear and ring gear separately, has higher value in PF case. This indicates the planet fault occurs in the PG. In addition, other components such as nf_{cr} and f_{pf} also show visible changes due to the effect of refitting the faulty gear. Nevertheless, these amplitudes are of lower amplitudes, compared with BL, indicating less vibration and improved condition associating these gears.

5. CONCLUSIONS

The proposed dual read head IAS estimation method has been investigated in theory and experiment. The encoder error model is presented and concluded that the eccentric and swash error will decrease the real IAS by generate a minus erroneous IAS oscillation at first two rotating frequency harmonics. Both simulation and experiment studies show the great capability in supressing the erroneous IAS caused by encoder imperfections, which exhibit mainly at rotating frequency harmonics, and thereby show the effectiveness and efficiency in monitoring the PG faults where the non-integer rotating harmonics are less influenced. Moreover, the fault signatures associated with the planet gear spalling damages can be identified successfully by the increased amplitudes at characteristic frequencies in the improved IAS signals.

ACKNOWLEDGMENT

The authors are grateful for the financial support provided by the National Natural Science Foundation of China under Contract No. 51475053 and China Scholarship Council No. 201606050039.

6. REFERENCES

1. Hahn, B., M. Durstewitz, and K. Rohrig, *Reliability of Wind Turbine–Experiences of 15 years with 1,500 WTs.* Institut für Solare Energieversorgungstechnik (ISET), Verein an der Universität Kassel eV, 2005. **34119**.
2. Wu, B., et al. *An approach to fault diagnosis of helicopter planetary gears.* in *AUTOTESTCON 2004. Proceedings.* 2004. IEEE.
3. Večeř, P., M. Kreidl, and R. Šmíd, *Condition indicators for gearbox condition monitoring systems.* Acta Polytechnica, 2005. **45**(6).

4. Ha, J.M., et al., *Autocorrelation-based time synchronous averaging for condition monitoring of planetary gearboxes in wind turbines.* Mechanical Systems and Signal Processing, 2016. **70**: p. 161-175.
5. Liang, X., M.J. Zuo, and L. Liu, *A windowing and mapping strategy for gear tooth fault detection of a planetary gearbox.* Mechanical Systems and Signal Processing, 2016. **80**: p. 445-459.
6. Samuel, P.D. and D.J. Pines. *Vibration separation methodology for planetary gear health monitoring.* in *Smart Structures and Materials 2000: Smart Structures and Integrated Systems*. 2000. International Society for Optics and Photonics.
7. Mark, W.D., et al., *A simple frequency-domain algorithm for early detection of damaged gear teeth.* Mechanical Systems and Signal Processing, 2010. **24**(8): p. 2807-2823.
8. Nguyen, K.D., et al., *A wearable sensing system for tracking and monitoring of functional arm movement.* IEEE/ASME Transactions on mechatronics, 2011. **16**(2): p. 213-220.
9. Liu, F., et al., *Error analyses and calibration methods with accelerometers for optical angle encoders in rotational inertial navigation systems.* Applied optics, 2013. **52**(32): p. 7724-7731.
10. Ovaska, S.J. and S. Valiviita, *Angular acceleration measurement: A review.* IEEE transactions on Instrumentation and Measurement, 1998. **47**(5): p. 1211-1217.
11. Mancini, D., et al. *Encoder system design: strategies for error compensation*. in *Telescope Control Systems III*. 1998. International Society for Optics and Photonics.
12. Li, Y., et al., *The measurement of instantaneous angular speed.* Mechanical Systems and Signal Processing, 2005. **19**(4): p. 786-805.
13. Qin, S., Z. Huang, and X. Wang, *Optical angular encoder installation error measurement and calibration by ring laser gyroscope.* IEEE Transactions on Instrumentation and Measurement, 2010. **59**(3): p. 506-511.
14. Filatov, Y.V., et al. *Laser goniometer systems for dynamic calibration of optical encoders*. in *Optical Measurement Systems for Industrial Inspection III*. 2003. International Society for Optics and Photonics.
15. Watanabe, T., et al. *Automatic high-precision calibration system for angle encoder (II)*. in *Recent Developments in Traceable Dimensional Measurements II*. 2003. International Society for Optics and Photonics.
16. Watanabe, T., et al., *An angle encoder for super-high resolution and super-high accuracy using SelfA.* Measurement Science and Technology, 2014. **25**(6): p. 065002.

Novel Bearing Fault Detection using Generative Adversarial Networks

S. Baggeröhr[1], W. Booyse[1], P.S. Heyns [1], D.N. Wilke[1]
[1] Centre of Asset Integrity Management, Department of Mechanical and Aeronautical Engineering, University of Pretoria, South Africa

ABSTRACT

Bearing fault detection and diagnosis (FDD) are important if one is to avoid the more catastrophic failure consequences of large rotating machinery. Faults usually manifest as marginal defects that intensify over time, allowing for well-informed preventative actions with early FDD. Detection of the fault begins with capturing signals from a machine in various forms, such as vibration response signals. Numerous methods exist to extract either time, frequency or time-frequency based diagnostic features from these signals. Traditionally, handpicked descriptive statistical features (mean, RMS, skewness, Kurtosis, etc.) or spectral diagrams are used for FDD. However, machine signals are usually generated under non-stationary operating conditions of varying loads and speeds, which makes FDD with these methods more challenging. A wealth of research has been invested towards machine learning based signal processing techniques to circumvent the problems associated with non-stationary signals. Many of these methods require vast amounts of historical data to train. Since a machine spends most of its life operating in a healthy condition, most historical data is occupied with data that comes from a healthy machine condition. Training these methods are difficult, due to the shortage of data from a machine running in an unhealthy condition. Generative Adversarial Networks (GANs) are proposed as a novel method to perform condition-based diagnostics on assets. This method has the advantage that it can be trained exclusively with healthy data. The GAN, when trained, learns a statistical manifold on which the healthy data lies. A deviation from this statistical manifold can indicate a fault within an asset. This paper presents a preliminary study on the use of GANs to detect and diagnose faults in bearings. It was found that GANs can detect faults in bearings. Also, by tracking the manifold describing the healthy data, one can locate and assess the type of fault. This method has the added benefit of training on the entire data set, as opposed to supervised learning methods, which require the data to be divided into training, validation and test sets.

Keywords: Condition monitoring, Fault detection and localization, Diagnosis, Bearings, Deep learning

Corresponding author: Stephan Baggeröhr (u12017095@tuks.co.za)

1. INTRODUCTION

Rotating machinery have become a staple in almost all areas of modern technology, for example in manufacturing or power generation, where high reliability has always been a big concern. Rotating machinery are generally supported by bearings, whose failure can lead to entire system shutdown. Consequently, bearings are considered one of an asset's most critical components. As a result, bearing fault detection and diagnosis (FDD) are important if one is to avoid the more catastrophic failure consequences of large rotating machinery. Faults usually manifest as marginal defects that intensify over time, allowing for well-informed preventative actions with early FDD. As such, FDD has become more popular amongst maintenance engineers to increase machine reliability and availability while reducing the costs associated with maintenance and unnecessary shutdowns.

FDD is traditionally broken down into a feature extraction phase and a fault classification phase. The feature extraction phase begins with acquiring signal data from a machine of interest. Signals are captured in various forms such as vibration or acoustic emissions for instance. These signals contain rich information about the condition of the machine, and numerous signal processing techniques are used to extract this information in the form of fault features. Features are generally extracted from the time or frequency domain or a combination of both. A good feature is one that is only sensitive to faults and performs well in signals with a low signal-of-interest (fault) to noise ratio. Statistical features such as Root Mean Square (RMS) magnitude, spectral skewness and Kurtosis are a popular choice amongst engineers [1]. Alternatively, in the frequency domain, spectral diagrams are used to manually track fundamental fault frequencies that can be calculated from the bearing geometry. Machines are, however, frequently operated under non-stationary conditions, such as varying speed or load. This makes extracting features with these methods more challenging, as they assume stationarity of a signal [2,3]. As a result, more advanced signal processing techniques are used in conjunction with order tracking to extract features in non-stationary conditions.

The second step of FDD involves classifying the fault based on the extracted features. Classification is done by comparing the values of the extracted features for healthy machines with those of unhealthy machines. The fault is identified by comparing the characteristics of the features to those of known fault modes. Once the fault is identified, a degradation metric or fault severity indicator (FSI) is selected and used to trend the fault over a predetermined monitoring interval. The resulting trend is then used to make a prediction about the remaining useful life (RUL) of the machine. The diagnosis is heavily dependent on the types of features extracted during the first phase. Features that are also sensitive to operating conditions may result in a misdiagnosis. The entire FDD process is not static, as feedback from each phase is used to update the process. For example, the signal processing method may be adjusted depending on the fault that is identified.

Knowing what types of fault features to use and what fault modes to expect requires expert knowledge or careful physics-based modelling of the system. Both of which can be costly or time-consuming to implement. As a result, data-driven approaches have gained increasing popularity over the years. These models have attempted to replace the second step of FDD by directly mapping the extracted features to fault modes. Because the second step of FDD is essentially a pattern recognition type problem, machine learning approaches have predominantly been used [4]. Majority of these methods require vast amounts of historical data to train. As a machine spends most of its life operating in a healthy condition, most historical data is occupied by data from a healthy machine condition. Training these methods is difficult because of this class imbalance. Machine learning algorithms also require good discriminative features to be extracted, and hence the careful selection of features and signal processing techniques are necessary to create a successful machine learning diagnostic algorithm, especially in non-stationary cases. A proposed framework can become very specific to the machines and faults it was developed for. This makes implementing the proposed framework more challenging when a diverse range of similar assets needs monitoring. More recently, advancements in deep learning allow algorithms more flexibility to find complex hierarchical representative features by exploiting deep network architecture. These algorithms can represent complex functions and as a result, they can extract more discriminative features over a broader range of signals.

Generative adversarial networks (GANs) are a type of generative model that exploits deep learning architecture to build a representation of a target distribution, such as the probability density of a vibration waveform from a healthy machine. In this paper, Generative Adversarial Networks have been proposed as a novel method to perform condition-based diagnostics on bearings. This method has the advantage that it can be trained solely on healthy data, requiring no preprocessing or advanced signal processing to extract features. The GAN, when trained, learns the probability density of the healthy data. This probability density can be considered as a low dimensional statistical manifold embedded in the high dimensional space of the problem. A deviation from this statistical manifold can indicate a fault within an asset. Unlike conventional statistical learning methods, such as Expectation Maximization or Variational Autoencoders, which require several assumptions regarding the shape of the data distribution (such as factorized gaussian distribution), a GAN can learn an implicit representation of the data which requires no prior assumptions of the structure of the data. As a result, no labelling of data is required. This paper presents a preliminary study on the use of GANs to detect and diagnose faults in bearings. This method has the added benefit of training on the entire data set, as opposed to supervised learning methods, which require the data to be divided into training, validation and test sets. The rest of the document is laid out as follows. The second section covers the theoretical framework for the proposed model. The third section summarizes the results obtained from preliminary studies on the application of the model.

2. GAN MODEL FOR BEARING FAULT DETECTION

2.1. GAN Framework

A generative adversarial network [5] is comprised of two subnetworks, namely a Generator network, $\boldsymbol{G}_{\theta}(\boldsymbol{z})$ and a Discriminator network, $\boldsymbol{D}_{\phi}(\boldsymbol{x})$, where θ and ϕ are trainable network parameters (such as weights and biases) and

$\boldsymbol{x}$ is a query sample. Training of the networks is achieved by setting the generator network against the discriminator network in a two player non-cooperative game expressed as the min max optimization problem V, shown in Eq. 1. The generator network passes random noise, $\boldsymbol{z}\sim U[0,1]$ or $\boldsymbol{z}\sim\mathcal{N}[0,1]$ through the network and produces a sample from a parametrized distribution, $\boldsymbol{Q}$ that is dependent on θ. The discriminator network tries to estimate the probability that the query sample was either produced by the generator or sampled from a target distribution, $\boldsymbol{P_{data}}$. Each sub network updates its own parameters with gradients derived from the cost function defined in Eq. 1 and back propagation of the error. The generator will improve the samples it produces based on the feedback it obtains from the discriminator. Theoretically, training is completed, when Nash-equilibrium is reached. In game theory, a Nash-equilibrium where a player has reached a point that leaves him/her in no better or no worse situation, no matter what his/her opponent decides to do. At this stage, the discriminator can no longer distinguish between samples produced by the generator and samples drawn from the target data distribution, and thus $\boldsymbol{Q} \approx \boldsymbol{P_{data}}$. The discriminator, now unsure of whether the sample is real or generated, outputs a probability of $\boldsymbol{D_{\phi}}(\boldsymbol{x}) \approx 0.5$.

$$\min_{\theta}\max_{\phi} V\left(\boldsymbol{D_{\phi}}(\boldsymbol{x}), \boldsymbol{G_{\theta}}(\boldsymbol{z})\right) = \mathbb{E}_{\boldsymbol{x}\sim p_{data}(\boldsymbol{x})}[\log \boldsymbol{D}(\boldsymbol{x})] + \mathbb{E}_{\boldsymbol{z}\sim p_{z}(\boldsymbol{z})}[\log(\boldsymbol{1} - \boldsymbol{D}(\boldsymbol{G}(\boldsymbol{x})))] \quad \cdots\cdots\cdots \quad (1)$$

For bearing diagnostics, a GAN is trained on purely healthy data. During the inference stage, a previously unclassified query sample is then passed to the discriminator. If the query matches the distribution of the training data ($\boldsymbol{Q} \approx \boldsymbol{P_{data}}$), the discriminator will output $\boldsymbol{D_{\phi}}(\boldsymbol{x}) \approx 0.5$, and the sample is classified as healthy. Instead, if the query sample results in $\boldsymbol{D_{\phi}}(\boldsymbol{x}) \neq 0.5$, the sample is believed to be anomalous from the healthy data and further investigation is performed to identify the cause of the anomaly. The loss functions for generator and discriminator are derived from Eq. 1. The loss function for the discriminator, shown in Eq. 2, when minimized ensures the discriminator correctly classifies samples as real or generated. The loss for the generator, shown in Eq. 3, is taken as the Kullback-Leibler (KL) divergence loss from [6] and ensures the generator produces realistic samples, whilst allowing both the discriminator and generator to reach an optimum. Optimization is done by RMSProp with a learning rate of 10^{-4} which is exponentially decayed at a rate of 0.98 every 4000 iterations. Where one iteration corresponds to one update of the trainable parameters using a batch of 64 random samples per iteration.

$$\mathcal{L}_D(\theta) = -\frac{1}{2}\log(\boldsymbol{D}(\boldsymbol{x})) - \frac{1}{2}\log(1 - \boldsymbol{D}(\boldsymbol{G}(\boldsymbol{z}))) \quad \cdots\cdots\cdots \quad (2)$$

$$\mathcal{L}_G(\phi) = -\log\left(\frac{\boldsymbol{D}(\boldsymbol{G}(\boldsymbol{z}))}{1 - \boldsymbol{D}(\boldsymbol{G}(\boldsymbol{z}))}\right) \quad \cdots\cdots\cdots \quad (3)$$

2.2. Architecture

For the diagnosis of faults in bearings, the architecture proposed by [7], known as a Deep Convolutional GAN (DCGAN), was used. The architecture was originally proposed to learn reusable feature representations in large unlabelled datasets, specifically in the context of computer vision. This makes it ideal for learning features from raw vibration signals. As vibration signals tend to be cyclo-stationary, certain features tend to repeat themselves, albeit at irregular intervals in the case of non-stationary operating conditions. The network is comprised of multiple layers of convolutions and thus the features that are learnt are somewhat invariant to scaling and transformations. This allows the network the flexibility to learn features even in non-stationary operating conditions. Some adjustments are done, to allow for the training on acceleration signals. These adjustments were:

- Batch Normalisation was *not* used in between layers.
- An additional fully connected layer was added after the last convolution layer in the discriminator.
- Exponential Linear Units (ELU) activation functions [8] were used in the discriminator in place of leaky Rectified Linear Units (RELU).

A visual representation of this architecture used in this study is presented in Fig. 1. The discriminator consists of 4 convolution layers with increasing number of filters on each layer. The stride is kept low on the first layer and then increased on the rest of the layers. The second last and last layers of the network are fully connected with an output size of 256 and 1 respectively. ELU activation is used on all layers except the output. In the generator network, random noise is passed through a fully connected layer, with an output size chosen and reshaped to ensure the generated samples match the dimension of the real samples. This is then passed through four deconvolution layers as in [7]. The final layer is passed through a tanh activation function. Therefore, training data is normalised between $[-1,1]$, before being presented to the discriminator. The GAN architecture is summarised as follows:

Generator architecture:
Input: $z = batch \times 100 \rightarrow$
Fully Connected: [4096], Reshape: [256 × 16], Relu →
Deconvolution 1: [128 × 25 × 1], Stride: [4 × 1], Relu →
Deconvolution 2: [64 × 25 × 1], Stride: [4 × 1], Relu →
Deconvolution 3: [32 × 25 × 1], Stride: [4 × 1], Relu →
Deconvolution 4: [1 × 25 × 1], Stride: [4 × 1], Tanh →
Output: Reshape: [4096 × 1]

Discriminator architecture:
Input: $x = batch \times 4096 \times 1 \rightarrow$
Convolution 1: [32 × 25 × 1], Stride: [2 × 1], Elu →
Convolution 2: [64 × 25 × 1], Stride: [5 × 1], Elu →
Convolution 3: [128 × 5 × 1], Stride: [5 × 1], Elu →
Convolution 4: [256 × 5 × 1], Stride: [5 × 1], Elu →
Fully Connected: [256], Elu →
Output: Fully Connected: [1], Sigmoid

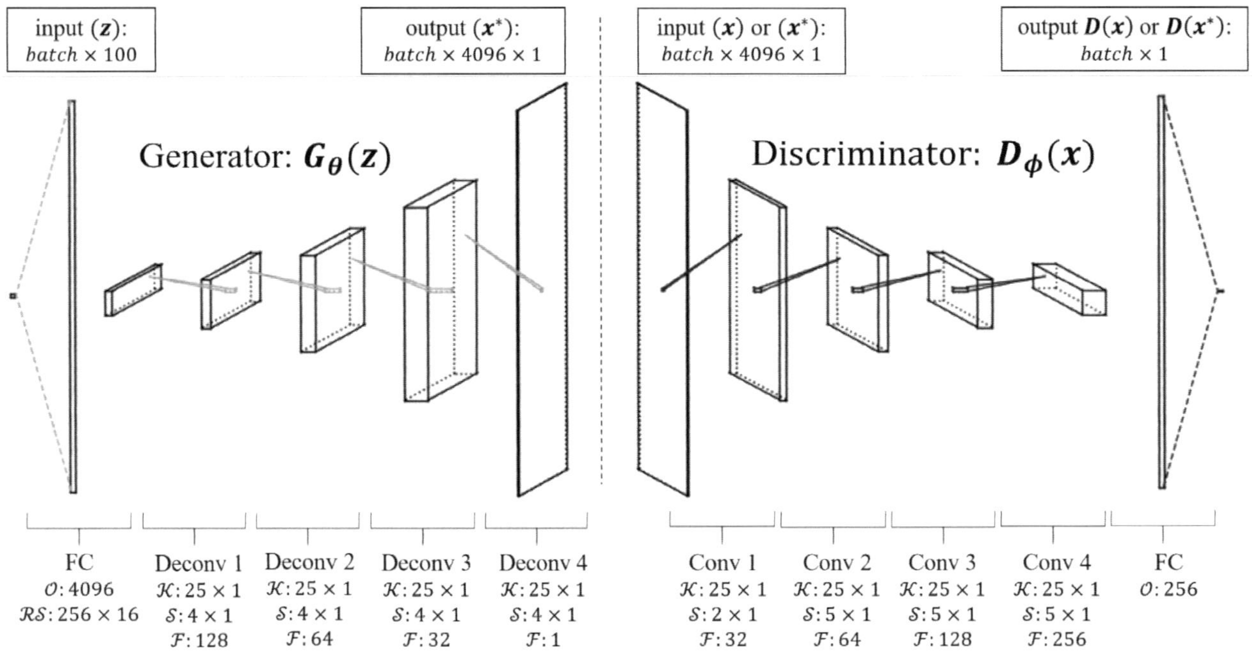

Figure 1: GAN Architecture for Discriminator and Generator networks, used for all investigations in this study.

2.3. Regularization for improved stability

The training of a GAN is a particularly difficult optimisation problem, since it involves finding the stationary points on a vector field defined by the gradients of the two loss functions (Eqs. 2 and 3). As opposed to conventional Artificial Neural Network training which only involves finding the stationary points of a vector field defined by only one loss function. When a vector field is defined by more than one function it becomes non-conservative. A non-conservative vector field is significantly more difficult to optimise as the optimisation process becomes path and starting point dependent. A vector field $\boldsymbol{F}$ is only conservative when there exists a function f such that $\nabla f = \boldsymbol{F}$ [9]. To improve the behaviour of the optimisation process researchers have recently proposed regularisation techniques to improve the behaviour of GAN training [10]. Two of these proposed methods were implemented in this paper, namely *consensus optimization* and *instance noise*.

2.3.1. Consensus optimization

The non-conservative vector field results in the optimization path that enters an orbit around a minimum instead of progressing towards it. Consensus optimization was introduced by [11] as a means of preventing this by modifying the loss functions of the generator and the discriminator. This is done by the addition of the term $L(\theta,\phi) = {}^{1}/_{2}\,\|\boldsymbol{v}(\theta,\phi)\|^2$, to both loss functions,

$$\tilde{\mathcal{L}}_D(\phi) = \mathcal{L}_D(\phi) + \gamma L(\phi,\theta) \text{ and } \tilde{\mathcal{L}}_G(\theta) = \mathcal{L}_G(\theta) + \gamma L(\phi,\theta), \quad (4)$$

where

$$\boldsymbol{v}(\theta,\phi) = \begin{pmatrix} \nabla_\phi \mathcal{L}_D(\phi) \\ \nabla_\theta \mathcal{L}_G(\theta) \end{pmatrix}, \quad (5)$$

is a vector containing the gradients of the two loss functions with respect to their trainable parameters and γ is a consensus factor. The initial γ was set to 3.5 for all experiments and then exponentially decayed at a rate of 0.5 every 8000 iterations.

2.3.2. Instance noise

At the start of training, the generator is essentially outputting noise. Therefore, the discriminator has an easy job of distinguishing between the real sample and the generated ones. It is highly unlikely for the parametric distribution of the generator to overlap with the distribution of the data distribution during early stages. As such [6] proposed the use of instance noise to overcome the problem. Instance noise ensures the parametric distribution and the data distribution initially overlap by adding Gaussian noise to both the generated and real samples. The noise has a mean of zero and a variance σ^2 of 0.2. As training progresses the variance of instance noise is linearly reduced to zero by iteration 10000.

3. EXPERIMENTAL VALIDATION

3.1. IMS Dataset

This dataset, developed by [12], contains the acceleration response of four Rexnord ZA-2115 double row bearings installed on a single shaft. The shaft was run by an AC motor kept at a constant speed of $2000RPM$ with a constant radial load of $6000lbs$. The dataset contains the results of three run-to-failure tests. Upon completion of test set No. 1, an inner race fault was discovered on bearing 3 and roller element fault was discovered on bearing 4. A GAN was trained using this test set No. 1. Only the acceleration signals of bearing 3 and 4 in the y direction were considered for this investigation and used for training. Two independent GANs were then trained for 30000 iterations (± 2 hours), separately on each of the bearing signals. The first 8% of data was taken as the reference condition (treated as healthy) and used to train the GAN. The training data corresponds to the blue highlighted portion of Fig. 2. In Fig 2a, the output of the GAN trained on the signal from bearing 3 is shown over a period of 35 days. Similarly, in Fig 2b, the output for the GAN trained on bearing 4 is shown. Also shown in both plots are the RMS and Kurtosis of each signal over the 35 days.

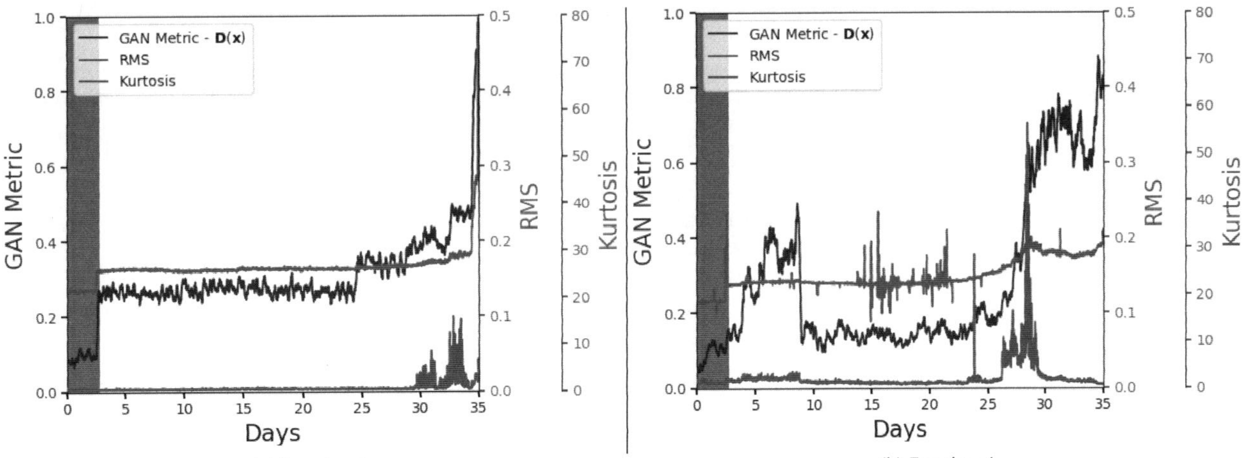

(a) Bearing 3 (b) Bearing 4

Figure 2: GAN metric of IMS dataset trended over 35 days. Blue highlighted portion represents first 8% of data used for training of the GAN. Also shown are traditional statistical features, RMS and Kurtosis of the same signal.

In Fig. 2a, the RMS and Kurtosis signals remain relatively unchanged during the first 30 days of operation, after which the RMS starts to increase, and the Kurtosis levels rise. This is in line with the observations of [13] and is typical behaviour of bearings with an inner race fault, where crack propagation and development occur over a short period of time. Looking at the GAN metric, the trend is relatively constant up until day 25, after which, a small jump is noticed. The GAN metric then remains constant until day 30, where it is seen to increase until the last day of observation. In Fig 2b, the GAN metric increases initially until day 10, after which it decreases and remains constant until day 25, where the rate increase until the end of sampling. A small increase in the Kurtosis is also noticed during the first 10 days. This initial increase and decrease are explained by a healing phenomenon whereby small surface cracks are smoothed by continuous rolling action. At a later stage, after day 25 when damage increases, the vibration levels increase once more. The jump that was noticed on day 25 of bearing 3 is explained by the increase in damage of bearing 4 around the same time. This shows the GAN's sensitivity to any deviation from conditions other than those that were used during training, treating any deviation as anomalous. Furthermore, the magnitude of the GAN metric during the first 10 days of bearing 4 indicate that it is more sensitive to the initialization of faults than the traditional statistical metrics of RMS and Kurtosis. This makes detecting faults during early fault development, far more accurate.

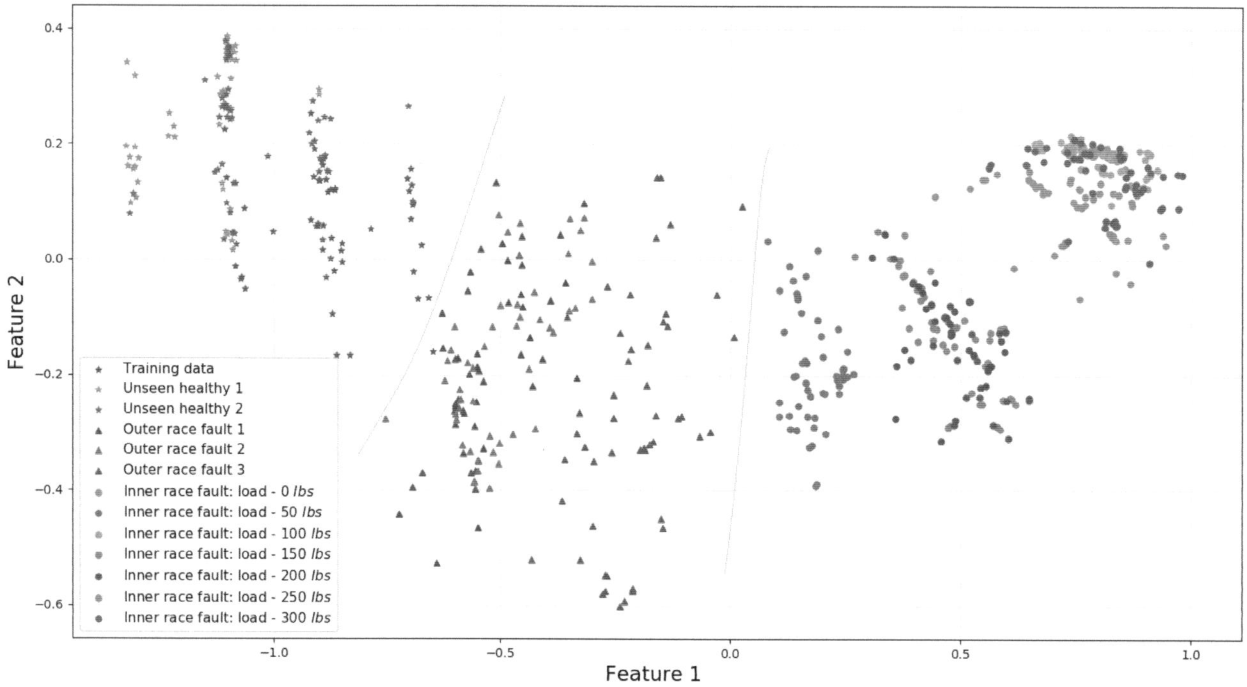

Figure 3: Scatter plot of samples with various fault modes presented to a trained GAN. Clustering of similar samples can be seen on the manifold

3.2. MFPT Dataset

The second dataset that was tested was a fault dataset released by the Society for Machinery Failure Prevention Technology (MFPT) [14] with an accompanying tutorial in bearing envelope analysis [15]. The dataset was released with the aim to facilitate research into bearing analysis. The dataset is comprised of several sets of test cases with data collected from a bearing test rig that was equipped with a NICE bearing that has 8 rolling elements. Also included were a few real-world cases, but these were not considered. The first dataset is a baseline condition that was sampled from a healthy bearing at a shaft speed of 25 Hz and a load of 270 lbs. The second set, contains data from a bearing with an outer race fault, operated at the same operating conditions as the baseline condition. The last of the datasets considered was that of a bearing with an inner race fault, sampled at varying loads between 0 and 300 lbs at a constant shaft speed of 25 Hz. Consequently, the dataset represents both stationary and non-stationary conditions in terms of load. A GAN was trained for 70000 iterations ($\pm$4 hours), solely on data from the one of the healthy baseline conditions. Once trained, samples from the remaining sets of data where shown to the discriminator for inference. The second last layer of the discriminator was used to visualise the manifold on which the data was mapped. ISOMAP was used to further reduce the dimension of the second last layer from 256 to 2 features for plotting as shown in Fig. 3. It can be seen in Fig. 3, that healthy samples are clustered together towards the top left corner of the plot. Even healthy samples that were not used for training, are clustered together with samples used for training. Similarly, samples with an outer race fault and the inner race fault are clustered towards the centre and right of the plot respectively. This shows that the features learnt by the discriminator are representative enough to separate samples from various fault modes, even when those samples have not been observed before. It also shows that samples of similar features are mapped to certain regions within the statistical manifold. Demonstrating that the approach may be viable in a semi-supervised diagnostic framework with sparse labels.

3.3. HSBD Dataset

High-speed bearing data (HSBD) was collected from a $2MW$ commercial wind turbine [16] and was used as the final dataset for this investigation. This real-world data set is inherently non-stationary, due to the varying wind speeds and loads found in nature. The measurements in this dataset were captured over a period of 50 consecutive days. The bearing from the wind turbine had an inner race fault that progressively deteriorated across the 50 days. The data was captured at a sampling rate of $97656\ sps$ for $6s$. The first 4 days (8%) of raw data was taken as the reference set and used to train a GAN for 70000 iterations (± 4 hours). Again, the output of the discriminator was used as a fault metric and trended over the period of 50 days and can be seen in Fig. 4. Also shown in Fig. 4, is the inner race envelope energy (measured in Gs as per [16]) of the raw and resampled data over the 50 days. The GAN metric is constant with small fluctuations until day 25, where it starts to increase until the last day of observations. Small fluctuations can be described by varying weather conditions. No resampling of the raw signal was required when training a GAN. If it is assumed that the training data contains samples from all possible operating conditions, the GAN learns how the healthy data looks at these varying operating conditions and any deviation can be explained by faults. Even when this assumption is weak, as was the case here, the overall trend of the fault can still be recovered. The small fluctuations present during the early stages of observations can be attributed to operating conditions not seen by the GAN during training. The duration of the peaks that are present in the early stages of observation are relatively short when compared to the duration over which the GAN metric increases. This indicates that the condition that caused the increase in the GAN metric is permanent, like a fault condition, and the condition that caused the small fluctuation is temporary, like an unseen operating condition.

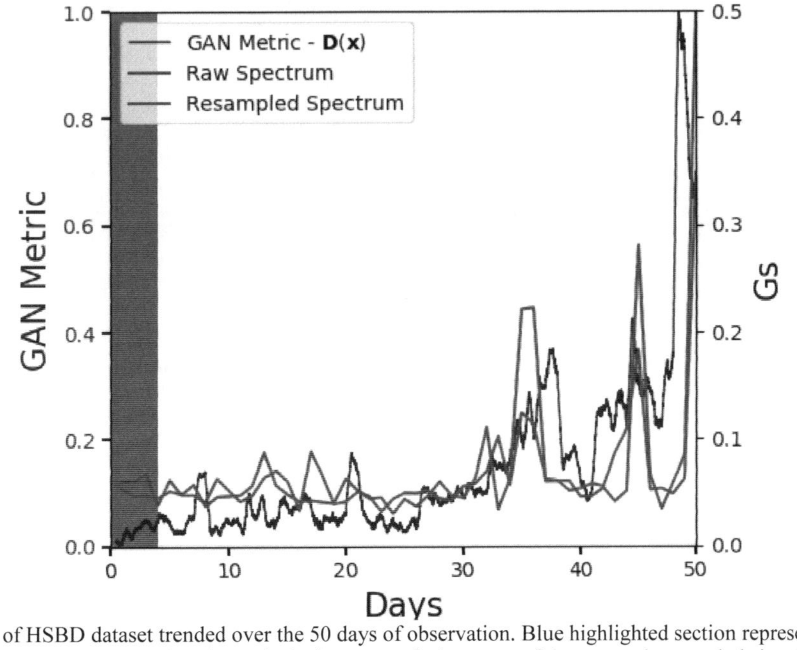

Figure 4: GAN metric of HSBD dataset trended over the 50 days of observation. Blue highlighted section represents the first 4 days of data used for training of the GAN. Also shown is the inner race fault energy of the raw and resampled signal measured in Gs.

4. CONCLUSION

This paper presents a novel technique that uses Generative Adversarial Networks (GANs) to perform bearing diagnostics. The issue of class imbalance, that is prevalent with diagnostics and typical machine learning algorithms, is solved by leveraging the GAN's ability to learn a statistical manifold on purely healthy data. Once the manifold is learnt, the discriminator of the network can be used to perform inference on unclassified samples. Faults can be trended by measuring how much the sample deviates from the statistical manifold. Furthermore, it is shown that the network automatically clusters samples that contain similar features on this learnt manifold. This clustering can be used for unsupervised classification of multiple fault modes. The framework is robust to non-stationary signals and requires no advanced signal processing or order tracking to perform.

REFERENCES

[1] N. Tandon, "A comparison of some vibration parameters for the condition monitoring of rolling element bearings," *Measurement*, vol. 12, no. 3, pp. 285–289, 1994.

[2] P. Boskoski and D. Juricic, "Fault detection of mechanical drives under variable operating conditions based on wavelet packet Renyi entropy signatures," *Mech. Syst. Signal Process.*, vol. 31, pp. 369–381, 2012.

[3] R. Zimroz, W. Bartelmus, T. Barszcz, and J. Urbanek, "Diagnostics of bearings in presence of strong operating conditions non-stationarity - A procedure of load-dependent features processing with application to wind turbine bearings," *Mech. Syst. Signal Process.*, vol. 46, no. 1, pp. 16–27, 2014.

[4] R. Liu, B. Yang, E. Zio, and X. Chen, "Artificial intelligence for fault diagnosis of rotating machinery: A review," *Mech. Syst. Signal Process.*, vol. 108, pp. 33–47, 2018.

[5] I. Goodfellow *et al.*, "Generative Adversarial Nets," *Adv. Neural Inf. Process. Syst. 27*, pp. 2672–2680, 2014.

[6] C. K. Sønderby, J. Caballero, L. Theis, W. Shi, and F. Huszár, "Amortised MAP Inference for Image Super-resolution," *Mach. Vis. Appl.*, vol. 25, no. 6, pp. 1423–1468, Oct. 2016.

[7] A. Radford, L. Metz, and S. Chintala, "Unsupervised Representation Learning with Deep Convolutional Generative Adversarial Networks," pp. 1–16, 2015.

[8] D.-A. Clevert, T. Unterthiner, and S. Hochreiter, "Fast and Accurate Deep Network Learning by Exponential Linear Units (ELUs)," pp. 1–14, 2015.

[9] J. Stewart, *Calculus: Early Transcendentals*, 7th ed. Cengage Learning, 2010.

[10] L. Mescheder, A. Geiger, and S. Nowozin, "Which Training Methods for GANs do actually Converge?," 2018.

[11] L. Mescheder, S. Nowozin, and A. Geiger, "The Numerics of GANs," 2017.

[12] H. Qiu, J. Lee, J. Lin, and G. Yu, "Wavelet filter-based weak signature detection method and its application on rolling element bearing prognostics," *J. Sound Vib.*, vol. 289, no. 4–5, pp. 1066–1090, 2006.

[13] T. Williams, X. Ribadeneira, S. Billington, and T. Kurfess, "Rolling element bearing diagnostics in run-to-failure lifetime testing," *Mech. Syst. Signal Process.*, vol. 15, no. 5, pp. 979–993, 2001.

[14] E. Bechhoefer, "Fault Data Sets – Society For Machinery Failure Prevention Technology." [Online]. Available: http://mfpt.org/fault-data-sets/.

[15] E. Bechhoefer, "A quick introduction to bearing envelope analysis."

[16] E. Bechhoefer, B. Van Hecke, and D. He, "Processing for Improved Spectral Analysis," *Annu. Conf. Progn. Heal. Manag. Soc.*, pp. 1–6, 2013.

Condition Monitoring of a Fan using Neural Networks

K. Passi[1], B. Zhang[1], M. Timusk[2]
[1]Department of Mathematics and Computer Science, Laurentian University, Sudbury, Ontario, Canada
[2]Bharti School of Engineering, Laurentian University, Sudbury, Ontario, Canada

ABSTRACT

Fans are widely used in various industrial fields and it plays a key role in cooling the machinery. For the machinery to work properly, the fan system should remain in stable and error-free condition. Condition monitoring is introduced as a maintenance tool for the failure diagnosis of a fan system. Some methods used in condition monitoring are vibration monitoring and thermal monitoring. Vibration monitoring method was used in this experiment. A fan system based on Machinery Fault SimulatorTM (MFS) was used to simulate different conditions of a fan in the laboratory. An accelerometer was installed on top of the bearing housing. It was used to detect the vibration signal of a running fan. A data acquisition program designed in LabVIEW was used to record and preprocess the raw vibration signal. The collected data was used to detect the condition of the fan system. Neural Network was used for the fault diagnosis. The raw vibration signal is a one-dimensional time domain series data, while the neural network requires multidimensional features as input data. Therefore, it is important to preprocess the raw vibration signal data. Two different preprocessing methods, time-domain features and Auto Regressive (AR) model features were used to preprocess separately. The neural network model was trained by these two methods respectively. The results show that the AR model gave better features than the time domain features method. The condition monitoring system consisted of the following parts: data acquisition, data storage, data preprocessing and the display of results. Some methods were programmed in Matlab, which were called by Matlab scripts in the LabVIEW software. The hybrid programming method helped to generate an efficient program which provided high accuracy of fault diagnosis.

Keywords: Fan, Failure diagnosis, Vibration analysis, Neural Network, LabVIEW, Matlab

Corresponding author: Kalpdrum Passi (kpassi@cs.laurentian.ca)

1. INTRODUCTION

Fans are widely used in various industrial fields. In factory premises, the operation of equipment generates enormous amounts of heat and its operation requires the maintenance of certain temperature conditions. High temperature causes overheating, which may break the machines. Good heat dissipation ability helps the machines and equipment to be in good working condition for a long time. Air-cooled radiator, which has the fan as a key component, exudes a tremendous amount of heat. It is used to maintain machine's temperature. Wind energy is a renewable, widely distributed and non-polluting source of energy [11] and fan blades are the key components of wind turbines.

Fans working at high-speed for a long time in harsh working conditions with inappropriate maintenance and excessive demands may cause faults. Therefore, monitoring the fan's condition is critical.

Fault Diagnosis is an emerging discipline that is based on reliability theory, information theory, cybernetics theory and systems theory [7]. It uses modern test instruments and computer technology as technical means to study a variety of diagnostic objects' features. Neural network is a highly complex large nonlinear adaptive system, which is made up of a large number of simple processing units [3]. Artificial neural networks can perform massively parallel processing. In this paper, we apply neural network for condition monitoring of a fan.

1.1. Fan's failure types

A rotating fan is basically combined of three parts: the drive device, such as the motor, transmission device, such as the shaft and the blades. Usually in order to control the rotating speed of the blades, electronic control unit or gearbox is added. Fan's parameters include the following features: wind velocity, air volume, wind pressure, lifetime and noise. These parameters impact the performance of the fan. Based on these parameters, researchers are able to make condition monitoring plans. Typical failures of rotating machinery include unbalance, misalignment, static and dynamic friction. Fan blades unbalance, rotor unbalance and bearing failure are the main reasons of fans' failures, which causes abnormal vibration in the equipment. These abnormal vibrations shorten the life of the equipment and cause some other unpredictable damages.

2. EXPERIMENT SET-UP

In order to test and validate a prototype condition monitoring system, the ideal situation would be to collect data and perform experiments on a fan in a plant exhibiting all of the fault conditions of interest. However, there are some limitations that make it impractical to do this. This test required various data sets collected from different fault conditions of a fan. For this reason, the experiments required were conducted in the laboratory using a machinery diagnostics simulator with a fan attachment.

The apparatus employed for the experiments had three capabilities. The first capability was that it was able to simulate the fan's normal operation. The second capability was that it was possible to seed faults on the fan and to operate it under these faulty conditions. The last capability was it was able to collect data from an accelerometer installed on the fan's bearing housing.

The laboratory platform that was used is a commercially available Machinery Fault Simulator^{TM} (MFS) manufactured by Spectra Quest of Richmond, Virginia. This MFS is designed to reproduce multiple types of machine faults such as shaft or coupling misalignment, mechanical looseness, motor faults, rolling element bearing faults, rotor unbalance and pump faults. In the case of the present work, the test bench used in fan's fault simulation and its schematic is shown in Figure 1.

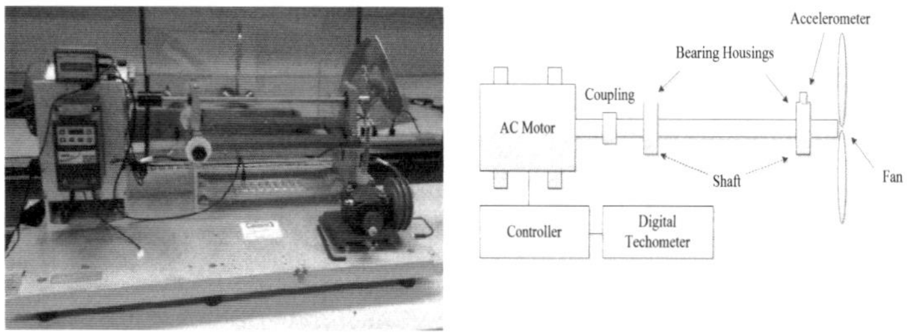

Figure 1. Photo of test bench and Schematic of the test apparatus

The equipment used to control the rotating speed of the motor was a Lenze AC Tech Controller. The test bench was fitted with a digital tachometer to measure and display the rotating speed of the motor. There are six buttons on the keypad, which are used to start or stop the drive, change the speed, and change the rotation direction. In the case of this work the motor was operated in three states: low speed, medium speed and high speed corresponding to 10Hz, 20Hz and 30Hz respectively.

The unbalance condition is the most common type of fan fault. In order to simulate the blade unbalance condition, a small mass was added to one of the three blades. In order to simulate the rotor unbalance, a cocked bearing housing was used. The lower surface of the cocked bearing housing that contacted with the rotor base was designed to achieve about a 0.5-degree tilt. The last fault type examined was a rolling element bearing fault. Due to the limited condition in the laboratory, in this case the damped bearing is selected as the failure bearing.

Platinum Low-cost Industrial ICP ® Accelerometer, which is made by ICP ® Sensor, was used for measuring the vibration signal in the test [10]. It is now widely used for general industrial purpose all over the world. A piezoelectric element is used by the accelerometer to convert the mechanical motion into an electrical signal. The value of the electrical signal reflects the amplitude of the vibration of the object that the accelerometer is attached to. A data acquisition module manufactured by National Instruments (model number PXIe-4492) was used to collect the accelerometer data. This data acquisition module is specifically designed for sound and vibration applications [8].

National Instruments LabVIEW (Laboratory Virtual Instrument Engineering Workbench) is a graphical programming language, similar to other traditional programming languages. The LabVIEW program written for this work performed the following functions: acquisition of the vibration signal, display of the vibration signal, display of the preliminary analysis of the vibration data, and storing the vibration signal data for further analysis and training of fault detection programs. In this experiment, the task was to collect fan's vibration signals in different states of health. The NI PXIe-4492 module would pick out the samples by the sample rate that was set up at 2000 Hz in the DAQ Assistant function. The samples were saved, processed and analyzed in LabVIEW for fan's test conditions of healthy, blade unbalance, rotor unbalance and bearing fault at rotating speeds 10 Hz, 20 Hz and 30 Hz, respectively.

3. NEURAL NETWORK DESIGN FOR CONDITION MONITORING

A neural network consists of an input layer, one or more hidden layers and an output layer. The input layer of neural network model needs multidimensional data, but the raw vibration signal is a one-dimensional time series data. So, the fan's vibration signal should be preprocessed to fit the input parameters of the neural network. The raw vibration signal needs to be converted to multiple features as input nodes representing healthy and faulty machine operations. Two different data preprocessing methods, time domain features and auto regressive (AR) model, were compared in the training process.

Using the raw time domain vibration signal for analysis is a popular, simple and effective method [9]. The original vibration signal can directly reflect fan's working condition. Thus, feature extraction is needed for processing the raw vibration data that can reduce the size of the training set while preserving features that correspond to the condition of the fan. Six statistical features were used to describe the raw vibration data: Root Mean Square Value, Peak to Peak Value, Kurtosis, Crest factor, Impulse factor and Energy in time domain [9]. These six features are selected as the input features of the neural network. In each data set, the sampling rate is 2000 Hz, which means there are 10000 samples in each segment of time series data. Based on the sample rate and the rotating speed of the fan system, 400 samples were selected as a period to calculate their features. The reason for choosing 400 samples as a segment of time series data is that they represent two complete revolutions of the rotating fan (10Hz rotating frequency). Matlab was used to calculate the features. The feature set values are the input data for the Back Propagation (BP) neural network module as shown in Figure 2.

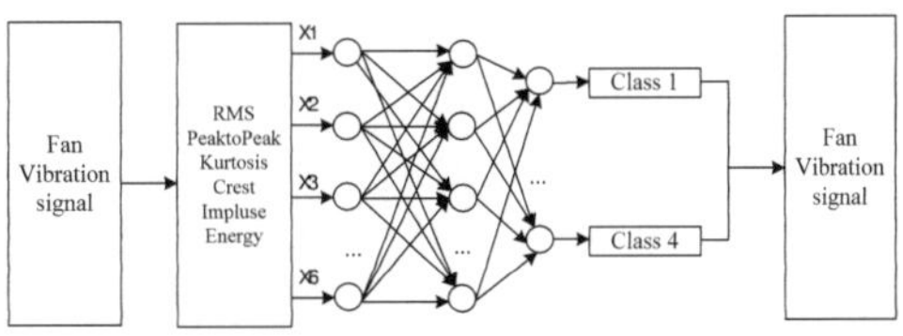

Figure 2. Neural network with time domain features as input nodes

There are 6 input nodes based on the fan's vibration signal feature vector and 4 output nodes based on the fan's conditions. The number of hidden layers can be calculated by the inequalities $l < \sqrt{(m+n)} + a$ and $\ell > log_2 n$ where n is the number of input nodes; m is the number of output nodes; a is a constant number between 1 and 10. Thus the number of nodes in the hidden layer range from 4 to13. Using Matlab modeling of the BP Neural Network, different number of nodes in the hidden layer were tried and it was found that 12 nodes gave the best accuracy. The fan vibration signal features representing four different conditions are labeled as 1, 2, 3, 4. Based on the labels, the target output values were set as 1000, 0100, 0010, 0001 representing healthy fan, blades unbalance, rotor unbalance and bearing fault, respectively. Neural Network Pattern Recognition Tool in Matlab was used to train the neural network. In the toolbox, 70% sample set (210 samples) was used as a training set, 15% sample set was used for validation, and the remaining 15% sample set was used for testing. The network adjusts the weights according to the classification error based on the training samples. Figure 3 shows the training results for four conditions with different speeds.

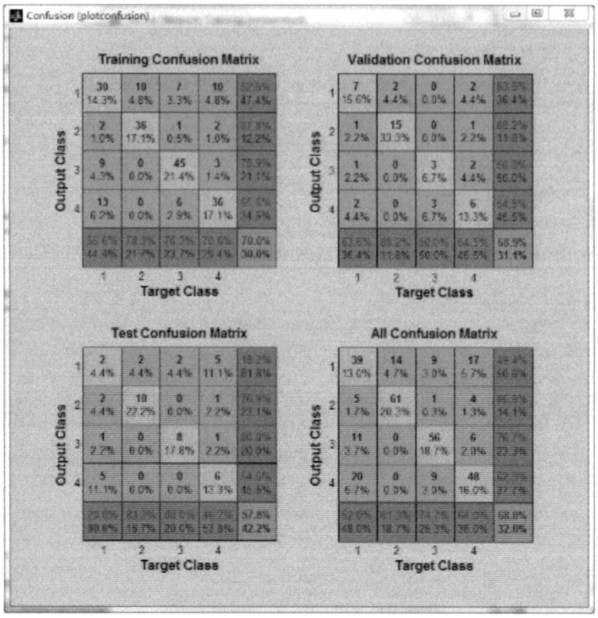

Figure 3. Training results for four conditions with different speeds

The first three matrices show the results of the training set, validation set and testing set and the fourth matrix shows the results of the combined data set. The columns in the matrices show the target classes, while the rows show the output classes. For example, in the Training Confusion Matrix, 30 samples (14.3% sample of training set) of output class 1 were correctly classified as the target class 1 (green background). Meanwhile, 2 samples (1% samples of training set) of target class 1 were misclassified as output class 2 (red background). The percentages shown in green color in the last row (grey background), columns 1 – 4 show the number of samples in the trained results that are correctly classified to the target class, and the percentages shown in red color show the number of samples in the trained results that are not classified to the target class. For example, in the Training Confusion Matrix, 55.6% of the samples were correctly classified as target class 1, and 44.4% of the samples were misclassified as other classes. The last column (grey background) are the results of the output classes. The percentages in green show in each output class how many samples were classified correctly, and the percentages in red show how many samples were classified incorrectly. For example, 52.6% samples of the output class 1 were classified correctly, the rest 47.4% samples in output class 1 were classified incorrectly. The last row-column (blue background) are the test results. For example, in the Training Confusion Matrix 70% of the samples in training set were classified correctly and 30% of the samples in the training set were classified incorrectly. The last matrix shows the test results of all the features. It is clear that 68% of the samples were classified correctly, which means for all of the 300 feature vector samples, 96 samples were classified to the wrong class.

From this test, it was obvious that using only one accelerometer to collect data, enough information was not available for training the BP neural network module. One accelerometer just gave us the information from the top of the bearing house. If we want to obtain more vibration signal information, we need more accelerometers which should be installed in different positions in order to get data from other dimensions.

Auto Regressive (AR) model [6] is a time series analysis method and its model parameters contain all the important features about the machinery's condition. Accurate AR model reflects the objective rules of dynamic system. AR model's coefficients are very sensitive to state changes in a dynamic system. Therefore, the actual AR coefficients were used as the feature vector to analyze the state of the dynamic system.

AR model is a widely used in practical work, and the function is a set of linear equations. Based on the knowledge of digital signal processing, a P-order AR model is equivalent to a P-order linear predictor. Based on the output of the AR model, the prediction parameters are calculated, which is the linear prediction coefficient. This linear prediction coefficient was used in voice coding at first, so it is named Linear Prediction Coding (LPC) [2]. The AR model's coefficients can be calculated by linear prediction analysis.

Many algorithms are used to calculate AR model's parameters, such as Burg's lattice-based method, Forward-backward approach, Geometric lattice approach, Least-squares approach and Yule-Walker approach. In practical work, Forward-backward approach is often used to calculate AR model's parameters [5]. Once the signal has been analyzed by AR model into discrete parameters, these parameters can be used for classification. Every set of parameters is able to show the state of fan in a time interval. This process is able to simplify the data and convert one-dimensional data into multidimensional data, which can be the input to the neural network model.

When using AR model to process vibration signal, users choose the number of AR model parameters. However, it is important to find the right number. If AR model is used to fit the time series, there is eventually a number that fits best, and this number is the optimal choice for the AR model. Two optimal value selection methods were used to find the number of parameters: Final Prediction Error (FPE) [1] and An Information Criterion (AIC) [4]. The FPE and AIC curves were plotted and the number of parameters were determined to be 20.

Minimizing the prediction error in the least squares sense, Linear Prediction Coding (LPC) is used to determine the coefficients of the forward linear predictor. It is used in filter design, eigenvalue extraction and speech coding. As the number of the parameters has been determined to be 20, fan's vibration signal is used as the AR (20) model's output. The parameters calculated from LPC by every 20 fan vibration signal samples can be used as the fan vibration features.

In each original fan vibration signal data file, there are 10,000 acceleration samples, and as the number of parameters is 20, every 400 samples are used to calculate their eigenvalues. The feature vector sample is the concentrated expression of the selected 400 vibration signal samples. It includes the features of the 400 samples, which can be used in classification algorithm. Feed-forward Neural Network with input parameters selected as eigenvectors from AR modeling is shown in Figure 4.

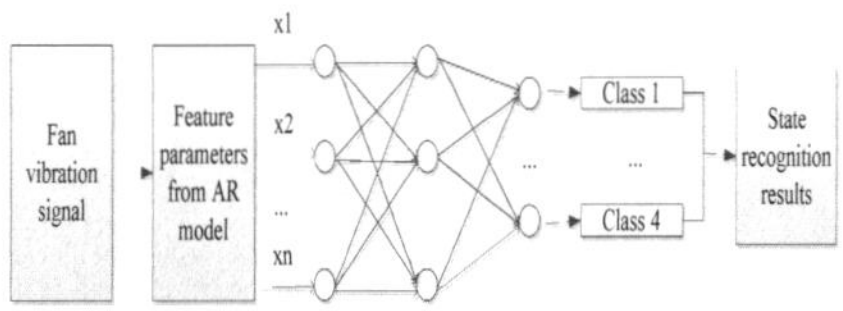

Figure 4. Neural network with AR model parameters as input nodes

There are 20 input nodes based on the fan vibration signal feature vector. According to the inequalitties discussed earlier, the number of nodes in the hidden layer range from 5 to19. Using Matlab modeling of the BP Neural Network, we try these different numbers as the number of nodes in the hidden layer and select 18 nodes in the hidden layer for the test. The structure of the BP neural network is 20-18-4, which means the input layer has 20 nodes, the hidden layer has 18 nodes and the output layer has 4 nodes. After all of the input vectors, output vectors, number of hidden layers are known, the Neural Network Pattern Recognition Tool is used to train the network. This step is similar to the training process used the input features Root Mean Square Value, Peak to Peak Value, Kurtosis, Crest factor, Impulse factor and Energy in time domain. Figure 5 shows the test results. The correct classification rate of the trained neural network model was 100%, which means for the entire 300 feature vector samples all of them were classified to the correct class. This shows that preprocessing by the AR model gives accurate results for classification by the neural network. The 100% result is achieved in the ideal conditions in the lab. It may not give 100% results in an industrial setting.

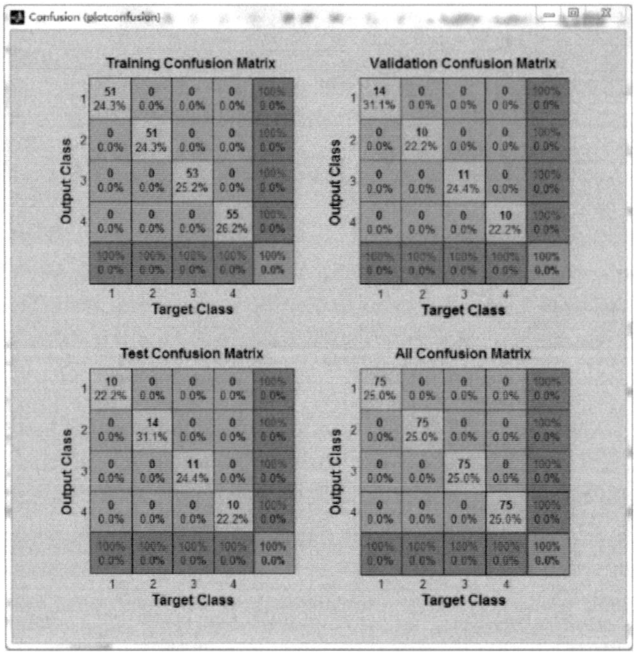

Figure 5. Training results for the four conditions

Two different methods were used for processing the original data and the processed data was used as input to train the neural network. The results show that the features that were calculated by time domain features do not train the neural network well. Enough information cannot be obtained for the classification by a traditional method with limited equipment available in the experiments. The comparison of the results is shown in Figure 6.

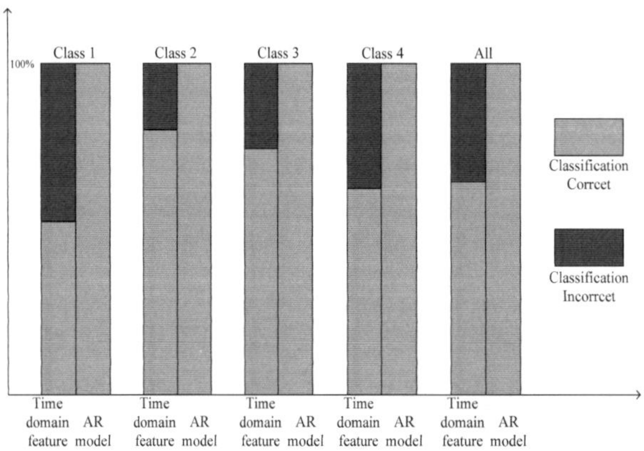

Method	Classification	Class 1	Class 2	Class 3	Class 4	All
Time domain features	Correct	52%	81.3%	74.7%	64.0%	68%
	Incorrect	48%	18.7%	25.3%	36.0%	32%
AR model	Correct	100%	100%	100%	100%	100%
	Incorrect	0%	0%	0%	0%	0%

Figure 6. Comparison of Time Domain Features and AR Model feature vector

Based on the above results, AR model was selected to process the original fan vibration signal.

3.1. Classifying the test data on the trained neural network

The raw vibration signal samples were used to do the test the neural network. The data in one-dimensional waveform signal was converted into a matrix form to use on Matlab. The raw vibration signal was converted to AR model feature vector consisting of 20 features. AR model feature vector set was used as the input to the trained BP neural network. The output of the BP neural network classification function is a one-dimensional array which is close to the class label. The class label is normalized by rounding the number to the nearest integer, which is 0 and 1. The output of the BP neural network model is the class label. As mentioned before, class label 1000, 0100, 0010 and 0001 represents healthy fan, fan blades unbalance, rotor unbalance and bearing fault, respectively. Since the class labels are listed by columns, number of 1's is calculated in each column in order to get the number of samples classified to their target class.

The result count is shown in a graphical display window as shown in Figure 7. The white bar shows the number of samples belonging to class 1. The test data used in this process was healthy fan's vibration signal. The classification result was correct.

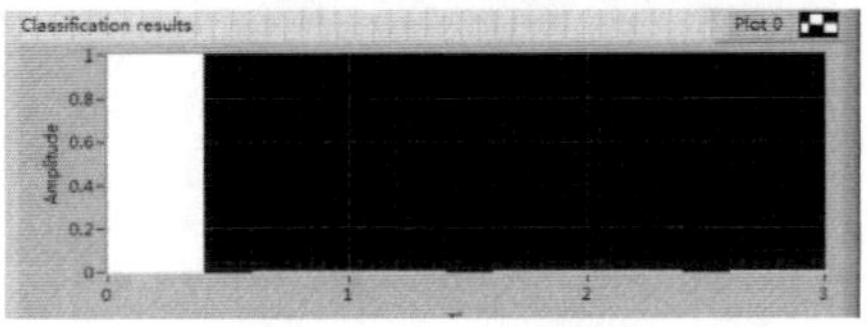

Figure 7. Results display in LabVIEW

4. CONCLUSION AND FUTURE WORK

This paper explored the application of neural networks for failure diagnosis of a fan. LabVIEW program was used to collect and process data and call Matlab functions for the classification. The results from the experiment show that the AR model is the appropriate preprocessing method for calculating the input data for the neural network. The powerful computing ability of Matlab and the data acquisition ability of LabVIEW can be perfectly combined together. The system designed in LabVIEW works well for fault diagnosis. After the fan vibration signal is collected, the system shows the class results in a short time. The results show that the system is a good fault diagnosis tool for fan condition monitoring.

Since the equipment used for the experiments was limited, there are still some other problems that have not been discussed in this paper. In a future study, other types of faults such as gearing box faults and motor faults may be considered. All of the faults tested in the project were in extreme cases; however, faults normally get worse gradually. Less extreme faults will be tested in the future. Extra accelerometers can be used to get more vibration details and database can be used in the program for data storage. Other classification techniques such as rough sets can be tested and compared with the neural network for fault detection. It would be useful to detect faults in the industrial setting.

REFERENCES

[1] Akaike, H. Fitting autoregressive models for prediction. (1969) Annals of the Institute of Statistical Mathematics. 243-247.
[2] Atal, B. S., (2006). The history of linear prediction. IEEE Signal Processing Magazine, 23(2). 154-161.
[3] Demuth, H., and Beale, M., (1993). Neural network toolbox for use with MATLAB.
[4] Fogel, D. B., (1991). An information criterion for optimal neural network selection, IEEE Transactions on Neural Networks, 2(5). 490-497.
[5] Hyndman, R. J., (1993). Yule-Walker estimates for continuous time autoregressive models, Journal of Time Series Analysis, 14(3): 281-296.
[6] Junsheng, C., Dejie, Y., and Yu, Y., (2006). A fault diagnosis approach for roller bearings based on EMD method and AR model, Mechanical Systems and Signal Processing, 20(2): 350-362.
[7] Mobley, R., K., (2012). An introduction to predictive maintenance," 2nd edition, Toronto, Ontario, Canada, Elsevier Science.
[8] National Instruments, (2012). NI PXIe-4492 Introduction, USA, http:// sine.ni.com/nips/cds/view/p/lang/en/nid/209541.
[9] Patel, J., Patel, V, and Patel, A., (2013). Fault diagnostics of rolling bearing based on improve time and frequency domain features using artificial neural networks, International Journal for Scientific Reaearch & Development, Vol. 1, Issue 4.

[10] IMI Sensors, (2012). Platinum Low-cost Industrial ICP ® accelerometer introduction.
[11] https://www.imi-sensors.com/Industrial_Accelerometers/Low-Cost _for_Permanent_Installation/Low-cost_2-pin_MIL_Connector/603C01.aspx.
[12] Smith, J.L., (2009). World oil: market or mayhem?. Journal of Economic Perspectives, 3(23):145-164.

Time-frequency domain analysis of varying speed vibration response of dual-rotor system

Yizhou Yang, Chao Liu, Dongxiang Jiang and Wenguang Yang
State Key Lab of Control and Simulation of Power Systems and Generation Equipment, Department of Energy and Power Engineering, Tsinghua University, Beijing, China

ABSTRACT

Dual-rotor is a common form of rotor system where two rotors are coupled via an inter-shaft bearing and the speeds of the rotors are usually different and variable. When the vibration signal of a dual-rotor system is non-stationary, traditional frequency-domain signal processing methods are not suitable. In this paper, variable speed vibration responses of a dual-rotor system dynamic simulation model is acquired and analyzed by a time-frequency domain signal analysis approach called frequency slice wavelet transform (FSWT). The dual-rotor system model is established under an inertia coordinate using finite element method. Then the rotor system model is solved by the explicit Newmark-β method to acquire the unbalance response under a fixed rotor speed or a varying speed, which could be movement of uniform angular acceleration or randomly fluctuating speed. By choosing the suitable scale coefficient, FSWT could achieve a good balance between time and frequency resolution, thus reflecting the detailed features in the vibration response. Meanwhile frequency components of the complex signal could be reconstructed by a FSWT representation.

Keywords: dual-rotor system, varying speed, time-frequency domain analysis, advanced signal processing, frequency slice wavelet transform.

Corresponding author: Dongxiang Jiang (jiangdx@tsinghua.edu.cn)

1. INTRODUCTION

Many rotating machinery often work under varying speed conditions, resulting in aperiodic characteristics in their vibration. Thus traditional methods of signal processing based on Fourier transform are not suitable for such non-stationary vibration signals [1]. The frequency components in the signal are variable, which could be analyzed by time-frequency domain signal methods [2]. Dual-rotor is a common form of rotor system in rotating machinery like aero-engines. The high-pressure rotor and the low-pressure rotor are coupled via an inter-shaft bearing, and the speed of the rotors are different and variable, making the vibration signals more complex than general rotating machinery.

Time-frequency signal processing methods are implemented on non-stationary signals in order to obtain the variation of the frequency components of a signal over time. The short-time Fourier transform (STFT) is one of the most basic forms of time-frequency analysis [3]. STFT divides a signal into shorter segments of equal length and compute the Fourier transform separately on each shorter segment then the changing spectra as a function of time is plotted. But once the window of STFT is fixed the time-frequency resolution is fixed so the ideal time-frequency resolution is hard to get. The wavelet transform obtains a scalable window function by selecting the scale factor and the shift factor, thus has a good time-frequency localization characteristic [4]. The ratio between the bandwidth and the analysis frequency remains unchanged, so that the wavelet function can automatically adapt to changes in the frequency components of the signal. The short-time window is used for high-frequency components and the long-time window is used for low-frequency components. Frequency slice wavelet transform (FSWT) is a kind of time-frequency signal analysis method by means of extension of STFT defined directly in frequency domain [5, 6]. The adjustable scale parameter makes the time-frequency resolution of FSWT decomposition controllable. The inverse transform of FSWT no longer rely on wavelet functions like wavelet transforms, so time-frequency domain segmentation could be performed flexibly in time-frequency space, and the desired signal components could be separated by signal reconstruction.

Besides of the measurement of an actual rotor system or an experimental rotor system, modelling and simulation is also an important way to acquire different kinds of rotor vibration signals. Researchers have built different types of models when dealing with dynamic problems of rotor systems [7-9]. The finite element model turn the continuous flexible structure into finite discrete elements, obtains rich internal information, and are also convenient to add nonlinear forces, which is suitable for numerical calculation [10]. Processing the flexible parts as finite elements and the parts of large stiffness as discrete lumped masses could reduce the complexity of the system model and ensure the accuracy of simulation results at the same time, making it a more efficient modeling approach.

In this paper, a finite element dual-rotor system model is established under an inertia coordinate. The rotors are modelled by beam elements and the bearings are using a nonlinear multi-body dynamic model. Unbalance responses under a fixed rotor speed or a varying speed are acquired by solving the system model using the explicit Newmark-β method. The varying rotor speed is realized by solving the rotation equations coupled with the system dynamic equations. Vibration displacement signals acquired from the dual-rotor system are then analyzed by frequency slice wavelet transform to obtain the variation of the main frequency components over time. Time-frequency contours of the transient signals are plotted on which the time-frequency domain segmentation could be based. The desired signal components are reconstructed using the inverse transform of FSWT.

2. DYNAMIC MODEL OF THE DUAL-ROTOR SYSTEM

The dual-rotor system consists of two rotors connected by an inter-shaft bearing. Each rotor is composed of a flexible shaft, rigid discs and bearings. The schematic of the dual-rotor system is shown in Figure 1.

The shafts are modelled with finite element method using the 3-D elastic beam element. The beam element has 2 nodes and each node has 6 degree of freedoms (DOFs), which are translations in the X, Y and Z directions and rotations around the X, Y and Z directions. The DOFs and the motion equation of the beam element are:

$$q^e = \left[u_{x1}, u_{y1}, u_{z1}, \theta_{x1}, \theta_{y1}, \theta_{z1}, u_{x2}, u_{y2}, u_{z2}, \theta_{x2}, \theta_{y2}, \theta_{z2}\right]^T \quad (1)$$

$$M^e \ddot{q}^e + (-\omega G^e)\dot{q}^e + K^e q^e = 0 \quad (2)$$

M^e is the consistent mass matrix, G^e is the gyroscopic matrix, K^e is the stiffness matrix and ω is the rotational speed.

The rigid discs is modelled as discrete lumped masses using the 3-D mass element, which has one node with 6 DOFs. The DOFs and the motion equation of the mass element are:

$$q^e = \left[u_x, u_y, u_z, \theta_x, \theta_y, \theta_z\right]^T \quad (3)$$

$$M^e \ddot{q}^e + (-\omega G^e)\dot{q}^e = Q^e \quad (4)$$

Q^e is the external force vector and could be unbalance forces or other types of excitation.

Use the element assembly method, the motion equations of the rotors could be derived as

$$M_i^r \ddot{q}_i^r + (C_i^r - \omega_i G_i^r)\dot{q}_i^r + K_i^r q_i^r = Q_i^r, \; i = 1,2 \quad (5)$$

C^r is the damping matrix, here using Rayleigh damping which is in the form of proportional damping.

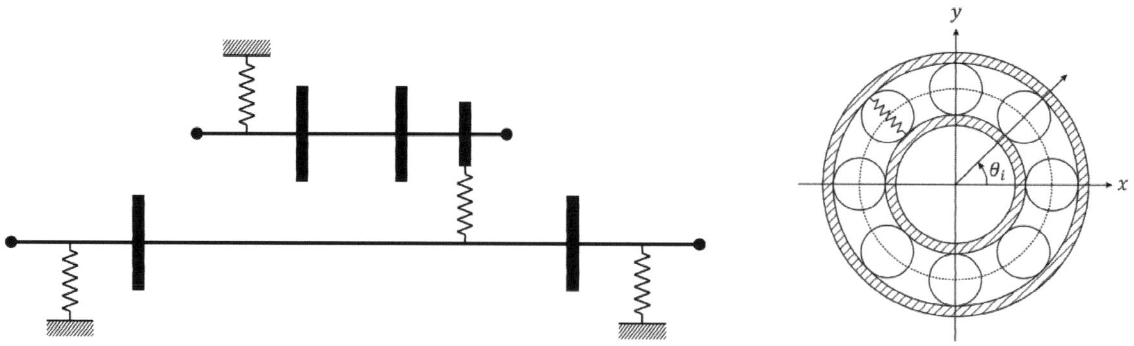

Figure 1. Dual-rotor system and nonlinear multi-body dynamic model of rolling element bearing.

The rolling element bearings supporting the rotors are modelled using a kind of nonlinear multi-body dynamic model, which is assumed as nonlinear springs with no mass and evenly arranged along the rolling element raceway [11, 12] (Figure 1). When the rotation position of the rotor is $\theta(t)$, the cage position is $\theta_c(t)$ and the angular position of the ith rolling element is $\theta_i(t)$.

$$\theta_c(t) = \frac{\theta(t)}{2}\left(1 - \frac{d}{D}\cos\alpha\right) \tag{6}$$

$$\theta_i(t) = \frac{2\pi(i-1)}{Z} + \theta_c(t) + \theta_0 + (0.5 - rand) \times \theta_{slip},\ i = 1,2,\dots,Z \tag{7}$$

θ_0 is the initial position of the first rolling element. $(0.5 - rand) \times \theta_{slip}$ is the phase deviation caused by slippage. If the displacement of the shaft (x_s, y_s) and the outer race (x_o, y_o) is known, the contact deformation for the ith rolling element is

$$\gamma_i = (x_s - x_o)\cos\theta_i + (y_s - y_o)\sin\theta_i - \delta \quad i = 1, 2, \dots Z \tag{8}$$

δ is the initial clearance between the rolling elements and the bearing races. The total bearing force could be derived in the form of external forces by summing the contact force of each rolling element.

$$F_{Bx} = -k_b \sum_{i=1}^{Z}(\lambda_i\gamma_i)^n \cos\theta_i\,,\ F_{By} = -k_b \sum_{i=1}^{Z}(\lambda_i\gamma_i)^n \sin\theta_i \tag{9}$$

k_b is the contact stiffness factor. λ_i is a coefficient for judging whether the rolling element deformation is positive or not. n is the exponent of Hertz contact, taking 3/2 for ball bearings and 10/9 for roller bearings.

The rotor system model is established by assembling all the finite elements and adding the external forces. The generalized external force vector of the system contains the unbalance forces, the weight forces and the nonlinear forces of the rolling element bearings.

$$M^s\ddot{q}^s + (C^s - \omega_1 G_1^r - \omega_2 G_2^r)\dot{q}^s + K^s q^s = Q^s \tag{10}$$

$$Q^s = F_u + W + F_B \tag{11}$$

The unbalance force on the disc is the main excitation of this dynamic model. When the rotation position of the rotor is $\theta(t)$, the rotation speed is $\omega(t) = \dot{\theta}(t)$, then unbalance forces will be

$$F_{uiy} = m_d e[\dot{\theta}(t)]^2 \cos[\theta(t)]\,,\ F_{uiz} = m_d e[\dot{\theta}(t)]^2 \sin[\theta(t)] \tag{12}$$

m_d is the disc mass and e is the disc eccentricity.

If the rotor speed is constant, $\theta(t) = \omega t$. But if the rotating speed changes over time, the angular velocity and displacement of the rotors could be obtained by solving the rotation equations.

$$I_i^r\ddot{\theta}_i + c_i^r\dot{\theta}_i + k_i^r\theta_i = T_i^r,\ i = 1,2 \tag{13}$$

I_i^r is the total moment of inertia of a rotor, c_i^r the rotational damping and k_i^r the rotational stiffness. T_i^r is the load torque, which could be controlled to control the movement of the rotor.

3. FREQUENCY SLICE WAVELET TRANSFORM

For a signal $f(t)$, $\hat{p}(\omega)$ is its frequency slice function (FSF) and the Fourier transformation of $p(t)$. The frequency slice wavelet transform (FSWT) of $f(t)$ is

$$W(t,\omega,\lambda,\sigma) = \frac{1}{2\pi}\lambda\int_{-\infty}^{+\infty}\hat{f}(u)\hat{p}^*\left(\frac{u-\omega}{\sigma}\right)e^{iut}du \tag{14}$$

where the scale $\sigma \neq 0$ and energy coefficient $\lambda \neq 0$ are constants or functions of ω and t. $\hat{p}^*(\omega)$ is the conjugate of $\hat{p}(\omega)$. Define another scale $\kappa = \omega/\sigma$, which is not directly related to ω and u, and can be adjusted to tune the transform to be more sensitive to frequency or more sensitive to time. Then FSWT is in the form as

$$W(t,\omega,\lambda,\kappa) = \frac{1}{2\pi}\lambda\int_{-\infty}^{+\infty}\hat{f}(u)\hat{p}^*\left(\kappa\frac{u-\omega}{\omega}\right)e^{iut}du \tag{15}$$

It is impossible to have high level of resolution in time and frequency domain at the same time due to the Heisenberg uncertainty principle. So two eclectic parameters of a signal are defined to estimate the scale parameter κ and have a compromise between time and frequency resolution. One is the frequency resolution ratio

$$\eta = \Delta\omega/\omega \tag{16}$$

The other is the expected response ratio of amplitude

$$0 < \upsilon \leq 1 \tag{17}$$

It is recommended to use the inequality below to estimate the scale parameter κ:

$$\kappa \geq \sqrt{(1/\upsilon)^2 - 1}/\eta \tag{18}$$

Since the time and frequency behaviors of a signal are not independent, the information of FSWT is redundant. There are various forms of inverse transforms to rebuild the original signal, and one of the simplest forms is often used and shown below. If the frequency slice function $\hat{p}(\omega)$ satisfies $\hat{p}(0) = 1$, the original signal $f(t)$ could be reconstructed as

$$f(t) = \frac{1}{2\pi\lambda}\int_{-\infty}^{\infty}\int_{-\infty}^{\infty} W(t,\omega,\lambda,\sigma)e^{i\omega(t-\tau)}d\tau\, d\omega \tag{19}$$

It can be seen that the inverse transform has no relation with the frequency slice function $p(t)$ or $\hat{p}^*(\omega)$. In any specific time-frequency region $(t_1, t_2, \omega_1, \omega_2)$, the signal component could be reconstructed [13] as

$$f_x(t) = \frac{1}{2\pi\lambda}\int_{\omega_1}^{\omega_2}\int_{t_1}^{t_2} W(t,\omega,\lambda,\sigma)e^{i\omega(t-\tau)}d\tau\, d\omega \tag{20}$$

For FSWT is used for processing discrete signals here, the discrete FSWT is

$$W(\cdot,k) = \lambda F^{-1}\{F\{f\}.*P_k\},\ k = 0,1,\dots,N-1 \tag{21}$$

P_k is the discrete FSF and could be seen as a series of filters. F is the Fourier transform and F^{-1} the inverse Fourier transform, which could be calculated by FFT and IFFT. Correspondingly, the discrete inverse transform of FSWT is

$$f(\cdot) = \frac{1}{\lambda N}F^{-1}\{\textstyle\sum_{l=0}^{N_r-1} W(l,k)e^{-i(2\pi/N)lk},\ k = 0,1,\dots,N-1\} \tag{22}$$

4. SIMULATION RESULT

The nonlinear equation of the dual-rotor system model is solved by the explicit Newmark-β method. This numerical integration approach is suitable for obtaining the dynamic response of a nonlinear system. Then the displacement, velocity and acceleration at each node of the finite element model can be obtained.

In a dual-rotor system, the high-pressure rotor and the low-pressure rotor both have unbalanced excitation forces. Usually the rotating speeds of the two rotors are different, and there is a certain relationship between them. Thus the excitation frequencies of the two rotors are also different. Here the speed relationship is taken as a fixed speed ratio of 1.5, either at simulations of constant rotor speed or variable rotor speed.

Calculate the steady-state unbalance response of the dual-rotor system at a fixed speed. The rotating speed is (60, 90) Hz, and the unbalanced excitation of the high-pressure rotor and the low-pressure rotor exist at the same time. The vibration displacement waveform of the low-pressure rotor's second disc in the horizontal direction and its frequency spectrum are shown in Figure 2. No clear nonlinear phenomenon appears in the shaft displacement. Both frequencies of the excitations are reflected on the same node due to the coupling effect of the inter-shaft bearing.

If the load torque of the rotor fluctuates, the rotor speed will fluctuate accordingly. Figure 3 shows the situation when the rotor load torques occur random fluctuations, which leads to changes in the amplitude and frequency of the vibration signal. The peaks of the rotational frequencies in the spectrum are wider than ones in the fixed-speed vibration spectrum.

Figure 4 shows an experimental signal acquired from a dual-rotor test rig [14], whose rotating speed is about (40.5, 54.0) Hz. Due to the mechanical dimensional errors and the speed fluctuate of the test rig, the experimental signal is not as smooth as the fixed-speed simulated signal. But the two main frequencies are clear, and have relatively wide peaks in the spectrum, which is similar to the simulated signal of fluctuating speed.

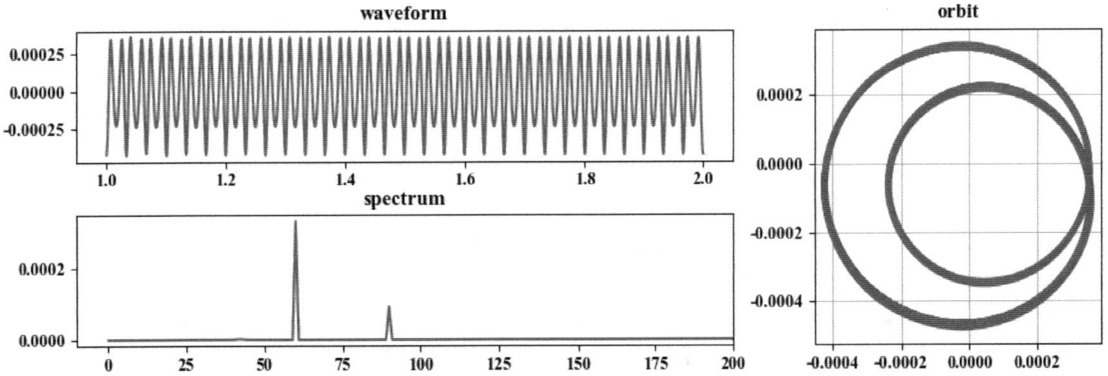

Figure 2. Steady-state unbalance response of the dual-rotor system.

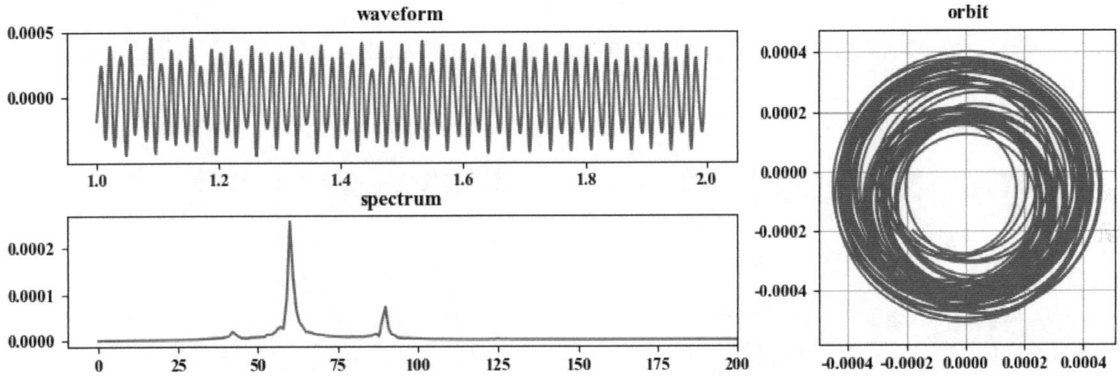

Figure 3. Unbalance response of the dual-rotor system with fluctuating rotor speed.

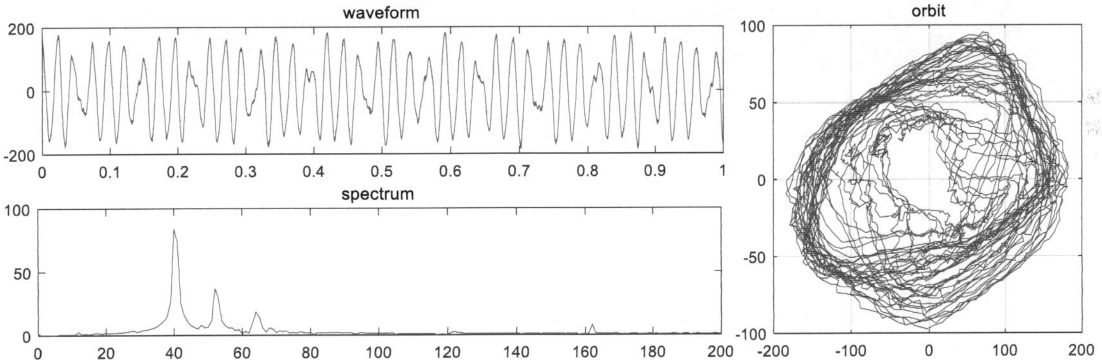

Figure 4. Experimental signal from a dual-rotor test rig.

The accelerating rotation process with a fixed angular acceleration is simulated, with the angular acceleration of (20, 30) Hz/s. Figure 5 is the simulated displacement signal of the dual-rotor system with only the unbalance excitation of the low-pressure rotor, and the one with only the unbalance excitation of the high-pressure rotor. It can be seen that different natural frequencies of the system are excited by the unbalance force of the low-pressure rotor or the high-pressure rotor. When the rotor is passing a natural frequency, known as a critical speed of the rotor system [15], the vibration amplitude first increase gradually, then decrease quickly after the peak amplitude. After the rotor passes the critical speed, the vibration at the natural frequency is in the decaying stage, and is synthesized with the higher frequency vibration, which results in the occurrence of the beat vibration.

Time-frequency contours of the speed-up signals in the region of (0s, 5s, 0Hz, 150Hz) drawn using FSWT are displayed in Figure 6. Take $\eta = 0.1$ and $\upsilon = 0.5$, and it is recommended that $\kappa \geq 10\sqrt{3}$. Since there is much uncertainty in the selection of the scale parameter κ, it should be adjusted based on the FSWT result to achieve a balance between the time and frequency resolution. Take $\kappa = 30$, and the FSWT results show the relationship between the signal frequency components and time clearly. The white line in the image shows the increasing of the main vibration frequency during the acceleration process. The red peaks show the rotating speed passing a critical speed.

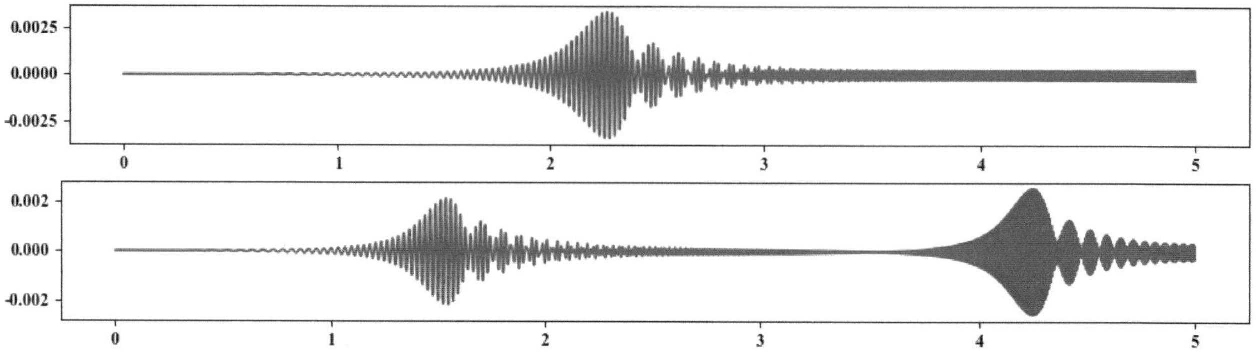

Figure 5. Simulated displacement signals with the unbalance excitation of the low-pressure rotor or the high-pressure rotor.

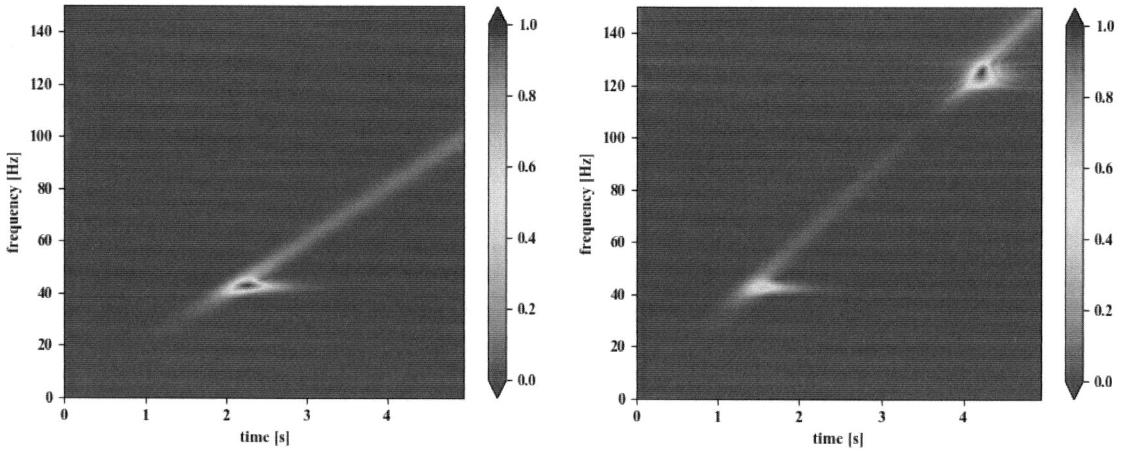

Figure 6. Time-frequency contours of the speed-up signals with the unbalance excitation of one rotor.

Figure 7 is the simulated displacement signal of the dual-rotor system with the unbalance excitation of both the low-pressure rotor and the high-pressure rotor. So natural frequencies of the both rotors could be excited. The FSWT contour of the signal is shown in Figure 8, where the time-frequency and the scale parameter remain the same. Two white lines show the two main frequencies changing at the same time, while the three red peaks show the first three orders of critical speeds of the dual-rotor system.

Short-time Fourier transform (STFT) is used for time-frequency analysis and its contour is also shown in Figure 8. Comparing with FSWT, although the frequency trend can be seen, the time-frequency resolution of the STFT output is much lower. This shortcoming might become a bigger problem when the frequency components become more complex.

Based on the FSWT result, the vibration signal could be decomposed and reconstructed. Split the time-frequency contour into two parts and each part contains the main frequency of one rotor like in Figure 6. By setting the FSWT values of one part all to zero, the other part could be rebuilt by the inverse transform of FSWT. The two reconstructed sub-signals are shown in Figure 9, which are very close to the signals excited by one rotor unbalance above. Adding the two reconstructed sub-signal as one signal, which is shown in Figure 9 and is very close to the original signal. It is noticed that although the reconstruction requires a square time-frequency region, any shape of time-frequency part could rebuild a signal by setting the unwanted part to zero. So the decomposition and reconstruction in the time-frequency domain of a signal using FSWT is quite flexible.

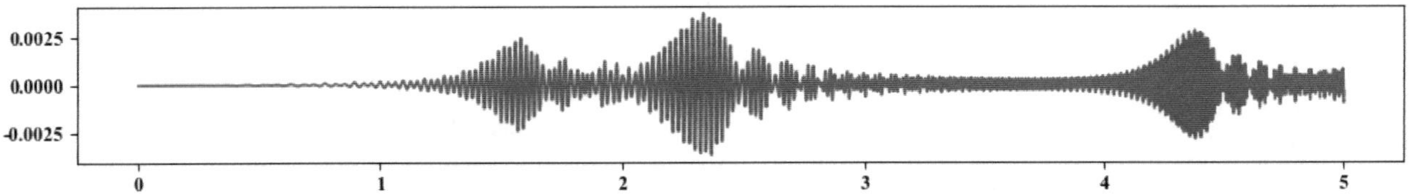

Figure 7. Simulated displacement signal with the unbalance excitation of two rotors.

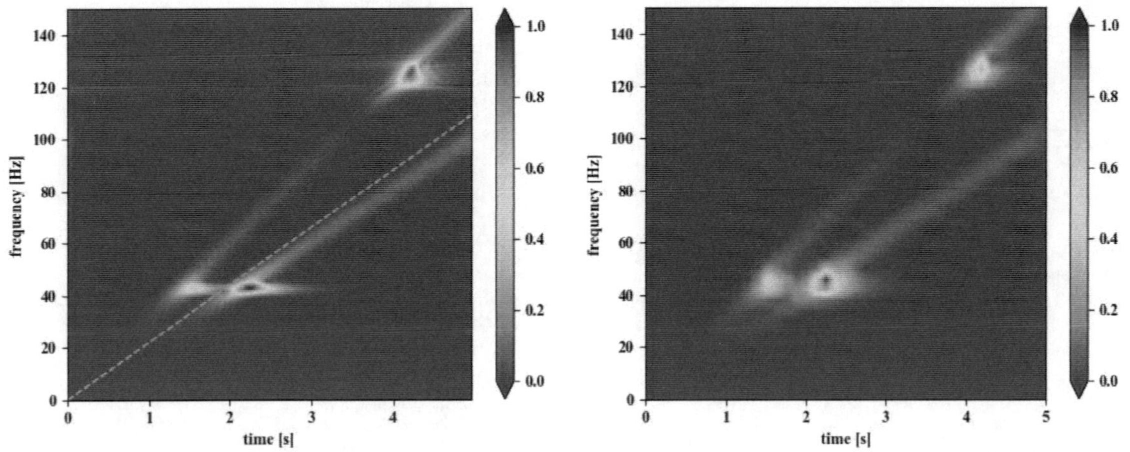

Figure 8. Time-frequency contours of the speed-up signal with the unbalance excitation of two rotors using FSWT and STFT.

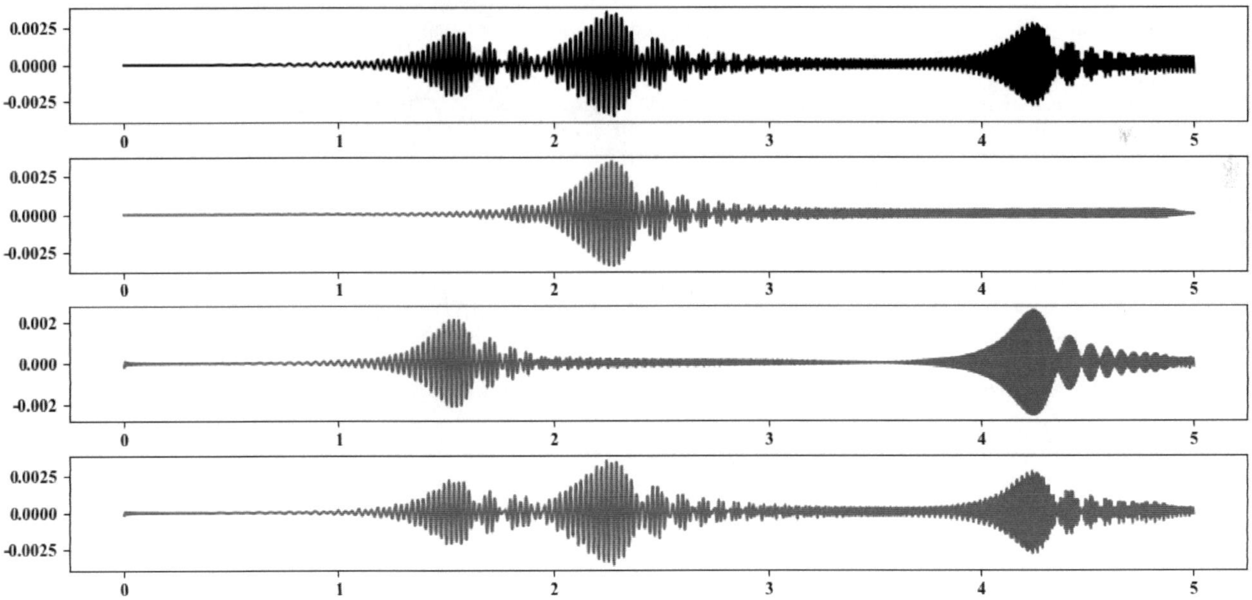

Figure 9. Signal decomposition and reconstruction using FSWT. (1) Origin signal; (2) rebuilt sub-signal of low-pressure rotor; (3) rebuilt sub-signal of high-pressure rotor; (4) rebuilt signal.

5. CONCLUSION

This paper presents a dual-rotor system model built by finite element method and a time-frequency method to analyze the varying speed vibration response of the system. In the system model, two rotors are supported on rolling element bearings and coupled by an inter-shaft bearing, where the shafts are modelled by beam elements and the bearings are using a nonlinear multi-body dynamic model. Unbalance responses under a fixed rotor speed or a varying speed are simulated and the varying speed is realized by solving the rotation equations of the rotors. The acceleration process will pass through the first few critical speeds, some of which are excited by the low-pressure unbalance and the others are excited by the high-pressure unbalance.

A time-frequency analysis approach called frequency slice wavelet transform is utilized to obtain the variation of the main frequency components over time in a transient signal. The time-frequency contours plotted by FSWT have a higher time-frequency resolution than the traditional short-time Fourier transform. Upon the FSWT coefficients time-frequency domain segmentation could be performed flexibly and desired signal components could be reconstructed by the inverse transform of FSWT. For the speed-up signal of the dual-rotor system, the vibrations of natural frequencies excited by different rotors are separated and rebuilt with a higher degree of reduction.

REFERENCES

[1] Z. M. Lin J, "Dynamic signal analysis for speed-varying machinery: A review (in Chinese)," *Sci Sin Tech,* vol. 45, pp. 669 - 686, 2015.

[2] Z. Feng, M. Liang and F. Chu, "Recent advances in time - frequency analysis methods for machinery fault diagnosis: A review with application examples," *Mechanical Systems and Signal Processing,* vol. 38, pp. 165-205, 2013.

[3] A. K. S. Jardine, D. Lin and D. Banjevic, "A review on machinery diagnostics and prognostics implementing condition-based maintenance," *Mechanical Systems and Signal Processing,* vol. 20, pp. 1483-1510, 2006.

[4] C. Torrence and G. P. Compo, "A Practical Guide to Wavelet Analysis," vol. 79, pp. 61 - 78, 1998.

[5] Z. Yan, A. Miyamoto and Z. Jiang, "Frequency slice wavelet transform for transient vibration response analysis," *Mechanical Systems and Signal Processing,* vol. 23, pp. 1474-1489, 2009.

[6] Z. Yan, A. Miyamoto, Z. Jiang, and X. Liu, "An overall theoretical description of frequency slice wavelet transform," *Mechanical Systems and Signal Processing,* vol. 24, pp. 491-507, 2010.

[7] G. Chen, "Vibration modelling and verifications for whole aero-engine," *Journal of Sound and Vibration,* vol. 349, pp. 163-176, 2015.

[8] P. M. Hai and P. Bonello, "An impulsive receptance technique for the time domain computation of the vibration of a whole aero-engine model with nonlinear bearings," *Journal of Sound and Vibration,* vol. 318, pp. 592-605, 2008.

[9] J. J. Sinou, "Non-linear dynamics and contacts of an unbalanced flexible rotor supported on ball bearings," *Mechanism and Machine Theory,* vol. 44, pp. 1713-1732, 2009.

[10] H. D. Chiang, C. Hsu, W. Jeng, S. Tu, W. Li, A. S. O. M. Engineers, and Null, "Turbomachinery Dual Rotor-Bearing System Analysis," in *American Society of Mechanical Engineers(ASME) Turbo Expo 2002 v.4 pt.B:Ceramics Industrial and Cogeneration Structures and Dynamics* Amsterdam,The Netherlands, 2002, p. 8.

[11] N. Sawalhi and R. B. Randall, "Simulating gear and bearing interactions in the presence of faults," *Mechanical Systems and Signal Processing,* vol. 22, pp. 1924-1951, 2008.

[12] K. Kappaganthu and C. Nataraj, "Nonlinear modeling and analysis of a rolling element bearing with a clearance," *Communications in Nonlinear Science and Numerical Simulation,* vol. 16, pp. 4134-4145, 2011.

[13] D. Chendong, G. Qiang and X. Xianfeng, "Generator Unit Fault Diagnosis Using the Frequency Slice Wavelet Transform Time-frequency Analysis Method," *Proceedings of the CSEE,* vol. 33, p. I0014, 2013.

[14] N. Wang, D. Jiang and K. Behdinan, "Vibration response analysis of rubbing faults on a dual-rotor bearing system," *Archive of Applied Mechanics,* vol. 87, pp. 1891-1907, 2017.

[15] M. Guskov, J. J. Sinou, F. Thouverez, and O. S. Naraikin, "Experimental and Numerical Investigations of a Dual-Shaft Test Rig with Intershaft Bearing," *International Journal of Rotating Machinery,* vol. 2007, pp. 1-12, 2007.

Remaining Useful Life Prediction and Uncertainty Modelling with Bayesian Deep Learning

C.J. Louw, P.S. Heyns
University of Pretoria, South Africa

ABSTRACT

The linearly decreasing remaining useful life modeling strategy was successfully implemented on the FD001 NASA turbofan engine degradation simulation data set time series regression problem, with 4 different Bayesian Deep Learning network architectures. This included a Dense Network, Simple Recurrent Network, Gated Recurrent Unit Network and Long Short-Term Memory Network, with Monte Carlo Dropout for remaining useful life prediction and uncertainty modeling. The Long Short-Term Memory Network was the most accurate at predicting the remaining useful life from the condition monitoring information for the turbofan engine examples in the testing set, due to its advanced sequence modeling architecture. The complex distributions and the model uncertainty associated with remaining useful life predictions were successfully and efficiently modeled with Monte Carlo Dropout. This made it possible to better understand the model uncertainty associated with remaining useful life predictions, which was previously given extremely limited attention in prognostics and health management research with Deep Learning. The Long Short-Term Memory Network model uncertainty drastically decreased, and model prediction accuracy drastically increased close to failure for all the turbofan engine examples in the testing set. The Long Short-Term Memory Network could on average very accurately predict the remaining useful life for all the turbofan engine examples 50 cycles before failure in the testing set, with a mean absolute error of only 1.954 cycles.

Keywords: Bayesian Deep Learning, Prognostics, Remaining Useful Life Modelling, Condition Monitoring, Pattern Recognition

Corresponding author: Carel Johannes Louw (u12032876@tuks.co.za)

1. INTRODUCTION

The accurate prediction and uncertainty modeling of remaining useful life (time to failure) for complex physical assets is extremely important for minimizing downtime and maintenance costs, and maximizing production output, availability, reliability and profitability [1]. The linearly decreasing remaining useful life modeling strategy [1] [2] can generally be used for the convenient remaining useful life modeling of complex physical assets, with numerous machine learning modeling techniques as a supervised time series regression problem. For this strategy the condition monitoring information time series inputs are modeled (mapped) directly on to the linearly decreasing remaining useful life time series output, for multiple examples that the complex physical asset was run to failure. This strategy does however assume that historic run to failure condition monitoring information is available and that degradation is trendable with respect to the measured condition monitoring information. The definition of failure (threshold) with respect to the measured condition monitoring information must also be consistent between examples. There are however extremely limited machine learning techniques that can accurately model long-term sequence information in condition monitoring information time series. Deep Learning with Simple Recurrent, Gated Recurrent Unit, Long Short-Term Memory Networks has achieved state-of-the-art performance on the majority of supervised machine learning sequence modeling problems [3]. There has however been extremely limited previous research done on applying Simple Recurrent, Gated Recurrent Unit, Long Short-Term Memory Networks for remaining useful life prediction and uncertainty modeling, which is crucial for asset management decision making.

2. DEEP LEARNING BACKGROUND

2.1. Deep Learning Models

Supervised Deep Learning regression models consist of numerous stacked layers, where each layer has trainable parameters and a nonlinear activation function. These layers can include Dense, Simple Recurrent, Gated Recurrent Unit, Long Short-Term Memory and Regression Layers. The layers in Deep Learning regression models are stacked. This means that the output vector of a previous layer is used as the input vector for the next layer, with the objective of modeling complex input to output mappings [4]. For this paper the inputs were condition monitoring information time series and the output was a remaining useful life time series of multiple turbofan engine examples that were run to failure.

2.2. Dense Layer

The Dense Layer [4] is very simple and can model general input tot output mappings, but cannot model sequence information from previous time steps. The Dense Layer output vector $h_i^{(t)}$ is calculated as shown in equation (1). Where $W_{i,j}$ is the weight matrix, $x_j^{(t)}$ is the input vector and b_i is the bias vector.

$$h_i^{(t)} = \mathrm{ReLU}\left(\sum_j W_{i,j} x_j^{(t)} + b_i\right) \tag{1}$$

The rectified linear unit activation function $\mathrm{ReLU}(z_i)$ for any vector z_i is calculated as shown in equation (2).

$$\mathrm{ReLU}(z_i) = \begin{cases} z_i \text{ for } z_i \geq 0 \\ 0 \text{ for } z_i < 0 \end{cases} \tag{2}$$

2.3. Simple Recurrent Layer

The Simple Recurrent Layer [4] can model sequence information from previous time steps, but suffers from the vanishing and exploding gradient problem [5] during training which limits its capacity to model long-term sequence information. The Simple Recurrent Layer output vector $h_i^{(t)}$ is calculated as shown in equation (3). Where $W_{i,j}$ is the weight matrix, $x_j^{(t)}$ is the input vector, $R_{i,j}$ is the recurrent weight matrix, $h_j^{(t-1)}$ is the output vector from the previous time step and b_i is the bias vector.

$$h_i^{(t)} = \tanh\left(\sum_j W_{i,j} x_j^{(t)} + \sum_j R_{i,j} h_j^{(t-1)} + b_i\right) \tag{3}$$

The hyperbolic tangent activation function $\tanh(z_i)$ for any vector z_i is calculated as shown in equation (4).

$$\tanh(z_i) = \frac{e^{z_i} - e^{-z_i}}{e^{z_i} + e^{-z_i}} \tag{4}$$

2.4. Gated Recurrent Unit Layer

The Gated Recurrent Unit Layer [6] is significantly more complex, but manages the vanishing and exploding gradient problem [5] during training with numerous gating operations that drastically increases its capacity to model relevant long-term sequence information. The Gated Recurrent Unit Layer candidate cell state vector $s_i^{(t)}$ is calculated as shown in equation (5). Where $W_{i,j}^s$ is the candidate cell state weight matrix, $x_j^{(t)}$ is the input vector, $R_{i,j}^s$ is the candidate cell state recurrent weight matrix, $r_j^{(t)}$ is the reset gate vector, $h_j^{(t-1)}$ is the output vector from the previous time step and b_i^s is the candidate cell state bias vector.

$$s_i^{(t)} = \tanh\left(\sum_j W_{i,j}^s x_j^{(t)} + \sum_j R_{i,j}^s r_j^{(t)} h_j^{(t-1)} + b_i^s\right) \tag{5}$$

The Gated Recurrent Unit Layer update gate vector $u_i^{(t)}$ is calculated as shown in equation (6). Where $W_{i,j}^u$ is the update gate weight matrix, $x_j^{(t)}$ is the input vector, $R_{i,j}^u$ is the update gate recurrent weight matrix, $h_j^{(t-1)}$ is the output vector from the previous time step and b_i^u is the update gate bias vector.

$$u_i^{(t)} = \sigma\left(\sum_j W_{i,j}^u x_j^{(t)} + \sum_j R_{i,j}^u h_j^{(t-1)} + b_i^u\right) \tag{6}$$

The sigmoid activation function $\sigma(z_i)$ for any vector z_i is calculated as shown in equation (7).

$$\sigma(z_i) = \frac{1}{1 + e^{-z_i}} \tag{7}$$

The Gated Recurrent Unit Layer reset gate vector $r_i^{(t)}$ is calculated as shown in equation (8). Where $W_{i,j}^r$ is the reset gate weight matrix, $x_j^{(t)}$ is the input vector, $R_{i,j}^r$ is the reset gate recurrent weight matrix, $h_j^{(t-1)}$ is the output vector from the previous time step and b_i^r is the reset gate bias vector.

$$r_i^{(t)} = \sigma\left(\sum_j W_{i,j}^r x_j^{(t)} + \sum_j R_{i,j}^r h_j^{(t-1)} + b_i^r\right) \tag{8}$$

The Gated Recurrent Unit Layer cell state vector $c_i^{(t)}$ is calculated (updated) as shown in equation (9). Where $u_i^{(t)}$ is the update gate vector, $s_i^{(t)}$ is the candidate cell state vector and $c_i^{(t-1)}$ is the cell state vector from the previous time step.

$$c_i^{(t)} = u_i^{(t)} s_i^{(t)} + \left(1 - u_i^{(t)}\right) c_i^{(t-1)} \tag{9}$$

The Gated Recurrent Unit Layer output vector $h_i^{(t)}$ is then finally calculated as shown in equation (10). Where $c_i^{(t)}$ is the cell state vector.

$$h_i^{(t)} = c_i^{(t)} \tag{10}$$

2.5. Long Short-Term Memory Layer

The Long Short-Term Memory Layer [5] is even more complex, but also manages the vanishing and exploding gradient problem [5] during training with numerous gating operations that drastically increases its capacity to model relevant long-term sequence information. The Long Short-Term Memory Layer candidate cell state vector $s_i^{(t)}$ is calculated as shown in equation (11). Where $W_{i,j}^s$ is the candidate cell state weight matrix, $x_j^{(t)}$ is the input vector, $R_{i,j}^s$ is the candidate cell state recurrent weight matrix, $h_j^{(t-1)}$ is the output vector from the previous time step and b_i^s is the candidate cell state bias vector.

$$s_i^{(t)} = \tanh\left(\sum_j W_{i,j}^s x_j^{(t)} + \sum_j R_{i,j}^s h_j^{(t-1)} + b_i^s\right) \tag{11}$$

The Long Short-Term Memory Layer update gate vector $u_i^{(t)}$ is calculated as shown in equation (12). Where $W_{i,j}^u$ is the update gate weight matrix, $x_j^{(t)}$ is the input vector, $R_{i,j}^u$ is the update gate recurrent weight matrix, $h_j^{(t-1)}$ is the output vector from the previous time step and b_i^u is the update gate bias vector.

$$u_i^{(t)} = \sigma\left(\sum_j W_{i,j}^u x_j^{(t)} + \sum_j R_{i,j}^u h_j^{(t-1)} + b_i^u\right) \tag{12}$$

The Long Short-Term Memory Layer forget gate vector $f_i^{(t)}$ is calculated as shown in equation (13). Where $W_{i,j}^f$ is the forget gate weight matrix, $x_j^{(t)}$ is the input vector, $R_{i,j}^f$ is the forget gate recurrent weight matrix, $h_j^{(t-1)}$ is the output vector from the previous time step and b_i^f is the forget gate bias vector.

$$f_i^{(t)} = \sigma\left(\sum_j W_{i,j}^f x_j^{(t)} + \sum_j R_{i,j}^f h_j^{(t-1)} + b_i^f\right) \tag{13}$$

The Long Short-Term Memory Layer output gate vector $o_i^{(t)}$ is calculated as shown in equation (14). Where $W_{i,j}^o$ is the output gate weight matrix, $x_j^{(t)}$ is the input vector, $R_{i,j}^o$ is the output gate recurrent weight matrix, $h_j^{(t-1)}$ is the output vector from the previous time step and b_i^o is the output gate bias vector.

$$o_i^{(t)} = \sigma\left(\sum_j W_{i,j}^o x_j^{(t)} + \sum_j R_{i,j}^o h_j^{(t-1)} + b_i^o\right) \tag{14}$$

The Long Short-Term Memory Layer cell state vector $c_i^{(t)}$ is calculated as shown in equation (15). Where $u_i^{(t)}$ is the update gate vector, $s_i^{(t)}$ is the candidate cell state vector, $f_i^{(t)}$ is the forget gate vector and $c_i^{(t-1)}$ is the cell state vector from the previous time step.

$$c_i^{(t)} = u_i^{(t)} s_i^{(t)} + f_i^{(t)} c_i^{(t-1)} \tag{15}$$

The Long Short-Term Memory Layer output vector $h_i^{(t)}$ is then finally calculated as shown in equation (16). Where $o_i^{(t)}$ is the output gate vector and $c_i^{(t)}$ is the cell state vector.

$$h_i^{(t)} = o_i^{(t)} \tanh\left(c_i^{(t)}\right) \tag{16}$$

2.6. Regression Layer

The Regression Layer [4] is the last layer in a supervised Deep Learning regression model and is just a Dense Layer with a linear activation function. The Regression Layer output vector $y_i^{(t)}$ is calculated as shown in equation (17). Where $W_{i,j}$ is the weight matrix, $x_j^{(t)}$ is the input vector and b_i is the bias vector.

$$y_i^{(t)} = \sum_j W_{i,j} x_j^{(t)} + b_i \tag{17}$$

2.7. Deep Learning Optimization Algorithms

There are numerous optimization algorithms that can be used to train the weight matrices, recurrent weight matrices and bias vectors of Deep Learning models on training set examples. The most popular optimization algorithms in Deep Learning were investigated, which included Stochastic Gradient Descent with Momentum, AdaGrad, RMSProp and Adam [4]. The Adam optimization algorithm generally produced the best results for the investigated FD001 NASA turbofan engine degradation simulation data set time series regression problem.

2.8. Dropout Regularization

Deep Learning models are unfortunately prone to overfitting on training set examples. This means that trained Deep Learning models can generally very accurately model previously seen training set examples, but can generally not generalize for previously unseen testing set examples. Dropout [7] is a stochastic regularization technique which reduces overfitting and increases the generalization ability of the Deep Learning models for more accurate predictions on previously unseen testing set examples. With Dropout regularization a fraction of elements in numerous output vectors in different layers of a Deep Learning model are randomly deactivated or dropped out (by zero multiplication) during a training iteration. This greatly reduces overfitting and increases the generalization ability of the Deep Learning model. Dropout regularization can also be interpreted as an efficient form of ensemble learning, since the Deep Learning model changes with each training iteration. This greatly improves modeling performance and generalization on the training and testing set.

2.9. Bayesian Deep Learning

Deep Learning models can generally not model the complex distributions and model uncertainty associated with predictions. Monte Carlo Dropout [8] however is a new and efficient Bayesian Deep Learning technique that approximates the complex distributions and model uncertainty associated with predictions. With Monte Carlo Dropout a Deep Learning model is trained with the stochastic Dropout regularization technique in order to reduce overfitting. Dropout is generally only used during model training and turned off during model testing. With Monte Carlo Dropout however, Dropout is kept on during model testing and multiple stochastic predictions are made for the same input, which results in a complex output distribution. The average of the complex distribution can then be calculated in order to make a prediction. More importantly, the complex distribution can then be used in order to understand the model uncertainty associated with a prediction by calculating confidence intervals.

3. DATA SET DESCRIPTION

The FD001 NASA turbofan engine degradation simulation data set time series regression problem [9] was investigated for this paper, because of its high popularity in prognostics and health management research [1], high complexity and general comparability to other engineering systems. The objective of the NASA turbofan engine degradation simulation data set time series regression problem, was to model the remaining useful life time series output from 24 condition monitoring information time series inputs, for multiple turbofan engine examples that were run to failure in the training set. The training set therefore represents previously seen historic run to failure condition monitoring information for multiple turbofan engine examples. The model then had to predict the remaining useful life time series output from 24 unseen condition monitoring information time series inputs for multiple turbofan engine examples that were run to failure in the testing set. The testing set therefore represents previously unseen real world operational run to failure condition monitoring information for multiple turbofan engine examples. The 24 condition monitoring information time series inputs that were measured for each of the turbofan engines that was run to failure in the training and testing set, included numerous temperatures, pressures, fan speeds, efficiencies (ratios), flow rates and other measurements. The condition monitoring information time series inputs were averaged (by the data set authors) after each operational cycle for each of the turbofan engines that was run to failure in the training and testing set. The training and testing set included 80 and 20 turbofan engine examples that were run to failure respectively. The high complexity of the problem is mainly due to unknown amounts of initial wear and the relatively high variance in failure times of the turbofan engine examples that were run to failure. The mean, standard deviation, minimum and maximum failure times of the turbofan engine examples were 206.31, 46.34, 128 and 362 cycles respectively. The condition monitoring information measurements were also extremely noisy which made modeling the data set even more challenging.

4. MODELING STRATEGY

The linearly decreasing remaining useful life time series modeling strategy [1] [2] was used for the FD001 NASA turbofan engine degradation simulation data set time series regression problem. The condition monitoring information time series inputs were therefore trained (mapped) directly on the linearly decreasing remaining useful life time series output with numerous Bayesian Deep Learning models, for each turbofan engine example that was run to failure in the training set. The linearly decreasing remaining useful life time series modeling strategy does however make the important and often overlooked assumption that the definition of failure (threshold) with respect to the measured condition monitoring information is consistent for all the training and testing set turbofan engine examples. The linearly decreasing remaining useful life time series modeling strategy was implemented with 4 different Bayesian Deep Learning models. This included a Feedforward Network, Simple Recurrent Network, Gated Recurrent Unit Network and Long Short-Term Memory Network. The Feedforward Network (FN) architecture consisted of the following layers from input to output: 4 Dense Layers and 1 Regression layer. The Simple Recurrent Network (SRN) architecture consisted of the following layers from input to output: 1 Dense Layer, 2 Simple Recurrent layers, 1 Dense Layer and 1 Regression layer. The Gated Recurrent Unit Network (GRUN) architecture consisted of the following layers from input to output: 1 Dense Layer, 2 Gated Recurrent Unit layers, 1 Dense Layer and 1 Regression layer. The Long Short-Term Memory Network (LSTMN) architecture consisted of the following layers from input to output: 1 Dense Layer, 2 Long Short-Term Memory layers, 1 Dense Layer and 1 Regression layer. The models were all trained on the training set with the Adam optimization algorithm in TensorFlow and regularized with Dropout in each layer, except the regression layer. The remaining useful life distribution at each time step for each turbofan engine in the testing set, was obtained from 10,000 Monte Carlo Dropout predictions. The remaining useful life distribution at each time step for each turbofan engine in the testing set, was then averaged in order to make a remaining useful life (RUL) prediction, and the 5% and 95% percentiles calculated for lower and upper confidence intervals (CI) respectively.

5. RESULTS

The remaining useful life prediction performance of the 4 different Bayesian Deep Learning models were compared by calculating the mean squared error and mean absolute error, between the actual and predicted remaining useful life for all the turbofan engine examples in the testing set. Table 1 compares the testing set mean squared error performance in cycles^2 and the testing set mean absolute error performance in cycles for the 4 different Bayesian Deep Learning models.

Table 1: The testing set mean squared error performance in cycles^2 and the testing set mean absolute error performance in cycles for the 4 different Bayesian Deep Learning models

Deep Learning Model	**Testing Set Mean Squared Error [Cycles^2]**	**Testing Set Mean Absolute Error [Cycles]**
FN	2303.315	35.704
SRN	1229.455	26.495
GRUN	955.236	22.123
LSTMN	716.467	17.871

The remaining useful life prediction performance of the 4 different Bayesian Deep Learning models were visualized by plotting the predicted remaining useful life versus time in cycles for 4 turbofan engine examples in the testing set. Figure 1 shows the predicted remaining useful life versus time in cycles of the 4 different Bayesian Deep Learning models for 4 turbofan engine examples in the testing set.

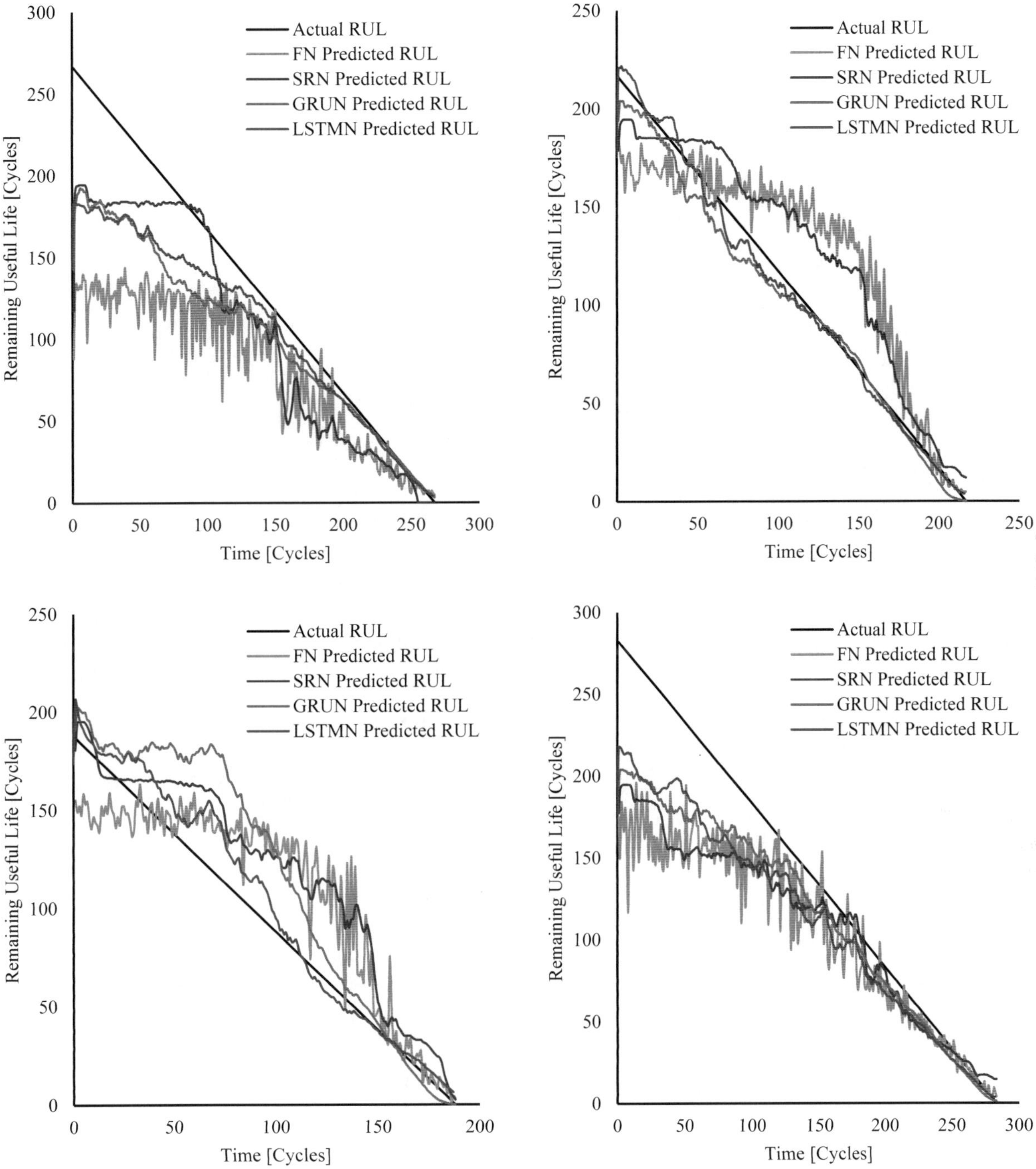

Figure 1: The predicted remaining useful life versus time in cycles of the 4 different Bayesian Deep Learning models for 4 turbofan examples in the testing set

The remaining useful life prediction with confidence intervals of the 4 different Bayesian Deep Learning models were also visualized by plotting the remaining useful life with confidence intervals versus time in cycles for another turbofan engine example in the testing set. Figure 2 shows the predicted remaining useful life with confidence intervals versus time in cycles of the 4 different Bayesian Deep Learning models for a turbofan engine example in the testing set.

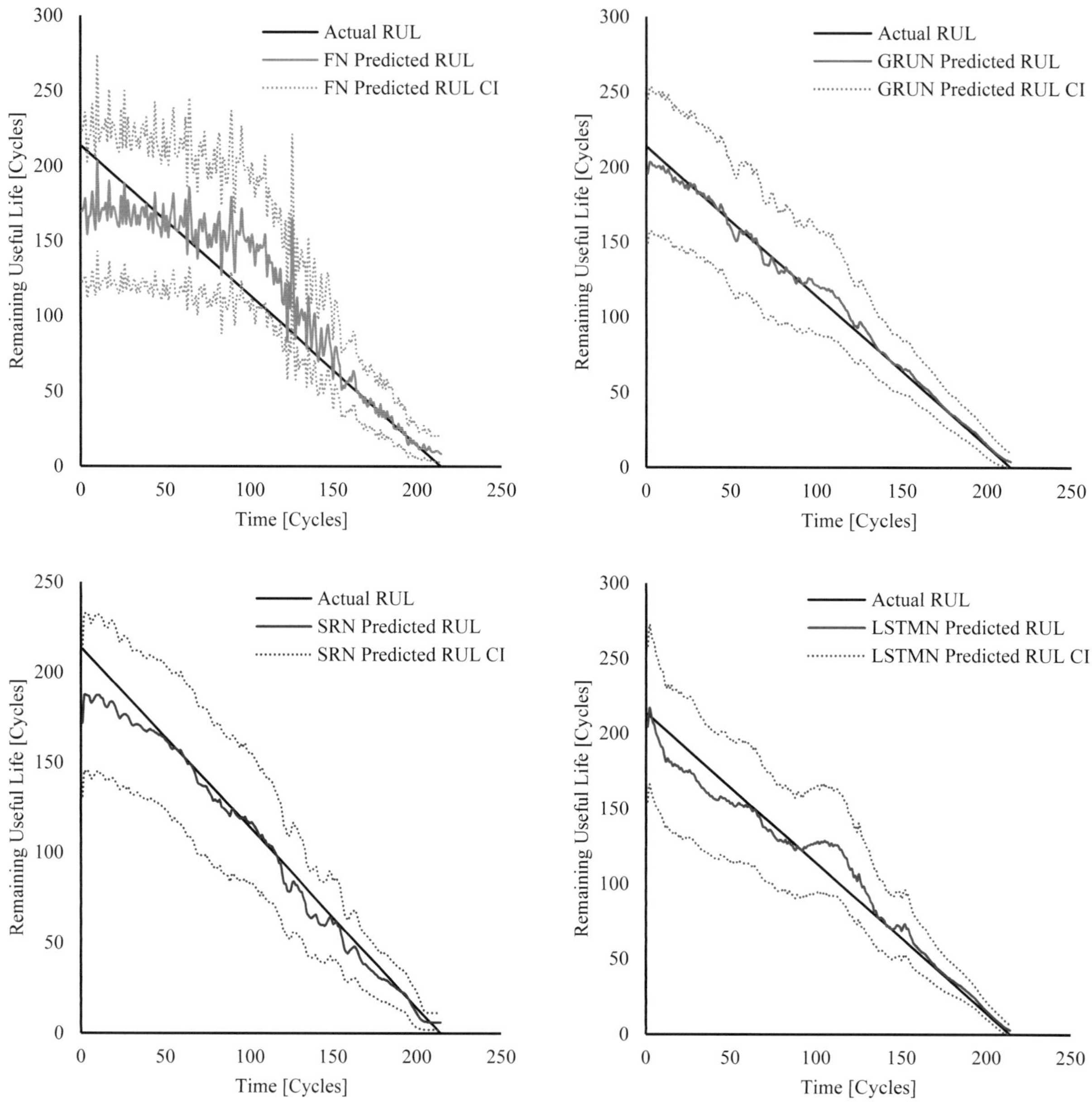

Figure 2: The predicted remaining useful life with confidence intervals versus time in cycles of the 4 different Bayesian Deep Learning models for a turbofan engine example in the testing set

The turbofan engine degradation with respect to the condition monitoring information generally had significantly increased trendability 50 cycles before failure and significantly increased remaining useful life prediction performance. The remaining useful life prediction performance of the 4 different Bayesian Deep Learning models were therefore also compared by calculating the mean squared error and mean absolute error, between the actual and predicted remaining useful life for all the turbofan engine examples in the testing set 50 cycles before failure. Table 2 compares the testing set mean squared error performance in $cycles^2$ and the testing set mean absolute error performance in cycles for the 4 different Bayesian Deep Learning models 50 cycles before failure.

Table 2: The testing set mean squared error performance in cycles^2 and the testing set mean absolute error performance in cycles for the 4 different Bayesian Deep Learning models 50 cycles before failure

Deep Learning Model	Testing Set Mean Squared Error [Cycles^2]	Testing Set Mean Absolute Error [Cycles]
FN	47.061	6.561
SRN	18.082	3.597
GRUN	11.474	2.253
LSTMN	9.577	1.954

6. CONCLUSION

The linearly decreasing remaining useful life modeling strategy was successfully implemented on the FD001 NASA turbofan engine degradation simulation data set time series regression problem, with 4 different Bayesian Deep Learning network architectures. This included a Dense Network, Simple Recurrent Network, Gated Recurrent Unit Network and Long Short-Term Memory Network, with Monte Carlo Dropout for remaining useful life prediction and uncertainty modeling. The Long Short-Term Memory Network was the most accurate at predicting the remaining useful life from the condition monitoring information for the turbofan engine examples in the testing set, due to its advanced sequence modeling architecture. The complex distributions and the model uncertainty associated with remaining useful life predictions were successfully and efficiently modeled with Monte Carlo Dropout. This made it possible to better understand the model uncertainty associated with remaining useful life predictions, which was previously given extremely limited attention in prognostics and health management research with Deep Learning. The Long Short-Term Memory Network model uncertainty drastically decreased, and model prediction accuracy drastically increased close to failure for all the turbofan engine examples in the testing set. The Long Short-Term Memory Network could on average very accurately predict the remaining useful life for all the turbofan engine examples 50 cycles before failure in the testing set, with a mean absolute error of only 1.954 cycles.

7. BIBLIOGRAPHY

[1] Y. Lei, N. Li, L. Guo, N. Li, T. Yan and J. Lin, "Machinery health prognostics: A systematic review from data acquisition to RUL prediction," *Mechanical Systems and Signal Processing,* no. 104, p. 799–834, 2018.

[2] F. O. Heimes, "Recurrent Neural Networks for Remaining Useful Life Estimation," *International Conference On Prognostics And Health Management,* pp. 1-6, 2008.

[3] F. Chollet, Deep Learning with Python, 2017.

[4] I. Goodfellow, Y. Bengio and A. Courville, Deep Learning, MIT Press, 2016.

[5] S. Hochreiter and J. Schmidhuber, "Long Short-Term Memory," *Neural Computation,* vol. 9, no. 8, pp. 1735-1780, 1997.

[6] K. Cho, "Learning Phrase Representations using RNN Encoder–Decoder for Statistical Machine Translation," *Universite de Montreal,* 2014.

[7] N. Srivastava, G. Hinton, A. Krizhevsky, I. Sutskever and R. Salakhutdinov, "Dropout: A Simple Way to Prevent Neural Networks from Overfitting," *Journal of Machine Learning Research,* no. 15, pp. 1929-1958, 2014.

[8] Y. Gal and Z. Ghahramani, "Dropout as a Bayesian Approximation: Representing Model Uncertainty in Deep Learning," *University of Cambridge,* 2015.

[9] A. Saxena and K. Goebel, "Turbofan Engine Degradation Simulation Data Set, NASA Ames Prognostics Data Repository," 2008. [Online]. Available: https://ti.arc.nasa.gov/tech/dash/groups/pcoe/prognostic-data-repository/. [Accessed 3 March 2018].

Rolling Element Bearings Prognostics Using High-Frequency Spectrum of Offline Vibration Condition Monitoring Data

Mehdi Behzad[1], Hesam Addin Arghand[2] and Abbas Rohani Bastami[3]

[1] School of Mechanical Engineering, Sharif University of Technology, Tehran, Iran
[2] School of Mechanical Engineering, Sharif University of Technology, Tehran, Iran
[3] Mechanical and Energy Engineering Department, Shahid Beheshti University, Tehran, Iran

ABSTRACT

Remaining useful life (RUL) prediction of rolling element bearings (REBs) with offline condition monitoring (CM) data is the purpose of this paper. A data driven algorithm based on feedforward neural network (FFNN) is proposed for this aim. Since, usually the number of offline measurements are not enough, the generalized Weibull failure rated function is used for producing the auxiliary points that are employed for training. Considering the physics of the bearing degradation, vibration level in the high-frequency bandwidth of the spectrum is used as a feature and its performance in REB prognostic problem is compared with that of using popular recommended features in the diagnostic standards.

Bearing accelerated life test data as well as data from two industrial bearings are used to investigate the purpose of this study. The results show that using the high-frequency vibration level features rather than the proposed frequency bandwidth in guidelines and standards for recording the vibration of rotating machines produce more accurate prediction of RUL.

Keywords: Prognostics, Offline Measurement, Neural Network, High-Frequency Vibration Level, Remaining Useful Life.

Corresponding author: Mehdi Behzad (m_behzad@sharif.edu)

1. INTRODUCTION

Rolling element bearings (REB) are highly used components in the rotating machines. Therefore, the accurate remaining useful life (RUL) prediction of these components leads to considerable enhancement of reliability in this kind of equipment. Condition based maintenance (CBM) is the advanced method which has been developed in recent decades. Therefore, many researchers have focused on the methods for RUL prediction of REBs.

In this regard, developing the data-driven approaches have been the goal of many studies in last two decades. An et al. [1] reviewed data-driven approaches as well as physics based methods. Then, they proposed some recommendations for choosing the appropriate method. Generally, the data-driven approaches are classified to two main categories: statistical methods and artificial intelligence (AI) methods [2]. The base of all the AI methods includes two main steps. At the first step, the condition monitoring (CM) data of one or more run-to-failure tests are used for training an AI model. Next, in the second step the trained AI model is used for estimating the RUL by using the CM history data [3]. Two main challenges in developing the data-driven approaches includes introducing new features and introducing more powerful algorithms. Both challenges have been of interest for researchers in this field of study. Qiu et al. [4] proposed a robust method that uses the wavelet filter for feature extraction combining with self-organized map in order to do prognostics of REBs. Gebraeel and Lawley [5] employed the vibration level in the defective frequency and its six harmonics as the features for neural network (NN) modelling. Mahammad et al. [6] used feedforward neural network (FFNN) algorithm for RUL prediction problem. They employed the Weibull hazard rate function for fitting the features to overcome their fluctuations. Tian [7] employed generalized Weibull hazard rate function for the same purpose. This fitting function improved the performance and the accuracy of the FFNN in prognostic problem. Ben Ali et al. [8] developed an algorithm based on NN classification combining with statistical features for RUL prediction of REBs. In another research, Van and Kang [9] proposed a two-stage method in order to extracting rich and more accurate features for REBs degradation. Chouri et al. [10] developed an algorithm based on supporting vector machine (SVM). They employed eight features in the time domain and two features in the frequency domain as the inputs of their proposing algorithm. Benkedjouh et al. [11] used the energy of signal in different levels as the features. Additionally, they combined the supporting vector regression (SVR) and isometric feature mapping reduction in order to make a data-driven model. In another research, Liu et al. [12] used a data driven framework with statistical features in the time domain. Zhao et al. [13] employed the time-frequency domain features and linear regression model for RUL prediction of the REBs. Behzad et al. [14] employed vibration level in high-frequency bandwidth for tracking the degradation and RUL estimation with an AI method.

Some guidelines and standards recommend to measure and record the level of the vibration in the frequency bandwidth [10,1000] Hz for monitoring in rotating machine. Therefore, this feature is used for many industrial applications. However, selecting high-frequency bandwidth for vibration condition monitoring (VCM) of REBs is more reliable. In this article, the comparison between vibration level in the mentioned frequency interval and the vibration level in the high frequency bandwidth for prognostic of REBs is studied. An appropriate algorithm for offline VCM records is developed and the proposing features capability for RUL prediction are investigated. The represented results verify the better performance of high-frequency bandwidth vibration level for REBs prognostics.

2. FEATURE SELECTION

A significant challenge in data-driven prognostic methods is finding appropriate features. Since the degradation of REB is a monotonic process, a proper feature is expected to have a monotonic trend during the life of REB. In addition, it should be an indicator for degradation level of the REB.

Many researchers have employed different type of features for bearing RUL prediction. However, it is still a challenge to find features with the best accuracy for the prognostics purpose. In this article, an idea based on physics of bearing degradation is used for tracking the fault growth. Behzad et al. [15] studied the vibration of defective bearing with a physics-based model. They explained that when the defect starts in the contacting metal surfaces of the REB components, the roughness of surfaces increases in some regions. Therefore, the micro-geometric characteristics of the surface change gradually. Then, the impacts between two contacting surfaces increase. Employing the impulse train model, they showed the degradation of the surfaces causes the increase of vibration level in the high-frequency bandwidth.

Based on this idea, tracking the vibration in the high-frequency bandwidth is expected to indicate the defect earlier rather than vibration level in the lower frequencies. Therefore, the RMS of the spectrum in the high-frequency bandwidth [a,b] is considered as a feature and it is shown with HFRMS(a,b). The frequency interval [a,b] is selected based on spectrum study of REB. The natural frequency regions are also considered for selecting this interval. This feature is previously employed for online VCM data in [14]. In this article, the similar approach is developed for offline VCM data.

3. AI MODEL

Artificial neural network (ANN) is one of the commonly used methods in data-driven approaches. In this article, a FFNN is employed for modelling the prognostic problem. A very popular FFNN structure with one hidden layer and one output layer has been used by many researchers for modelling engineering nonlinear problems successfully. This structure which has been depicted in figure 1, is used in this study too.

Since the degradation of REBs is a dynamic problem, the features in two steps (the current step and a step in previous measurement point) are considered for inputs of this structure.

In the online monitoring system, usually the distance between measurement points is constant and relatively short. Therefore, there are plenty of measurement values in online condition monitoring system. However, in the offline system, the gap between measurement values are not necessarily same and the measurement values number is much less than the online data. The data-driven algorithm that is used in this study, has been developed for offline CM data. To this aim, at the first step an appropriate function is fitted to the trend of selected feature in the training process. The generalized Weibull hazard rate function is selected for fitting at this step. This function which has been represented in equation (1), includes three parameters that are tuned in fitting step.

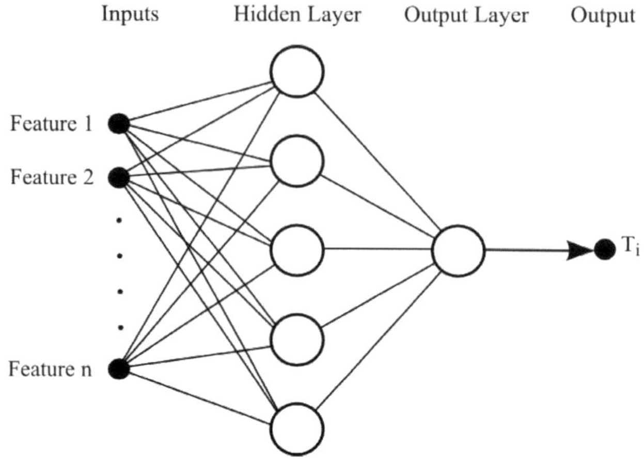

Figure 1. FFNN structure for modelling the prognostic of REBs

$$V_f = A + Bt^C \tag{1}$$

In the equation (1) t is the time, A is the fitted value at $t = 0$, B is the scale parameter, C is the shape parameter, and V_f is the fitted function.

Since the number of measurements during the offline condition monitoring of a machine may be limit, it is proposed to use some auxiliary points. To this aim, a set of auxiliary equidistant points in $[0, t_{final}]$ is considered where t_{final} is the whole life of an REB. The values of the features in these points are calculated by fitted function. The number of 2 to 5 times of original measurement points number is proper for auxiliary points number. Then, in the training process, the auxiliary points are used for training and the original points are used for validation. This is due to preventing the overfitting in training stage. Next, the trained FFNN is used for prediction. The target in training stage and the output in prediction stage are life ratio and estimation of life ratio respectively. The life ratio (T_i) is defined as the life of the REB in the current time (t_i) per the final life (t_{final}) of the REB:

$$T_i = \frac{t_i}{t_{final}} \tag{2}$$

At the prediction stage, first the features are calculated from the measurement data. Then, applying the calculated features to the trained NN, the life ratio estimation is calculated. Finally, the estimation of RUL at point i is calculated as:

$$\overline{RUL_i} = \bar{t}_{final} - t_i = \frac{t_i}{\overline{T_i}} - t_i \tag{3}$$

where, $\overline{RUL_i}$ is the estimation of RUL at point i, $\bar{t}_{final}$ is estimation of final life, and $\overline{T_i}$ is the estimation of life ratio at point i. The discussed algorithm is shown in a flowchart in Figure 2. In the next section, more details of using this algorithm with actual CM data will be discovered.

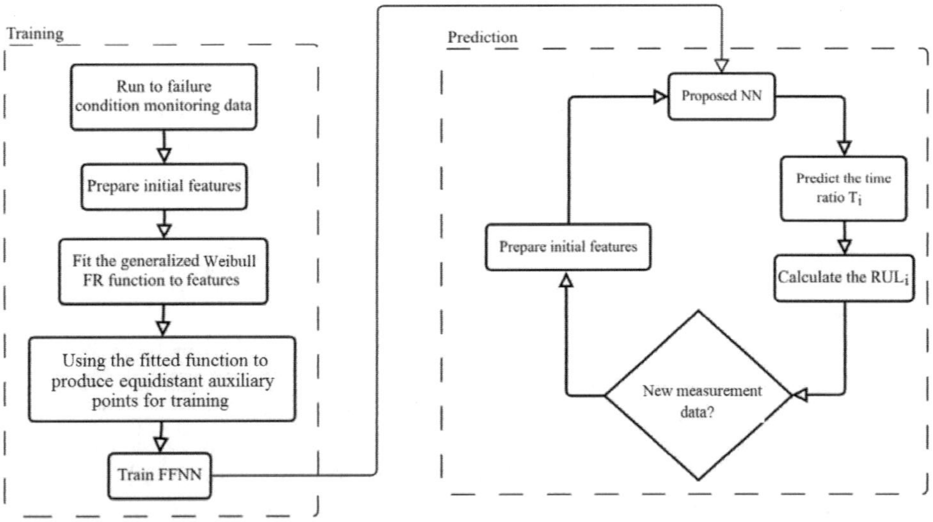

Figure 2. Flowchart of the proposed prognostic algorithm for offline CM data

4. RESULTS

An important challenge of using data-driven approaches is providing enough data for training stage. Usually in these methods, it is required to assume that the operating condition of machine in training data and prediction data have been the same. Therefore, in actual industrial cases, it may be difficult to provide enough data for training in many applications. However, implementing the accelerated run-to-failure tests of REBs in the laboratory is an alternative for checking the performance of algorithms.

In this article, the proposed method is used for prediction of both experimental accelerated life test data and actual industrial data. The published data of PRONOSTIA experiment [16] is used as laboratory data. In addition, data of two industrial case-studies provided by Behravesh Vibration Engineering Company have been analyzed with the proposed algorithm.

4.1. PRONOSTIA Experiment data

PRONOSTIA experiments data which published in PHM 2012 conference [16] has been used by many researchers for prognostic study of REBs. Figure 3 shows the details of this experimental test rig.

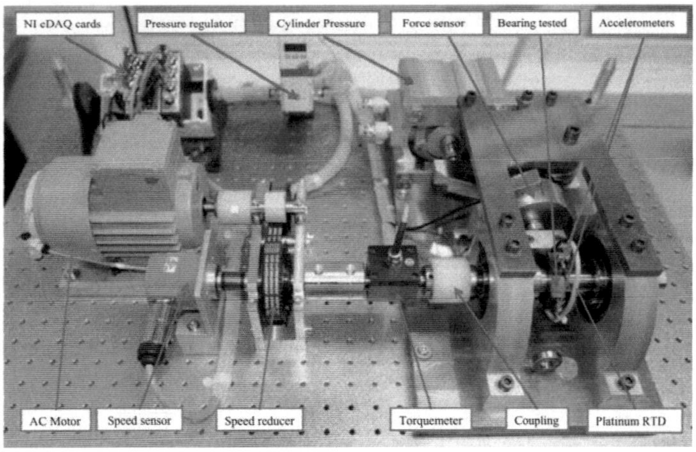

Figure 3. overview of the Prognostia test rig (16)

This set of data includes 17 full run-to-failure online vibration measurement data which are in three different operating conditions. The data of the first operating condition (shaft speed 1800 rpm and radial load 4000 N) is employed for the purpose of this article. Totally, there are seven series of data in this operation condition. Based on the initial study of these data, the kurtosis of the vibration records of the second bearing is fluctuating in a very large range. Even in the healthy condition of this REB, the kurtosis touches very high thresholds (over 100) [14]. Therefore, it is assumed that there existed a problem in the measurement or installation of the second bearing. Due to this reason, the second test is excluded from the analysis and the other six tests in the first operating condition are considered for the analysis in this part.

The inner diameter and the outer diameter of the test bearing in these experiments are 20 mm and 32 mm respectively. Additionally, the dynamic load capacity has been reported 4000 N. Vibration signals are recorded every 10 seconds with the 25.6 kHz sampling frequency during whole life of the bearing and the length of each signal is 0.1 second. Since, the proposed method in this article is going to be tested with offline data, one measurement in every 25 minutes (1500 sec) is selected for analysis. The selected signals are not equidistant in time and a random selection algorithm is used for choosing the offline points. More details about the using data is represented in table 1. Comparing the average of the mean error in the numerical tests, shows the better performance of the selected features in numerical test 4. Beside the frequency range [8,10] kHz which has produced the most accurate results, the results of numerical tests 1, 3 and 5 are very close to each other.

Here, sample prediction result plots for bearing 1_7 in all numerical tests are depicted in Figure 4. In each prediction, the vibration condition monitoring data of five other bearings are used for training. These results verify a better prediction accuracy of the features employed in numerical test 4.

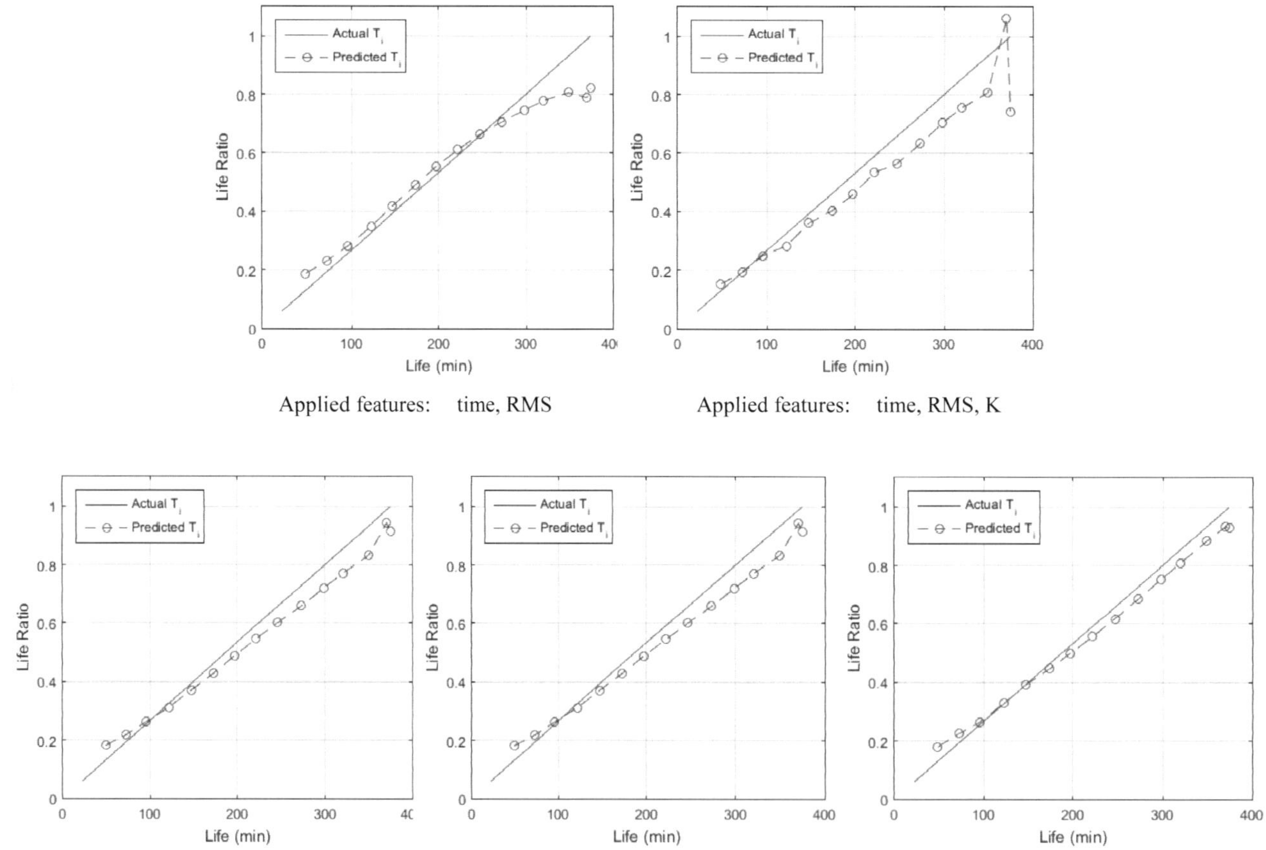

Applied features: time, RMS

Applied features: time, RMS, K

Applied features: time, HFRMS(1,10)

Applied features: time, HFRMS(8,10)

Applied features: time, RMS(0,1)

Figure 4. Prediction results on Bearing 1_7

Table 3. Numerical test results for laboratory data

	error	B 1_1	B1_3	B1_4	B1_5	B1_6	B1_7	Average error
Test 1	*e*	0.115	0.137	0.286	0.045	0.076	0.114	**0.129**
	E	0.203	0.391	0.375	0.130	0.213	0.184	**0.249**
Test 2	*e*	0.072	0.092	0.245	0.096	0.179	0.117	**0.133**
	E	0.198	0.189	0.455	0.345	0.594	0.257	**0.340**
Test 3	*e*	0.170	0.075	0.312	0.064	0.058	0.099	**0.130**
	E	0.609	0.292	0.408	0.183	0.111	0.167	**0.295**
Test 4	*e*	0.093	0.043	0.218	0.068	0.044	0.041	**0.085**
	E	0.307	0.094	0.374	0.104	0.082	0.080	**0.174**
Test 5	*e*	0.096	0.111	0.350	0.064	0.075	0.076	**0.128**
	E	0.176	0.315	0.467	0.167	0.161	0.228	**0.252**

4.2. Industrial data

The results of proposed algorithm in this article on the bearing accelerated life test VCM data analyzed in previous part. On the other hand, the two industrial cases are going to be studied in this part for the same purpose. The first case-study is for a bearing on the drive-end of a fan in a car manufacturing company. The bearing has been monitored regularly over its full life. From August 2012 to August 2013, sixteen vibration measurements have been recorded for this bearing. The second case-study is for a bearing of a sanding machine in a wood-based products manufacturer company. The vibration measurement of this bearing has been recorded from Feb 2015 to Feb 2017 during the whole life of the bearing. Totally, twenty-three measurements have been recorded during this period of time and the bearing has gone to failure at the end of this period.

The frequency bandwidth [10,1000] Hz is selected for vibration measurement of these machines based on recommendations of guidelines and standards for vibration condition monitoring of rotating machines. Besides that, the vibration in a wide frequency bandwidth [10,8000] Hz is also recorded for all measurements.

The offline records in both sets of data are not equidistant in time. Four numerical tests with different feature groups is used for this part of study. Table 4 presents the features corresponding to each numerical test.

Table 4. numerical tests along with applied features for industrial case study

Numerical Test No.	Applied features
Test 1	t_{i-1} , t_i , RMS_{i-1} , RMS_i
Test 2	t_{i-1} , t_i , RMS_{i-1} , RMS_i , K_{i-1} , K_i
Test 3	t_{i-1} , t_i , $HFRMS(1,8)_{i-1}$, $RMS(1,8)_i$
Test 4	t_{i-1} , t_i , $HFRMS(0,1)_{i-1}$, $RMS(0,1)_i$

Each numerical test has been repeated three times. In each repetition, the best result between five consecutive attempts is considered as the final result. The corresponding prediction plots of all the numerical tests for the both industrial bearings are depicted in Figure 5 and Figure 6. Finally, the average values of three final results of each numerical test has been calculated and reported in Table 5.

Comparing the results in Table 5, shows the better performance of the high-frequency bandwidth features. This comparison verifies the importance of the high-frequency vibration investigation in bearing degradation analysis.

Although the frequency bandwidth [10,1000] Hz commonly is used for condition monitoring and diagnostics of the machines, the results depicted in this study represents how the analysis of high-frequency bandwidth of the spectrum concludes more accurate estimation in REBs prognostics.

Table 5. Numerical test results comparison

	error	**Bearing 1 (Car Manufacturing Co.)**	**Bearing 2 (Wood-based products Manufacturing Co.)**
Test 1	*e*	0.037	0.122
	E	0.143	0.449
Test 2	*e*	0.277	0.415
	E	1.010	1.940
Test 3	*e*	0.045	0.070
	E	0.170	0.419
Test 4	*e*	0.209	0.161
	E	0.753	0.610

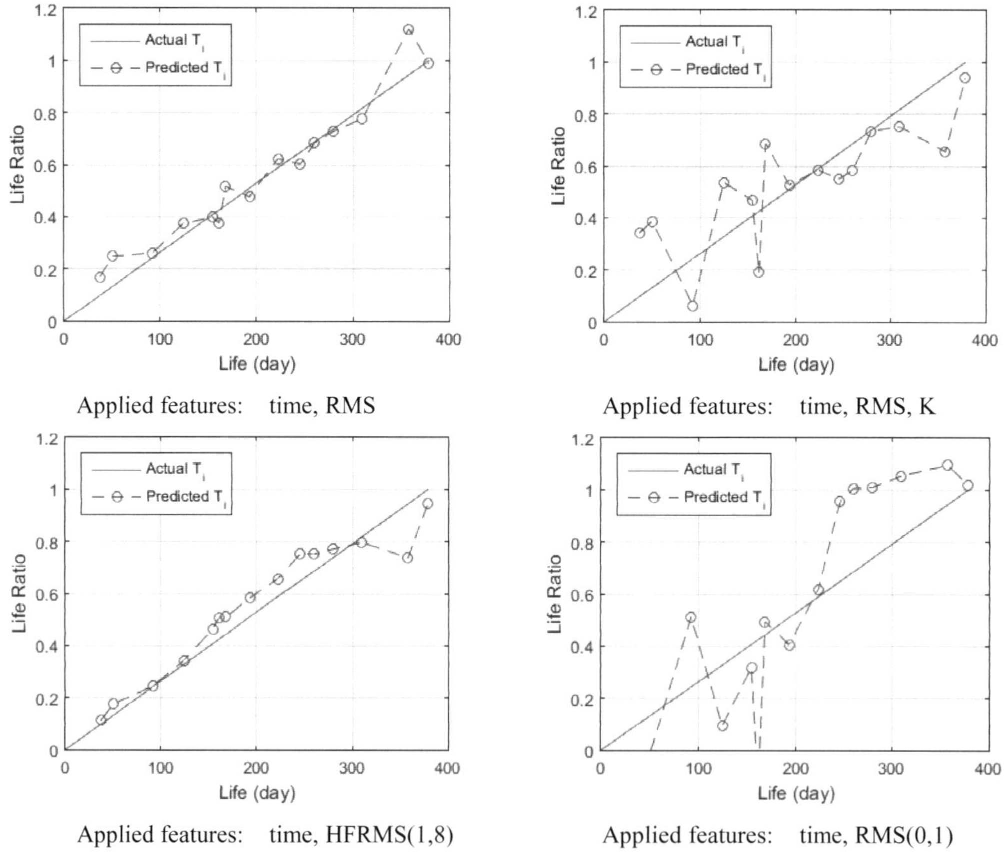

Figure 5. Prediction results on the industrial bearing 1 regarding to the Fan in car manufacturing company.

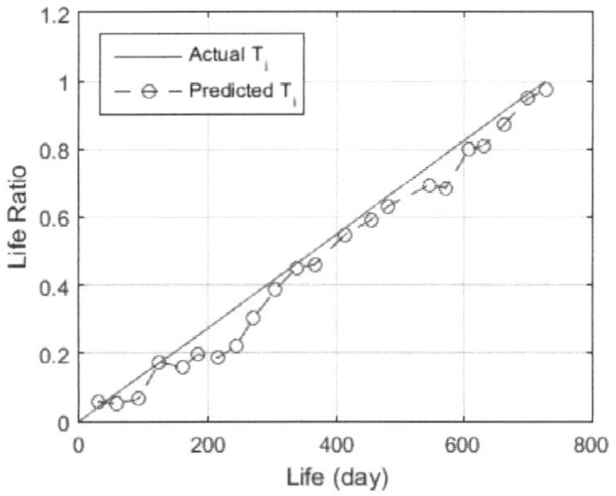

Applied features: time, RMS

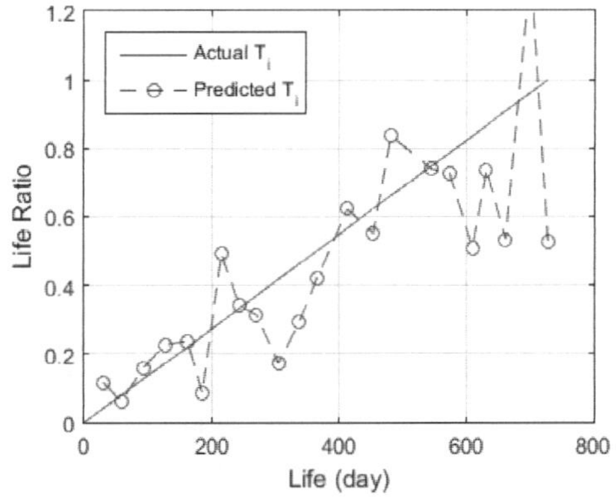

Applied features: time, RMS, K

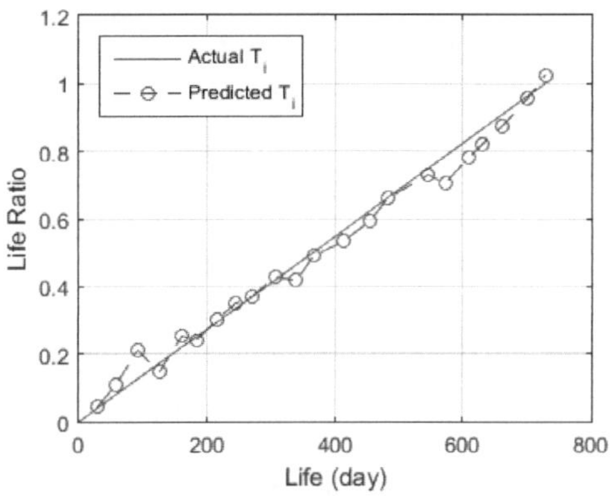

Applied features: time, HFRMS(1,8)

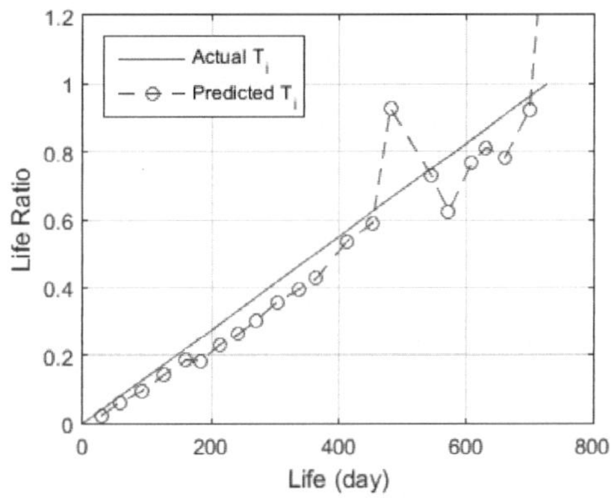

Applied features: time, RMS(0,1)

Figure 6. Prediction results on the industrial bearing 1 regarding to the sanding machine in wood-based product manufacturing Co.

5. CONCLUSION

In this article, an algorithm based on FFNN approach has been proposed for prognostics of REBs by using offline VCM data. Since the number of offline measurements during the whole life of the equipment are not enough, this algorithm produces some auxiliary points for training. Features in two different time steps are used as inputs of the NN model due to dynamic behavior of degradation. RMS of vibration spectrum in various frequency bandwidths was considered as a feature. In addition, time and kurtosis were two other features that have been combined with RMS in different numerical tests.

The proposed algorithm has been applied to a set of bearing accelerated life run-to-failure tests data as well as two industrial cases data. According to the results of this study, the RMS of high-frequency bandwidth in vibration spectrum is a better feature for tracking the degradation. Therefore, using these features for RUL estimation has produced more accurate predictions.

ACKNOWLEDGEMENT

The financial support of the ministry of science, research and technology is gratefully acknowledged for this study. In addition, the authors greatly appreciate the help of Behravesh Vibration Engineering Co. for providing the industrial machine vibration data.

REFERENCES

[1] An, D., Kim, N. H., & Choi, J. H. (2015). Practical options for selecting data-driven or physics-based prognostics algorithms with reviews. Reliability Engineering & System Safety, 133, 223-236.

[2] Peng, Y., Dong, M., & Zuo, M. J. (2010). Current status of machine prognostics in condition-based maintenance: a review. The International Journal of Advanced Manufacturing Technology, 50(1-4), 297-313.

[3] Huang, H. Z., Wang, H. K., Li, Y. F., Zhang, L., & Liu, Z. (2015). Support vector machine based estimation of remaining useful life: current research status and future trends. Journal of Mechanical Science and Technology, 29(1), 151-163.

[4] Qiu, H., Lee, J., Lin, J., & Yu, G. (2003). Robust performance degradation assessment methods for enhanced rolling element bearing prognostics. Advanced Engineering Informatics, 17(3-4), 127-140.

[5] Gebraeel, N., Lawley, M., Liu, R., & Parmeshwaran, V. (2004). Residual life predictions from vibration-based degradation signals: a neural network approach. IEEE Transactions on industrial electronics, 51(3), 694-700.

[6] Mahamad, A. K., Saon, S., & Hiyama, T. (2010). Predicting remaining useful life of rotating machinery based artificial neural network. Computers & Mathematics with Applications, 60(4), 1078-1087.

[7] Tian, Z. (2012). An artificial neural network method for remaining useful life prediction of equipment subject to condition monitoring. Journal of Intelligent Manufacturing, 23(2), 227-237.

[8] Ali, J. B., Chebel-Morello, B., Saidi, L., Malinowski, S., & Fnaiech, F. (2015). Accurate bearing remaining useful life prediction based on Weibull distribution and artificial neural network. Mechanical Systems and Signal Processing, 56, 150-172.

[9] Van, M., & Kang, H. J. (2016). Two-stage feature selection for bearing fault diagnosis based on dual-tree complex wavelet transform and empirical mode decomposition. Proceedings of the Institution of Mechanical Engineers, Part C: Journal of Mechanical Engineering Science, 230(2), 291-302.

[10] Chouri, B., Monteiro, F., Tabaa, Mohamed, Dandache, A., Komarasamy, G., Wahi, A., ... & Huerta, L. (2013). Residual useful life estimation based on stable distribution feature extraction and SVM classifier. Journal of Theoretical and Applied Information Technology, 55(3), 299-306.

[11] Benkedjouh, T., Medjaher, K., Zerhouni, N., & Rechak, S. (2013). Remaining useful life estimation based on nonlinear feature reduction and support vector regression. Engineering Applications of Artificial Intelligence, 26(7), 1751-1760.

[12] Liu, Z., Zuo, M. J., & Qin, Y. (2016). Remaining useful life prediction of rolling element bearings based on health state assessment. Proceedings of the Institution of Mechanical Engineers, Part C: Journal of Mechanical Engineering Science, 230(2), 314-330.

[13] Zhao, M., Tang, B., & Tan, Q. (2016). Bearing remaining useful life estimation based on time–frequency representation and supervised dimensionality reduction. Measurement, 86, 41-55.

[14] Behzad, M., Arghand, H. A., & Rohani Bastami, A. (2017). Remaining useful life prediction of ball-bearings based on high-frequency vibration features. Proceedings of the Institution of Mechanical Engineers, Part C: Journal of Mechanical Engineering Science, 0954406217734885.

[15] Behzad, M., Bastami, A. R., & Mba, D. (2011). A new model for estimating vibrations generated in the defective rolling element bearings. Journal of Vibration and Acoustics, 133(4), 041011.

[16] Nectoux, P., Gouriveau, R., Medjaher, K., Ramasso, E., Chebel-Morello, B., Zerhouni, N., & Varnier, C. (2012, June). PRONOSTIA: An experimental platform for bearings accelerated degradation tests. In IEEE International Conference on Prognostics and Health Management, PHM'12. (pp. 1-8). IEEE Catalog Number: CPF12PHM-CDR.

Investigation of Infrared Thermography as a Dual Online Diagnostic Tool for Dynamic Structural Health Monitoring

D.A. Desai[1], S.M Talai[1], P.S. Heyns[2]

[1] Department of Mechanical Engineering, Mechatronics & Industrial Design, Tshwane University of Technology, Pretoria 0001, South Africa

[2] Department of Mechanical & Aeronautical Engineering, University of Pretoria, Hatfield 0028, South Africa

ABSTRACT

It is well known that infrared thermography (IRT) has fully matured as a tool for temperature condition monitoring. However in this paper, an experimental investigation of IRT as a dual online diagnostic tool in terms of temperature and vibration frequency characterization of dynamic structures is carried out, as such studies have not as yet been adequately investigated. Tests were conducted on both healthy and defect-induced cantilever beam-like structures coupled with a lacing wire subjected to forced excitations. Thermal images were acquired at the frictional interfaces, employing two infrared cameras. The analyzed frequencies obtained from the frictional temperature evolution time domain waveform using a Matlab FFT algorithm compared well with alternate accelerometer data acquired with the maximum relative error being 0.30 %. Moreover, the beam with the induced defect showed a temperature increase of 8.1 $^{\circ}C$ compared to the healthy one and exhibited multiple spectral peaks around the dominant frequency, hence revealing successful discrimination between them. This study in essence proves that IRT is capable of reliably describing structural vibration behavior in terms of frequency. These findings are particularly useful in overcoming many limitations inherent in some of the current contact vibration measuring techniques operating in turbulent, unclean environments.

Keywords: Frictional heat generation, structural health indicators, condition monitoring, structural defect, non-destructive testing and evaluation (NDT&E)

Corresponding author: Dawood Desai (desaida@tut.ac.za)

1. INTRODUCTION

In today's mechanical and aerospace engineering communities, the need for enhanced ability to monitor dynamic structures and detect potential damages at the earliest possible stage for effective structural health monitoring (SHM) is ever increasing [1].

The online temperature measurement performed in a non-contact way using Infrared thermography (IRT) facilitates early diagnosis of probable faults and proper preventive measures adopted to avoid major shut downs. Furthermore, the non-destructive testing and evaluation (NDT&E) techniques based on IRT are the most advanced techniques available today and has attracted much research attention [2]. Titman [3] explained that they represent the most promising method of structural defect detection using the principle that all bodies emit infrared radiation with the absolute temperature greater than 0 K. Bagavathiappan et al. [4] reported that since the abnormal temperature pattern is an indication of the unhealthy state; it's widely used in temperature condition monitoring. Vibrothermography which is a NDT&E technique is used to find surface and near surface defects such as cracks and delaminations through observations of vibration induced heat generation [5]. It involves exposing the specimen to a sonic or ultrasonic vibration, then observing the induced frictional heating with an infrared camera. Montanini and Freni [6] investigated the correlation between the vibrational mode shapes and viscoelastic heat generation in vibrothermography. The authors found that there exists a minimum threshold of vibration amplitude that allows reliable flaw detections for a specific geometry. The results showed that exciting the specimen at lower modes is sufficient to generate heat flux for flaw identification. Gao et al. [7] described a statistical method for conducting a good vibrothermography inspection test for crack detection in aircraft engine fan blades. They showed that use of the Bayesian method enables the development of confidence intervals for the probability of detecting cracks. Furthermore, Dimarogonas and Syrimbeis [8] successfully predicted the vibration signature from heat generation due material damping by considering a vibrating rectangular plate. Most recently, Talai et al. [9] developed a technique for structural vibration characteristics measurement using an IRT approach by considering healthy cantilever beam-like structures coupled with a lacing wire.

For decades, however, vibration monitoring has been utilized to assess SHM for predicting potential failures that, in turn, enhances reliability and availability [10-14] where the time domain waveform signals are transformed into frequency domain using Fourier's transformation. Nevertheless, the available measurement techniques such as strain gauges have a shorter sensor life span due to fatigue, turbulent and unclean environments coupled with centrifugal forces at high temperatures and erosion as reported by Al-Bedoor [15] in power plant applications. Also, there are times, when one can't have easy access to a machine because of its location, or desires the additional confidence of having another diagnostic tool available. For this reason, a non-contact diagnostic tool with ability of monitoring dual indicators of structural health is of great benefit.

Therefore, this paper presents a practical application of IRT as a dual online diagnostic tool for dynamic SHM in terms of temperature monitoring and spectral vibration frequency. To achieve this, both healthy and defect induced cantilever beam-like structures coupled with a lacing wire subjected to mechanical excitation in the laboratory environment are employed with infrared cameras focused on the frictional interfaces at the lacing wire zone for the thermal imaging. Results using this technique were successfully validated against miniature accelerometer acquired measurements. Hence, this study has great significance for the effective SHM of dynamic structures, thus, reducing down-time and maintenance costs, leading to increased plant efficiency.

2. MATERIALS AND EXPERIMENTAL METHOD

2.1. Materials

Both the beams and lacing wires were manufactured of AISI 304 stainless steel due their low thermal conductivity, hence, generate heat with slight frictional effect [16]. The geometric and material properties are given in table 1 and table 2, respectively.

Table 1. Beam and lacing wire geometric dimensions.

Description	Dimension
Beam mass	0.10 *kg*
Length	300 *mm*
Width	25 *mm*
Thickness	2 *mm*
Lacing wire diameter	5 *mm*
Beam hole diameter [17] [tolerance grade: *F8*/*js7*]	$5^{+0.422}_{+0.412}$ *mm*
Beam-lacing wire hole location from fixed end	250 *mm*
Location of the exciters	290 *mm*

Table 2. AISI 304 steel material properties [18].

Material properties	Parameters
Density, ρ	7740 kg/m^3
Young modulus, E	200 GPa
Poisson ratio, v	0.33
Static friction coefficient, μ_s ($\mu_k = 0.75\ \mu_s$) [19]	0.15
Thermal properties	
Thermal conductivity, k	16.5 W/mK
Specific heat capacity, c	500 J/kgK

2.2. Experimental procedure

The experimental setup used in the work of Talai et al. [9] was employed in this study as depicted in figure 1. A pre-load force of 15 *N* applied at the lacing wire throughout the experiments. Three excitation patterns of the beams were considered. Firstly; both beams at the same forcing frequency of 20 *Hz* , secondly; Beam 1 at 20 *Hz* and Beam 2 at 25 *Hz* , finally; both beams at 25 *Hz* but with Beam 1 induced with structural defect. These excitation frequencies were considered based on Shannon's sampling rate theorem of 0.4 times the rated resolution of the IR cameras used. The thermal images were recorded continuously for 150 *s*. The analyzed frequencies from interface temperature time domain waveform were subsequently validated against the corresponding frequencies acquired concurrently using miniature Deltatron accelerometers attached to the respective beams.

3. RESULTS AND DISCUSSIONS

In case of both beams excited at the same frequency value of 20 *Hz* , it was expected that the thermal imaging will attain the same maximum frictional temperature. However, this was not the case practically. The thermal imaging exhibited temperature difference of 0.9 °*C* (figure 2) that was attributed partly to the emanating beam's dynamics leading to the non-uniform interface boundary motion and the different thermal sensitivities of the IR cameras used in this study. Accordingly, the frequencies evaluated from frictional temperature evolution for Beam 1 and 2 were 19.9885 *Hz* (figure 3a) and 19.9973 *Hz* (Figure 3b), respectively. These were in good agreement to the acquired accelerometer frequencies (figure 3a and 3b) with relative errors being 0.26 % and 0.21 % of the former and latter, respectively.

Montanini and Freni [6] explained that the amount of frictional heat generated depends on several parameters, which includes coefficient of friction, normal applied force and the sliding relative speed. However, this study kept other factors constant while the forcing frequencies varied. As expected, the structure under higher excitation frequency allowed the frictional interface to slide against each other more often (as opposed to lower frequencies) leading to greater frictional heat generation resulting in a temperature difference of 2.2 °*C* depicted in figure 4 in the case of Beam 1 and Beam 2 excited at 20 *Hz* and 25 *Hz*, respectively. Moreover, Mabrouki et al. [16] observed similar findings while investigating frictional heating for efficient use of vibrothermography. Hence, the conclusion was that frictional heat increases with the interface frequency. On the other hand, the analyzed frequencies from thermal imaging were 19.9832 *Hz* (figure 5a) and 25.0025 *Hz* (figure 5b) for Beam 1 and Beam 2, respectively. Likewise, good agreement was exhibited when compared to the accelerometer measured frequencies, the relative errors being 0.28 % (figure 5a) for the former and 0.19 % (figure 5b) for the latter.

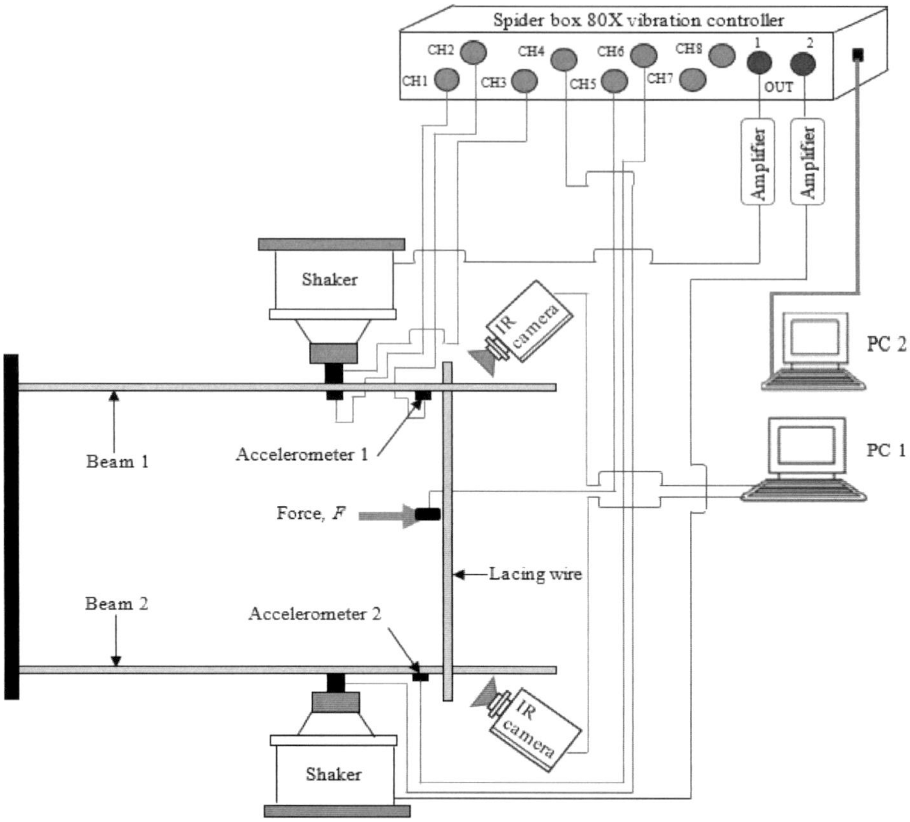

Figure 1. Laboratory experimental setup.

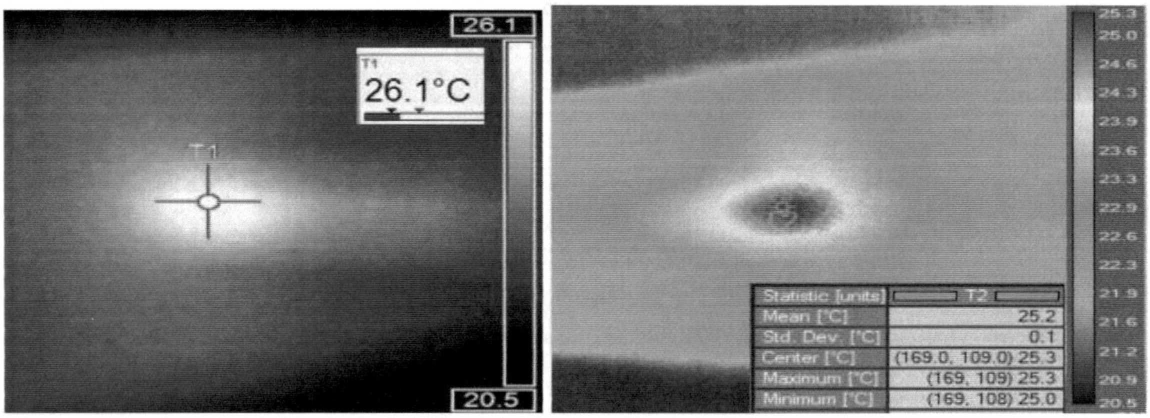

Figure 2. Thermal images for both beams excited at 20 *Hz* : 150 *s* (a) Beam 1 and (b) Beam 2.

It is known that the presence of structural defects reduce stiffness, hence, larger displacement variations occur resulting in increased frictional heat generation due to interfacial metallurgical boundary local deformations. This was evident by the abnormal temperature difference of 8.1 $^{\circ}C$ among the beams investigated with a forcing frequency of 25 *Hz* with Beam 1 induced with a structural defect (figure 6) compared to the healthy beam. The analyzed frequencies from frictional temperature generation for Beam 1 around the forcing frequency were 24.9748 *Hz*, 24.2328 *Hz* and 25.7605 *Hz* (figure 7a), while the frequencies for Beam 2 had a clear spectral peak at 25.0013 *Hz* (figure 7b). Based on the dominant frequencies, the exhibited relative errors were 0.30 % and 0.19 % of Beams 1 and 2, respectively; compared to the accelerometer measured values. It was interesting to observe the unhealthy beam being clearly distinguished by its abnormal temperature generation and vibration frequency spectral peaks, both of which were monitored online using IRT.

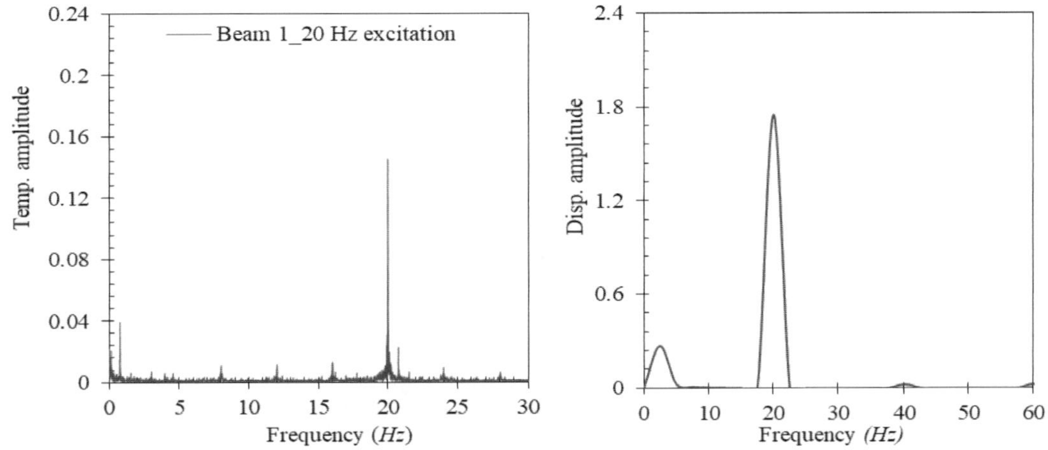

(a)

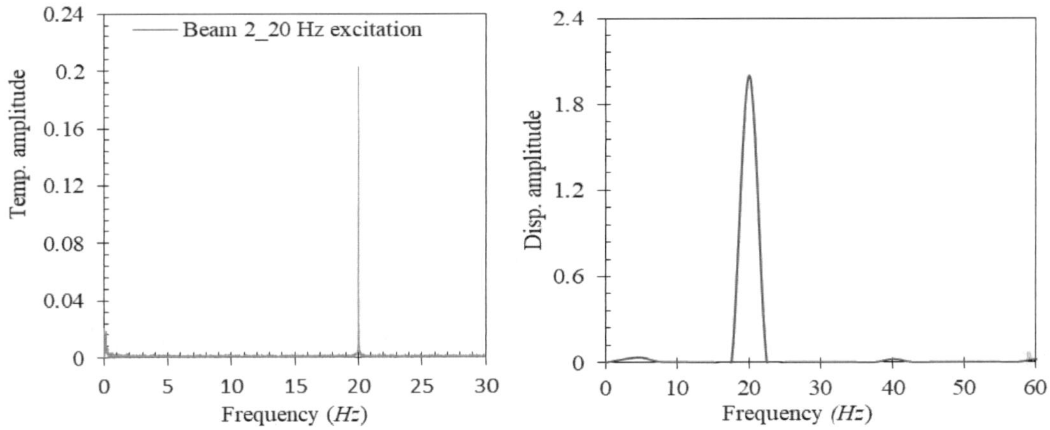

(b)

Figure 3. FFT of temperature evolution and displacement: both beams at 20 *Hz* (a) Beam 1 and (b) Beam 2.

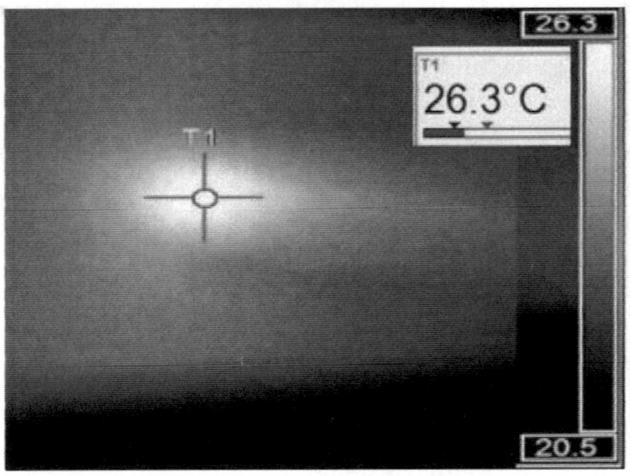

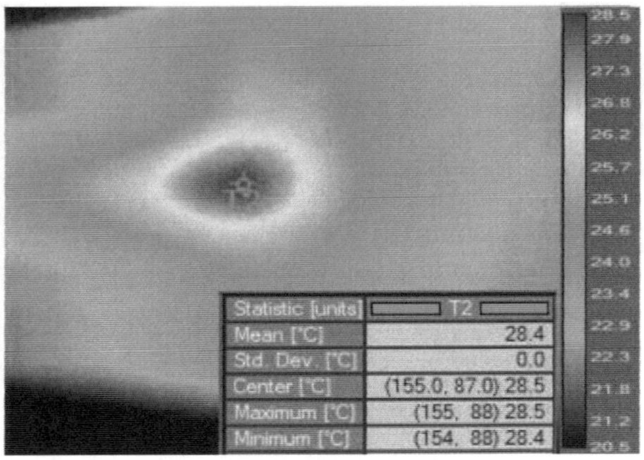

(a) *(b)*

Figure 4. Thermal images for excitation: 150 *s* (a) Beam 1 at 20 *Hz* and (b) Beam 2 at 25 *Hz*.

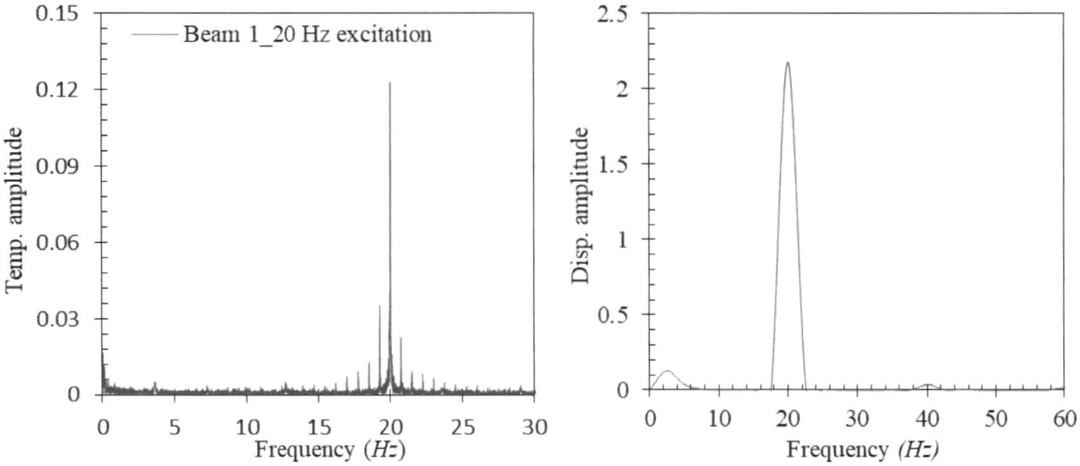

(a)

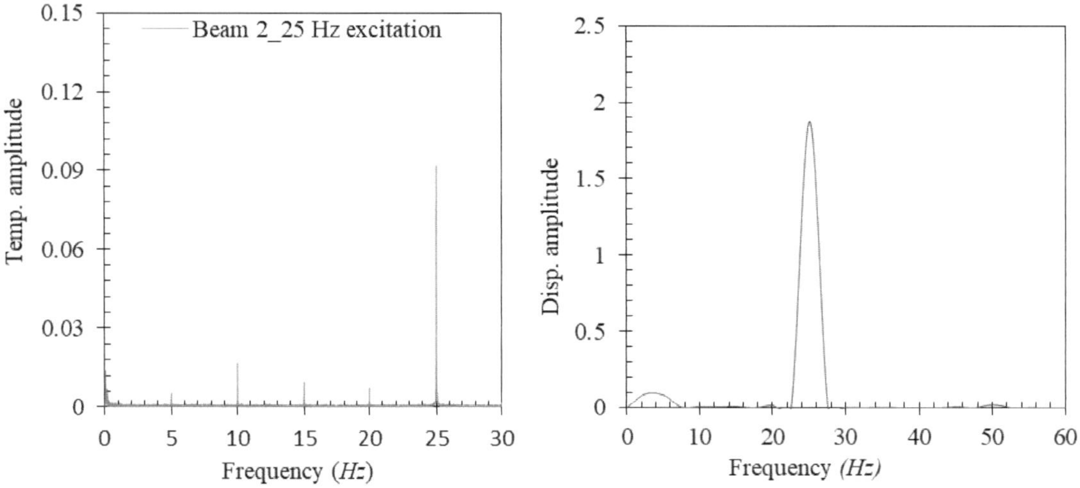

(b)

Figure 5. FFT of temperature evolution and displacement (a) Beam 1 at 20 *Hz* and (b) Beam 2 at 25 *Hz*.

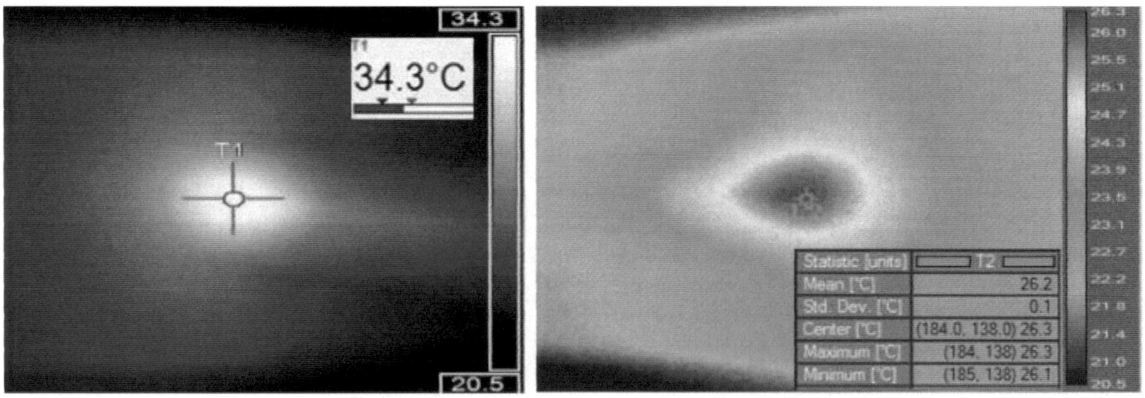

(a) (b)

Figure 6. Thermal images for both beams excitated at 25 *Hz* : 150 *s* (a) defect induced and (b) healthy.

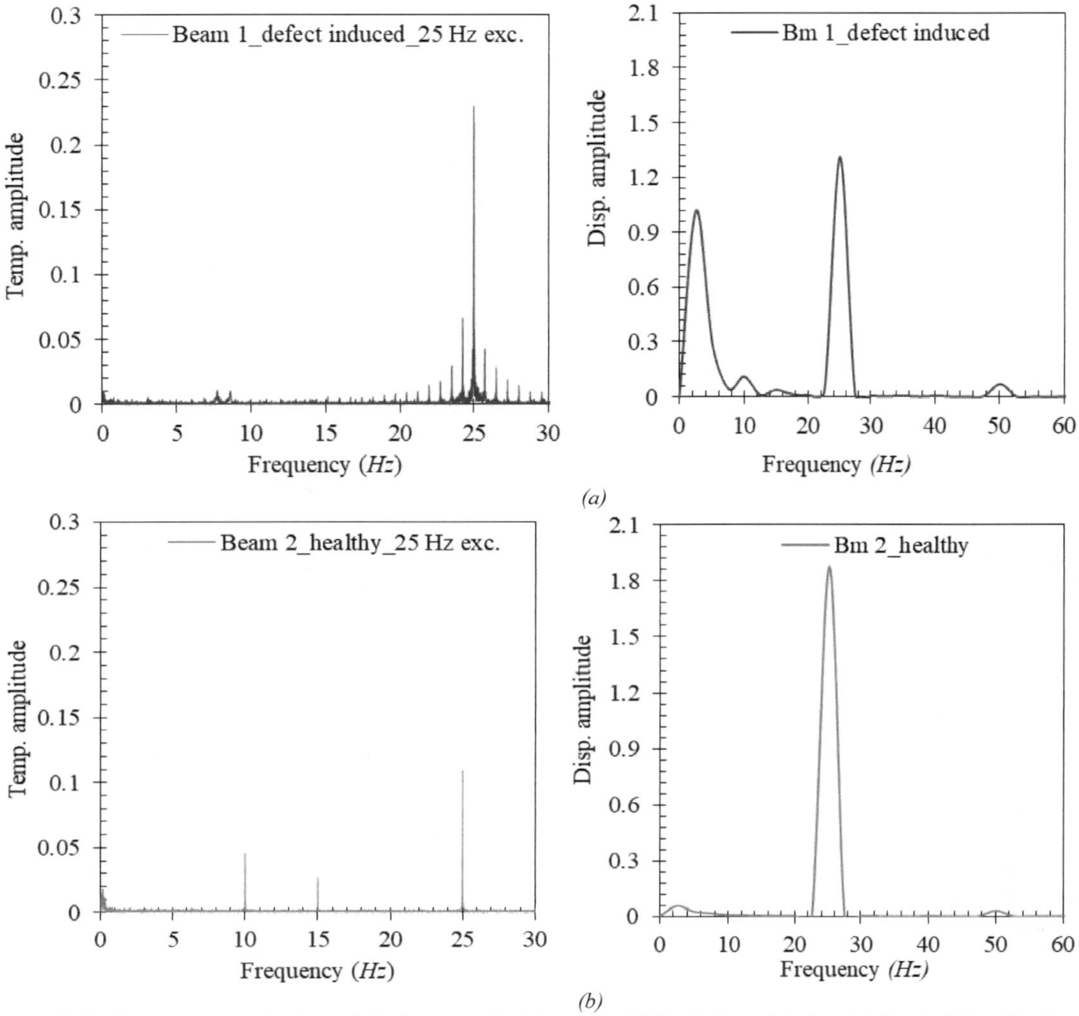

Figure 7. FFT of temperature evolution and displacement: both beams at 25 *Hz* (a) Beam 1 induced defect & (b) healthy Beam 2.

4. CONCLUSION

All the frequencies predicted through the IRT approach agreed well with those acquired by miniature accelerometer sensors, with the maximum relative error being 0.30 % thus, validating experimental measurements using IRT. Also, the unhealthy beam was clearly discriminated based on the abnormal temperature generation and multiple spectral frequency peaks around the forcing frequency. Therefore, it can be concluded that IRT is reliable as a dual online diagnostic tool for SHM as evident by this research work.

5. ACKNOWLEDGEMENT

The authors greatly appreciate the support of Tshwane University of Technology, University of Pretoria and Eskom Power Plant Institute (South Africa) in funding this research.

REFERENCES

[1] Le, T. T. H., Point, N., Argoul, P., Cumunel, G. (2016) Structural changes assessment in axial stressed beams through frequencies variation, International Journal of Mechanical Sciences 110: 41-52.

[2] Meola, C., Carlomagno, G. M. (2004) Recent advances in the use of infrared thermography, Measurement Science and Technology 15:(9) 27-58.

[3] Titman, D. J. (2001) Applications of thermography in non-destructive testing of structures, NDT&E International 34: 149-154.

[4] Bagavathiappan, S., Lahiri, B. B., Saravanan, T., Philip, J., Jayakumar, T. (2013) Infrared thermography for condition monitoring – A review, Infrared Physics & Technology 60: 35-55.

[5] Renshaw, J., Chen, J. C., Holland, S. D., Bruce, T. R. (2011) The sources of heat generation in vibrothermography, NDT&E International 44:(8) 736-739.

[6] Montanini, R., Freni, F. (2013) Correlation between vibrational mode shapes and viscoelastic heat generation in vibrothermography, NDT&E International 58: 43-48.

[7] Gao, C., Meeker, W. Q., Mayton, D. (2014) Detecting cracks in aircraft engine fan blades using vibrothermography nondestructive evaluation, Reliability Engineering and System Safety 131: 229-235.

[8] Dimarogonas, A. D., Syrimbeis, N. B. (1992) Thermal signatures of vibrating rectangular plates, Journal of Sound and Vibration 157:(3) 467-476.

[9] Talai, S. M., Desai, D. A., Heyns, P. S. (2016) Vibration characteristics measurement of beam-like structures using infrared thermography, Infrared Physics & Technology 79: 17-24.

[10] Mukhopadhyay , N. K., Chowdhury, S. G., Das, G., Chattoraj, I., Das, S. K., Bhattacharya, D. K. (1998) An investigation of the failure of low pressure steam turbine blades, Engineering failure analysis 5:(3) 181-193.

[11] Wei, F., Pizhong, Q. (2010) Vibration-based Damage Identification Methods: A Review and Comparative Study, Structural Health Monitoring 10:(1) 83-111.

[12] Rao, S. J. (1991) Turbo-machine blade vibration (first edition), New Delhi: New age international publishers 7-20.

[13] Jaiswal, B. L., Bhave, S. K. (1994) Experimental evaluation of damping in a bladed disk model, Journal of sound and vibration 177:(1) 111-120.

[14] Torshizi, M. S. E., Nikravesh, Y. S. M., Jahangiri, A. (2009) Failure analysis of gas turbine generator cooling fan blades, Engineering Failure Analysis 16:(5) 1686-1695.

[15] Al-Bedoor, B. O. (2002) Discussion of the available methods for blade vibration measurement, In: ASME 2002 Pressure Vessels and Piping Conference. Vancouver, BC: Canada. 1561: 53-61.

[16] Mabrouki, F., Thomas, M., Genest, M., Fahr, A. (2009) Frictional heating model for efficient use of vibrothermography, NDT&E International 42:(5) 345-352.

[17] Sanders, W. P. (1996) Turbine steam path engineering for operations and maintenance staff (first edition), Richmond Hill, Ontario: Canada 236-239.

[18] Madhusudana, C. V. (1999) Thermal conductance of cylindrical joints, International Journal of Heat and Mass Transfer 42: 1273-1287.

[19] Oden, J. T., Martins, J. A. (1985) Models and computational methods for dynamic friction phenomena, Computer methods in applied mechanics and engineering 52: 527-634.

Experimentally validated numerical simulation of prediction of structural vibration frequencies from interfacial frictional temperature signature

S.M. Talai[1], D.A. Desai[1], P.S. Heyns[2]
[1] Department of Mechanical Engineering, Tshwane University of Technology, South Africa
[2] Department of Mechanical & Aeronautical Engineering, University of Pretoria, South Africa

ABSTRACT

Friction studies have shown that the contact interface of dynamic structural components generally results in a temperature increase due to periodic motion of the boundary. Furthermore, the well-known heat conduction equation yields the temperature distribution. Subsequently, the aim of this study was to investigate the prediction of vibration frequency from the interfacial frictional temperature signatures of structures under excitation. In this work, an experimentally validated finite element (FE) model was developed using the commercial software ABAQUS. Tests were done on AISI 304 steel of healthy cantilever beam-like structures coupled with a friction interface and subjected to forced excitations. A coupled temperature-displacement transient methodology was utilized to predict the frictional heat generation and subsequently the vibrational frequencies extracted from the temperature wave form, which in turn were validated by comparing them to the experimental results employing infrared thermography. The results confirmed the feasibility of using the interface frictional temperature signature to describe the vibration frequency with good accuracy and may be used to overcome some limitations encountered with conventional instrumentation such as strain gauge failures due to fatigue.

Keywords: Frictional heat generation, structural health indicator, condition monitoring, vibration frequency, infrared thermography.

Corresponding author: Stephen M. Talai (stevetalai@gmail.com)

1. INTRODUCTION

Temperature monitoring of dynamic components aids early diagnosis of the potential damages and proper preventive measures adopted to avoid major breakdowns [1-3]. Titman [4] explained that it represent the most promising method of structural defect detection using the principle that all bodies emit infrared radiation with the absolute temperature greater than 0 K.

For decades, however, vibration monitoring has been utilized to assess SHM for predicting potential failures that, in turn, enhance the reliability and availability [5-9]. Several techniques are applied for the online structural vibration frequency monitoring. Strain gages bonded to the structure, experience the same motion (strain) as the structure and hence its change in resistance gives the strain applied to the structure. Even though this method has the advantage of performing measurement of individual structure, it has several disadvantages such as shorter sensor life span due to fatigue. Laser Doppler Vibrometry (LDV) with an Eulerian approach allows overcoming most of the limitations mentioned in the use of strain gauges [10]. Nonetheless, Castellini et al. [11] reported that its main limitations are the speckle effects and poor signal-to-noise ratio when measuring vibration on the low diffusive surface. Also, interferometry has the ability to provide traceability of vibration displacement measurement. However, its accuracy is limited by the environment, thus it is often not practical to perform vibration measurements in turbulent environments like in the case of power plants [12].

There are times, however, when one cannot have access to a machine because of its location, or desires the additional confidence of having another diagnostic tool available. For this reason, a diagnostic tool with the ability of monitoring dual indicators of structural health is of benefit. This paper presents an experimentally validated numerical simulation of predicting vibration frequency from the temperature wave form. To achieve this, both healthy and defect induced cantilever beam-like structures were coupled with a friction interface subjected to mechanical excitation in a laboratory environment, while infrared cameras focused on the frictional interfaces for the thermal imaging. This study is significant for the effective SHM of dynamic structures, thus, reducing the down-time and maintenance cost, leading to increased plant efficiency.

2. FE MODELLING AND SIMULATION

FE modelling and simulations were carried out using the commercial software ABAQUS/CAE version 6.13-1. This software is well suited for performing non-linear simulations [13]. The models were developed as follows:

2.1. FE geometric model

The geometric model composed of 3D deformable solid parts consisting of two straight rectangle cantilever beam coupled with lacing wire. Similarly, Petreski [14] used a group of cantilever beams to numerically study the natural frequencies and mode shapes due to changes of lacing wire such as diameter, elasticity and location on the LP steam turbine blades. The geometric dimensions and material properties are as used in the work of Talai, Desai and Heyns [15] on vibration characteristics measurement of beam-like structures using infrared thermography.

2.1.1. Assembly, loading, boundary condition and meshing

The model assembly is depicted in figure 1.

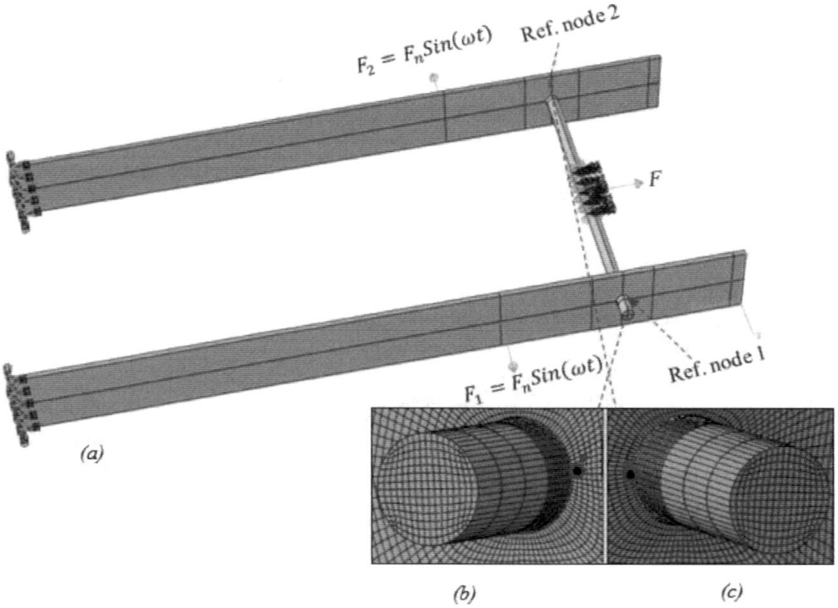

Figure 1. FEM model (a) assembly, boundary conditions and loading (b) Beam 1 interface mesh with reference node (c) Beam 2 interface mesh with reference node.

The sinusoidal loads were defined on the beams' surface in order to simulate the mechanical excitations while a lateral force to the lacing wire to mimic the centrifugal force. The forced excitations of beams were achieved through the formulation of a sinusoidal load, $F(\omega, t)$ using a varying force function expressed as

$$F(\omega,t) = F_n Sin(\omega t), \qquad (1)$$

where F_n is the excitation force which varies sinusoidally with time, t at a frequency of ω *cycle/second*. The assumptions made when defining the loads and boundary conditions were [16]:

i. Beams and lacing wire initially isothermal at 22 °C.
ii. The heat losses due to convection effect are of convection coefficient of $30\ W/m^2.K$ at 22 °C ambient temperature.
iii. The radiation heat losses negligible compared to convection.
iv. The frictional heat transfer obeys the heat partition model of the ratio of thermal conductivity.

The Bauschinger effect which is the changing material properties under cyclic loading is generally involved in frictional contacts [17, 18]. For instance, loading of a specimen beyond its yield limit in a given direction lowers the yield strength in the opposite direction. This leads to non-linear structural material properties over the entire period and for the case of numerical simulation, it usually results in higher computational cost [19, 20].

Accordingly, higher loading of the FEM at the expense of greater computational efforts leads to more frictional heat generation. However, since this study entailed an investigation, it considered small loads to shorten the computational time. Therefore, the transverse sinusoidal force of amplitude of 0.10 N and lateral force of 0.001 N defined to idealise the beams harmonic loading and lacing wire lateral loading, respectively. The simulations were carried as follows: first; both beams at 20 Hz, secondly; both beams at 40 Hz, thirdly; beam 1 at 20 Hz while beam 2 at 40 Hz and finally; both beams at 25 Hz, but with beam 1 induced with a defect. These frequencies were considered based on Shannon's sampling rate theorem on rated optical resolution of the IR cameras used in this study.

An encastre boundary condition defined for FE beams end, while asymmetric boundary condition at the centre of the lacing wire i.e. $U1 = U2 = UR3 = 0$ *(allows the free lateral movement)*. Also, a reference node *(beam 1: Ref. node1 and beam 2: Ref. node 2)* created at each beam-lacing wire interface as a set in the modelling facilitated a precise data acquisition for the frictional temperature evolution and displacement spectrum, hence, analysis.

The model was meshed using element type C3D8RT. However, the beam interface hole and the lacing wire contact region meshed with highly dense elements since it was the main focus of the study, for the obvious reason of increasing the accuracy of the results. The mesh seed size around the beam interface was 0.157 mm, while the rest of the model 2.5 mm. On the other hand, lacing wire composed of 0.302 mm seed size. The total elements generated for the healthy FEM used in the simulation was 100 320.

2.1.2. Friction interaction properties

The interaction properties define the contact properties between the master and slave surfaces. ABAQUS considers the master surface as the surface that undergoes larger motion and vice versa for the slave. In this regard, the beam was assigned as master and the lacing wire surface as the slave. The following interaction properties were defined:

i. The friction formulation of static-kinetic exponential decay as tangential behaviour.
ii. Heat distribution according to the partition ratio given in equation.
iii. Thermal conductance using clearance dependent data based on material conductivity (table 1).

Table 1. Interface clearance dependent data.

Conductance $(W/mm^2.K)$	Clearance (mm)
0.0165	0
0.0165	1
0	100

2.1.3. Analysis procedure

The static analysis type simulated the lateral force acting on the lacing wire. The contact frictional heat, however, is strongly dependent on the relative motion [16, 20]. The ABAQUS/CAE solver uses coupled temperature-displacement analysis of simulations involving thermo-mechanical analysis [13]. It takes into account the initial ambient temperature and the relative interaction properties for calculating the transient temperature distribution. Wenya et al. [20], carried out a numerical modelling and simulation of friction welding processes based on ABAQUS using coupled temperature displacement transient. The temperature results were in good agreement with experimentally measured using thermocouples. Correspondingly, this study utilised similar analysis type in simulating the interface frictional thermal generation.

2.2. Experimental Setup

The laboratory experimental setup is presented in figure 2. The infrared cameras and all sensors were as in the work of Talai, Desai and Heyns [15]. During the forced excitation, first both beams were excited at 20 Hz, secondly, one beam was excited at 20 Hz and the second one at 40 Hz, to facilitate the validation of the FEA results. During the entire excitation period, the thermal images were recorded continuously for 150 s.

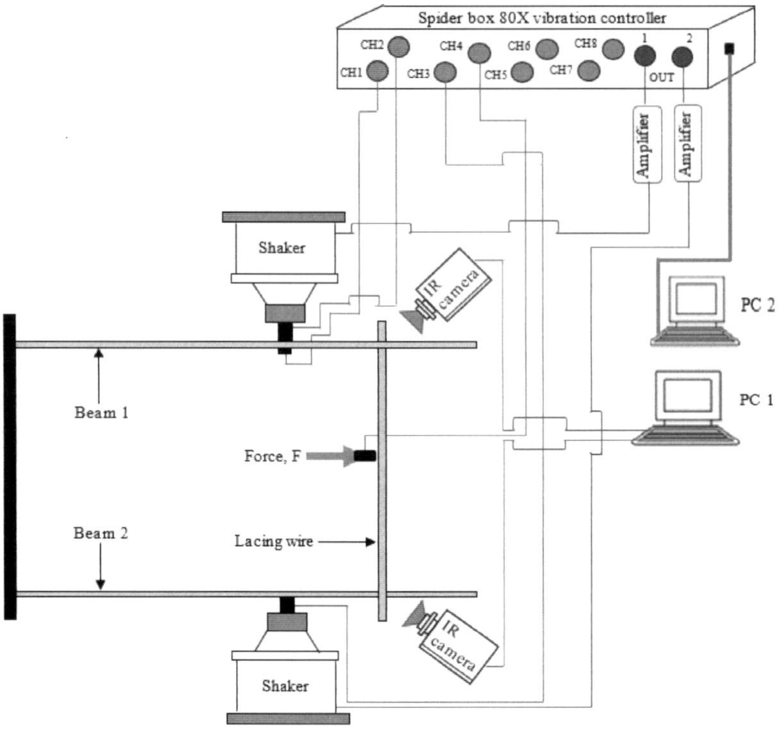

Figure 2. Laboratory experimental setup.

3. RESULTS AND DISCUSSION

A PC with the 64-bit operating system was used for the simulation. It consisted of four processors each with 3.20 GHz and RAM of 32 GB. The total simulation wall clock time was 272 437 s. Since the forcing frequencies were periodic, whether the beams were excited in phase or out of phase, the analysed frequencies will be of the same value regardless of the phase. Based on this knowledge, the study considered out of phase excitations.

Primarily, it is known that frictional heating concentrates within the real area of contact between two bodies in relative motion [21]. In addition, according to the Coulomb law of friction, frictional heat generation occurs only if there exists a relative motion between the interacting structures [22]. This means that the frictional temperature signature is consistent with the displacement pattern. Thus, the structure under higher excitation frequency allows the frictional interface to slide against each other more times; unlike at lower frequencies, hence, leading to greater frictional heat evolution as seen with temperature increase from 0.11 $°C$ (figure 3a) to 0.20 $°C$ (figure 4a) for 20 Hz and 40 Hz, respectively. Moreover, Mabrouki et al. [22] observed similar findings on investigation of the frictional heating model for efficient use of vibrothermography. Hence, the conclusion was that frictional heat increases with the interface frequency. The analyzed frequencies using a MATLAB FFT algorithm are presented in figure 3b and figure 4b. The observed increasing temperature peak and vice versa for displacement with increasing excitation frequency was attributed to the increasing temperature as prior explained and shortening periodic time considering the same amplitude of displacement, respectively. Successfully, the frequencies analyzed from the frictional temperature evolution were in good agreement to that acquired from the corresponding displacement as summarized in table 2.

Table 2. Frequencies for both beams excited at same frequency values.

Forcing frequency for both beams (Hz)	Results	Frequency predicted	
		Temp. signature	Disp. curve
20	Figure 3a	19.99 Hz	19.99 Hz
40	Figure 4b	39.99 Hz	39.99 Hz

In the case of beams excitation at different forcing frequencies of beam 1 at 20 Hz and beam 2 at 40 Hz; the results revealed a frictional temperature increase of 0.24 and 0.32 °C in the former and latter, respectively (figure 5a). Interestingly, the analyzed frictional temperature evolution from incredibly predicted the vibration frequency (figure 5b) just as displacement curves as shown in table 3. Also, the values agreed well with those of corresponding excitations prior obtained for same excitation frequencies (table 2). This implies that the irrespective of the structural excitations, the analysis of the interface frictional heat evolution is capable enough of predicting its vibration frequency as evidenced by these findings.

Table 3. FEA frequencies with different excitation frequencies.

Forcing frequency	Results	Frequency predicted	
		Temp. signature	Disp. curve
Beam 1: 20 Hz	Figure 5b	19.99 Hz	19.99 Hz
Beam 2: 40 Hz		39.97 Hz	39.97 Hz

FEM simulation results (figure 3a and figure 4a) showed that exciting beams at the same value of frequencies and time period, yields equal maximum frictional temperature. However, this was not the case experimentally. The thermal imaging exhibited a temperature difference of 0.9 °C for both beams excited at 20 Hz (figure 6) at the ambient temperature being 22.4 °C. The frequencies evaluated from the thermal imaging (figure 6) post analysis of beam 1 and 2 were 19.9885 Hz (figure 7a) and 19.9973 Hz (figure 7b), respectively. These were in good agreement with the corresponding FEA predicted frequencies (table 2) with relative errors being 0.0075 % and 0.0365 % for the former and latter, respectively, thus, validating the FEM simulation results. Furthermore, beam 1 possessed several smaller frequencies that were associated with the beam multi-dynamics due to the periodic loading.

It is known that the existence of a crack reduces the stiffness of a structural member due to local flexibility [23]. Yang et al. [24] analytically examined the forced vibration of a cracked cantilever beam. The authors found that dynamic deflection increases due to the crack, thus increasing frictional heat evolution as evidenced by the similar contour interface temperature for healthy FE model (figure 8a) and vice versa for a model with an induced defect (figure 8b). This finding implied that a beam induced with a defect can be positively identified from the observation of contact interface thermal variation. Importantly, it is worth pointing out that this method is currently being used for dynamic structural condition monitoring as reported by Bagavathiappan et al. [1]. Even although the frequencies analysed from the frictional temperature evolution and displacement were in agreement for both models (25.00 Hz) as seen in figure 9, significant variations in peaks existed. The frequency peak evaluated from frictional temperature signature increased in the case of beam induced with a defect and vice versa for the healthy one. This may be attributed to the increased frictional temperature due to increased displacement in respect to an induced structural defect, while there was less frictional heat in the case of the healthy beam due to less resulting displacement.

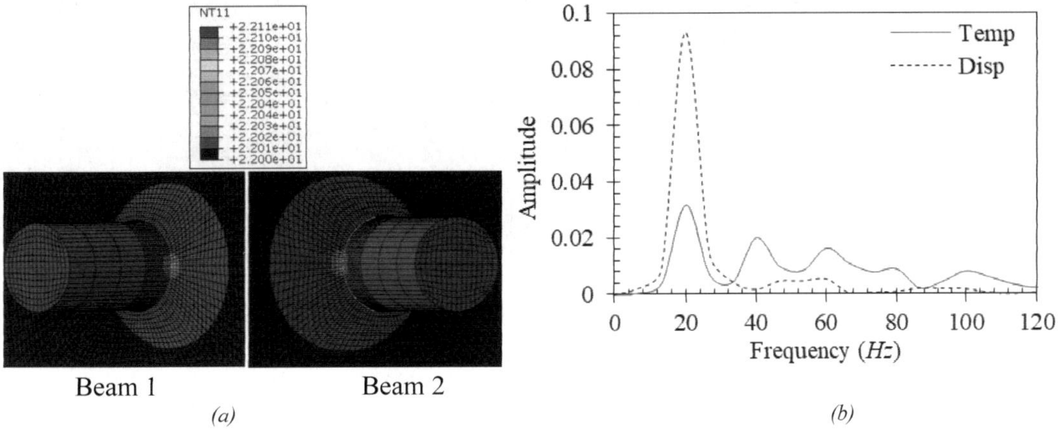

Figure 3. Both beams excited at 20 Hz (a) interface temperature contour after 0.1 s and (b) FFT of temperature evolution and displacement.

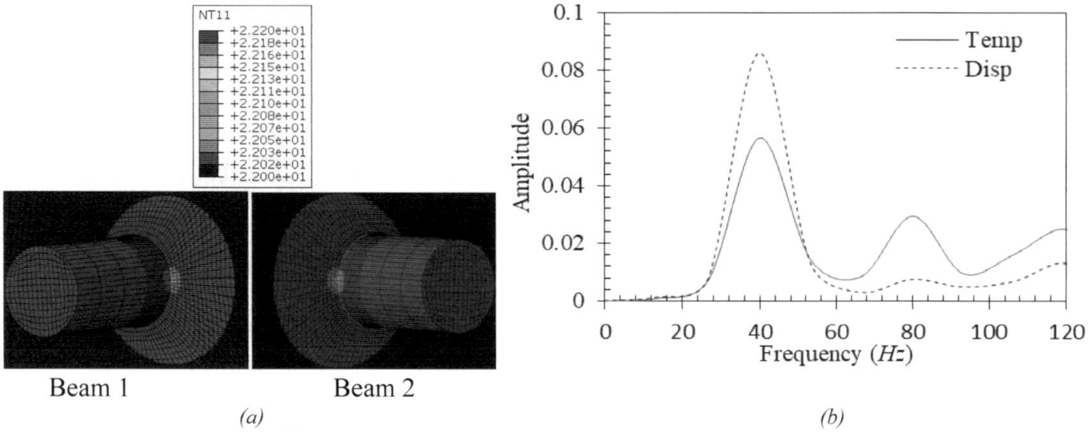

Figure 4. Both beams excited at 40 Hz (a) interface temperature contours after 0.05 s and (b) FFT of temperature evolution and displacement.

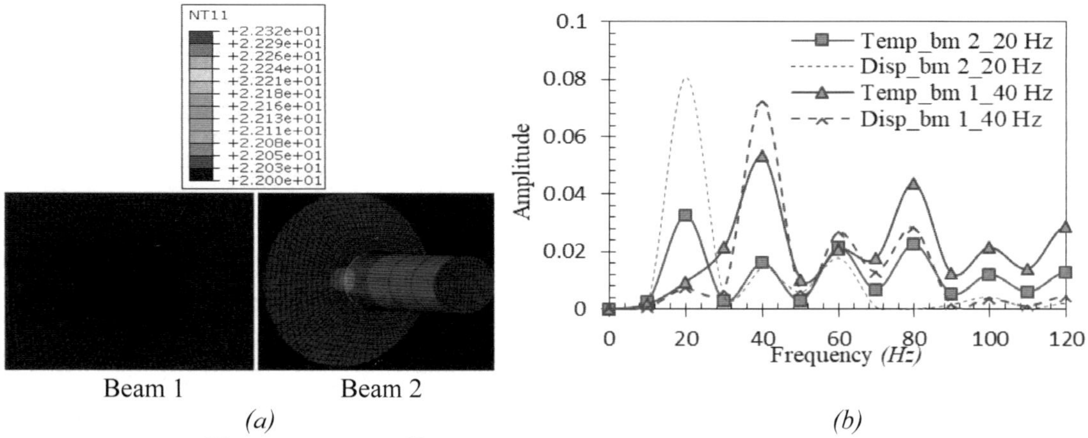

Figure 5. Beam 1 excited at 20 Hz and beam 2 at 40 Hz (a) interface temperature contours after 0.1 s and (b) FFT of temperature evolution and displacement.

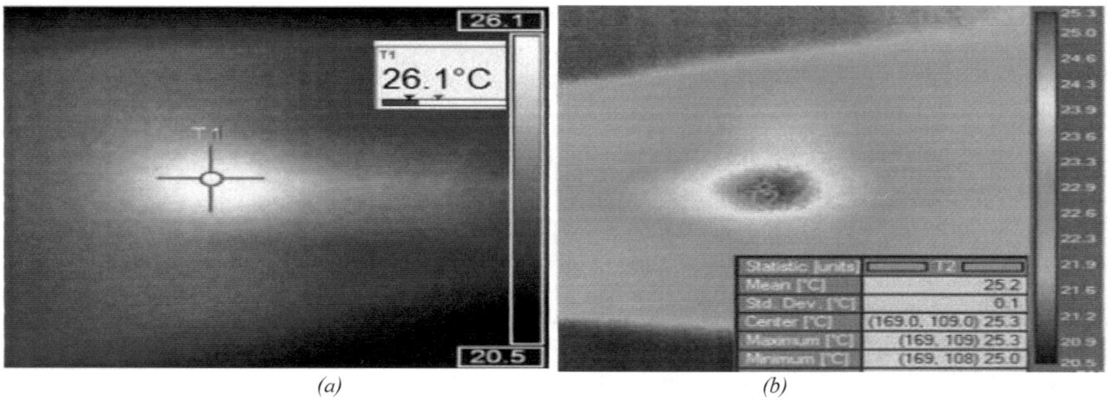

Figure 6. Thermal images for both beams excited at 20 Hz.

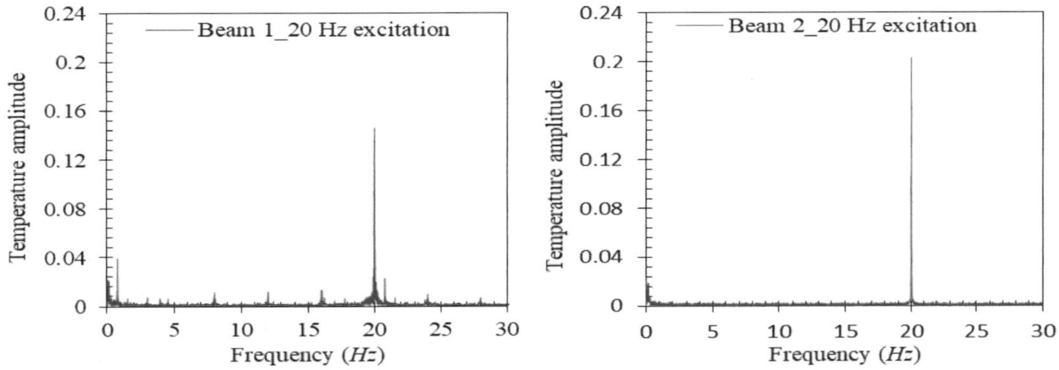

(a) *(b)*

Figure 7. FFT of temperature evolution for both beams excited at 20 ***Hz*** (a) beam 1 and (b) beam 2.

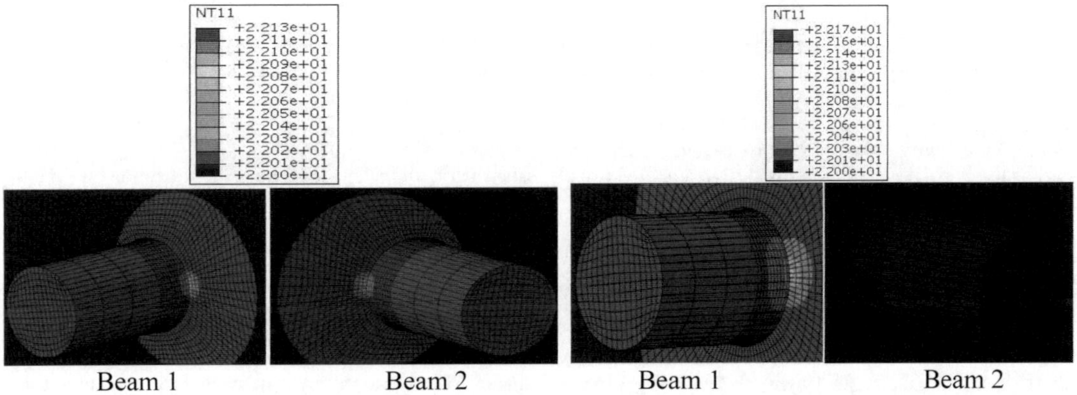

(a) *(b)*

Figure 8. Interface temperature contour for both beams excited at 25 *Hz* (a) Both beams are healthy (b) Beam 1 induced with defect while Beam 2 is healthy.

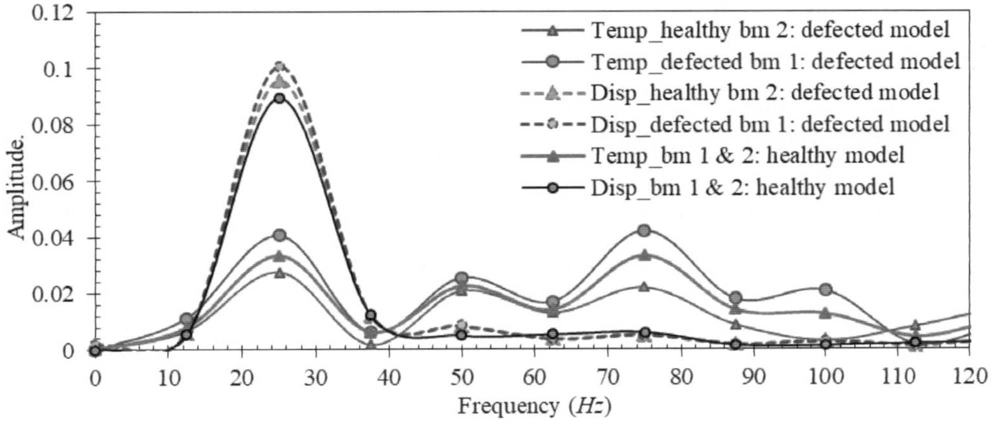

Figure 9. FFT of temperature evolution and displacement for both healthy and defected FE beam models.

4. CONCLUSION

In conclusion, the analysed frictional temperature time-domain wave form obtained from the simulation of 3D FEM formulated confirmed the ability of predicting the structural vibration frequency of mechanical components. Furthermore, the approach developed forms the basis for online monitoring of vibration frequency, hence, effective SHM employing infrared thermography as a dual diagnostic tool.

5. ACKNOWLEDGEMENT

The authors greatly appreciate the support of Eskom Power Plant Engineering Institute (Republic of South Africa), University of Pretoria and Tshwane University of Technology for funding this research.

REFERENCES

[1] Bagavathiappan, S., Lahiri, B. B., Saravanan, T., Philip, J., Jayakumar, T. (2013) Infrared thermography for condition monitoring – A review, Infrared Physics & Technology 60: 35-55.

[2] Lahiri, B. B., Bagavathiappan, S., Soumya, C., Mahendran, V., Pillai, V. P. M., Philip, J., Jayakumar, T. (2014) Infrared thermography based defect detection in ferromagnetic specimens using a low frequency alternating magnetic field, Infrared Physics & Technology 64: 125-133.

[3] Meola, C., Carlomagno, G. M. (2004) Recent advances in the use of infrared thermography, Measurement Science and Technology 15:(9) 27-58.

[4] Titman, D. J. (2001) Applications of thermography in non-destructive testing of structures, NDT&E International 34: 149-154.

[5] Mukhopadhyay , N. K., Chowdhury, S. G., Das, G., Chattoraj, I., Das, S. K., Bhattacharya, D. K. (1998) An investigation of the failure of low pressure steam turbine blades, Engineering failure analysis 5:(3) 181-193.

[6] Wei, F., Pizhong, Q. (2010) Vibration-based Damage Identification Methods: A Review and Comparative Study, Structural Health Monitoring 10:(1) 83-111.

[7] Rao, S. J. (1991) Turbo-machine blade vibration (first edition), New Delhi: New age international publishers 7-20.

[8] Jaiswal, B. L., Bhave, S. K. (1994) Experimental evaluation of damping in a bladed disk model, Journal of sound and vibration 177:(1) 111-120.

[9] Torshizi, M. S. E., Nikravesh, Y. S. M., Jahangiri, A. (2009) Failure analysis of gas turbine generator cooling fan blades, Engineering Failure Analysis 16:(5) 1686-1695.

[10] Oberholster, A. J., Heyns, P. S. (2011) Eulerian laser Doppler vibrometry: Online blade damage identification on a multi-blade test rotor, Mechanical Systems and Signal Processing 25:(1) 344-359.

[11] Castellini, P., Martarelli, M., Tomasini, E. P. (2006) Laser Doppler Vibrometry: Development of advanced solutions answering to technology's needs, Mechanical systems and signal processing 20: 1265-1285.

[12] Brock, N., Hayes, J., Kimbrough, B., Millerd, J., North-Morris, M., Novak, M., Wyan, C. J. (2005) Dynamic Interferometry, *In:* Proceedings of SPIE 5875.

[13] ABAQUS/CAE (2013) Analysis User's Manuel, Dassault Systèmes.

[14] Petreski, Z. (2009) Natural frequencies of a blade group with a lacing wire, Mechanical scientific engineeering 28:(1) 1-5.

[15] Talai, S. M., Desai, D. A., Heyns, P. S. (2016) Vibration characteristics measurement of beam-like structures using infrared thermography, Infrared Physics & Technology 79: 17-24.

[16] Miller, S. F., Shih, A. J. (2007) Thermo-Mechanical Finite Element Modeling of the Friction Drilling Process, Journal of manufacturing science and engineering 129: 531-538.

[17] Papuga, J., Růžička, M., Meggiolaro, M. A., Castro, J. T. P. d., Wu, H. (2015) Computationally-efficient Non-linear kinematic Models to Predict Multiaxial Stress-strain Behavior under Variable Amplitude Loading, *In:* 3rd International Conference on Material and Component Performance under Variable Amplitude Loading, VAL 2015: Procedia Engineering 101: 285-292.

[18] Claeys, M., Sinou, J. J., Lambelin, J. P., Todeschini, R. (2016) Modal interactions due to friction in the nonlinear vibration response of the "Harmony" test structure: Experiments and simulations, Journal of Sound and Vibration 376: 131-148.

[19] Arrazola, P. J., Özel, T. r. (2010) Investigations on the effects of friction modeling in finite element simulation of machining, International Journal of Mechanical Sciences 52:(1) 31-42.

[20] Wenya, L., Shanxiang, S., Feifan, W., Zhihan, Z., Tiejun, M., Jinglong, L. (2012) Numerical simulation of friction welding processes based on ABAQUS environment., Journal of engineering science and technology review 5:(3) 10-19.

[21] Mabrouki, F., Thomas, M., Genest, M., Fahr, A. (2009) Frictional heating model for efficient use of vibrothermography, NDT&E International 42:(5) 345-352.

[22] Simar, A., Bréchet, Y., de Meester, B., Denquin, A., Gallais, A., Pardoen, T. (2012) Integrated modeling of friction stir welding of 6xxx series Al alloys Process, microstructure and properties, Progress in Materials Science 57: 95-187.

[23] Dimarogonas, A. D. (1996) Vibration of cracked structures: A state of the art review, Engineering Fracture Mechanics 55:(5) 831-857.

[24] Yang, J., Chen, Y., Xiang, Y., Jia, X. L. (2008) Free and forced vibration of cracked inhomogeneous beams under an axial force and a moving load, Journal of Sound and Vibration 312:(1-2) 166-181.

Corrosion monitoring with Acoustic Emission of steel embedded in concrete that is subjected to different environmental conditions

T. Sigoba[1], M. Howse[2], Q.I. Sikakana[3]
[1, 3] Department of Non-Destructive Testing and Physics, Faculty of Applied and Computer Sciences, Vaal University of Technology, South Africa
[2] SSA Acoustic & Specialised Inspections, Vereeniging, South Africa

ABSTRACT

The use of steel reinforcement in concrete structures is an established building code requirement for structural integrity. The Acoustic Emission (AE) technique has been used in industry to evaluate corrosion in reinforced concrete since the late 1970's and early 1980's [1,2]. The AE technique is well suited for limited access applications as the waves travel along entire structures.

This study utilized five samples of mild steel imbedded in concrete subjected to different environmental conditions and the AE activity and intensities were recorded and compared. The environmental conditions were to simulate actual locations where concrete structure are built. The five samples consisted of uncoated mild steel bars embedded in concrete, which were then submerged in, drinking water; salt water; salt water (insulated steel); hydrochloric acid; salt water and copper forming a galvanic cell. The samples were kept at ambient temperature and monitored over a seven week period utilizing AE sensors with a resonant frequency of 30 kHz.

The hydrochloric acid sample exhibited the highest cumulative AE energy, followed by the galvanic cell, salt water, drinking water and lastly salt water (insulated steel). The cumulative number of AE activity followed the same trend except for the saltwater and galvanic cell samples obtaining the fewer hits than the drinking water. These findings were confirmed, with the exception of the galvanic cell sample, by visual examination of the surface corrosion present on the samples post concrete removal.

Keywords: accelerated corrosion of steel; acoustic emission; unloaded reinforced concrete; environmental conditions; structural health monitoring

1. INTRODUCTION

1.1 Corrosion of Steel Reinforcement in Concrete

Corrosion effects of steel reinforcement in concrete structures must be considered in the design phase. It is a fact that corrosion of steel in concrete is accelerated when the structures are located in environments with high salt levels or where the temperatures are constantly high with humidity. Active corrosion in steel in concrete is for the most part, not externally visible, and it may take years after initiation of cracks that it surfaces. Concrete acts, as an electrolyte and the reinforcement will develop a potential. Corrosion is an electrochemical process in which electric current flows between the cathodic and anodic (steel) sites through the concrete.

In new concrete, steel is protected from corrosion due to the formation of a protective, passive film on the surface of the embedded steel in the highly alkaline surroundings of hydrated cement (>pH 12.5). Eventually a pH reduction caused by carbonation or by ingress of chloride causes the passive film to degrade, allowing the reinforcement to corrode in the presence of oxygen and moisture. To minimise the risk of corrosion in steel reinforced concrete; the concrete must be of appropriate quality for the intended application and that the depth of cover to the steel reinforcement must meet the ACI 318 code [3]. Proper compaction and curing of the fresh concrete is essential, so as to have dense concrete with low permeability. Porous concrete on the other hand, has high permeability increasing the risk of corrosion. The cement mixing ratios are therefore crucial, particularly the water-to-cement ratio.

It is well documented that corrosion of steel reinforcement in concrete on existing structures can be assessed by different methods such as mentioned in the review by Ha-Wong Song et al. [4].

1.2 Acoustic Emission Monitoring of Reinforced Steel Bars in Concrete

Acoustic Emission (AE) is the elastic energy released from materials which are undergoing deformation. The rapid release of elastic energy, the AE event, propagates through the structure to arrive at where a transducer is mounted. By analysis of the resultant waveform in terms of feature data such as amplitude, energy and time of arrival, the severity and location of the AE source can be assessed.

One study of the applicability of AE technique for detecting corrosion of reinforced-steel bar in concrete was performed by Zongjin et al. [5]. Their results demonstrated that there is a clear relationship between the AE events and reinforced-steel bar corrosion in concrete.

A comparative study of the rate of corrosion of five samples of mild steel bars embedded in concrete were done. The analysis of the characteristics of the AE waveforms emanating from the steel-concrete interface where the chemical reactions for corrosion takes place are presented. The purpose of this research project is thus to advance the knowledge of AE as an evaluation tool for static structures subjected to different environmental conditions in the absence of dynamic stresses (unloaded). This study also contributes to determining steel reinforced concrete structural durability and its deterioration [6-8]. Figure 1, is a standard three stage graph depicting the deterioration process due to salt attack [9].

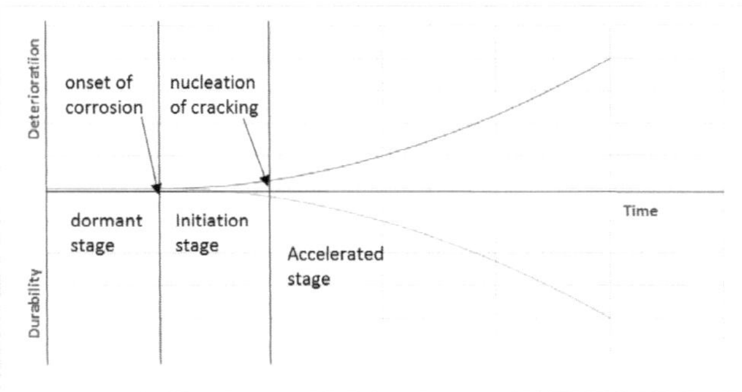

Figure 1: Deterioration / Durability process of steel reinforced concrete due to corrosion

Application of AE in concrete has its limitation, for the following reasons [10]:

a) Each application has a unique set of parameters, resulting in different results influenced by sample size, type of transducer, path of signal, etc.
b) Strength of the signal is dependent on the energy content of the source.
c) Susceptibility to noise, that is, external noise sources may skew results.

Nonetheless, AE from unloaded steel reinforced concrete has its own 'signature' as compared to those from other materials reinforced concrete, like pre-stress and post-tension reinforced concrete, stainless steel in concrete, epoxy-coated steel in concrete, galvanizes steel reinforcement , fibre, etc. One such case are the amplitudes of the events from steel reinforced concrete tend to be larger than those without steel. Thus, this study of inspection for corrosion in steel reinforced concrete subjected to different environmental conditions using AE aims to optimise the procedures.

The source location of the acoustic signal was not the objective of this study, since only one sensor per sample was used.

1.3 The operational basics of an AE system

The typical operational setup for an AE system consists of one to several AE sensors, computer-based data acquisition devices and a couplant.

The role of the AE sensor is acting as a receiver to convert detected dynamic displacements or sound waves into electric signals. There are two categories of sensors usually applied in the AE testing: resonant and broadband sensors. Resonance type sensors, which are usually applied in the material with high attenuation like concrete, gain higher sensitivity but lower frequency range than the broadband sensors.

The choice of couplant is crucial to the test sensitivity. The couplant layer should be even and it is essential to get rid of bubbles and make couplant layer as thin as possible to guarantee good acoustic transmission.

1.4 The AE signal feature descriptions

In the AE testing, count, amplitude, duration, rise time, and energy are the five most frequently used parameters. Other parameters like threshold, RA value, frequency, to name a few are also widely employed in crack classification [11,12]. These parameters are shown in figure 2.

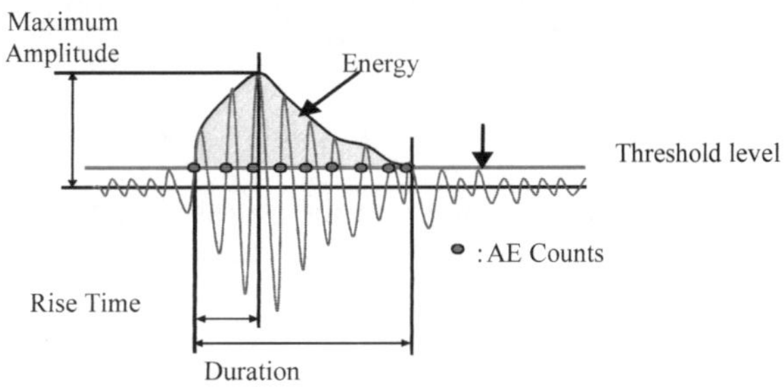

Figure 2. Acoustic Emission parameters for a hit

Although in this experiment the samples were not subjected to loading, it is noteworthy that the RA value; defined to be the ratio of rise time and amplitude, is now a standard value used to identify the classification of fractures [13,14]. For example, a decreasing RA value suggests the tendency of tensile fracture [15].

The features utilised as defined by the AE acquisition system's user manual [16] were:

- Hit: the detection and measurement of an AE Signal on a Channel
- ASL (Average Signal Level): measure of the average amplitude signal received on a channel
- Absolute Energy: this is derived from the integral of the squared voltage signal divided by the reference resistance (10 kΩ) over the duration of the AE waveform packet. Measured in Attojoules. Its time driven dataset is calculated as follows;

$$Absolute\ Energy\ [aJ] = \sum \left(\frac{(Sensor\ Voltage)^2 \times 10^{-6} sec}{10\ k\Omega} \times \frac{1\ Joule}{WattSec} \right) \tag{1}$$

2. THE PENCIL LEAD BREAK TEST VERIFICATION

The purpose of verification is twofold. First, it ensures that the transducers are in good acoustic contact with the part being monitored. Generally, the lead breaks should register amplitudes of at least 90 dB for a reference voltage of 1 mV and a total system gain of 90 dB. Second it checks the accuracy of the source location setup. This last purpose involves indirectly determining the actual value of the acoustic wave speed for the object being monitored.

In this experiment, source location determination could not be used due to only one sensor per sample utilised. Thus, the pencil lead breaks were conducted by breaking 0.5 mm 2H lead approximately 25 mm on the steel bars from each sensor; exclusively to verify coupling sensitivity. Pencil / lead breaks conducted confirmed an amplitude response of 90 dB ± 2 dB.

3. SAMPLE PREPARATION

Ratios of concrete mix used (in kilograms) were, Cement: aggregate: sand = 1 : 3 : 3. A proportionate amount of water was added to this concrete mix which is medium to high strength concrete for civil structures. When cured after 28-days it has a compressive strength of about 3000 psi. (21 MPa).

Tins were used for setting the concrete mix blocks. The 200 mm long mild steel bars were placed vertically in the middle of each concrete block. All five set-ups samples were exposed to the sun for three days, to allow the concrete to partially dry. Holes were drilled in each set-up sample to reach the mild steel bar so as to accelerate the corrosion rate and to simulate water ingress into the concrete structure. The partially dry concrete blocks were removed from the tins, and allowed to completely dry. The concrete blocks were then each placed in plastic containers and numbered according to the different conditions they were to be subjected to as described in table 1. 22 days was allowed for further curing of the concrete, as this produces noise because of material friction during shrinkage. Data acquisition was thereafter initiated.

TABLE 1: Sample Description

Sample Number	Sample Type	Composition	Description
1	Salt water mixture (control)	35g salt in 1000 ml of water	Insulated steel embedded in concrete and submerged in salt water; concentration in 500 ml
2	Tap (drinking) water	500 ml of tap water	Non insulated steel embedded in concrete and submerged in water
3	Salt water mixture (brine)	35g salt in 1000 ml of water	Non insulated steel embedded in concrete and submerged in salt water
4	Hydrochloric Acid (HCl)	500 ml of commercial hydrochloric acid	Container is placed in a well-ventilated area. Non insulated steel embedded in concrete and submerged in acid (hydrochloric acid)
5	Galvanic cell	500 ml of salt saturated tap water with a copper wire inserted vertically	Non insulated steel embedded in concrete and submerged in salt water And copper wire (galvanic cell)

Figure 3 shows a cross-sectional view of a steel bar embedded in the concrete block with a transducer attached.

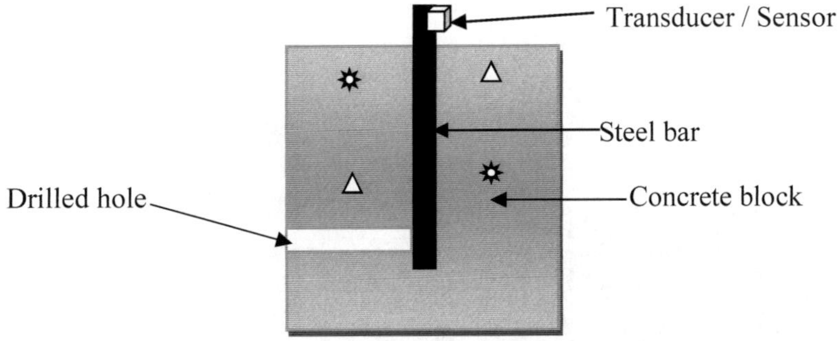

Figure 3: Schematic diagram of a steel bar in concrete with transducer attached

Table 2 summarises the chemical element composition intensities of the mild steel bars. Three runs for intensity counts per second (cps) were done and the average relative standard deviation (RSD) calculated. The RSD showed big variances for some elements, implying that dilution for the analysed sample was not consistent. Nonetheless, the elements identified were carbon, zinc, iron, cobalt, nickel, copper, aluminium, chromium, vanadium, manganese and lead. These chemical elements are consistent with the family of mild structural steels.

TABLE 2: Chemical element composition of mild steel bars

Intensities report

9/14/2017 11:03:55 AM

LabBook Steel samples scan.imexp

Sample Mild steel [redacted]

Analysis Date 9/14/2017 10:51:50 AM

Amount Final Quantity Dilution Factor 1

	C 247.856 {136} (Axial)	Zn 213.856 {457} (Axial)	Fe 259.940 {130} (Axial)	Co 228.616 {447} (Axial)	Ni 231.604 {445} (Axial)	Cu 324.754 {104} (Axial)
Intensity average	159,396 cps	31,308 cps	1,560,415 cps	5,533 cps	15,032 cps	72,423 cps
Intensity per Run 1	19,952 cps	31,501 cps	1,911,039 cps	5,554 cps	15,098 cps	2,330 cps
Intensity per Run 2		31,427 cps	1,168,109 cps	5,549 cps	15,096 cps	
Intensity per Run 3	298,840 cps	30,996 cps	1,602,098 cps	5,497 cps	14,901 cps	142,516 cps
Intensity RSD	123.7 %	0.9 %	23.9 %	0.6 %	0.8 %	136.9 %

	Al 396.152 {85} (Axial)	Cr 283.563 {119} (Axial)	V 310.230 {109} (Axial)	Mn 257.610 {131} (Axial)	Pb 216.999 {455} (Axial)
Intensity average	1,452,133 cps	66,696 cps	5,605 cps	1,050,116 cps	3,049 cps
Intensity per Run 1	90,637 cps	4,290 cps	1,218 cps	181,106 cps	3,064 cps
Intensity per Run 2			7,334 cps		3,049 cps
Intensity per Run 3	2,813,629 cps	129,102 cps	8,263 cps	1,919,126 cps	3,033 cps
Intensity RSD	132.6 %	132.3 %	68.3 %	117.0 %	0.5 %

4. EXPERIMENTAL SET_UP

Figure 4 is a picture of the five samples already connected to the AE system.

Figure 4: Set-up of the Five Test Samples

The system set-up – including the five samples – has the acquisition system (Mistras Group micro express 8 system) connected in-line with the 20 dB pre-amplifiers, which are connected to the Mistras R3I (30kHz resonant frequency) sensors. Couplant was applied to each of the sensors and attached to the samples.

The 20 dB pre amplifier were then switched on and data acquired on all five samples for between 45 minutes to 24 hours and then normalised to 1 hour for comparison and trend analysis.

Table 3 shows the Software Acquisition Settings for the experiment. Sample 4's threshold was increased to 44 dB due to data rate saturation during acquisition.

TABLE 3: AEwin for Express 8 v5.70 Software Acquisition Setup

Sample Number	**Threshold [dB]**	**Analogue Frequency Filter [kHz]**	**Sampling Rate [k Samples per second]**	**Peak Definition Time [µs]**	**Hit Definition Time [µs]**	**Hit Lockout Time [µs]**
1,2,3,5	40	20 - 100	10000	2500	5000	5000
4	44	20 - 100	10000	2500	5000	5000

5. RESULTS AND DISCUSSIONS

5.1 Acoustic Emission Data

The corrosion process occurs at a steady rate and hence, extraneous noise - such as mechanical bangs from nearby activities – was identifiable and the associated time removed before analysis of the data.

The normalised data was trended in figures 5a and 5b. It was observed that, for all five samples, the absolute energy remained relatively constant during the course of the experiment (Figure 5a), while the number of hits decreased with time (Figure 5b).

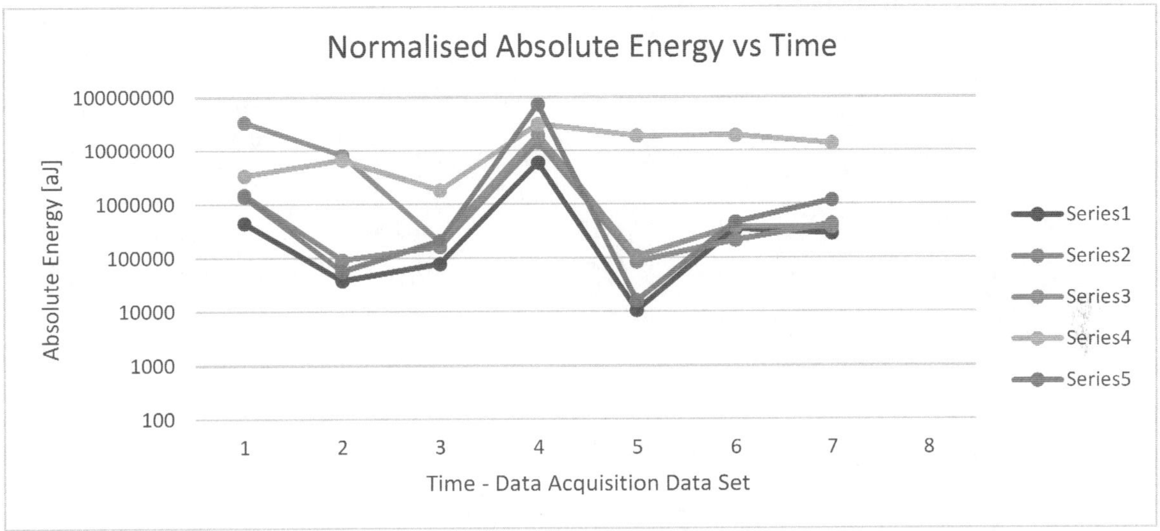

Figure 5a: Normalised Absolute Energy vs Time

The decreasing rate of hit with time was attributed to the build-up of by-products on the mild steel, and thereby reducing the effective surface area available for the reactions, hence impeding the reaction rate.

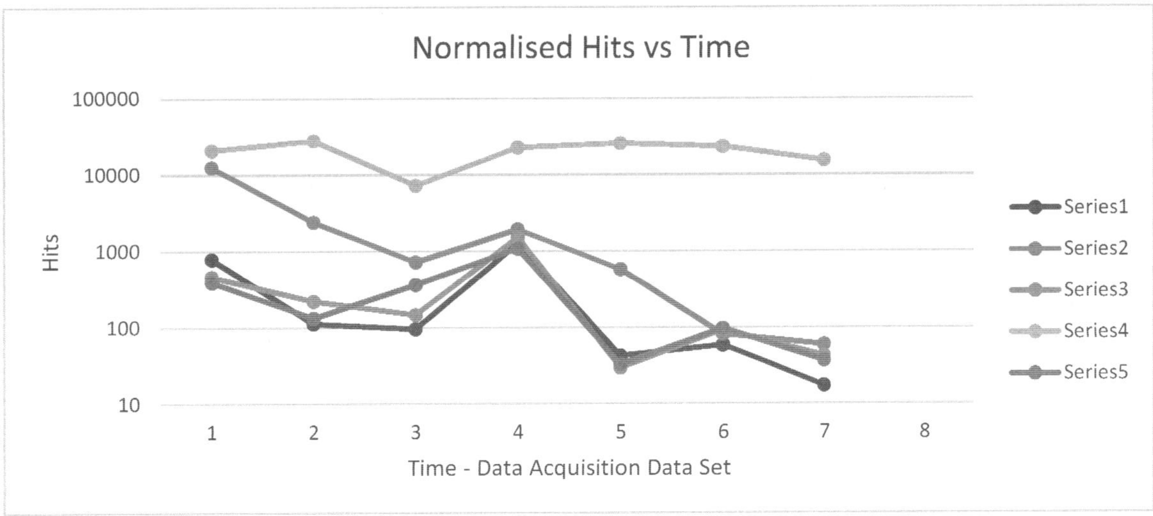

Figure 5b: Normalised Hits vs Time

The cumulative hit and absolute energy per sample set-up are captured in the graph of figure 6.

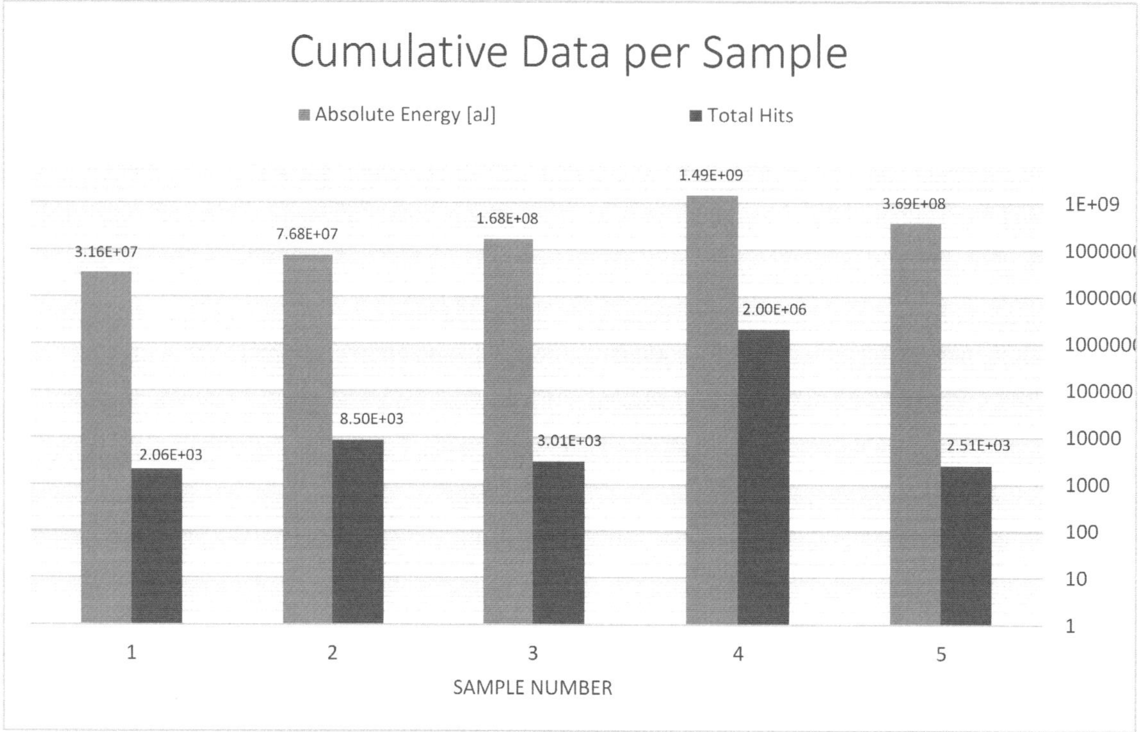

Figure 6: Cumulative Hit & Energy Data per Sample

The graph extracted numerical cumulative hits and absolute energies for each sample set-up are tabulated in descending order in tables 4a & 4b, respectively.

TABLE 4a: Sample Cumulative Hits

Sample	Hits
4 (HCL)	2.00E+06
2 (Water)	8.50E+03
3 (Brine)	3.01E+03
5 (Galvanic Cell)	2.51E+03
1 (Control)	2.06E+03

TABLE 4b: Sample Cumulative Energy

Sample	Absolute Energy
4 (HCL)	1.49E+09
5 (Galvanic Cell)	3.69E+08
3 (Brine)	1.68E+08
2 (Water)	7.68E+07
1 (Control)	3.16E+07

The cumulative energy (Table 4b) behaved as anticipated for the five samples and correlated with visual observations conducted at the end of the experiment. The hydrochloric acid sample had the greatest energy release, followed by the galvanic cell sample, then the brine sample, then the drinking water sample and finally the insulated control sample.

The cumulative number of hits, however, revealed an alternative order of decreasing activity as follows: hydrochloric acid, drinking water, brine, galvanic cell, insulated control. The drinking water sample and galvanic cell sample exchanged positions. To explain this behaviour, it is postulated that the brine sample and galvanic cell sample produced a residue build-up which slowed down the hit rate but not the overall energy released.

5.2 Visual Observations

The samples steel bars were removed from the concrete 6 weeks post-termination of the data acquisition phase. Each sample steel bar was cleaned via a nail brush to reveal any surface damage. The visual observations are described in table 5 below, with the photographs of the corresponding steel bars shown in figure 7.

TABLE 5: Visual Findings Summary

Sample	Condition of Material Embedded in Concrete	Condition of Material at Concrete to Air Interface & Above
1 (Control)	Minor corrosion in areas where paint failure occurred	Minor corrosion in areas where paint failure occurred
2 (Water)	Localised pits near drilled hole locations	Light general pitting at concrete to air interface
3 (Brine)	Deep pit (+-40% loss) at drilled hole location (bottom right of photograph) Highly localised damage in areas	Minor corrosion at concrete interface Localised damage in a number of areas
4 (HCL)	Light general scaling present	Concentrated pitting and scaling present from interface to top of sample
5 (Galvanic Cell)	Minor localised pits near drilled hole locations	Light general pitting at concrete to air interface below residue layer

Figure 7: Varied corrosion of sample steel bars post-termination of data acquisition

Comparison between the AE data and visual observation results of the condition of the sample steel bars is summarised and ordered as follows:

Area of general corrosion (greatest to least) : Sample 4, Sample 2, Sample 3, Sample 5, Sample 1
Corrosion pits (deepest to shallow) : Sample 3, Sample 4, Sample 2, Sample 5, Sample 1
Overall Condition (worst to best) : Sample 4, Sample 3, Sample 2, Sample 5, Sample 1

6. CONCLUSION

The AE system was utilised to detect the rate of corrosion / material loss on five samples of mild steel bars embedded in concrete.

The hydrochloric acid sample (Sample 4) had the most extensive corrosion damage, followed by the brine sample (Sample 3), then the drinking water sample (Sample 2), then the galvanic cell (Sample 5) with the insulated steel (Sample 1) only having minor damage where paint failure occurred.

There was a correlation between the visually observed areas of corroded surface with the total number of AE hits from the sample. This correlates with the fact that the larger the surface area, the higher the chemical reaction rate which is then observed as an increase in the number of AE hits / activity. Thus, the rate of corrosion activity is reflected in the amount of AE activity (number of hits) emanating from a sample.

The cumulative absolute AE energy released from the samples coincides with observations on the number of deep pits on the samples. The only exception being sample 5 (galvanic cell) which did not exhibit particularly deep pits but possessed the second highest energy released. This could be due to the energy release during the formation of salt crystals on the outer surface of the sample during the experiment.

The correlation between energy detected and severity of pitting on the samples may be explained by the greater energy released during the breaking of bonds deeper into the lattice structure.

The decrease in AE hit rate during the experiment coupled with the absolute AE energy remaining relatively constant points towards each source event (spalling of corrosion) releasing more energy as time passes. This phenomenon is often observed in the industry where "older" corrosion with deeper pits produce higher amplitude hits (and higher energy) than "fresh" corrosion which usually only exhibit minor surface pitting.

Noting that the areas of corrosion in this study were within 200 mm of the sensors; future experimentation to determine the 'effective' AE monitoring distance for optimal corrosion detection of reinforcement steel in concrete is envisaged.

Acknowledgement

The authors hereby acknowledge SSA Acoustic & Specialised Inspections for the use of their equipment and premises for the duration of this project, particularly, Gavin McCabe (MD) and the assistance given by Bradley Fagan (AE level 1).

REFERENCES

[1] McCabe, R., Koerner, R.M. and Lord, A.E., "Acoustic Emission Behavior of Concrete Laboratory Specimens." ACI J. (July 1976).

[2] Li, L. and Poorooshasb, H.B., "Characteristics of Acoustic Emission from Reinforced Concrete," Progress in Acoustic Emission III, The Jap. Soc. of NDI, 522-528 (1986).

[3] ACI Committee 318, Building Code Requirements for Structural Concrete, AC 318-02, American Concrete Institute, Farming Hills, Michigan, 2002.

[4] Song H.W., and Saraswathy V., "Corrosion Monitoring of Reinforced Concrete Structures - A Review," Int. J. Electrochem. Sci. **2**, 1-28 (2007).

[5] Zongjin Li., Faming Li., Zdunek A., Landis E. and Shah S., "Application of acoustic emission technique to detection of rebar corrosion in concrete," ACI Mater J. **95**, 68–76 (1998).

[6] Goszczyńska, F.B., "Analysis of the process of crack initiation and concrete with acoustic emission testing," Archives of Civil and Mechanical Engineering **14**, 134-143 (2014).

[7] Goszczyńska, B., Świt, G., Trąmpczyński, W., Krampikowska, A., Tworzewska, J. and Tworzewski, P., "Experimental verification of the acoustic emission (AE) method for the cracking process determination and location in concrete elements," in: IABSE Rotterdam Conference Report **99**, 150–151 (2013).

[8] Goszczyńska, B., Świt, G., Trąmpczyński, W., Bacharz, K., Godowska, M. and Krampikowska, A., "Identification of acoustic emission signals in unloaded concrete," in: Proceedings 58th Annual Conference on Scientific Problems of Civil Engineering, Krynica – Rzeszów, 202–203 (2012).

[9] JSCE (2001) Standard Specification for concrete and concrete structures on maintenance, JSCE, Tokyo.

[10] Maji, A.K., and Sahu, R., "Acoustic Emissions from Reinforced Concrete," Experimental Mechanics **34**, 379 – 388 (December 1994).

[11] Kawasaki, Y., Wakuda, T., Kobarai, T. and Ohtsu, M., "Corrosion mechanisms in reinforced concrete by acoustic emission," Construction and Building Materials **48**, 1240–1247 (2013).

[12] Pollock, A.A., "Acoustic Emission Inspection." Metals Handbook **17**, 278- 294 (1989).

[13] Berkovits, A. and Fang, D., "Study of fatigue crack characteristics by acoustic emission." Engineering Fracture Mechanics **51**(3), 401-409 (1995).

[14] Grosse, C., Reinhardt, H. and Dahm, T., "Localization and classification of fracture types in concrete with quantitative acoustic emission measurement techniques," NDT & E International **30**(4), 223-230 (1997).

[15] Soulioti, D., Barkoula, N.M., Paipetis, A., Matikas, T.E., Shiotani, T. and Aggelis, D.G., "Acoustic emission behavior of steel fibre reinforced concrete under bending," Construction and Building Materials **23**(12), 3532-3536 (2009).

[16] Express-8 AE System User manual Rev 0, Mistras Group Incorporated (2014).

Variations in vibration responses of an ice-going vessel during wave slamming

C.M van Zijl[1], A. Bekker[1]

[1]Sound and Vibration Research Group, Department of Mechanical and Mechatronic Engineering, Stellenbosch University, South Africa

ABSTRACT

A polar supply and research vessel was instrumented with an accelerometer network to determine full-scale dynamic responses of the operational ship structure. The vessel in question is pre-disposed to problematic stern slamming. In a recent voyage to the Marginal Ice Zone, the vessel encountered extremely rough weather, including storm in a field of brash ice. During this voyage wave slamming occurred and ice debris was flung onto deck spaces. This work investigates the variation in vibration and modal responses, which result from harsh operational conditions. The aim is to examine the feasibility of modal tracking towards potential long-term structural health monitoring and inverse force estimation. Three case studies are identified from operational data. The vibration responses are subjected to conventional time history analyses and newly developed slamming response analyses. LMS PolyMax automatic parameter selection is used to identify the changes in modal frequencies under stormy conditions. These values are reflected against results from a finite element analysis and earlier measurements in a harbor environment. In the investigated cases, the fundamental modal frequencies varied by 0.1 Hz in different operational conditions. It is found that rotational speed of the propulsion shaft corresponds to the modal frequency of the first bending mode. As such, this mode is not ideal for modal tracking because of harmonic contamination. The excitation of higher order bending modes could be an indicator of wave slamming, as these modes are not significantly excited in moderate seas.

Keywords: Operational modal analysis, ice-going vessel, modal tracking

1. INTRODUCTION

Antarctic research institutes and their scientists have relied on polar supply and research vessels to supply bases and serve as floating laboratories. The need to predict tipping points in climatology is adding momentum to polar and oceanographic research in attempts to increase our understanding of the sparsely explored oceans (Moedas, Pandor and Kassab, 2017).

South Africa's polar supply and research vessel, the SA Agulhas II, is a state-of-the-art ship, which has been designed to meet the harsh operational requirements of the stormy Southern Ocean and icy conditions of Antarctica. The full-scale ship responses from this first-in-class vessel are highly relevant to influence the design of future research vessels. As a result a consortium was established to conduct full scale measurements on-board the SAA II to create a legacy of data to enhance the scientific basis for ice-going ships (Bekker *et al.*, 2018).

The particular focus of the present work includes an investigation into the potential use of accelerometer network to track variable vessel responses as a result of varying environmental conditions or long term (20 years) modal tracing to assess decaying structural integrity. Modal tracking is enabled through operational modal analysis (OMA), which is an output only modal analysis technique. In OMA, structural responses to ambient excitation (by ice and water) during the normal operating conditions are used to achieve dynamic deformation of the structure (Brincker and Ventura, 2015). This deformation is influenced by engineering design parameters (such as mass, stiffness and damping properties) and boundary conditions that constrain and load the structure (ice, water, vessel draft, temperature).

A further application of modal tracking includes the potential of inverse force estimation, which could indicate novel methods of estimating ship ice resistance or wave slamming forces (Soal, 2018). Outside the ambit of engineering, vibration signatures could potentially be indicative of specific environmental conditions. In the context of research vessels such phenomena could enable the automated detection and sampling of ice thickness or wave height (Sharkh *et al.*, 2014), which would add localized environmental observations towards the increased fidelity and interpretation of satellite images.

2. METHODS

2.1. Vessel and voyage

The SAA II can carry up to 100 passengers in 46 cabins and is operated by a crew of 45 members. She is equipped with two shaft-lines with four-bladed variable pitch propellers, which are powered electric motors. The diesel generators comprise 6-cylinders, which are run at a constant speed of 750 rpm. The general specifications of the SAA II are reported in Table 1.

Table 1. Specifications of the SA Agulhas II.

Gross tonnage	12 897 tons	Diesel engine type	6L32
Length	143 m	Electric motor type	N3 HXC 1120 LL8
Breadth	22 m	Speed at MCR*	140 rpm
Classification	DNV	Power at MCR*	4500 kW
Class notation	1A1 PC-5/ICE-10	Nominal torque	307 k(Sharkh *et al.*, 2014)Nm
Yard / Year	STX Finland /2012	Propeller maker	Rolls Royce
Main engine maker	Wärtsilä	No. of blades / Diameter	4 / 4.3 m

* MCR – Mean Continuous Rating

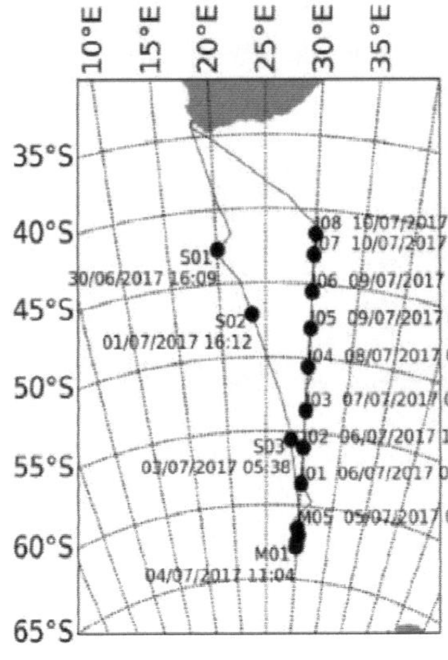

Figure 1 A map showing the voyage route of the SAA II on her Winter Cruise in July 2017. The selected case studies are indicated: Case 1 (Blue), Case 2 (Red) and Case 3 (Green).

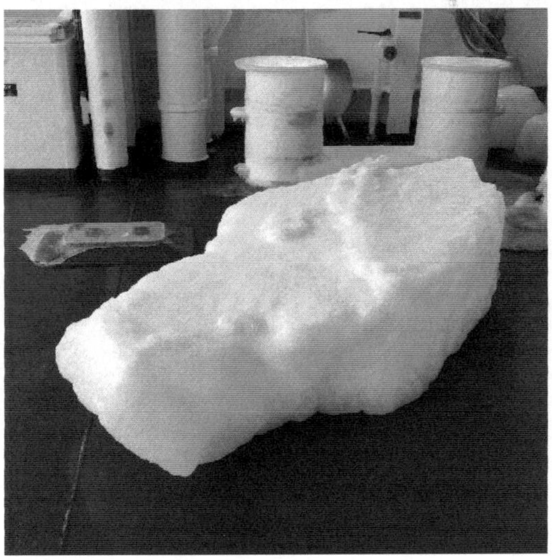

Figure 2. The violent wave action during a storm in brash ice, caused ice debris to be flung onto the deck.

The SAA II is prone to stern slamming, which has led to passenger complaints and equipment damage on board (Omer and Bekker, 2018). Slamming is the exposure of a vessel structure to wave impacts, which give rise to large forces for a short duration of time (Kapsenberg, 2011) caused by water entry. The lightly damped nature of many vessel structures result in a global vibratory response, known as whipping, which follows a slamming event (Dessi, 2014). Concurrent with the findings of Mouton (1991) and Kapsenberg *et al.* (2011), the slamming-prone dynamics of the SAA II are attributed to the flat, extended transom design which originated from the desire to create additional deck space at the stern. Owing to the sparse availability of full-scale wave slamming data, the long-term effects of excessive slamming on the structure of the SAA II are yet unknown.Full-scale measurements were performed on a designated scientific Winter Cruise of SAA II in July 2017 as shown in the voyage track in Figure 1. The focus of scientific operations was set on investigations of the ice and waves in the Marginal Ice Zone (MIZ). This voyage is of particular interest as wave slamming was encountered in open water as well as the MIZ.

The SAA II departed from Cape Town harbour on 28 June 2017 with an ambitious scientific agenda, which suffered owing to the encounter of several storms and motion sickness of the newly sea-faring scientists. In order to regain lost time the vessel speed was increased to 17 knots by switching the engine into "ice mode" on 1 July. The drop keel was raised to reduce resistance. During the next two days, the operations on board mostly consisted in underway sampling and navigation in stormy waters. Weather forecasts indicated a substantial frontal system approaching the vessel in the area where sea ice was expected on 4 July. At 4:45am GMT on 4 July, the first brash sea ice was encountered. The vessel entered a large field of pancake ice with floes smaller than 3 m, and an average thickness of 40 to 60 cm. Long waves with estimated periods of more than 10 s were persistently high and were experienced by the ship which was proceeding at a sustained speed of 16 to 17 knots. The progress of scientific stations was hindered by several large waves and slams that interrupted the work. The ship continued to head NW heaving into the swell in wind conditions exceeding 20 m/s and finally reached stormy open water. Snow was falling heavily. Wind speed reached 30 m/s with stronger gusts. In search of sea ice, the vessel course was changed in a SE direction. After the change in heading, the pancake floes increased in size. The ice surface detected by satellite images indicated 100%, although the encountered ice was not consolidated, with estimated significant wave height of up to 7 m. The vessel voyaged North and ice conditions changed rapidly to a more mobile ice surfaces.

The stormy conditions decreased in the morning of 5 July, which enabled work and inspection of the deck. Several large floes were washed on deck as shown in Figure 2. An ice impact had damaged a mast on the ship's bow, which was taken down. Calmer weather enabled scientific work to resume. Another storm was encountered on 7 July. The vessel speed was reduced and scientific stations were again interrupted by rough weather conditions, which persisted until 8 July. The weather improved on 9 July enabling the voyage to proceed as planned. The vessel finally returned to Cape Town harbour on 13 July.

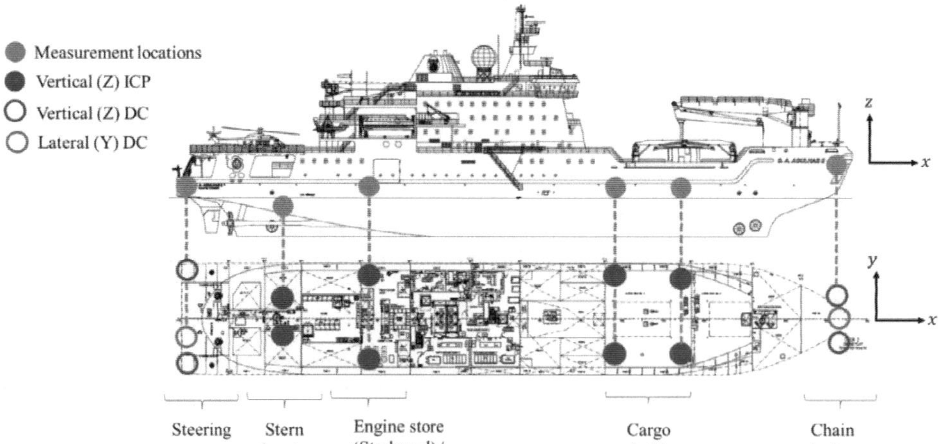

Figure 3. A diagram showing the measurement locations of the accelerometer network on the SAA II.

2.2. Full-scale measurements

Table 2. Sensor layout and location, types of sensors, cable routing, visual observations.

Point reference	Channel	Axis	Deck	Side	Description	Type	Sensitivity [mV/m.s-^{2}]
1	9	+Z	4	Port	Bow chain locker	DC	20.00
2	10	+Y	4	Middle	Bow chain locker	DC	20.39
3	11	+Z	4	Starboard	Bow chain locker	DC	20.36
4	14	-Z	3	Port	Cargo hold fore	ICP	9.96
5	15	-Z	3	Starboard	Cargo hold fore	ICP	10.05
6	16	-Z	3	Port	Cargo hold aft	ICP	10.25
7	17	-Z	3	Starboard	Cargo hold aft	ICP	10.31
8	18	-Z	2	Port	Fresh water room	ICP	10.44
9	19	-Z	2	Starboard	Engine store	ICP	10.36
10	20	+Z	2	Port	Stern thruster	ICP	10.43
11	21	+Z	2	Starboard	Stern thruster	ICP	10.72
12	26	+X	2	Starboard	Steering gear aft	DC	20.19
13	27	+Y	2	Starboard	Steering gear aft	DC	20.22
14	28	+Z	2	Starboard	Steering gear aft	DC	20.13
15	29/25	+Z	2	Port	Steering gear aft	DC	20.25
16	5	+Y	4	Starboard	Aft stairwell	ICP	10.60
17	6	+Z	4	Starboard	Aft stairwell	ICP	10.18
18	7	+Y	7	Starboard	Passenger accommodation	ICP	10.01
19	8	+Z	7	Starboard	Passenger accommodation	ICP	10.56
20	12	+Y	8	Starboard	Chef's quarters	ICP	10.03
21	13	+Z	8	Starboard	Chef's quarters	ICP	10.20
22	3	+Y	9	Starboard	Bridge	DC	20.29
23	4	+Z	9	Starboard	Bridge	DC	20.46
24	22	+Z	9	Port	Bridge	ICP	10.10

Measurements from 24 accelerometers were utilized in the present analysis. The hull instrumentation comprised six pairs of accelerometers on starboard and portside locations for vertical acceleration measurements and two lateral sensors, one on the starboard side at the stern and one on the vessel centre line in the bow chain locker (see Figure 3). The remaining sensors monitored accelerations in the super-structure, closer to accommodation areas, recreational spaces and the bridge. The sensor network includes both DC and ICP accelerometers as summarized in Table 2. Data was acquired using an LMS SCADAS master-slave system and LMS.TurbineTesting acquisition software. Acceleration was sampled at 2048 Hz.

2.3. Case studies

Three case studies were selected for analysis from the operational measurements as detailed in Table 3. Each case comprised 30 minutes of data. With vibration monitoring and modal tracking in mind, the cases were selected to highlight the differences in operational conditions that may be experienced from calmer to extreme conditions. Case 1 and Case 2 reflect instances where slamming was reported in the voyage notes in water close to 0 °C. Case 1 represents data from an open water storm, whereas Case 2 reports on the rare occurrence of a storm with slamming in brash ice. Case 3 represents ship responses as experienced in some of the calmest conditions experienced on the voyage in warmer water. Operational conditions were recorded from the ship logbook. The Beaufort scale is an empirical measure, which relates wind speed to observed conditions at sea on a scale of zero (Calm) to 12 (Hurricane force).

Table 3 Operational conditions during the three selected case studies.

	● Case 1	● Case 2	● Case 3
Date	7 July 2017	5 July 2017	12 July 2017
Time (UTC)	03:00 – 3:30 AM	01:00 – 1:30 AM	08:00 – 8:30 AM
Conditions	Storm (Open water)	Storm (MIZ)	Calm water
Ship true heading [°]	0°	322 °	307 °
Swell height [m] & direction	3.5 WNW	6.0 WNW	2.0 SW
Beaufort scale	8	7	5
Ice conditions [-]	None (open water)	Brash ice (floes < 3m)	None (open water)
Hull temperature [C]	0.6 °C	-1.8 °C	18 °C
Ship speed	27.8 – 29.0	11.4 – 15.0	13.9 – 25.6

3. RESULTS

3.1. Acceleration time histories

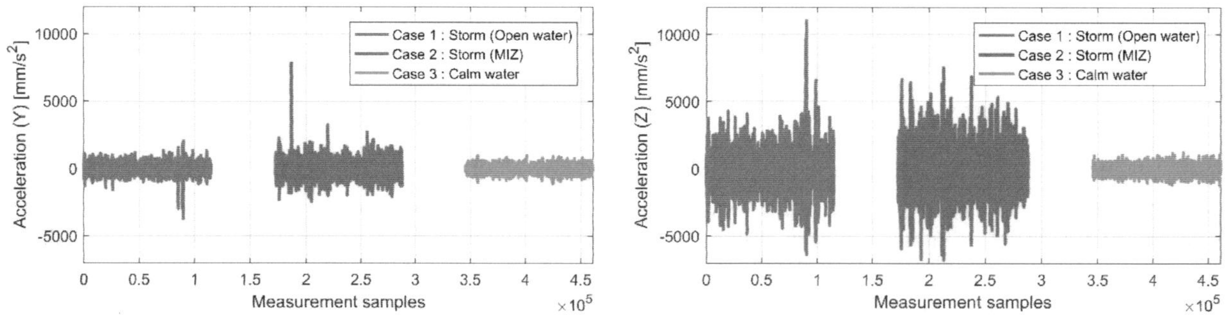

(a) Lateral acceleration. (b) Vertical acceleration.

Figure 4. Acceleration time histories as measured at the bow and stern for three case studies.

Figure 4 presents the de-trended and high-pass filtered time-series of acceleration measurements in the lateral (Y) and vertical (Z) directions from sensors at the bow and stern during the three respective cases. Levels of vibration are significantly higher in the vertical direction than the lateral direction for the two storm cases while vibration levels in the two directions are similar for the calm water case. The time-histories of the two storm cases include several impulsive events followed by decaying vibration amplitudes. Impulsive events followed by decaying vibrations are characteristic of wave slamming, followed by whipping of the global vessel structure. The frequency and magnitude of impulsive events more frequently appear in the MIZ storm case (Case 2). The extreme peak in the lateral acceleration measurements (Figure 4a) is believed to be attributed to impact of the hull with floating ice debris. The maximum peak vibration occurs during the open water storm (Case 1). The time history for the calm water case appears stationary with small impulsive events, which are not prominent compared to the general vibration floor. Some focus is afforded to the global distribution of peak metrics for the fourteen hull sensors to shed light on the possible effects of wave slamming (see Figure 5). Peak acceleration values are lower in the mid-ship owing to the stiffening presence of the superstructure as seen in Figure 5b. In Figure 5c there is a clear trend when considering the maximum crest factor, which shows that the most impulsive acceleration was excited during the open water storm.

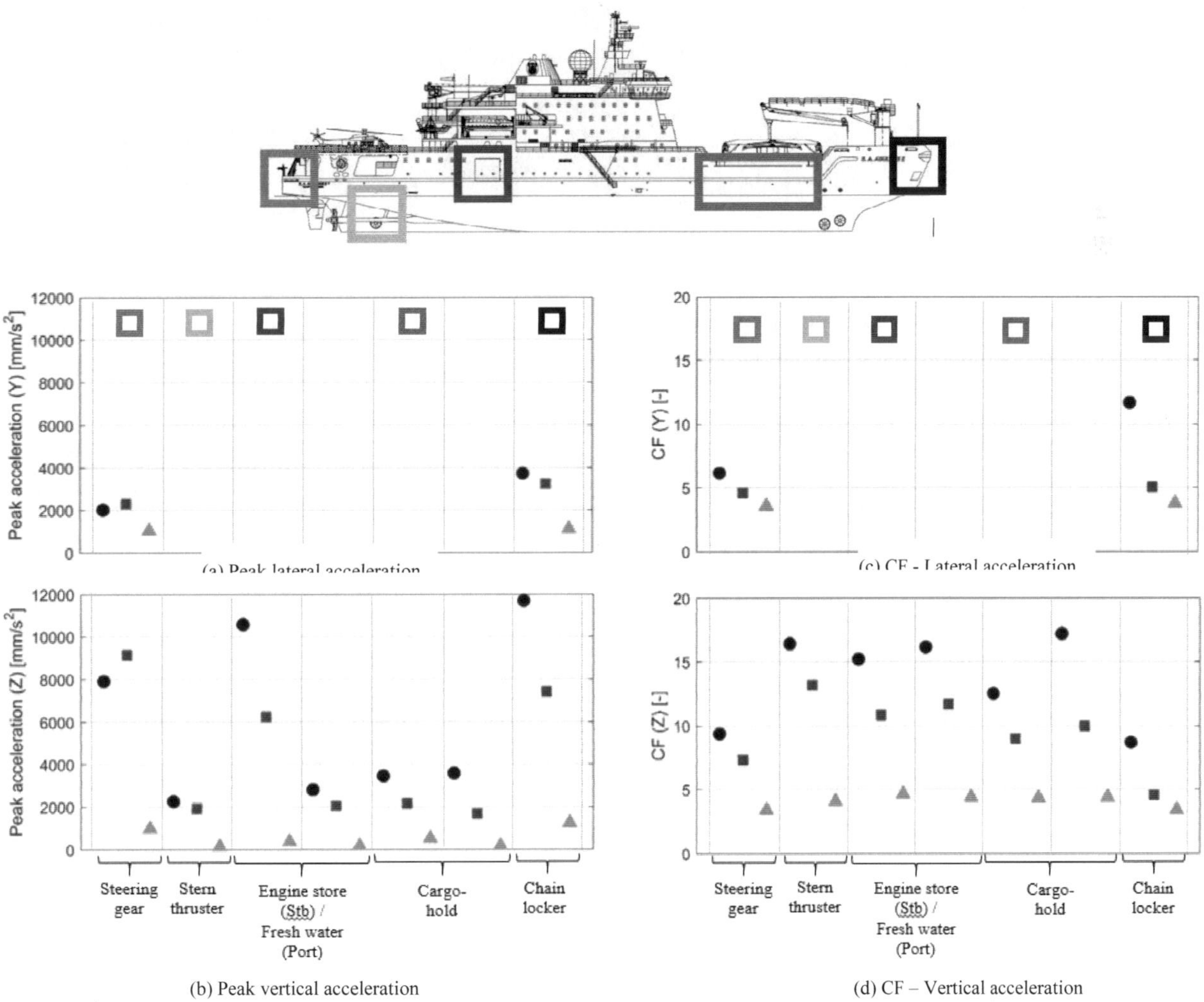

Figure 5 A diagram showing the global distribution of vibration peak values and crest factors throughout the hull structure (● Case 1, ■ Case 2, ▲ Case 3).

3.2. Slamming analysis

An in-house slamming detection algorithm (Bekker, Zijl and Saunders, 2018) was used to extract impulsive peaks caused by unpredictable, non-linear wave slamming. The slamming location was decided based on the sensor on which the peak occurred first (Omer and Bekker, 2016). The number of detected slams as well as significant slams (defined to have CF > 6 (Bekker, Zijl and Saunders, 2018)) and their impact sites are shown in Figure 6. For Case 1 the vessel is cruising at full speed into 3.5 m waves. Structural excitation is dominated by bow slams, with more slams recorded at the starboard bow shoulder, although waves are approaching from the port side. Slamming for Case 2 is most frequent in the stern when the vessel is cruising at medium speed, almost directly into 6 m swell. Small slams frequently occur all around the vessel in oblique moderate seas, with no significant slams recorded.

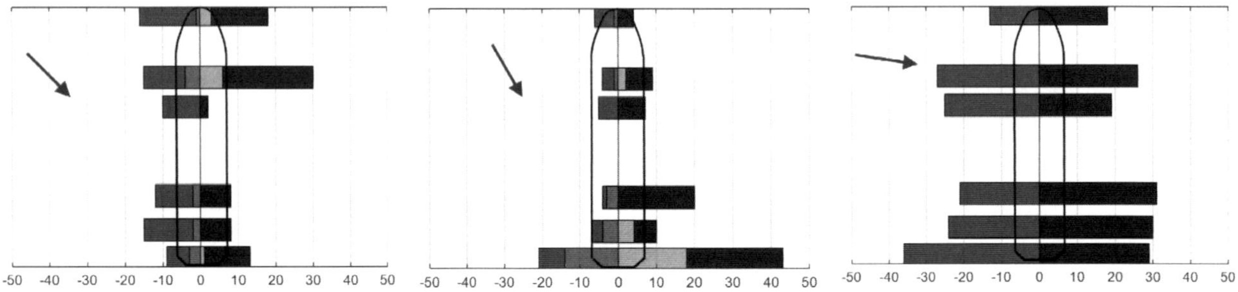

(a) Case 1 (b) Case 2 (c) Case 3

Figure 6 A summary of wave slamming impacts sites as recorded for the different case studies. The ship outline (top view) is indicated as well as the incident wave angle (blue arrow). The detected slamming counts are indicated by ■ on the starboard side and ■ on the port side. Larger slams with a crest factor > 6 are indicated by ■ on the starboard side and ■ on the port side.

3.3. Power spectral densities

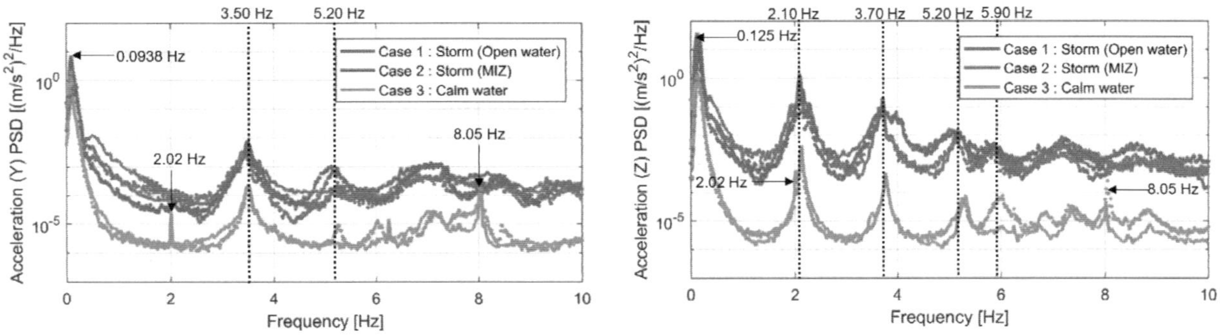

(a) Lateral acceleration. (b) Vertical acceleration.

Figure 7 Auto-power spectral densities of the acceleration signals in the bow (solid line) and stern (dashed line) for the case studies.

Auto-power spectral densities (ASPDs) of the un-filtered DC acceleration measurements at the bow and stern are presented for the three cases in Figure 7. Low frequency peaks in the ASPDs of all three cases occur in the lateral direction and vertical direction at 0.094 Hz and 0.125 Hz, respectively. These are thought to represent the rigid body motion of the vessel, which is also the most significant for Case 2, with the greatest significant wave height. Sharp, narrow frequency peaks are discernible in the ASPDs of the calm water case at 2.02 Hz and 8.05 Hz. These peaks are likely harmonic excitation from the first and fourth blade pass frequency of the four-bladed propeller. The propeller mostly operates at a speed of 140 rpm (2.33 Hz) and the pitch is adjusted to achieve different levels of thrust. Higher levels of ambient vibration during the storm cases mask propulsion harmonics in the respective APSDs. Two possible modes occurring at 3.50 Hz and 5.20 Hz may be identified from the ASPDs of the lateral direction measurements. The peak at 3.50 Hz is present in measurements from both the stern and bow and may represent a lateral bending mode. The ASPDs of measurements in the vertical direction present four peaks of interest: 2.10 Hz,

3.70 Hz, 5.20 Hz and 5.90 Hz. The peak at 5.90 Hz is present only in measurements recorded at the stern of the vessel, while the other three peaks of interest are visible in the APSDs of bow and stern measurements.

3.4. Operational modal analysis

LMS Test.Lab 11A Operational PolyMax was used to perform an operational modal analysis on decimated data (128 Hz) from all 24 acceleration measurements on the SAA II. Cross-powers were calculated with 32,768 NFFT points and a 50% overlap exponential window was applied, resulting in a frequency resolution of 0.0390 Hz. The vertical direction accelerometer in the engine store room (Point 19) was selected as the reference sensor since this resulted in the clearest stabilization diagrams. The automatic pole selection add-in of Operational PolyMax was used to select poles from the stabilization diagrams. Selected modes were validated using the MAC criteria (Pastor, Binda and Harčarik, 2012) as shown in Figure 8. The identified modal frequencies for the case studies are shown in Table 4.

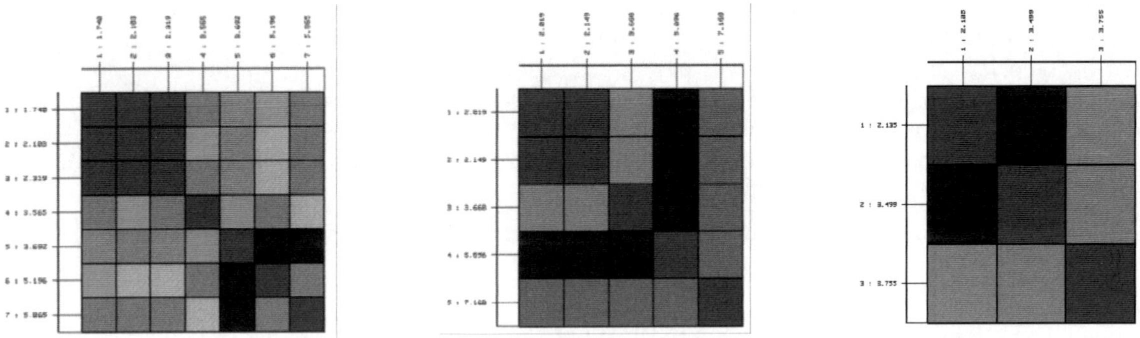

(a) Case 1 – Storm (Open water). (b) Case 2 – Storm (MIZ). (c) Case 3 – Calm water.

Figure 8 A comparison of the MAC matrices for the three case studies according to LMS PolyMAX automatic parameter selection.

Table 4. A comparison of the modal frequencies identified through operational modal analysis compared to earlier investigations.

Mode description	STX Finland (2010)	Harbour (K. Soal, Bienert and Bekker, 2015)	Case 1	Case 2	Case 3
First vertical bending	2.09	1.94	2.10	2.15	2.14
First lateral bending	3.19		3.57	[redacted]	3.50
Second vertical bending	3.76	3.37	3.69	3.67	3.76
Second torsional	4.93		-	-	-
Third vertical bending	5.37	4.72	5.20	5.01	[redacted]
Combined torsion and bending	6.52		-	-	-
Second lateral bending	6.93		-	-	-

4. DISCUSSION

The SAA II encountered stormy conditions in open water as well as the MIZ, which has resulted in the recording of unique acceleration measurements of wave slamming on a ship structure in open water and in ice. Wave slamming in open water and in ice causes increased excitation of the vessel modes, but also raises the general vibration level. This raised vibration floor challenges the detection of wave slamming occurrences. The first bending mode dominates the flexural response of the vessel. The use of this mode for modal tracking is challenged as the modal frequency coincides with the harmonic excitation from shaft rotation. In general the identified modal frequencies in the present investigation are higher, than those determined by Soal et al. (2015) in earlier harbor measurements. The determined frequencies do however correlate well with the finite element analysis by STX Finland (2010). Modal parameter variations could be attributed to multiple factors including vessel speed, water temperature, and ship draft, to name a few. At 5.2 Hz, wave slamming excites significant responses in lateral acceleration measurements at the stern. The automated modal identification by LMS Polymax does not succeed in the identification of the first lateral bending mode for Case 2 nor the third vertical bending mode for Case 3.

5. CONCLUSIONS & RECOMMENDATION

The rotational speed of the propulsion shaft corresponds to the modal frequency of the first bending mode. This mode is not ideal for modal tracking as a result of harmonic contamination. The monitoring of lateral motion in the stern could be an indicator of wave slamming, especially at frequencies around 5.2 Hz. The meaningful interpretation of modal tracking on a polar supply and research vessel requires high quality data of environmental conditions (wave conditions, ice conditions, temperature) as well as the ship state (draft, ship speed, shaft speed). Automated modal tracking will require algorithms that ensure robust modal identification. The use of the MACXP criterion (Vacher, 2010) has been successfully used by Soal (Soal, 2018) to clear up spurious modes for modal tracking of a polar vessel.

ACKNOWLEDGEMENTS

Funding through the National Research Foundation (NRF) SANAP programme is thankfully recognized.

REFERENCES

[1] Bekker, A. *et al.* (2018) 'From data to insight for a polar supply and research vessel', *Ship Technology Research*, pp. 1–17. doi: 10.1080/09377255.2018.1464241.

[2] Bekker, A., Zijl, C. M. Van and Saunders, C. F. W. (2018) 'The detection of wave slamming from vibration measurements on a polar supply and research vessel', in *Submitted to the 14th International Conference on Vibration Engineering and Technology of Machinery*. Lisbon, Portugal.

[3] Brincker, R. and Ventura, C. (2015) *Introduction to Operational Modal Analysis*. First Edit. John Wiley & Sons Ltd. doi: 10.1002/9781118535141.

[4] Dessi, D. (2014) 'Whipping-based criterion for the identification of slamming events', *International Journal of Naval Architecture and Ocean Engineering*, 6(4), pp. 1082–1095. doi: 10.2478/ijnaoe-2013-0232.

[5] Finland), (STX (2010) *FINNSAP finite element analysis of the PSRV NB1369*. Rauma, Finland.

[6] Kapsenberg, G. K. *et al.* (2002) 'Whipping loads due to aft body slamming', in *24th Symposium on Naval Hydrodynamics*. Fukuoka, Japan.

[7] Kapsenberg, G. K. (2011) 'Slamming of ships: where are we now?', *Philosophical Transactions of the Royal Society A: Mathematical, Physical and Engineering Sciences*, 369(1947), pp. 2892–2919. doi: 10.1098/rsta.2011.0118.

[8] Moedas, C., Pandor, N. and Kassab, G. (2017) *Belem Statement on Atlantic Research Innovation Cooperation, 13 July*. Lisbon: Atlantic Ocean Research Alliance. Available at: https://ec.europa.eu/research/iscp/pdf/belem_statement_2017_en.pdf (Accessed: 21 July 2017).

[9] Moton, C. J. (1991) *Open-water resistance and seakeeping characteristics of ships with ice-breaking bows*. Annapolis, Maryland. Available at: http://www.dtic.mil/dtic/tr/fulltext/u2/a245643.pdf (Accessed: 18 March 2018).

[10] Omer, H. and Bekker, A. (2016) 'Detection of wave slamming sites from ship deflections', *Research and Development Journal of South Africa*, 32, pp. 50–57.

[11] Omer, H. and Bekker, A. (2018) 'Human responses to wave slamming vibration on a polar supply and research vessel', *Applied Ergonomics*. Elsevier Ltd, 67, pp. 71–82. doi: 10.1016/j.apergo.2017.09.008.

[12] Pastor, M., Binda, M. and Harčarik, T. (2012) 'Modal Assurance Criterion', *Procedia Engineering*, 48, pp. 543–548. doi: 10.1016/j.proeng.2012.09.551.

[13] Sharkh, S. M. *et al.* (2014) 'A Novel Kalman Filter Based Technique for Calculating the Time History of Vertical Displacement of a Boat from Measured Acceleration', *Marine Engineering Frontiers*, 2(August). Available at: www.seipub.org/mef.

[14] Soal, K. (2018) *System identification and modal tracking of ship structures*. Stellenbosch University.

[15] Soal, K., Bienert, J. and Bekker, A. (2015) 'Operational modal analysis on the polar supply and research vessel the S.A. Agulhas II', in *6th International Operational Modal Analysis Conference, IOMAC 2015*.

[16] Vacher, P. (2010) 'Extensions of the MAC Criterion to Complex Modes.', in *International Conference on Noise 330 and Vibration Engineering (ISMA)* .

Remote Monitoring of Wind Turbine Blades based on High-speed Photogrammetry

Miaoshuo Li1, Xianghong Wang2, Fulong Liu1, Xiaoli Tang1, Fengshou Gu1 & Andrew D. Ball1
[1]School of Computing and Engineering, University of Huddersfield, Queens gate, Huddersfield, HD1 3DH, UK
[2]Schoole of Automobile and Mechanical Engineering, Changsha University of Science and Technology, 410004, China

ABSTRACT

The wind turbine energy industry has been growing rapidly in recent years. Due to extreme operating environments, condition monitoring of wind turbine blades is a challenging task and received extensive studies. The conventional methods based on contact transducers such as ultrasonic or fibre-optical strain gauge are expensive and complex to install and the measurement range is also limited to discrete points. To overcome this issue, this paper presents a new way of condition monitoring for the wind turbine blades based on a high-speed photogrammetry in combination with a scheme a non-contact excitation. The research progress is reviewed regarding to the condition monitoring techniques of wind turbine blades and development on photogrammetry. A photogrammetry system is built to identify the vibration modes of wind turbine blades without mounting any sensors on blades. An air impulse excitation system that consists of a solenoid valve and high-pressure air from air compressor is used to produce the artificial impulse excitation. A high-speed industrial camera captures the image and measures the dynamic displacement of the blades. The results obtained by photogrammetry is accurate and reliable, compared with that of conventional accelerometer-based measurements. Moreover, the faults caused by snow or ice covering are simulated by changing mass of blades, likewise fault of loosening the screws was also tested. The modal parameters extracted from baseline and faulty conditions show significant distinctions, which has affirmed that this vision method is effective and sensitive to blade faults.

Keywords: condition monitoring, wind turbine blades, photogrammetry, modal analysis, high-speed camera.

Corresponding author: Fengshou Gu(F.Gu@hud.ac.uk)

1. INTRODUCTION

Due to the harsh service environment and complex working conditions, faults of wind turbines are inevitable and frequent, which will lead to long downtimes and high maintenance costs [1]. As the core component of the wind turbine to capture wind energy, the blades are often operating under a high stress and the blade faults always cause secondary accident [2]. Therefore, detecting faults of wind turbine blades timely are significant for improving operational reliability and reducing maintenance costs [3].

At present, the testing methods for the condition monitoring of wind turbine blades [4-6] mainly include stress method, acoustic emission method, impedance method, ultrasonic method, vibration method, and infrared detection method. Apart from the infrared detection method, the other methods require to install many sensors that is necessary to provide information at limited discrete points. However, the complex dynamic response of complex blades cannot be easily measured using a limited set of sensors. Placing sensors on a utility-scale wind turbine blade is labor intensive, affects the normal operation of the blades, and can also introduce electrical noise to the measured signal due to the extensive wiring. Infrared detection is a type of non-contact detection, with features such as a large area and remote detection. However, its detection results are easily affected by environments such as external temperature, and the identification and qualitative analysis of defects are still difficult to be achieved.

Due to the limitations of these methods, the researchers had used photogrammetry technology [7, 8] to detect the blade state. The technology combines the advantages of the vibration method and the visual method to measure the vibration of the object in a non-contact method. Corten et al. [9] used photogrammetry technology for the first time to measure the vibration of a wind turbine with a diameter of 10m. The measurement error was 0.043% of the visual field range. Johnson et al. [10] developed a point projection stereography system consisting of a digital projector and a binocular camera. The system was used to collect surface vibrations during the static and fatigue tests of the blades and captured submillimeter-level deformation and torsion. Baqersad et al. [11] and Lundstrom et al. [12] conducted dynamic experimental studies on small wind turbine blades and identified their modal shapes. Ozbek et al. [13] applied the three-dimensional point tracking to a 2.5 MW wind turbine and the modal parameters of the rotating utility-scale wind turbine blade were identified using stereo-photogrammetric measurement.

The above studies are static, or fatigue results conducted in a laboratory environment. Moreover, these studies mainly focused on the measurement of blade deformation and vibration, and did not further analyze the correspondence between deformation, vibration and faults such as blade cracks and unbalanced failures.

To make more advancement of this promising technology, the research of condition monitoring for a 2 kW wind turbine blades was carried out based on the high-speed photogrammetry in this paper. Firstly, a marked point tracking method is put forward to detect the motion of markers on the blade. Secondly, the effectiveness of photogrammetry inspection on vibration is verified by comparing with the results of the acceleration detection. Thirdly, the experiment and results of faults identification is acquired. Finally, the conclusions are given. This current paper paves the way for developing an optical condition monitoring system that can be used to perform full-sized structural evaluation of a real-world wind turbine blades.

2. METHODOLOGY

The measurement of vibration displacement of wind turbine blades using photogrammetry technology includes the following key steps: 1) preprocessing, including strobe filter, which removes the interference noise generated by white-bright lamps, and image preprocessing, which is mainly used to enhance the brightness and contrast of the image and realize image inversion; 2) measurement of vibration displacement, mainly recognizing the feature points and obtaining the actual vibration displacement; and 3) modal identification using a stochastic subspace method. The specific theories are detailed in this section in order to provide a basis for experimental studies.

1) Preprocessing

The traditional fluorescent lamp is directly used for 50 Hz alternating current, and its strobe frequency is 100 Hz. When the digital camera collecting, the water ripple will appear due to strobe frequency of current of lamp. To get the stable lightning condition in the laboratory, a filter of the video is implemented to cut out most fluorescent ambient light.

At the same time, to highlight the feature of point on the blades, the invert and adjustment of brightness and contrast of images is implemented. In this study, the brightness and contrast are adjusted by

$$g(i,j) = \alpha \cdot f(i,j) + \beta \tag{1}$$

where $f(i,j)$ is the pixel of source image point (i, j), and $g(i,j)$ is the adjusted pixel $I(i, j)$ indicate that the pixel is in row i and column j. α and β controls the contrast and brightness of image, respectively.

2) Measurement of vibration displacement

The first step of measuring vibration is to track the feature points on the wind turbine blades. Because Harris corner detector has the features of strong invariance to rotation, scale, illumination variation and image noise [14], the Harris corner detector is adopted to recognize the feature points in the wind turbine blades and get a good performance. This algorithm measures the local gray level changes of an image with patches shifted by a small amount in different directions based on a local auto-correlation function of the image.

Given the pixel gray level as $f(i,j)$, the gray level intensity changes $E_{x,y}(u,v)$ of each pixel *(u, v)* of the image moves a very small (*x, y*) can be described as follows:

$$\begin{aligned} E_{x,y}(u,v) &= \sum_{x,y} W_{u,v} * (f(u+x, v+y) - f(u,v))^2 \\ &= \sum_{x,y} W_{u,v} * [f_x u + f_y y + o(u^2, v^2)]^2 \\ &\cong [x,y](\sum_{x,y} W_{u,v} * \begin{bmatrix} f_x^2 & f_x f_y \\ f_x f_y & f_y^2 \end{bmatrix}) \begin{bmatrix} x \\ y \end{bmatrix} \end{aligned} \tag{2}$$

where W is the low pass filtering by a Gaussian function, and $W_{u,v}$ is the coefficient that Gauss filter in location *(u, v)*. $f_x = \frac{\partial f}{\partial x}$, $f_y = \frac{\partial f}{\partial y}$ is the respective horizontal and vertical derivative of image $f(x,y)$, respectively. $*$ denotes convolution.

Denoting $M = \sum_{x,y} W_{u,v} * \begin{bmatrix} f_x^2 & f_x f_y \\ f_x f_y & f_y^2 \end{bmatrix}$, so Eq. (2) can be described as

$$E_{x,y}(u,v) = [x \quad y] M \begin{bmatrix} x \\ y \end{bmatrix} \tag{3}$$

The eigenvalues of M can describe the extreme curvature of the image gray level autocorrelation function. And a corner point can be judged according to the size of eigenvalues. Therefore, a corner response function (CRF). R is defined to determine whether the pixel *(u, v)* is a corner point by the size of R. R is described as:

$$R(x,y) = \det[M] - k(\text{trace}\, M)^2 \tag{4}$$

where $\det M = \lambda_1 \lambda_2, \text{trace}\, M = \lambda_1 + \lambda_2$, λ_1 and λ_2 are the eigenvalues of M. $k = 0.04 \sim 0.06$ is an empirical constant.

Therefore, the pixel displacement of feature points on the blades can be obtained. Then the homographic transformation [15] is used to obtain the vibration displacement of each feature point.

3) Modal identification

Since the input of air impulse excitations is hardly measured precisely. A type of output-only modal identification is adopted. Specifically, the covariance-driven stochastic subspace identification (Cov-SSI) algorithm [16], which can be implemented on line due to less computational cost, was applied to the multi-point vibration responses and then the modal parameters, including natural frequencies, damping ratio and mode shapes will be identified, which is crucial to distinguish healthy and faulty conditions.

3. EXPERIMENT

3.1. Experimental Rig

In this research, an experimental rig consists of a set of blades from a 2kW wind turbine and an air excitation system. The model of wind turbine is iSTA Breeze I-2000, it has 3 composite blades with length of 107 cm and weight 650 gr. To obtain vibration modal properties (frequency, damping ratio and shape) of these lightweight blades in laboratory, the blades is fixed on a frame shown in Figure 1 and excited by air pulses. In this way, the measurement was carried out which are more similar to the real operating conditions. In addition, 7 black markers are made on the trailing edge as the objective to be tracked by a remote high-speed camera for obtaining the flap-wise vibration as illustrated in Figure 1.

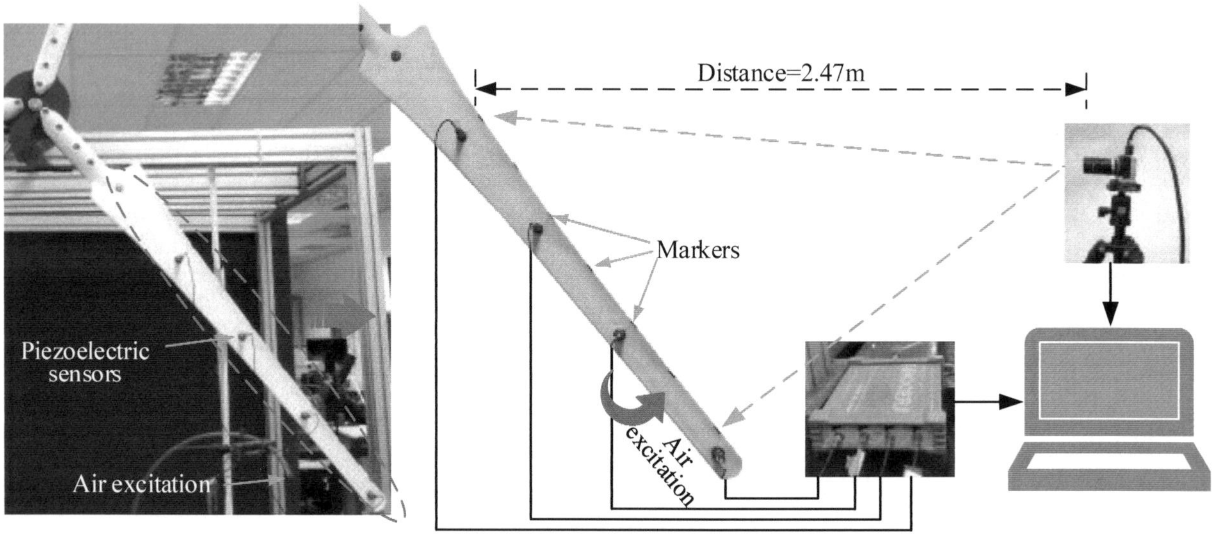

Figure 1. Schematic diagram of experimental rig and instrumentation of the high-speed camera and the accelerometers

The air excitation system is made up of a compressor to generate high pressure air, an air pressure regulator gauge, and a solenoid valve. This system can produce an air impulse to excite blades, being like natural excitation due to a strong wind turbulence. In combination of the high-speed camera set at 2.47 metres away from the blades, this test system can measure the vibration modes in a remote way and expected to be more accurate compared conventional contact measurements.

3.2. Instrumentation

The video measurement system consists of a high-speed industrial camera with a telescopic lens. The high-speed camera uses a CMOS sensor, the resolution is 640×480 pixel and the maximum fps is at 600. In this experiment, frames per second was set to 200 fps for low computational cost and found is sufficient to identify the vibration properties. The captured videos then are processed in Python & OpenCV and MATLAB using method mentioned in previous sections.

Meanwhile, for the verification of the photogrammetric results obtained, acceleration responses were also collected by 4 piezoelectric accelerometers. These sensors are mounted at points near location of markers. The signals are acquired by a four-channel dynamic measurement system which is hosted by laptop (Figure 1).

3.3. Experimental Procedure

The first step, to verify the accuracy of optical measurement. While the camera is capturing the video, piezoelectric accelerometer is collecting acceleration at the same time. Because the traditional contact sensor is precise and frequency range is wide, the vibration signals collected from accelerometers can be regarded as a good reference to verify the error of the photogrammetry method.

The second step, the baseline condition is obtained. All used piezoelectric accelerometers are removed from the blades. since mass of the sensors inevitably has impact on the dynamic characteristics of the blades. So that using camera we can obtain the more real dynamic characteristics of healthy condition,

The third step, according to other researchers' previous research [17], the trailing edge deformation and real load distribution at root end are two important factors leading to failure. Thus, at the non-destructive condition, loosening the screws at the end root and adding mass near the trailing edge are implemented to simulate the typical fault on the wind turbine blades. The aim is clear that judge the fault using photogrammetry.

4. RESULT ANALYSIS

4.1. Verification in Time Domain and Frequency Domain

Unlike accelerometers measuring the acceleration straightforwardly, photogrammetry need to detect and track markers in the video, get the trajectory of markers' motion, and then convert trajectory of pixel in successive images into physical displacement in space, finally solve the accelerations through differential calculation. Figure 2 shows 5 acceleration responses on 5 monitored markers. In this figure, periodicity of signals is clear, and the amplitudes increase with respect to the distance of blade root.

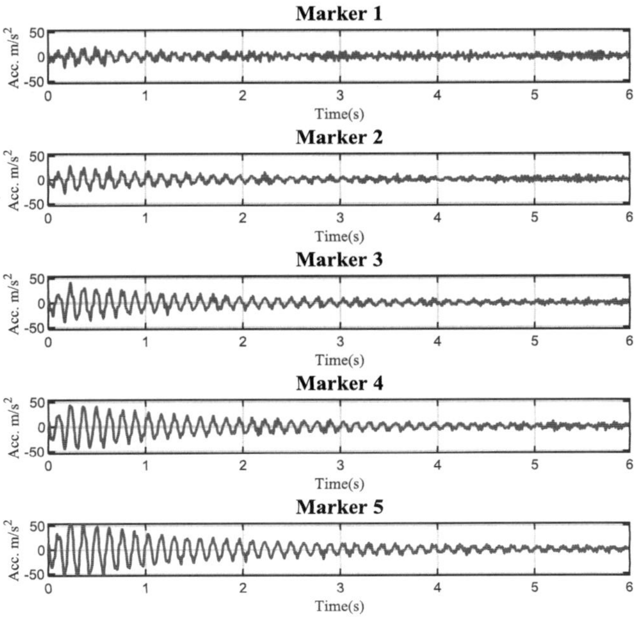

Figure 2. Acceleration of 5 markers acquired from photogrammetry

The comparison between photogrammetry signals and accelerometer signals in the time domain and frequency domain is shown in Figure 3. Sampling rate of piezoelectric sensor is 1.5k Hz, that is significantly higher than photogrammetry (200 Hz). Apparently more high frequency components can be acquired by the accelerometers. Meanwhile, the resolution of CMOS sensor also has the influence on the result. Despite these, the amplitude and frequency values are well agreeable. Given that large-scale wind turbine blades have lower natural frequency, in this case the photogrammetry should be very effective.

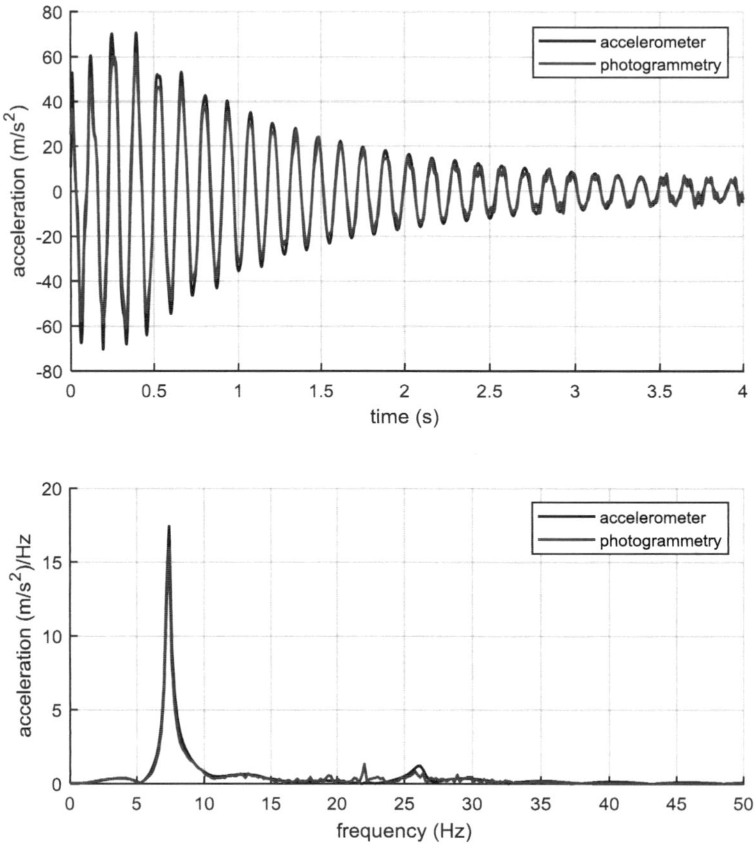

Figure 3. Comparison of comparison between photogrammetry signals and accelerometer signals in Time domain and frequency domain

4.2. Verification of Modal Identification

Having confirmed that the vibration response signals is reliable, the stochastic subspace identification is implemented to identify the natural frequency, damping ratio, and the modal shape in order to implement a condition monitoring method without the needs of baseline data.

To calibrate the results, the stabilization diagram and the results are provided in Figure 4 and Figure 5 respectively for the case of normal blade conditions. Although, there are different spurious modes in these two diagrams, both they successfully identify the same modes. In Figure 5, the differences for first and second natural frequencies is 0.31% and 0.1% respectively. However, the error of damping ration is more obvious, which is common problem in using output-only methods. Moreover, two mode shapes are consistent with each other for the two methods. These then confirms that the photogrammetry system shows acceptable performance.

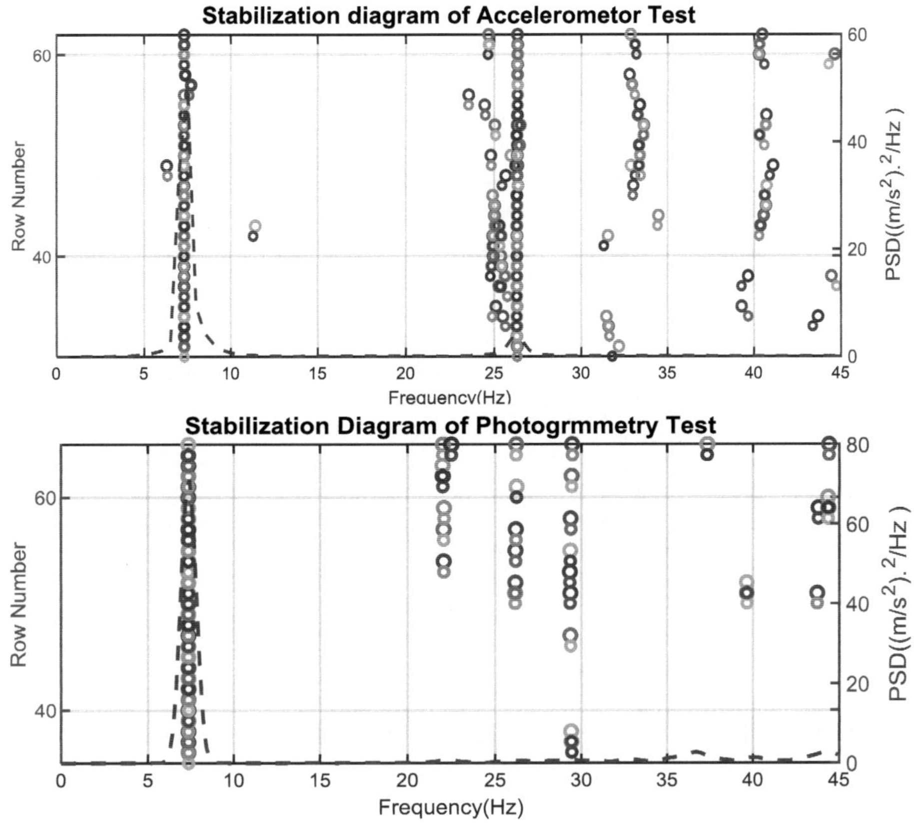

Figure 4. The stabilization diagram of the method of accelerometer and photogrammetry

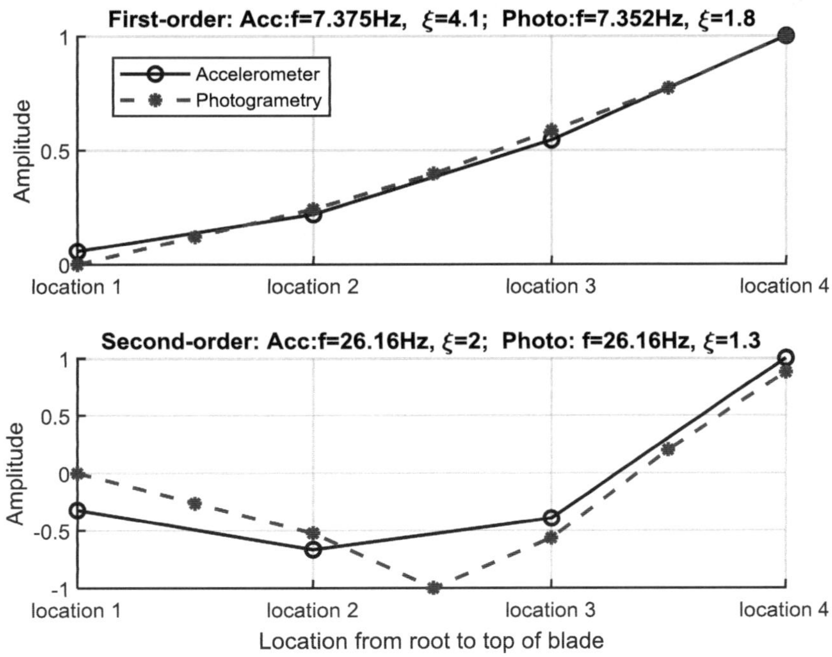

Figure 5. The modal parameters for the first and second modes

4.3. Condition Monitoring based on Modal Parameters

Because the modal parameters of wind turbine blades can be accurately extracted from photogrammetry. The condition monitoring test based on the photogrammetry can be implemented logically. The artificial faults are set to check out if this method is stable and sensitive to some common fault of the wind turbine blades. All in all, the change of natural frequency can indicate the faults, regulation of damping ratio is not very clear. Modal shapes keep feature of first and second order mode but cannot see a significant change.

At the beginning, all sensors are removed from the blade to test the parameters of baseline condition, so that the faulty cases could have a standard. Since part mass are subtracted, the natural frequency increase naturally, comparing with the above verification test.

After that, the sensors as 3 extra mass (11g per sensor) are gradually added on the blade, the change of frequency are discovered clearly, especially when the third mass block is added near the end of the blade, where often accumulates most snow or ice and cause most enormous imbalance and lead to fault easily.

Finally, the screw which fix blades on the shaft was loosened, this part always bears huge stress. Therefore, this fault is chosen to simulate early crack or loosening. From table 1 it can be seen that the change in natural frequency is distinctive, which allows for an assessment of blade conditions.

Table 1: change of modal parameters due to faults

		First order				Second order			
		Natural frequency	Change rate δf	Damping ratio	Change rate $\delta\xi$	Natural frequency	Change rate δf	Damping ratio	Change rate $\delta\xi$
Baseline case		8.79Hz	0%	1.4%	0%	29.73Hz	0%	0.9%	0%
Snow or frozen	1 mass	8.63Hz	-1.82%	1.0%	-28.57%	28.87Hz	-2.89%	1.3%	44.44%
	2 mass	8.34Hz	-5.12%	1.6%	14.29%	28.65Hz	-3.63%	1.2%	33.33%
	3 mass	7.50Hz	-14.68%	1.4%	0%	26.83Hz	-9.75%	1.0%	11.11%
Screw Loosen		8.61Hz	-2.05%	1.2%	-14.29%	29.26Hz	-1.58%	1.4%	55.55%

5. CONCLUSION

In this research, modal parameters identified using a photogrammetry method are validated by the result of the traditional piezoelectric sensor based method. The difference of the natural frequency is less than 0.31%. Furthermore, the healthy case and the faulty cases are differentiated by the image signals distinctly. The mass change of blades as small as 11g can be detected, which is shown more significantly when the added mass is more distant from the root.

In conclusion, as a novel non-contact method, photogrammetry combines with the air impulse excitation is an effective way to implement the remote condition monitoring of wind turbine blades.

REFERENCES

[1] Yang W, Tavner P J, Crabtree C J, et al. Wind turbine condition monitoring: technical and commercial challenges[J]. Wind Energy, 2014, 17(5): 673-693.

[2] C A Walford. Wind turbine reliability: Understanding and minimizing wind turbine operation and maintenance costs [R]. USA: Sandia National Laboratories,2006.
[3] M H Hansen. Aeroelastic instability problems for wind turbines[J]. Wind Energy, 2007, 10(6):551-577.
[4] W Yang, Z Peng, K Wei, et al. Structural health monitoring of composite wind turbine blades: challenges, issues and potential solutions[J]. Iet Renewable Power Generation, 2017, 11(4):411-416.
[5] Li D, Ho S C M, Song G, et al. A review of damage detection methods for wind turbine blades[J]. Smart Materials and Structures, 2015, 24(3): 033001.
[6] Ye G, Neal B, Boot A, et al. Development of an ultrasonic NDT system for automated in-situ inspection of wind turbine blades[C]//EWSHM-7th European Workshop on Structural Health Monitoring. 2014.
[7] E Mikhail, J Bethel, J C McGlone, Introduction to Modern Photogrammetry, New York:Wiley, 2001
[8] Poozesh P, Baqersad J, Niezrecki C, et al. Large-area photogrammetry based testing of wind turbine blades[J]. Mechanical Systems and Signal Processing, 2017, 86: 98-115.
[9] G P Corten, J C Sabel. Optical motion analysis of wind turbines [C]. In: Proceedings of European union wind energy conference. Goteborg, Sweden, 1996, 5: 20-24.
[10] J T Johnson, S Hughes, J V Dam. A stereo video grammetry system for monitoring wind turbine blade surfaces during structural testing[J]. ASME Early Career Tech J, 2009, 8:1-10.
[11] J Baqersad, J Carr, T Lundstrom. Dynamic characteristics of a wind turbine blade using 3D digital image correlation[C]. In: health monitoring of structural and biological systems, Proceedings of SPIE, San Diego, USA, 2012, 83:12-15.
[12] T Lundstrom, J Baqersad, C Niezrecki, et al.Using high speed stereo photogrammetry techniques to extract shape information from wind turbine/rotor operating data. In: Topics in modal analysis II[C]. Proceedings of the 30th IMAC, a conference on structural dynamics. New York: Springer, 2012, 6: 269-275.
[13] Ozbek M, Rixen D J. Operational modal analysis of a 2.5 MW wind turbine using optical measurement techniques and strain gauges[J]. Wind Energy, 2013, 16(3): 367-381.
[14] Lowe D G. Distinctive Image Features from Scale-Invariant Keypoints[C]. International Journal of Computer Vision. 2004:91-110.
[15] Hartley, Richard, Zisserman, Andrew. Multiple view geometry in computer vision[J]. Kybernetes, 2004, 30(9/10):1865 - 1872.
[16] Reynders E, Roeck G D. Reference-based combined deterministic-stochastic subspace identification for experimental and operational modal analysis[J]. Mechanical Systems & Signal Processing, 2008, 22(3):617-637.
[17] Zhou, H.F., et al., A review of full-scale structural testing of wind turbine blades[J]. Renewable and Sustainable Energy Reviews, 2014. 33: p. 177-187.

An improved resonance demodulation technique based on spectral kurtosis and fault characteristic harmonic-to-noise ratio

Chang Yan[1], Jing Lin [1,2], Ming Zhao[1], liang Zeng[1]
[1] School of Mechanical Engineering, Xi'an Jiaotong University, Xi'an, China
[2] School of Reliability and Systems Engineering, Beihang University, Xueyuan Road No. 37, Haidian District, Beijing, China

ABSTRACT

Rolling bearings are widely used in energy, construction machinery and high-speed railway because of its strong load-bearing capacity and small start-up torque. The faults of rolling bearings as the core component of the machine can lead to huge economic losses and even fatal accidents. The health monitoring and fault diagnosis of rolling bearings have attracted increasing attention over the past decades. Traditional spectral kurtosis (SK) method is commonly implemented by kurtogram to determine the resonance frequency band for demodulation. However, there exist problems when SK is applied to signals containing random strong interferences. As a result, envelope harmonic-to-noise ratio (EHNR) is proposed to reduce the influence of random strong interferences and strengthen the periodic components in the signal. Unfortunately, EHNR is not sensitive to fault impact when the periodic components in the signal are not related to the failure such as rotational frequency interferences and power frequency interferences. To overcome these limitations, this paper proposes an improved resonance demodulation technique (RDT) based on SK and fault characteristic harmonic-to-noise ratio (FCHNR). Compared with SK, the improved RDT has two distinct advantages. Firstly, the proposed method can be applied to the rolling bearings under the working condition with variable speed through integrating with order tracking technique. Secondly, the resonance frequency band is no longer solely determined by the maximum value of kurtosis or EHNR. Instead, the periodicity and impulsivity of rolling bearing fault can be taken into account simultaneously, thus the improved RDT has higher robustness. Finally, the improved RDT is validated by real fault data of railway bearings. The results show that the improved RDT can detect the fault of rolling bearing under variable speed with higher efficiency and superiority.

Keywords: spectral kurtosis; fault characteristic harmonic-to-noise ratio; resonance frequency band; order tracking

1. INTRODUCTION

With the advancement of modern industry, rolling element bearings are widely used in various rotating machinery as core components, such as high-speed railway, mine excavator, aircraft engine, etc. The faults of rolling bearings may lead to huge economic losses and even fatal accidents [1,2]. Hence, increasing attention has been paid on the health monitoring and fault diagnosis of rolling bearings over the past decades. Since the vibration signals contain rich information on health conditions of rotary machinery, vibration analysis is an effective approach for fault diagnosis of rolling bearings [3-6]. Once the localized faults occur on a rolling bearing, the periodic impacts caused by faults can excite the system resonant and generate amplitude modulation phenomenon. Therefore, how to accurately determine the resonance frequency band of vibration signals and realize faults feature extraction in the background of strong noise is crucial.

Various signal processing methods have been developed for the diagnosis of rolling bearings. Antoni J [7] proposed spectral kurtosis (SK) methodology to find the optimal filter for fault diagnosis. Y. Wang and M. Liang [8] proposed an adaptive SK technique to determine the bandwidth and center frequency for the fault detection of rolling element bearings. Y. Lei et al. [3] proposed an improved Kurtogram based on wavelet packet transform. Xu X et al. [9] introduced a new index, EHNR, to improve Kurtogram for periodic impulses detection.

Among all kinds of signal processing methods, SK and EHNR have been demonstrated as effective and efficient indexes for impact detection of rolling bearings. However, the kurtosis index is only sensitive to the impact of fault, but not the periodicity of fault. When the fault is weak, the impact of the fault is not obvious and easily submerged by random strong interference in the environment, resulting in the error of the resonance frequency band determined by the kurtosis index. On the other hand, EHNR is proposed to weaken the influence of random strong interference and enhance the periodicity component. However, the sensitivity of now available EHNR to impact caused by faults is insufficient, thus it is easy to be misled by the periodic component caused by non-fault factors when determining the resonance frequency band.

To overcome these drawbacks of SK or EHNR and take the impact and periodicity of the fault into account, an improved RDT is explored in this paper. The proposed method filters kurtosis value by calculating FCHNR, which is different from the existing EHNR, of the same frequency band. FCHNR has directivity to the fault impact. Finally, the effectiveness and robustness of the proposed improved RDT is validated by the data from a railway bearing test rig.

2. THEORETICAL BACKGROUND

2.1.SK and kurtogram

SK was first proposed by Dwyer to locate the resonance frequency to estimate the impulsivity. The theoretical basis of SK was further developed by J. Antoni [10,11] for real cases. Given time variable t; $X(t, f)$ is the complex envelope of signal $x(t)$ at frequency f. The 2n order spectral instantaneous moments of the signal $x(t)$ is defined as follows:

$$S_{2n}(t, f) = \langle |X(t, f)|^{2n} \rangle, \quad (1)$$

where $\langle \cdot \rangle$ is the mean operation over time. Thus, the SK of $X(t, f)$ can be described as follows:

$$K(t, f) = \frac{S_4(t, f)}{S_2(t, f)} - 2, \quad (2)$$

2.2. Fault characteristic harmonic-to-noise ratio

For rolling bearings, assuming that the theoretical ball pass order of outer race (BPOO), ball pass order of inner race (BPOI), fundamental train order (FTO) and roller spin order (RSO) are f_o, f_i, f_c and f_r, respectively. For input signal $x(t)$, the equal angle resampling signal can be denoted $y(\theta)$, of which the Fourier transform is $O(f)$. Due to the discrete error, the actual value of $O(f)$ at f_o, f_i, f_c and f_r are $O(\hat{f_o})$, $O(\hat{f_i})$, $O(\hat{f_c})$ and $O(\hat{f_r})$, respectively, and their high order harmonics are $O(\widehat{nf_o})$, $O(\widehat{nf_i})$, $O(\widehat{nf_c})$ and $O(\widehat{nf_r})$, where n=2,3,4$\cdots$. It's worth noting that $O(\widehat{nf})$ is not strictly equal to $O(n \cdot \hat{f})$. Let σ be the standard deviation of the $O(f)$. For any j, if $O(f_j) \geq 6 \cdot \sigma$, remove $O(f_j)$ from the array of $O(f)$. As a result, a new array $N(\tilde{f})$ can be obtained, whose length is less than the length of $O(f)$. The noise energy level can be estimated by $N(\tilde{f})$, of which $\tilde{\sigma}$ is the standard deviation. When two and three times the fault characteristic frequency occur on the spectrum clearly, it is generally believed that the bearing has a early failure. If there is only a clear fault characteristic frequency, we can't come to the same conclusion.

As a result, the FCHNR of outer race is defined as follows:

$$r_o = \frac{O(2\hat{f}_o)\cdot O(3\hat{f}_o)}{\tilde{\sigma}^2}, \quad (3)$$

The FCHNR of inner race is defined as follows:

$$r_i = \frac{O(2\hat{f}_i)\cdot O(3\hat{f}_i)}{\tilde{\sigma}^2}, \quad (4)$$

The FCHNR of fundamental train is defined as follows:

$$r_c = \frac{O(2\hat{f}_c)\cdot O(3\hat{f}_c)}{\tilde{\sigma}^2}, \quad (5)$$

The FCHNR of roller is defined as follows:

$$r_r = \frac{O(2\hat{f}_r)\cdot O(3\hat{f}_r)}{\tilde{\sigma}^2}, \quad (6)$$

2.3. An improved RDT

Due to the limited impact energy caused by early faults of rolling bearings, it is difficult to determine the resonant frequency band by the SK or the FCHNR alone. An improved RDT is proposed to integrate SK and FCHNR to improve the resonance demodulation technique as shown in figure 1. First, the SK is applied to find the impact components in the measured signal. And then FCHNR is employed to determine whether the impact components is related to the failure of outer race, inner race or the fundamental train of rolling bearings. The threshold can be the average of the FCHNRs of all bands or the median of the FCHNR array.

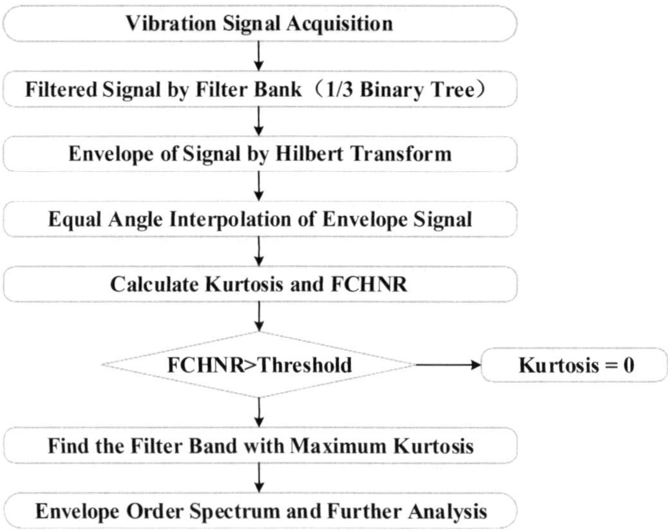

Figure 1. The flow chart of the proposed hybrid method.

3. EXPERIMENT

The experimental data of a railway rolling bearing test rig is utilized to illustrate the validity of the proposed method. The outer race of the bearings is fixed with the carriage and the inner race rotates with the wheel shaft, whose rotating speed is monitored by a tachometer. A PCB accelerometer is mounted on the top of axle box to pick up the vibration signal of the bearing. The geometric parameters of the bearing are listed in table 1, and the sampling frequency is 25600Hz. The original vibration signal with 0.5 seconds is shown in figure 2 and the rotating speed signal is shown in figure 3. Figure 4(a) (b) and (c) show the results of the original kurtogram, EHNR-based kurtogram and the proposed hybrid kurtogram, respectively. Obviously, they find different optimal resonance frequency bands. The filtered signals at those optimal resonance frequency bands are demodulated with Hilbert transform, and combined order tracking method, order envelope spectrum (OES) was obtained. Figure 5(a) shows the result of OES of the original vibration signal, from which we can't find obvious fault feature. There is also no fault harmonic in the results of OES based on SK or EHNR, which are shown in figure 5(b) and figure 5(c), respectively. Figure 5(d) shows the result of OES based on the proposed method, and we can see the obvious outer race fault characteristic harmonics. The diagnosis result is consistent with the actual fault.

Table 1. Geometric parameters of the Rolling bearings.

Number of rollers	Contact angle (degree)	Roller diameter (mm)	Pitch diameter (mm)
19	10	26	180

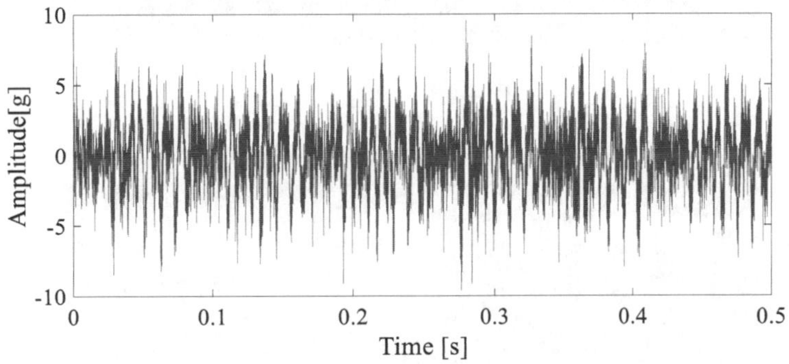

Figure 2. The original vibration signal

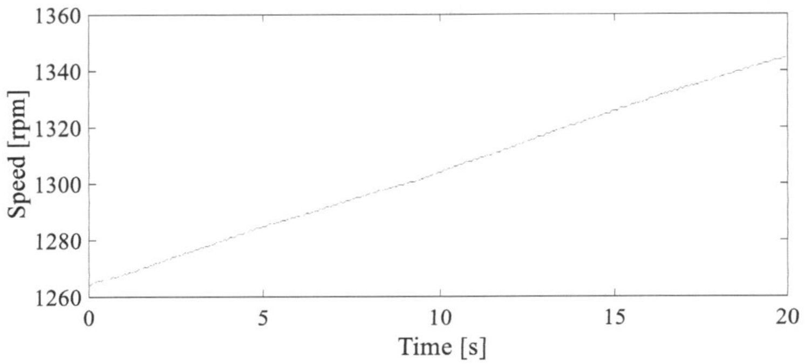

Figure 3. The rotating speed signal

(a)

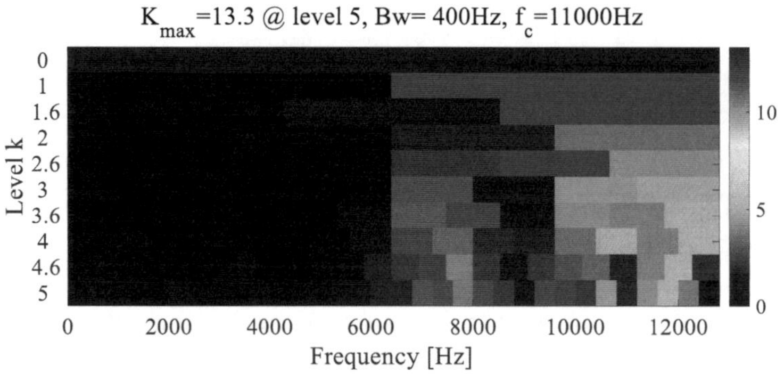

(b)

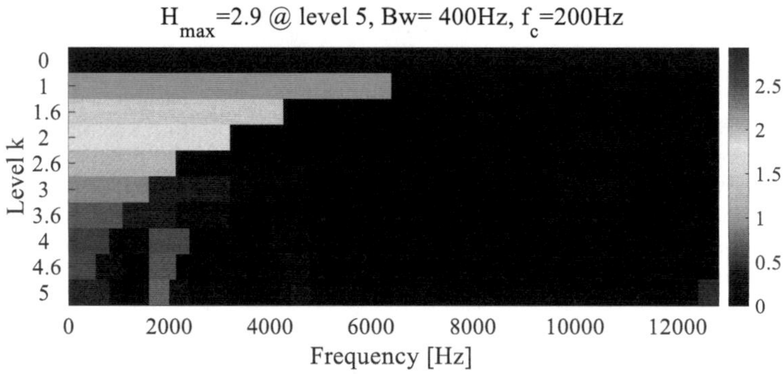

(c)

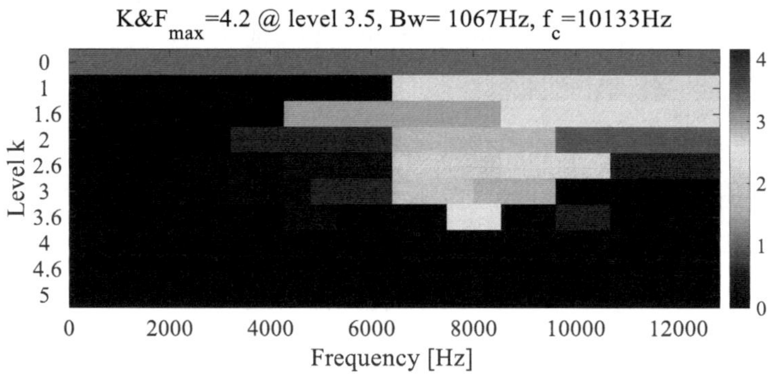

Figure 4. The result of kurtogram:

(a) The original kurtogram;(b) EHNR-based kurtogram;(c) HM-based kurtogram

(a)

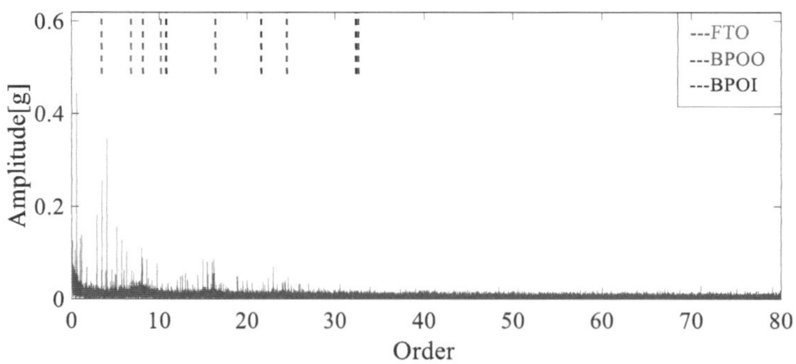

(b)

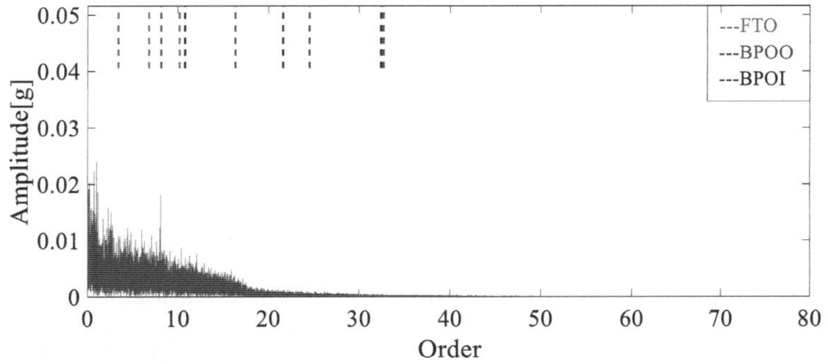

(c)

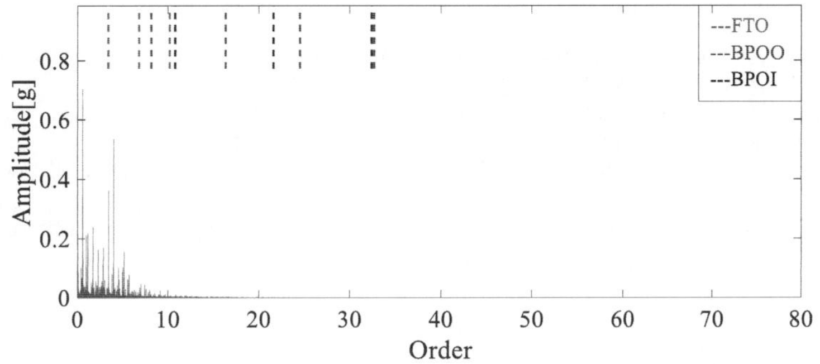

(d)

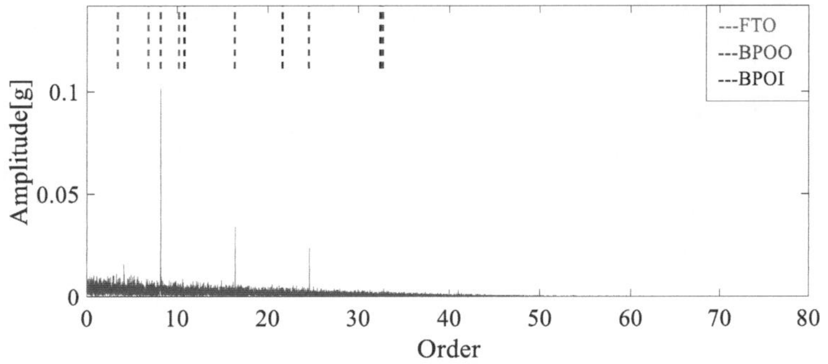

Figure 5. The order envelope spectrum: (a) OES of the original vibration signal;

(b) OES based on SK; (c) OES based on EHNR; (d) OES based on proposed method

4. CONCLUSION

Through the comparisons among the SK-based RDT, EHNR-based RDT and the improved RDT, it is appropriate to draw the following conclusions that the proposed hybrid method combined SK and FCHNR has higher robustness. It improved the traditional resonance demodulation technique and is proved effective for fault diagnosis of rolling bearing. SK is used to find the frequency band which is related to the impact. And the FCHNR is applied to determine whether the impact is related to failure of rolling bearing. The proposed method is demonstrated by data collected from a railway rolling bearing test rig. And it can detect the fault of rolling bearing under variable speed with higher efficiency and superiority.

5. ACKNOWLEDGMENT

This project is supported by the National Natural Science Foundation of China (Grant No. 51421004), which are highly appreciated by the authors.

REFERENCES

[1]. Randall R B. Vibration-based condition monitoring : industrial, aerospace and automotive applications[J]. Mechanisms & Machine Science, 2011, 3(4):431-477.
[2]. Wang D, Zhao Y, Yi C, et al. Sparsity guided empirical wavelet transform for fault diagnosis of rolling element bearings[J]. Mechanical Systems & Signal Processing, 2018, 101:292-308.
[3]. Lei Y, Lin J, He Z, et al. Application of an improved kurtogram method for fault diagnosis of rolling element bearings[J]. Mechanical Systems & Signal Processing, 2011, 25(5):1738-1749.
[4]. Ou L, Yu D, Yang H. A new rolling bearing fault diagnosis method based on GFT impulse component extraction[J]. Mechanical Systems & Signal Processing, 2016, 81:162-182.
[5]. Li Y, Xu M, Wang R, et al. A fault diagnosis scheme for rolling bearing based on local mean decomposition and improved multiscale fuzzy entropy[J]. Journal of Sound & Vibration, 2016, 360:277-299.
[6]. Janjarasjitt S, Ocak H, Loparo K A. Bearing condition diagnosis and prognosis using applied nonlinear dynamical analysis of machine vibration signal[J]. Journal of Sound & Vibration, 2008, 317(1):112-126.
[7]. Antoni J. The spectral kurtosis: a useful tool for characterising non-stationary signals[J]. Mechanical Systems & Signal Processing, 2006, 20(2):282-307.
[8]. Wang Y, Liang M. An adaptive SK technique and its application for fault detection of rolling element bearings[J]. Mechanical Systems & Signal Processing, 2011, 25(5):1750-1764.
[9]. Xu X, Zhao M, Lin J, et al. Envelope harmonic-to-noise ratio for periodic impulses detection and its application to bearing diagnosis[J]. Measurement, 2016, 91:385-397.
[10]. Antoni J, Randall R B. The spectral kurtosis: application to the vibratory surveillance and diagnostics of rotating machines[J]. Mechanical Systems & Signal Processing, 2006, 20(2):308-331.
[11]. Antoni J. Fast computation of the kurtogram for the detection of transient faults[J]. Mechanical Systems & Signal Processing, 2007, 21(1):108-124.

A vision of energy-based visualisation of large scale industrial systems for the purposes of condition monitoring

G. van Schoor[1], K.R. Uren[2]
[1] Unit for Energy and Technology Systems, Faculty of Engineering, North-West University, South Africa
[2] School of Electrical, Electronic and Computer Engineering, North-West University, South Africa

ABSTRACT

Most large scale industrial systems can be viewed as processes that convert energy from one form to another. Energy is a universal quality of which the distribution holds meaning across physical domains i.e. thermal-fluid, mechanical, electrical and chemical. This paper presents a vision of using energy information of a process to visualise or characterise what is happening in the process with regards to energy. This energy information is then represented in a form that retains the structural information, making it possible to relate patterns in the representation to specific locations in the process. The energy attributes considered include energy flow rates between components of the system and the change in exergy flow rates across components. An energy representation of a process can form the basis for fault detection and diagnosis (FDD) or optimal control through the definition of a reference energy representation. The use of energy information linked to specific locations in the system, classifies the approach to FDD in the hybrid class which represents a hybrid between a pure data-driven and model-based approach. Attributed graphs are considered in this work as a structure that can describe the system's energy attributes while retaining structural information. The usefulness of the energy representations for the purpose of condition monitoring is illustrated through a few case studies. The case studies include a heated two tank system, a gas to liquids process and a Brayton cycle. The results illustrate that energy-visualisation as a means to condition monitoring shows promise in terms of detecting and diagnosing typical faults in large scale industrial processes where energy conversion is the main concern.

Keywords: energy visualisation, condition monitoring, large scale industrial systems, attributed graphs

1. INTRODUCTION

Modern industrial plants are becoming more reliant on condition monitoring and fault tolerant control schemes to be more efficient and profitable. Also, strict safety and environmental regulations further motivate the need for advanced process monitoring schemes. For future monitoring schemes to be successful, aspects such as robustness, uncertainties and big data need to be addressed [1], [2]. Accordingly, new approaches need to consider a hybrid approach that combines various data-driven, analytical and knowledge-based techniques.

This paper shares the vision that energy as a unifying attribute can be used to characterise large scale industrial systems. The synthesis of any complex system into a geometric representative structure, with functional nodes inter-connected by process streams and mechanical, electrical or heat links and the subsequent energy characterisation of the structure, results in a data structure that poses very useful for the purposes of FDD and control.

The concept of energy visualisation as a means to characterise and assess the health of a system initiated from the study of [3], [4] on the health monitoring of a Brayton cycle using the enthalpy and entropy attributes of the system. In this work the enthalpy and entropy distribution across the different locations in the system are used as a basis for health assessment. This idea was further explored in a paper by van Schoor et al. presenting a survey on energy as basis for modelling, supervision and control [5]. This paper initiated the exploration of ways to visualize energy characteristics of components in industrial plants such as heat exchangers and compressors [6], [7].

Considering literature related to the analysis and characterisation of industrial systems from an energy perspective, it became evident that both the first and second laws of thermodynamics are important [8]. This implies that knowledge of both the quantity and quality of energy is necessary for a complete characterisation of a system. This involves the inclusion of exergy as a quality parameter for energy-based visualisation of systems. Exergy-based methods have their roots in exergetic and thermoeconomic analysis and their use for diagnosis and optimisation have been widely discussed in literature [9]–[11].

Since many industrial plants can be classified as complex and large-scale due to the large number of component interactions, a mathematical framework is needed to represent these component interactions and attributes. Also, this mathematical framework should be integrable in a fault detection and diagnosis scheme. Graph theory is considered a natural fit for this purpose. Graphical approaches toward fault detection and diagnosis are readily found in literature [12]–[14]. Smaili et al. [12], developed a fault monitoring framework for multi-energy systems using signed directive graphs. Directed graphs are used to capture the process topology and connectivity to show the causal relationships between variables. Graphs are then used for qualitative fault indicators in the form of fault signature matrices. Another study conducted by van Graan et al. considered a linear graph representation of a heat exchanger [15]. Exergy and energy flowrates were assigned as attributes to the nodes and links of the linear graph. This graph was then used for fault detection using a pattern recognition technique called graph matching [16]. Graph matching is mainly concerned with finding a correspondence between the node and link attributes of different graphs in an optimal way [17].

In this paper the goal is to discuss a novel approach and vision of representing industrial processes as attributed graphs containing energy information. These graphs will be called energy signatures which represent an energy visualisation of a process. The term visualisation can mean a visual picture of the condition of the plant, but may also take an abstract form such as a matrix. It will be shown how these energy signatures can be utilised for fault detection and diagnosis by means of attributed graph matching.

The paper is organized as follows. Section 2 shares the idea of energy visualisation as a means to characterise large scale industrial systems. In section 3 the concept of energy visualisation for fault detection and diagnosis is illustrated through a few case studies. The paper is concluded in section 4.

2. ENERGY VISUALISATION

In this section the nature of multi-domain systems is firstly discussed after which the idea of using graphs to visualise system energy attributes in a structured way is presented. Lastly the concept of using graph matching as a means to represent system fault information in a structured way is presented. Throughout this section the example of a simplified coal-fired power station is used to illustrate the applicable concepts.

2.1. Multi-domain systems

Energy is a universal quality of which the distribution holds meaning across physical domains i.e. thermal-fluid, mechanical, electrical and chemical [5]. In a coal-fired power station energy is transformed through a sequence of processes across physical domains. The source of energy would in this case be coal of a certain calorific value. The coal is transformed from chemical potential energy to thermal potential energy through a chemical process where chemical kinetics describe the transformation of energy. The thermal potential energy is transformed into mechanical energy through a thermal-fluid process that is described by thermodynamics. The mechanical energy is transformed to electrical energy through an electromechanical process described by electromagnetics. The product of the process in this case is electrical power.

The idea emphasised in this paper is that the energy distribution across the process with specific reference to structure reveals important information about the health of the system. Energy information in the form of energy stored, energy converted, and energy dissipated at the different points in the system will be of specific interest. Due to process irreversibility in the thermal-fluid and chemical domains also characterising these processes in terms of exergy lost or exergy change is of specific interest.

The energy interaction through a general multi-domain system component can be illustrated as shown in Fig.1. The component experiences thermal fluid or chemical domain interaction in the form of a fluid stream that enters and exits the component at energy rates of $\dot{Q}_{in}$ and $\dot{Q}_{out}$ respectively. This interaction stands in direct relation to the other energy interactions indicated i.e. electrical power in (P_{in}) and out (P_{out}), mechanical power in or out (P) and heat flow in or out ($\dot{Q}$). The change in exergy of a fluid stream through a component is denoted by ΔB and represents the change in the ability of the fluid to do work as it passes through the component. The general energy interaction portrayed can be made applicable to any system component by omitting the interactions that do not apply. In the case of a turbine for example the main interaction will be a fluid stream entering and exiting the component with mechanical power as the main output from the component. In the case of a heat exchanger the primary side can be represented by a fluid stream entering and exiting the component with heat flow in the direction of the secondary side of the heat exchanger. Similarly, the general multi-domain energy interaction can be applied to other process components e.g. a reactor, a pump, and an electrical transformer to name but a few.

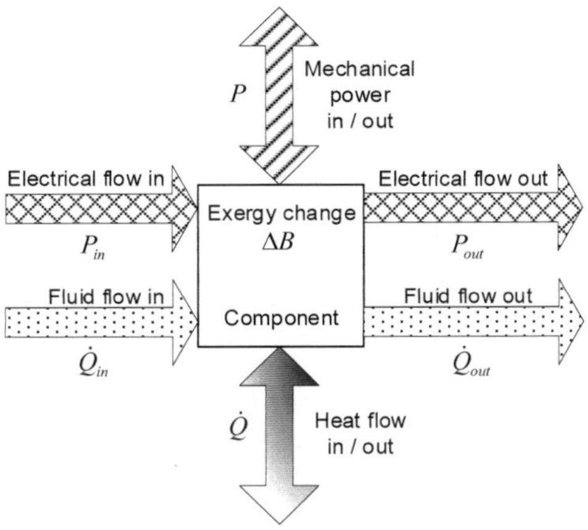

Figure 1. Illustration of energy interaction through a general multi-domain system component

The change in exergy of a fluid stream through a component constitutes components of kinetic, potential, physical and chemical exergy. For the case of a thermal fluid component, considering only the change in physical (ΔB_p) and chemical exergy (ΔB_c):

$$\Delta B = \Delta B_p + \Delta B_c = \pm \dot{W} + B_{destroyed}, \quad (1)$$

with $\dot{W}$ the nett energy flow rate in or out of the component. $\dot{W}$ could either be equal to P, the mechanical power in or out of the component or $\dot{Q}$, the nett heat flow in or out of the component. $B_{destroyed}$ represents the exergy destroyed across the component. The chemical exergy is important where the chemical composition of the stream entering the component differs from the one leaving. Furthermore, from the energy balance requirement:

$$\dot{W} = \dot{Q}_{in} - \dot{Q}_{out}, \quad (2)$$

where $\dot{Q}_{in}$ and $\dot{Q}_{out}$ are the heat flow in and out of the component due to the fluid stream.

The change in physical exergy of a fluid stream through a component at flow rate $\dot{m}$, flowing in at enthalpy and entropy values of h_{in} and s_{in} and out at enthalpy and entropy values of h_{out} and s_{out} respectively with a reference temperature of T_0, is given by:

$$\Delta B_p = \dot{m}(h_{in} - h_{out}) - \dot{m}T_0(s_{in} - s_{out}). \quad (3)$$

The change in chemical exergy of a fluid stream through a component at flow rate $\dot{m}$ is given by [18]:

$$\Delta B_c = \dot{m}\left(\textstyle\sum_i x_i \bar{b}_{c,i}\right)_{in} - \dot{m}\left(\textstyle\sum_i x_i \bar{b}_{c,i}\right)_{out}, \quad (4)$$

with x_i the fraction of the component in the substance and $\bar{b}_{c,i}$ the standard chemical exergy of the corresponding component.

From the similarity between (1) and the sum of (3) and (4) and the requirement for energy balance,

$$\dot{W} = \dot{m}(h_{in} - h_{out}) + \dot{m}\left(\textstyle\sum_i x_i \bar{b}_{c,i}\right)_{in} - \dot{m}\left(\textstyle\sum_i x_i \bar{b}_{c,i}\right)_{out} \text{ and} \quad (5)$$

$$B_{destroyed} = -\dot{m}T_0(s_{in} - s_{out}). \quad (6)$$

From (2) and (5) it follows that

$$\dot{Q}_{in} = \dot{m}h_{in} + \dot{m}\left(\textstyle\sum_i x_i \bar{b}_{c,i}\right)_{in} \text{and} \quad (7)$$

$$\dot{Q}_{out} = \dot{m}h_{out} + \dot{m}\left(\sum_i x_i \bar{b}_{c,i}\right)_{out} \quad (8)$$

2.2. Attributed graph as visualisation instrument

Graphs are very useful to represent complex system interactions in a structured way [19]. The use of a specific type of graph called an attributed graph gives the opportunity to capture both structural information and attributes associated with specific locations in the system structure [15], [20]. A graph representation of a system firstly requires the system to be represented by links and nodes. To make it an attributed graph attributes are associated with the links and the nodes. An attribute associated with a link is called a link attribute and an attribute associated with a node is called a node attribute. Different approaches to configuring energy attributed graphs could be taken. Firstly, the choice of what constitutes a link and what constitutes a node needs to be decided. Then what constitutes a link energy attribute and what constitutes a node energy attribute need to be decided. In this work a node is defined as a system component e.g. a compressor or turbine or reactor and a link is defined as a connection between two system components e.g. the connection between a compressor and a turbine. Node energy attributes are seen as energy changes that occur in a component e.g. energy losses in the component or in the case of a thermal-fluid component the exergy change or exergy lost in the component. In the case of a component where chemical energy transformation takes place both physical and chemical exergy changes would be node energy attributes of interest. In cases where energy is stored in a component, the stored energy would also be a node attribute of interest. Link energy attributes are seen as energy flow attributes. Here the power flow between components would normally be of interest as well as the direction of the power flow.

With these definitions in mind, it will now be illustrated how the simplified coal-fired power station as portrayed in Figure 2(a) can be represented as an energy attributed graph.

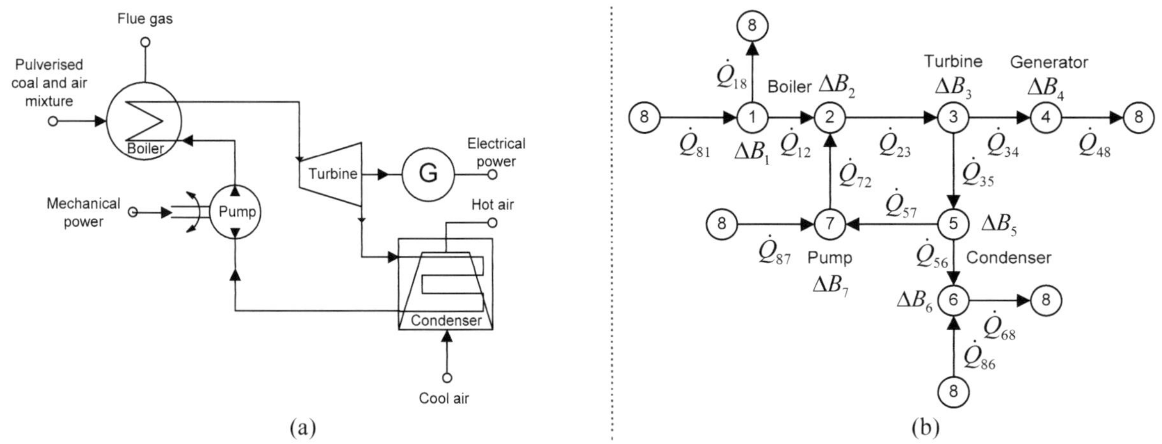

Figure 2. (a) Simplified representation of a coal-fired power station. (b) Attributed graph of a coal-fired power station

The simplified system constitutes a boiler, a power turbine, a generator (G), a condenser and a water pump as the main components. The boiler is represented as a heat exchanger with the pulverised coal and air mixture entering the primary side and the flu gas exiting the primary side. Heat is exchanged to the secondary side where water enters and super-heated steam exits. The super-heated pressurised steam then expands in the turbine that drives a generator that delivers electrical power as the system product. The expanded steam is cooled in the condenser which also acts as a heat exchanger. The hot steam enters the condenser on the primary side and water exits. Heat is exchanged to the secondary side that is typically a cooling tower with cool air entering and hot air exiting. To complete the water cycle, the water that exits the condenser is pumped to the boiler.

The next step to illustrate the concept of energy visualisation by means of a graph is to set up a graph from Figure 2(a) where each component represents a node and connections to and between components are represented by links. A heat exchanger is however assigned two nodes, one for the primary side and one for the secondary side with a link between the primary side node and the secondary side node that represents the heat exchanged from the one side to the other. Figure 2(b) shows the graph set up from Figure 2(a) with the node allocations as well as the attributes associated with each node indicated. In Figure 2(b) nodes 1 and 2 represent the boiler primary and secondary sides, nodes 3 and 4 represent the Turbine and Generator, nodes 5 and 6 represent the condenser primary and secondary sides and node 7 represents the pump. Node 8 is chosen as the environment and represents a node where all links to outside the system connect to.

The attribute $\dot{Q}_{81}$ represents the hydraulic and chemical power transfer from the environment (node 8) to the boiler primary side (node 1). All the energy flow rate attributes in Figure 2(b) should be similarly interpreted with the subscript designating the nodes between which the energy flow applies. ΔB_1 represents the change in chemical and physical exergy across the boiler primary side. The attributed graph is now mathematically described by the node signature based formulation given in [16]:

$$\mathbf{N}_s = \begin{bmatrix} \Delta B_1 & \vdots & \dot{Q}_{11} & \dot{Q}_{12} & \cdots & \dot{Q}_{1N} \\ \Delta B_2 & \vdots & \dot{Q}_{21} & \dot{Q}_{22} & \cdots & \dot{Q}_{2N} \\ \vdots & \vdots & \vdots & \vdots & \vdots & \vdots \\ \Delta B_N & \vdots & \dot{Q}_{N1} & \dot{Q}_{N2} & \cdots & \dot{Q}_{NN} \end{bmatrix} \quad \text{.........} \quad (9)$$

Each row of $\mathbf{N}_s$ describes the node in the graph corresponding to the row number. In this way row n describes all the attributes associated with node n. The node attributes are ordered in such a way that the structural information is retained. The first entry in every row represents the exergy change across the node corresponding to that row number. The attributes that follow for a particular row are the energy flows to and from the particular node. In the case of row n the attributes $[\dot{Q}_{n1} \quad \dot{Q}_{n2} \quad \cdots \quad \dot{Q}_{nN}]$ would represent all the energy flows from node n and in the order of the flow to node 1 first, followed by the flow to node 2 next.

2.3. Graph matching for condition monitoring

With the attributed structured graph $\mathbf{N}_s$ set up as described in section 2.2 it is now possible to use graph matching as a technique for condition monitoring. By defining a normal conditions graph $\mathbf{N}_{s,n}$ and comparing an operational conditions graph $\mathbf{N}_{s,o}$ with $\mathbf{N}_{s,n}$ it becomes possible, based on the difference between the two, to detect and diagnose faults. A first step in graph matching can be used to describe the difference between $\mathbf{N}_{s,n}$ and $\mathbf{N}_{s,o}$. The method as defined by [16] determines the so-called cost matrix to describe the difference between two graphs. It is based on determining the distance between every node of the two graphs being compared. The distance between two nodes is determined based on the difference between their node signatures.

The Heterogeneous Euclidean Overlap Metric (HEOM) for the distance between node i of $\mathbf{N}_{s,n}$ and node j of $\mathbf{N}_{s,f}$ is given by

$$HEOM(\mathbf{N}_{s,ni}, \mathbf{N}_{s,fj}) = \sqrt{\textstyle\sum_{a=1}^{k} \delta(\mathbf{N}_{s,ni}(a), \mathbf{N}_{s,fj}(a))^2}, \quad \text{.........} \quad (10)$$

where $\mathbf{N}_{s,ni}(a)$ refers to the (i, a) entry in the normal signature matrix, and $\mathbf{N}_{s,fj}(a)$ refers to the (j, a) entry in the faulty signature matrix. a refers to the a-th column entry of the rows considered, k is the length of the row and the function δ for the numeric attributes only case is given by

$$\delta(\mathbf{N}_{s,ni}(a), \mathbf{N}_{s,fj}(a)) = \frac{|\mathbf{N}_{s,ni}(a) - \mathbf{N}_{s,fj}(a)|}{range_a}, \quad \text{.........} \quad (11)$$

where $range_a$ is used to normalise the attributes, and is defined as

$$range_a = \max a - \min a\,, \quad \text{.........} \quad (12)$$

where $\max a$ and $\min a$ are the maximum and minimum values respectively observed in the a-th column entry of $\mathbf{N}_{s,n}$.

Next, a cost matrix, $\mathbf{C}$ can be compiled from the distances between the nodes of graphs $\mathbf{N}_{s,n}$ and $\mathbf{N}_{s,f}$. The cost matrix is a square matrix with dimension equal to the number of nodes.

Since the cost matrix represents the difference between two system conditions based on the change in the energy distribution in the system, analysing the characteristics of the cost matrix $\mathbf{C}$ to detect and diagnose faults is investigated. To this end the eigenvalues and eigenvectors of $\mathbf{C}$ are analysed.

For a system with N nodes, the cost matrix $\mathbf{C}$ will be an $N \times N$ matrix with N eigenvalues, $[\lambda_1 \quad \lambda_2 \quad \cdots \quad \lambda_N]$ and N eigenvectors, $[\mathbf{x}_1 \quad \mathbf{x}_2 \quad \cdots \quad \mathbf{x}_N]$. For the normal case the graph $\mathbf{N}_{s,n}$ is compared with itself to obtain the normal set of eigenvalues and eigenvectors $\boldsymbol{\lambda}_n$ and $\mathbf{x}_n$ respectively. For a fault case the corresponding set of eigenvalues and eigenvectors $\boldsymbol{\lambda}_f$ and $\mathbf{x}_f$ are obtained.

By applying a simple direction threshold to only the eigenvalues obtained, a simple $N \times r$ qualitative fault signature matrix $\mathbf{F}_q$ for r faults is obtained as follows:

$$\mathbf{F}_q = \begin{bmatrix} F_1 & F_2 & F_3 & \cdots & F_r \\ - & + & - & \cdots & - \\ + & - & - & \cdots & - \\ + & - & - & \cdots & - \\ \vdots & \vdots & \vdots & \cdots & \vdots \\ + & - & + & \cdots & + \end{bmatrix} \begin{matrix} \text{row} \\ 1 \\ 2 \\ 3 \\ \vdots \\ N \end{matrix} \quad \cdots\cdots (13)$$

where columns denoted $F_1, F_2, \cdots, F_r$ represent the eigenvalues for the respective r fault conditions. The size of the fault signature matrix can be extended $(N+1)$ fold by also applying the direction threshold to the eigenvectors. The extended fault signature for fault 1 can be given by

$$\mathbf{F}_{q1} = \begin{matrix} \text{eigenvalues} \\ \\ \\ \\ \\ \\ \\ \end{matrix} \begin{bmatrix} + & - & + & \cdots & + \\ \hline - & + & - & \cdots & - \\ + & - & - & \cdots & - \\ + & - & - & \cdots & - \\ \vdots & \vdots & \vdots & \cdots & \vdots \\ + & - & + & \cdots & + \\ \mathbf{x}_1 & \mathbf{x}_2 & \mathbf{x}_3 & \cdots & \mathbf{x}_N \end{bmatrix} \begin{matrix} \text{row} \\ 1 \\ 1+1 \\ 1+2 \\ 1+3 \\ \vdots \\ 1+N \end{matrix} \quad \cdots\cdots (14)$$

The + and – signs in (13) and (14) are fictitious to illustrate the idea. For k faults the extended $\mathbf{F}_q$ will take the form of $\mathbf{F}_q = [\mathbf{F}_{q1} \quad \mathbf{F}_{q2} \quad \cdots \quad \mathbf{F}_{qk}]$.

3. CASE STUDIES

The concept of energy visualisation is now illustrated through a few case studies, applying the method outlined in section 2. The first case study is the heated two tank system representing an open system with only the thermal-fluid domain of interest. Next a gas to liquids plant where the thermal-fluid and chemical domains are important is considered. The third case study is a Brayton cycle where a closed cycle with the mechanical and thermal-fluid domains of interest. Due to space limitations only a short summary of each case study is provided.

3.1. Heated two-tank system

A diagram of the system is shown in Fig. 3(a). The outlet flow from both tanks is proportional to the square root of the level in each tank. The outlet from the first tank flows into the second tank, and each tank has its own supply of cold water with a control valve to control the level of each tank. Each tank also exchanges heat with a hot water line. The temperature in the tanks is controlled using control valves on the hot water lines.

An energy-attributed graph for the two-tank system is depicted in Fig. 3(b). The change in exergy flow rate, ΔB, and the energy flow rate, ΔQ, are considered as the node and link attributes respectively. From Fig. 4(b) five nodes of interest are identified namely: (1) The hot water line transferring heat to tank1; (2) Tank 1; (3) The hot water line transferring heat to tank 2; (4) Tank 2; (5) The environment.

In this case study 5 possible faults were considered: two sensor faults, namely the tank level sensor, and the temperature sensor; two actuator faults, one related to the heat transfer from the hot water side to the tank, generally called fouling, and the other valve failure, and finally, a process fault in the form of tank leakage. These faults are induced one at a time, first for tank 1 and then for tank 2.

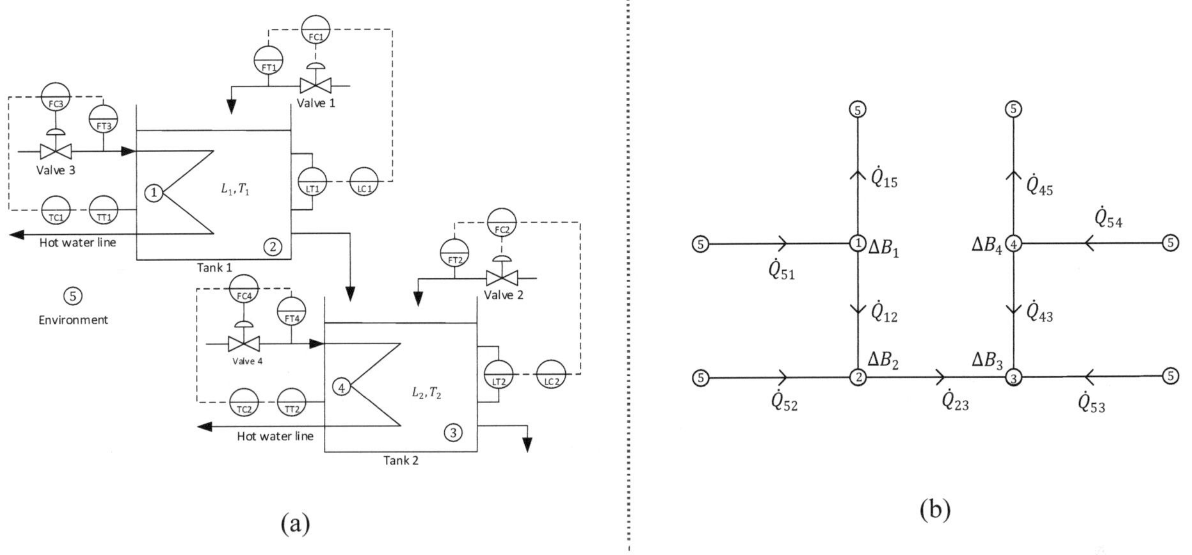

Figure 3. (a) Heated two-tank system. (b) Energy-based attributed graph

By applying a direction threshold on the fault eigenvalues with respect to the normal case eigenvalues, qualitative fault signature matrices for tank 1 and tank 2 are obtained as follows:

$$\mathbf{F}_{qtt1} = \begin{bmatrix} F_1 & F_2 & F_3 & F_4 & F_5 \\ + & + & + & + & + \\ + & + & + & - & - \\ + & + & + & - & + \\ - & - & - & - & + \\ - & - & - & - & - \end{bmatrix}, \text{ and } \mathbf{F}_{qtt2} = \begin{bmatrix} F_1 & F_2 & F_3 & F_4 & F_5 \\ + & + & + & + & + \\ + & + & + & + & + \\ + & + & + & - & + \\ - & - & - & - & - \\ - & - & - & - & - \end{bmatrix} \qquad (15)$$

Fault columns having the same pattern are considered not diagnosable using the current residual definition. From (15) it is clear that all faults are detectable. In terms of diagnosability fault types 4 and 5 of $\mathbf{F}_{qtt1}$ are diagnosable, but uniquely linked to a specific tank. For $\mathbf{F}_{qtt2}$ only fault type 4 is uniquely diagnosable.

3.2. Gas to liquids process

In this section, a gas to liquids process is considered. The modelled process, shown in Fig. 4, firstly comprises an autothermal reformer (ATR) which produces syngas when fed with specific ratios of feedstocks. The produced syngas is then fed into a Fischer-Tropsch reactor (FTR) in order to produce a variety of liquid products. The energy-attributed graph developed for this model is shown in Fig. 5. The nodes correspond to the units within the process as indicated on both diagrams. The attributes of the nodes being the change in exergy, ΔB, seen over the units. The links indicate how the units are connected to one another. The link attributes are described by the energy flow, $\dot{Q}$, of the material streams. The three faults that were considered are:

1) A +10% deviation of the molar flow of the methane feed stream, which is representative of an actuator fault.
2) A temperature deviation of the reactor feed stream of -10%. This would be the result of a faulty heater (node 5).
3) The third fault was induced within the recycle stream; a faulty compressor leading to a pressure drop of 300 kPa.

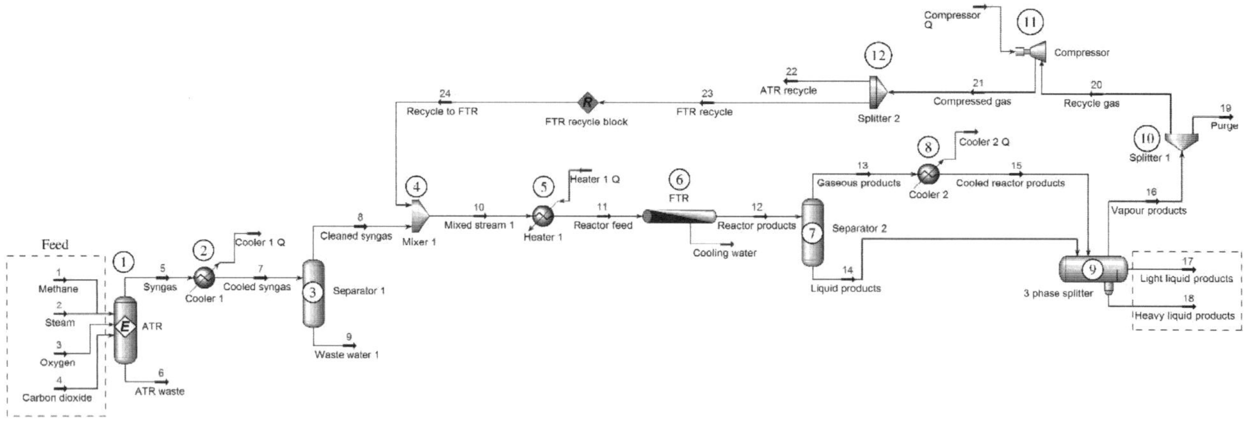

Figure 4. GTL plant layout

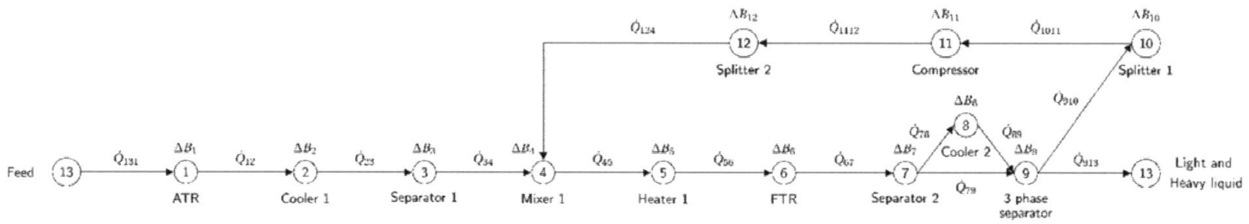

Figure 5. An energy-based attributed graph of the GTL plant

The qualitative fault matrix based on only the eigenvalues is given by

$$\mathbf{F}_{qGTL}{}^{\mathrm{T}} = \begin{bmatrix} F_1 & + & - & - & - & - & - & - & - & + & + & - & - & - \\ F_2 & + & - & - & - & - & - & - & + & + & + & - & - & - \\ F_3 & + & - & - & - & - & - & - & + & + & + & - & - & - \end{bmatrix} \quad (16)$$

As (16) reveals all three faults were detectable. Only F1 was found to be diagnosable though. One of the reasons would be that the recycle stream inevitably propagates the faults throughout the entire process.

3.3. Brayton cycle

The Brayton cycle considered in this section is portrayed in Fig. 6(a). In this closed-loop Brayton cycle the working fluid is compressed through the compressor (comp) before heat energy is add through a heat addition component. The working fluid is then expanded through a compressor turbine (CT) and then through a second turbine (PT) producing a nett work output. The working fluid is then cooled through a heat rejection component, before being re-introduced into the compressor. The compressor inlet temperature is controlled to be constant and the pressure is varied to produce the desired nett work output. The compressor turbine inlet is also controlled at a constant temperature.

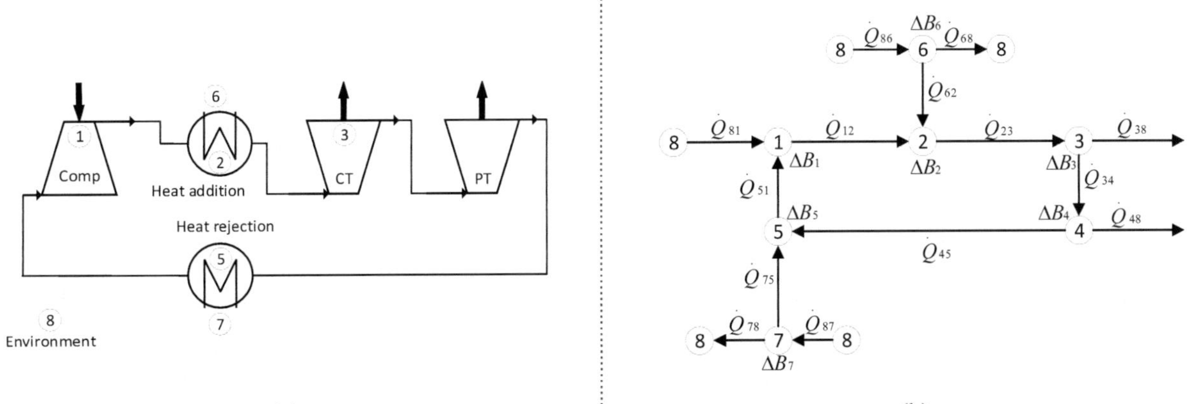

(a) (b)

Figure 6. (a) A Brayton cycle power conversion unit. (b) An energy-based attributed graph of the Brayton cycle

By applying the method described in section 2.2, the attributed graph of Fig. 6(b) is obtained. The attributed graph is constructed such that the work input to the compressor and the work output of the compressor turbine is not equal and therefore connected but rather connected to the environment node. The 8 nodes of interest are (1) the compressor, (2) the heat addition component secondary, (3) the compressor turbine, (4) the power turbine, (5) the heat rejection component primary, (6) the heat addition component primary, (7) the heat rejection component secondary and (8) the environment node.

The three fault conditions emulated are:

1) a change in the pressure ratio of the compressor,
2) a change in the pressure ratio of the power turbine,
3) a leak between the compressor and the heat addition component.

By comparing the eigenvalues and –vectors under fault conditions to that of the normal condition and applying a direction threshold, the following fault signatures are obtained.

Fault1					
Eigenvalues					
+	+	+	+	+	+
Eigenvectors					
-	-	-	-	-	+
-	-	-	+	+	-
+	-	+	-	-	-
-	+	+	-	-	+
+	+	-	+	-	-
-	+	-	+	+	+

(a)

Fault2					
Eigenvalues					
+	+	+	+	+	+
Eigenvectors					
+	+	-	-	-	-
+	-	-	-	+	-
+	+	+	+	-	+
+	+	+	+	-	-
+	-	-	+	+	+
+	+	-	-	+	-

(b)

Fault3					
Eigenvalues					
-	+	+	+	+	+
Eigenvectors					
+	-	-	+	+	+
+	-	-	-	-	-
+	+	+	+	+	+
+	-	+	-	-	-
+	-	-	+	+	+
+	+	-	-	-	+

(c)

Figure 7. Fault signatures for: (a) Fault 1. (b) Fault 2. (c) Fault 3.

From the fault signature matrices it is clear that when only considering the eigenvalues, fault diagnosis is not possible. By including the deviation in the eigenvectors fault diagnosis becomes possible.

4. CONCLUSION

The case studies discussed illustrate that the use of energy visualisation as a means of characterizing the health of industrial plants is viable. All faults emulated could be detected but not necessarily isolated. Extending the characterization to include the eigenvectors as part of detecting the deviation from the normal conditions, improves the diagnosis capability. Further work will include exploring different methods of compiling the node signature matrices to capture more node attributes per node while maintaining structural information. The simplicity of the mathematical formulation however presents a challenge. An important next step will be confirming the reliability of the energy visualisation techniques through appropriate sensitivity analyses. The vision of energy visualisation will be extended to include optimisation and control of large scale industrial systems.

5. ACKNOWLEDGEMENTS

This work is based on the research supported wholly or in part by the National Research Foundation (NRF) of South Africa (Grant numbers 91093, 103392). Opinions expressed, and conclusions arrived at are those of the authors and are not necessarily to be attributed to the NRF.

REFERENCES

[1] K. Severson, P. Chaiwatanodom, and R. D. Braatz, "Perspectives on Process Monitoring of Industrial Systems," *IFAC-PapersOnLine*, vol. 48, no. 21, pp. 931–939, 2015.

[2] M. S. Reis and G. Gins, "Industrial Process Monitoring in the Big Data/Industry 4.0 Era: From Detection, to Diagnosis, to Prognosis," *Processes*, vol. 5, no. 3, p. 35, 2017.

[3] C. P. du Rand and G. van Schoor, "Fault diagnosis of generation IV nuclear HTGR components – Part I: The error enthalpy–entropy graph approach," *Ann. Nucl. Energy*, vol. 40, no. 1, pp. 14–24, 2012.

[4] C. P. du Rand and G. Van Schoor, "Fault diagnosis of generation IV nuclear HTGR components - Part II: The area error enthalpy-entropy graph approach," *Ann. Nucl. Energy*, vol. 41, pp. 79–86, 2012.

[5] G. van Schoor, K. R. Uren, M. A. van Wyk, P. A. van Vuuren, and C. P. du Rand, "An energy perspective on modelling, supervision, and control of large-scale industrial systems: Survey and framework," in *IFAC World Congress*, 2014.

[6] L. B. Fouché, K. R. Uren, and G. van Schoor, "Energy-based visualisation of an axial-flow compressor system for the purposes of Fault Detection and Diagnosis," *IFAC-PapersOnLine*, vol. 49, no. 7, 2016.

[7] K. R. Uren and G. V. Schoor, "Energy-based visualisation of a counter-flow heat exchanger for the purpose of fault identification," *IFAC-PapersOnLine*, vol. 49, no. 7, 2016.

[8] V. Nikulshin, C. Wu, and V. Nikulshina, "Exergy efficiency calculation of energy intensive systems by graphs," *Int. J. Appl. Thermodyn.*, vol. 5, no. 2, pp. 67–74, 2002.

[9] H. Marais, G. van Schoor, and K. R. Uren, "The merits of exergy-based fault detection in petrochemical processes," *J. Process Control*, 2017.

[10] V. Verda, L. M. Serra, and A. Valero, "Zooming procedure for the thermoeconomic diagnosis of highly complex energy systems," *Int. J. Thermodyn.*, vol. 5, no. 2, pp. 75–83, 2002.

[11] V. Verda and R. Borchiellini, "Exergy method for the diagnosis of energy systems using measured data," *Energy*, vol. 32, no. 4, pp. 490–498, 2007.

[12] R. Smaili, R. El Harabi, and M. N. Abdelkrim, "Design of fault monitoring framework for multi-energy systems using Signed Directed Graph," *IFAC-PapersOnLine*, vol. 50, no. 1, pp. 15734–15739, 2017.

[13] M. R. Maurya, R. Rengaswamy, and V. Venkatasubramanian, "A systematic framework for the development and analysis of signed digraphs for chemical processes. 1. Algorithms and analysis," *Ind. Eng. Chem. Res.*, vol. 42, no. 20, pp. 4789–4810, 2003.

[14] G. Wu, L. Zhang, and J. Tong, "Online Fault Diagnosis of Nuclear Power Plants Using Signed Directed Graph and Fuzzy Theory," in *2017 25th International Conference on Nuclear Engineering*, 2017, p. V004T06A011--V004T06A011.

[15] S. van Graan, G. van Schoor, and K. R. Uren, "Graph matching as a means to energy-visualisation of a counter-flow heat exchanger for the purpose of fault diagnosis," *IFAC-PapersOnLine*, vol. 50, no. 1, pp. 2842–2847, 2017.

[16] S. Jouili, I. Mili, and S. Tabbone, "Attributed graph matching using local descriptions," in *International Conference on Advanced Concepts for Intelligent Vision Systems*, 2009, pp. 89–99.

[17] T. S. Caetano, J. J. McAuley, L. Cheng, Q. V Le, and A. J. Smola, "Learning graph matching," *IEEE Trans. Pattern Anal. Mach. Intell.*, vol. 31, no. 6, pp. 1048–1058, 2009.

[18] T. J. Kotas, *The exergy method of thermal plant analysis*. Elsevier, 2013.

[19] C. Godsil and G. F. Royle, *Algebraic Graph Theory*. Springer New York, 2013.

[20] M. Dehmer, F. Emmert-Streib, and Y. Shi, "Quantitative Graph Theory: A new branch of graph theory and network science," *Inf. Sci. (Ny).*, vol. 418–419, pp. 575–580, 2017.

A low-cost condition monitoring solution for industrial bakery equipment

H. Marais[1], J. Black[1]

[1] School of Electrical, Electronic and Computer Engineering, North-West University, South Africa

ABSTRACT

Artisanal bakeries perform a critical function in many societies all over the world. However, artisanal bakers struggle to compete with commercial bakeries' efficiencies due to the lack of technology deployment. One aspect that could lead to significant improvements of up to 36% are improved maintenance processes. Unfortunately the prohibitive costs associated with commercial condition monitoring equipment and a lack of proven efficiency improvements in the artisanal bakery hamper the adoption of such technologies. In this work a low-cost, remote, condition monitoring and measurement system is developed that makes use of commercial off-the-shelf technology. The prototype system provides immediate return on investment to the baker and maintenance staff while at the same time gathering high-resolution data for the development of more advanced diagnostic algorithms. The latter is especially of importance due to the lack of diagnostic algorithms capable of handling the variance introduced by the viscosity characteristics of dough and other mixer products.

Keywords: Practical, Low-Cost, Condition Monitoring

1. INTRODUCTION

The consumption of confectionary and other baked goods continues to rise leading to the impression that bakeries are lucrative high-profit industries. However, without careful management of bakery operations profit margins of up to 45% are quickly eroded [1]. Considering bakeries world-wide, a trend of increasing the degree of automation appears to be on the increase. Even in South Africa the use of technology such as robotic arms is not unheard of [2]. However, artisan bread (hand prepared) still dominate the African and Middle-East markets [1] and as such large commercial bakery operations compete with several smaller operations. Due to the technology dichotomy between commercial operators and smaller artisanal operators it is not atypical to find commercial operators employing the latest automation and process information to eke out even the smallest profits whilst artisans rarely employ technology beyond the machinery required to produce the relevant baked goods. This can largely be ascribed to the relative cost of commercial automation and monitoring technology. As such, the artisanal vendor is excluded from the advantages that such technologies could provide. In this work the development of a low-cost, scalable condition monitoring device for an artisanal bakery takes centre stage. In Section 2 some background on maintenance processes in artisanal bakeries are discussed, as well as how the changing technological landscape enables low-cost condition monitoring solutions to be developed. In Section 3, System Development, the architecture of the proposed solution is discussed in detail, and aspects such as scalability and data security receive special attention. Section 4 deals with the development of the monitoring algorithm. Section 5 provides initial results with a brief discussion of issues encountered during the prototype development. Finally, in Section 6 conclusions and future work is presented.

2. BACKGROUND

2.1 Bakery Equipment

Considering typical operations involved in the baking process the oven might spring to mind as being the most critical. However, the mixer plays an equally important role and arguably performs under harsher environmental conditions [3]. The typical components of an industrial 10 litre mixer is shown in Figure 1. Fundamentally a mixer is an electrical motor connected via a gearbox to a type of mixing attachment. The entire drivetrain is housed in a cast iron frame.

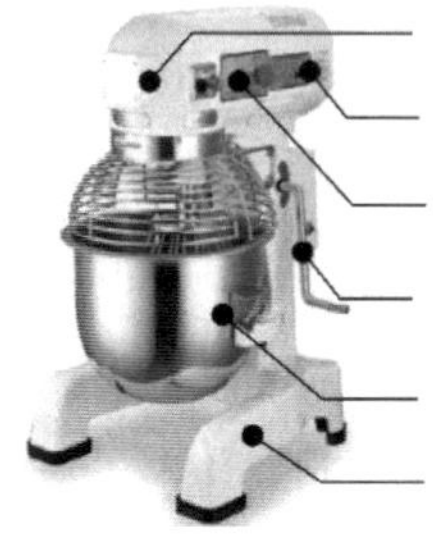

Figure 1: Typical components of an industrial mixer

2.2 Maintenance models

In a typical artisanal bakery environment the dominant maintenance *strategy* is run to failure (RTF) even though RTF has been shown to be one of the most costly strategies (mainly due to loss of operational time). Introduction of preventative maintenance (+18-25% overall financial improvement) and predictive maintenance strategies (+20-25% overall financial improvement) may be of significant interest to artisanal bakeries [4]. However, the operational environment in which mixers operate is harsh with high levels of atmospheric dust, long operational hours, and high temperatures. Although some advances have been made such as the use of sealed gearboxes the ingress of flour dust into drivetrain components remain a challenge. As such any monitoring or diagnostic system shall be required to be non-invasive. Non-invasive methods will allow minimal foreign matter ingress whilst also preserving any manufacturer provided guarantees.

2.3 Non-invasive diagnostic methods

The combination of the simplicity of a mixer in conjunction with typical operating procedures reduces the problem space to identification of motor and gearbox conditions. Methods such as vibrations monitoring [5] [6], temperature monitoring [7], and the monitoring of the electrical characteristics of the drive motor [6] [8], are commonly employed.

Amongst the methods employed, the analysis of motor current signature analysis (MCSA) is one of the most popular methods, and would be attractive from both a non-invasive perspective and the limited remote deployability the technique affords. Recent trends in literature [8] indicate that the signal processing aspect is receiving increasing attention [9]. Amongst techniques commonly deployed is the Fast Fourier Transform, the instantaneous power spectrum [10], and wavelet decompositions [11]. However, regardless of the specific signal processing the objective is to develop a unique representation for each of the faults of interest. This process is referred to as residual generation by Gertler or more commonly as signature generation.

It has been noted that large variance exist in the current signature of bearings presenting failure modes. Since there are several bearings present in the drivetrain and gearbox of the mixer it is realistic to expect similar results for the mixer. Additionally, and perhaps more importantly, the viscosity of dough is highly dependent on the properties of the mixture constituting the dough. Further complicating the matter is the variance in product for which the mixer is used. It is not uncommon for mixing operations to range from beating eggs through cake mixtures to kneading of bread dough. This situation further complicates matters of condition monitoring or fault detection in these machines.

As such, a theoretical analysis of techniques is of limited value. In practice the correlation of operational data with maintenance data, and sensor data would lead to an optimal solution. Unfortunately, this presents a conundrum with regards to the deployment of condition monitoring technology in artisanal bakeries. In this work the problem is addressed by the development of low-cost, flexible monitoring hardware as a driver for the adoption of condition monitoring and fault detection techniques in artisanal bakeries.

2.4 Remote measurement

Although there are several remote diagnostic architectures proposed in literature [12] [13], the rise of the internet of things (IoT) and the proliferation of low-cost computing hardware and sensors has seen the development of customised remote measurement solutions that are repurposed for remote diagnostics.

Primarily the difference lies in where the bulk of the computational load is carried. Simply retrofitting existing diagnostic systems with remote terminals leads to excessive data overhead being incurred while providing limited additional value. Traditionally, these diagnostic systems made use of high sampling frequencies with the advantages associated with that.

On the other hand systems driven primarily from the connectivity side is typically characterised by the means in which data is exchanged between the sensing node and the remote terminal. Unfortunately, naïve approaches yield systems with limited data usage and limited diagnostic usage. This scenario is presented schematically in Figure 2.

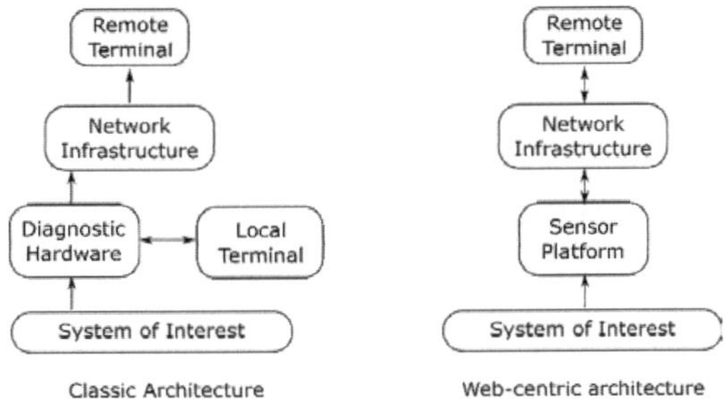

Figure 2: Comparison between classical and web-centric remote diagnostic system architecture

Neither of the architectures in Figure 2 make optimal use of resources. The classical approach leads to excessive data communication overhead, and the web-centric approach ignores the computational power that is located in the sensor platform; effectively centralising a decentralised problem. By leveraging the computing power in the edge of the network cascade failures introduced by the centralised nature of the web-centric approach can be avoided. This approach is referred to as fog computing by Cisco [14] and aims to optimally utilise computing power regardless of the location thereof.

2.5 Data security

One advantage afforded by the classic approach is that data security was less of a concern. However, the web-centric approach as well as an approach making use of fog computing creates a significant risk in terms of data security. It is reported that up to 45% of all IoT devices are vulnerable to data security risks. For this reason any data protocol that is used between the sensor platform (so-called edge nodes) and the central database should be designed for maximum security.

3. SYSTEM DEVELOPMENT

For the proposed remote diagnostic system the fog computing paradigm provides for the most flexibility in terms of deployment. Due to the large number of unknown factors a dedicated platform cannot be developed as diagnostic techniques have differing computational complexity. Secondly, due to the cost sensitive nature of the platform commercial off-the-shelf (COTS) hardware will be preferred where possible. Over the last two decades the cost of computing power has significantly decreased with microcontrollers becoming commonplace. The proliferation of single board computers (SBCs) has placed industrial computing power [15] within the means of even small business. Additionally, the rise of the Internet-of-Things (IoT) has seen significant advances with regards to the availability and affordability of sensors. Considering all of the above the following architecture is arrived at (see Figure 3). Such an architecture provides for various diagnostic algorithms to be implemented in an agile fashion whilst retaining the ability to derive a purpose made solution.

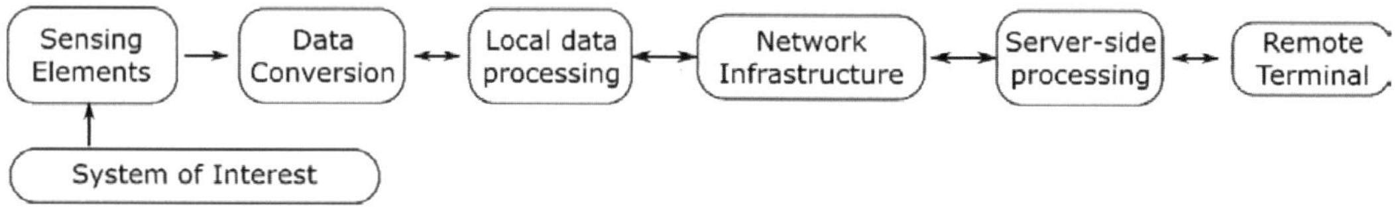

Figure 3: Remote diagnostic system with edge node computing

The system architecture in Figure 3 can be realised by making use of the Raspberry Pi SBC together with an analogue to digital conversion module. The Raspberry Pi Model 3+ already provides the capability to connect to network infrastructure via WiFi® and Ethernet (IEEE 802.03) thereby simplifying the deployment of the hardware. Additionally the Raspberry Pi provides more than adequate computational power (1GHz, 64-bit Quad-core architecture) thereby allowing for various diagnostic processing algorithms to be deployed on the edge node.

As mentioned in Section 2.5 data security is of primary concern but frugal use of data communication bandwidth remains important. Although there are several data protocols available that could satisfy either the security or overhead requirements the MQTT protocol was selected as the protocol of choice. This is not only due to the superior data security provided by means of transport layer security (TLS) but the low protocol overhead the MQTT protocol presents (having been designed for the IoT) [16].

Although the MQTT protocol is available as a free and open-source standard several commercial providers offer implementations. One such implementation is provided by Amazon Web Services (AWS) thereby removing the need for the system developer to implement this aspect of the system by hand. The system architecture in Figure 3 can be expanded as shown in Figure 4, in which the directions of the arrows indicate the expected data flows. It should be noted that although provision has been made for multiple types of sensor input only the current measurement and voltage measurement functions were implemented. In the case of the current measurement a low-cost, clamp-on, current transformer (CT) was selected due to cost considerations. The selected CT (YHDC SCT013) features a peak current rating of 15A_{rms} with a ratio of 30:1 although similar CTs can easily be accommodated. An example of the CT in question is depicted in Figure 5a. At this point the system's main components can either be procured from commercial vendors or from cloud computing platforms

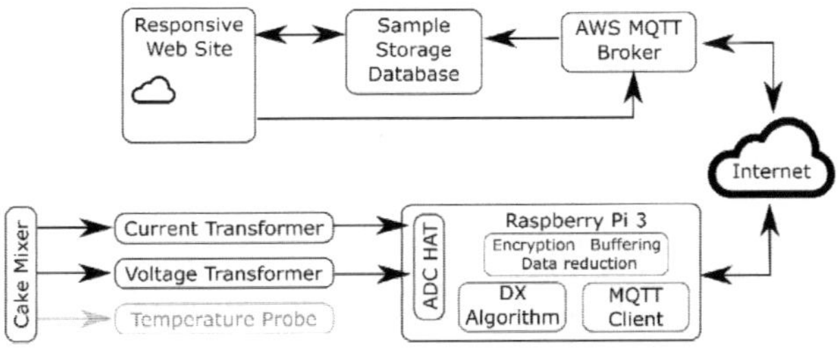

Figure 4: Physical System Architecture

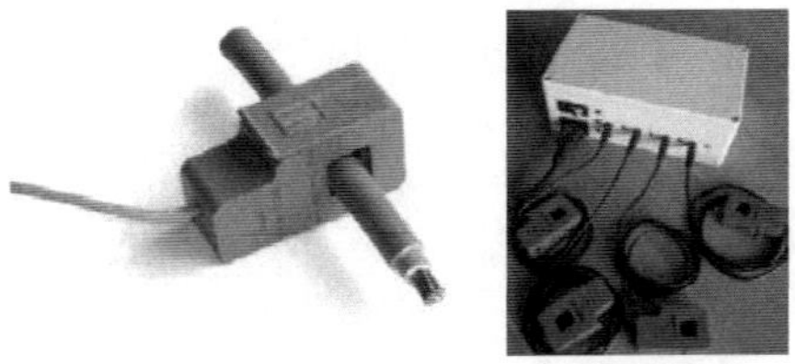

Figure 5: Depiction of the selected current transformer

Combining the Raspberry Pi 3, analogue to digital conversion (ADC) expansion board (HAT in Raspberry Pi parlance), an enclosure, and some CTs results in the prototype system depicted in Figure 5b. The prototype system can be developed for approximately US$80 (including a single CT) which is in line with the affordability constraints of the artisanal baker.

4. DIAGNOSTIC ALGORITHM

As mentioned in Section 2 MCSA algorithms are the most common form of non-invasive condition monitoring for induction machines and the systems to which these machines are connected. However, although the diagnostic environment in artisanal bakeries lends itself well to the application of condition monitoring techniques the identified challenges preclude the use of commercial solutions. Additionally, the lack of published literature complicates the business case presented to bakery owners. As such a two-pronged approach is followed in terms of the diagnostic algorithm.

An FFT-based approach has been identified as the most suitable and will provide sufficient information for future development endeavours. However, due to the large variance the mixer is subjected to, detailed diagnostic information is difficult to obtain reliably. In order to provide almost immediate commercial value to the bakery owner the existing maintenance process needs to be improved. Although the current maintenance process is largely RTF, economic improvements can be had if some operational information of the mixer were input to the maintenance scheduling activity. For this reason, the main diagnostic information gathered (and presented to the baker) concerns the running time of the mixer (operational hours) as well as the average and peak current draws of the mixer. The latter can be used by existing maintenance staff to aid in the diagnostic process.

From literature the start-up time of an induction machine can be indicative of incipient faults or simply wear and tear. As such, the algorithm should make provision for the identification of the following discrete events:

- Start-up;
- Peak starting current;
- Transition to operational mode;
- Machine shutdown.

A naïve approach would be to simply make use of threshold values which could be derived from the plate specifications of the machine. However, such an approach would not allow for natural variation between machines and, more importantly, would not allow for the commissioning of the system on used (or currently in-service) machines.

In order to accommodate machine variation derivatives are used in lieu of direct threshold values. With reference to Figure 6, the observed rms current clearly features the start-up, transient, running, and shutdown phases. However, in order to simplify matters a windowed rms function is deployed. Naturally the window length affects the transient detection resolution and thus the window should be kept as small as possible. In a South African power system the fundamental frequency of 50 Hz naturally lends itself to a window size of 20 ms. Although a crude approximation, this is similar to fundamental demodulation from [11].

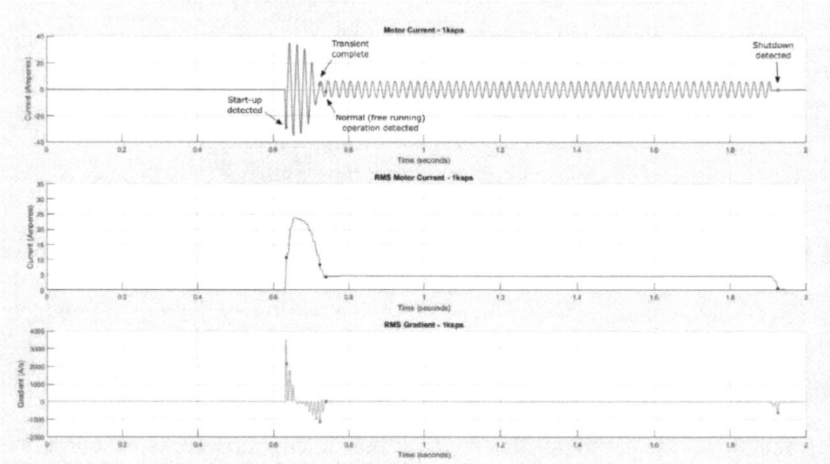

Figure 6: Up-time detection algorithm output

From the windowed rms current the inflection points can easily be identified by means of inspection. However, since inspection would be impractical for large-scale diagnostic deployments the use of current gradients and associated thresholds proved to be successful. A comment on the use of a threshold value is in order. Analytically speaking, setting the derivative of the current equal to 0 would provide the inflection points. However, since the differentiation is accomplished numerically the inflection points can not be guaranteed to be exactly 0. Rather than making use of tuned sensitivity parameters, setting gradient threshold values on the order of 750 A/s allows for natural machine variation while also accommodating the numerical variations expected. From Figure 6 it is clear that the proposed algorithm correctly identifies the various phases of machine start-up.

Although the raw data is stored locally on the Raspberry Pi's SD card, only the timestamps of the inflection points are transmitted via the MQTT protocol to the server. When the machine is running changes in operating current is handled in similar fashion, and should no changes be detected (steady-state operation), the average of the previous period is reported to the server.

5. INITIAL RESULTS

The application of the diagnostic algorithm in Figure 6 within the framework provided in Figure 4 results in the generation of a web-page that contains a visual representation of the mixer in question, some historic data including the total operational up-time of the machine, as well as the historic average current drawn.

The visual representation of the mixer is automatically updated to reflect the state of the machine, and the live diagnostic data is continually displayed as new field data is collected. A representation hereof is shown in Figure 8. The use of a responsive website rather than a dedicated application is that the website automatically allows for scalability across multiple platforms and devices.

Figure 7: Example of web front-end

Figure 8: Representation of live diagnostic data processing

The current implementation updates diagnostic data once per second at a total monthly cost of approximately US$8 (considering that 3G/LTE data is used). Due to the limitations imposed by only equipping a single cake mixer in a small artisanal bakery further exploration of the motor current data (or power spectra) can only be accomplished once sufficient maintenance data has been collected in addition to raw sensor data.

6. CONCLUSIONS AND RECOMMENDATAIONS

In this work the development of a prototype remote diagnostic and measurement system for artisanal bakeries were presented. It was demonstrated that a prototype system could be developed for approximately US$80 and feature affordable monthly operating costs. Although the current development is more focussed on the development of the hardware a return on investment is provided for by means of some operational use data that can be provided in real-time to the baker or maintenance staff. During the initial phase of the development detail diagnostic algorithms cannot be deployed due to the lack of operational and maintenance data. However, the system makes provision for the storage of up to 3 months' worth of raw data which would be suitable for the development of more sophisticated diagnostic algorithms.

Future work would include the development of more sophisticated diagnostic algorithms, specifically taking into account the variability in operational environment attributed to the characteristics of dough viscosity.

ACKNOWLEDGEMENTS

This work is based on the research supported, in part, by the National Research Foundation (NRF) of South Africa (grant nos. 91093 and 103392). Any opinion, finding and conclusion, or recommendation expressed in this material is that of the authors and not necessarily attributable to the NRF.

7. REFERENCES

[1] H. Zourides and T. Bolton, "Irresistibly better baking!," *Supermarket & Retailer,* pp. 12-15, January 2016.

[2] Vector Magazine, "Robotics in action in commercial bakeries," EE Publishers, 27 September 2016. [Online]. Available: http://www.ee.co.za/article/abb-robotics-proactive-connected-services-in-action-at-south-african-pioneer-foods-bakeries.html. [Accessed 15 February 2018].

[3] V. Bamford, "Smooth operators: maintain your mixers," *British Baker,* pp. 31-32, February 2018.

[4] G. P. Sullivan, H. Pugh, A. P. Melendez and W. D. Hunt, Operations & Maintenance Best Practices: A Guide to Achieving Operational Efficiency, Federal Energy Management Program, 2010.

[5] S. Mitra and C. Koley , "Vibration signal analysis of induction motors used in process control operation," in *2013 IEEE 1st International Conference on Condition Assessment Techniques in Electrical Systems*, Kolkata, India, 2013.

[6] S. Mitra and C. Koley, "Different measurement techniques for detection of bearing faults in industrial actuators — comparative study," in *2017 IEEE Calcutta Conference*, Kolkata, India, 2017.

[7] J. U. Porep, D. R. Kammerer and R. Carle, "On-line application of near infrared (NIR) spectroscopy in food production," *Trends in Food Science & Technology,* vol. 46(2) Part A, pp. 211-230, 2015.

[8] S. Karmakar, Induction motor fault diagnosis : approach through current signature analysis, Singapore: Springer, 2016.

[9] L. T. Ran, E. Kim and A. C. Tan, "A practical signal processing approach for condition monitoring of low speed machinery using Peak-Hold-Down-Sample algorithm," *Mechanical Systems and Signal Processing,* vol. 36, no. 2, pp. 256-270, 2013.

[10] M. Irfan, N. Saad, R. Ibrahim and V. Asirvadam, "Condition monitoring of induction motors via instantaneous power analysis," *Journal of Intelligent Manufacturing,* vol. 28, no. 6, pp. 1259-1267, 2017.

[11] D. Miljković,, "Brief review of motor current signature analysis," *CrSNDT Journal,* vol. 2, pp. 14-26, 2015.

[12] F. Campos, W. Mills and M. Graves, "A reference architecture for remote diagnostics and prognostics applications," in *AUTOTESTCON Proceedings*, Huntsville, AL, USA, 2002.

[13] W. Chen, Z. Chai and Q. Zhang, "A Real-Time Remote Diagnostic System Architecture Based on OSGi Platform," *Future Wireless Networks and Information Systems. Lecture Notes in Electrical Engineering,* 2012.

[14] Cisco, "Fog Computing and the Internet of Things: Extend the Cloud to Where the Things Are," Cisco Networks, USA, 2015.

[15] T. Ahmad, H. Studiawan and T. Ramadhan, "Developing a Raspberry Pi-based Monitoring System for Detecting and Securing an Object," *International Electronics Symposium (IES) 2014,* 2014.

[16] "MQTT," [Online]. Available: www.mqtt.org.

Online Performance Monitoring of Discrete Legs in a Convective Heat Exchanger of a Coal Fired Power Plant Boiler

G.T. Prinsloo, P.G. Rousseau, P. Gosai
Department of Mechanical Engineering, University of Cape Town, South Africa

ABSTRACT

The convective pass of a coal fired power plant boiler consists of several heat exchanger stages. Knowledge about the real-time performance of these heat exchangers is valuable for informing operational practices, preparing maintenance strategies and performing root cause analysis. For the specific boiler being studied, the steam in each stage is split into four legs that run in parallel across the boiler width. These legs alternate between the sides and across the centre of the boiler to disperse the effects of an uneven heat profile along the cross section of the flue gas flow path. Previous work has shown that the typical measurement setup in these boilers allows the performance parameters of each complete heat exchanger stage to be determined online using one-dimensional mass and energy balance equations [1][2]. However, flow and temperature maldistribution occurring in steam headers and the flue gas flow path leads to a non-uniform distribution of the performance of individual legs. We present and compare two methods of determining the performance of the individual heat exchanger legs online from process measurements. The first method employs soft sensors based on a thermo-fluid process model that not only accounts for the mass and energy balance, but also solves for momentum conservation. This allows the prediction of the flow distribution in common headers and ducts. The Flownex one-dimensional thermo-fluid network software is used for the modelling effort. The second method is based on the reconstruction-based contribution method which compares measured data to plant performance signatures that allows identification of deterioration of individual heat exchanger legs. These signatures are originally generated with the aid of the process model developed above for different fault scenarios. A case study of a final superheater stage is presented to assess the benefits and limitations of each method. These methods enable the performance of individual legs to be monitored online from process measurements and therefore provide higher resolution results within a heat exchanger stage.

Keywords: heat exchanger modelling, model-based performance monitoring, fault signatures, data reconstruction, diagnosis

Corresponding author: Gert Thomas Prinsloo (prnger003@myuct.ac.za)

1. INTRODUCTION

Coal fired power plants (CFPP) are the single largest generator of electrical energy worldwide [3]. In these plants, boilers transfer energy contained in coal to water/steam in the form of thermal energy. Optimal management of the steam generation process directly impacts on boiler reliability and the overall power plant thermal efficiency. The convective pass of coal fired boilers, where heat from the flue gas is transferred to tube surfaces mainly by convection, consists of several heat exchanger stages. A typical arrangement of these heat exchanger stages is shown in Figure 1. Knowledge about the real-time thermal performance of these heat exchangers is valuable for informing operational practices, preparing maintenance strategies and performing root cause analysis. Online performance monitoring is made possible by online measurements from which performance parameters can be deduced. In CFPP boilers, the flue gas temperature, abrasiveness and flow geometry inhibit representative online measurements on the flue gas side. Therefore, the conventional approach to online performance monitoring utilize soft sensors based on mass and energy balance (MEB) models to obtain flue gas parameters required to estimate heat transfer rates and overall heat transfer coefficients of the entire convective pass [4] or heat exchanger stages [1][2], from steam side process measurements. Other techniques such as heat flux measurements or strain gauged sling tubes are usually applied together with MEB models to corroborate results [5].

In this study we consider a specific case study boiler where each of the heat exchanger stages in turn consist of four legs running in parallel across the boiler width. These legs alternate between the centre and the sides of the boiler as the steam travels from one heat exchanger stage to the next. This is done to disperse the effects of an uneven flue gas heat profile along the cross section of the flue gas flow path. This configuration is the norm in modern boiler design [6]. Since the resolution to which the outcomes of online monitoring can be optimized is directly linked to the resolution of monitoring, we aim to monitor the performance of individual legs. We propose two methods for creating such a monitoring system. The first method, being the most intuitive approach to online monitoring, is the design of soft sensors to estimate unmeasured parameters of interest. We base the soft sensor design on a steady state thermo-fluid process model. The second method is similar to data-based monitoring, which classes patterns in measurement data according to the fault status. However, we obtain our labelled training data from process model simulation results rather than from historical process data. We compare the benefits and limitations of the two methods at the hand of actual plant data while assuming healthy sensors and actuators.

2. CASE STUDY BOILER DESCRIPTION

The case study boiler is a 618 MW two pass drum boiler with reheat. For this paper we focus on monitoring the final superheater (SH3). Figure 1 shows a diagram of this boiler with the location of SH3.

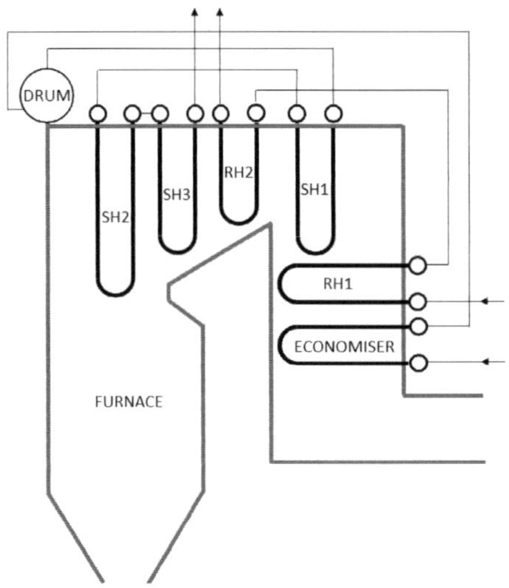

Figure 1: Side view of the case study boiler convective pass heat exchanger stage layout.

SH3 has a cross-parallel flow configuration and consists of 28 elements, which can be further broken down into four legs running in parallel, each containing seven elements. Throughout this paper, each of these legs will be represented as a single lumped heat exchanger as represented in Figure 2.

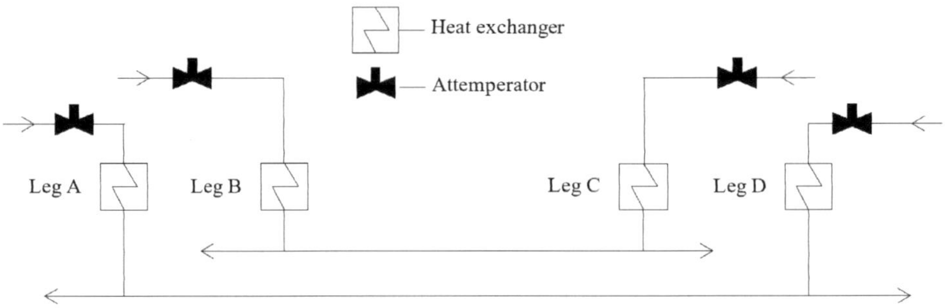

Figure 2: Final superheater steam flow layout.

On the steam side, each leg has a separate inlet connected to the separate outlets of SH2 by the final superheater spray water attemperator stations. The outlets of legs A and D are connected by a common outlet header. The steam flow can therefore be distributed between these outlets. This implies that the steam flow and temperature measurements on the left-hand side of the final outlet header, shown in Figure 3, do not necessarily only represent the average values of steam coming out of leg A. It can also include the effect of steam coming out of leg D, depending on how much mixing takes place within the header, since the flow will rarely be distributed symmetrically. This configuration and reasoning also holds for legs B and C.

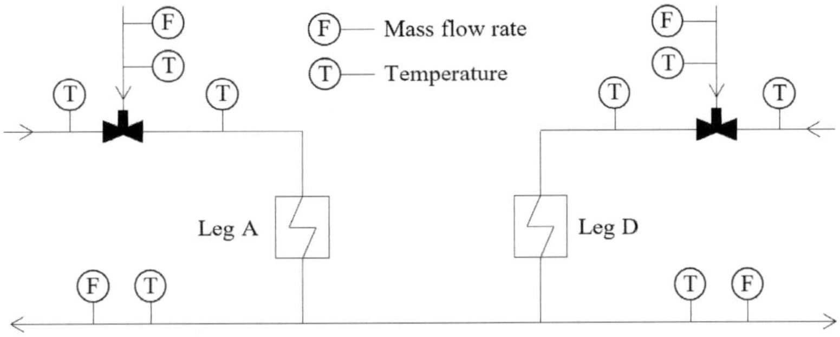

Figure 3: Final superheater measurement configuration. Note that the configuration shown for legs A & D also holds for legs B & C, which is not included in this figure for clarity.

Because of the abrasive, high-temperature combustion products it is difficult to obtain representative measurements on the flue gas side. Process measurements are therefore predominantly located on the steam side, taken on interconnecting piping where no significant heat is being transferred. Flue gas temperature measurements are located up- and downstream of the economiser heat exchanger, which is far downstream of the final superheater. Measurements of overall coal and air flow rate into the boiler are also available.

3. METHOD 1 – SOFT SENSOR DESIGN

In this method, a process model is developed using a first principles one dimensional approach to predict steam flow through each leg of the final superheater. The resultant soft sensors predicts values of unknown parameters of interest based on the limited measured values indicated in Figure 3.

3.1. Steam side model

The steam side is modelled according to the flow layout shown in Figure 2. Each heat exchanger leg is modelled as a pipe with uniform heat input along the length. The dimensions and other physical characteristics of the pipe and remaining flow connections are taken from plant design data to account for flow resistance as a function of flow characteristics. Using the measurements shown in Figure 3 as input boundary conditions, the model solves for the conservation of mass, energy and momentum. The results show the thermodynamic state of the steam at each location throughout the final superheater system with accompanying flow distributions. Of specific interest is the mass flow rate through each heat exchanger leg and the corresponding outlet steam temperatures, from which the heat transferred to the steam flowing through each leg is determined.

3.2. Flue gas side model

The flue gas path is modelled as four streams linked to each leg in the final superheater as shown in Figure 4. Each stream is modelled as a pipe with the assumption that mixing between streams within the heat exchanger stage is negligible. The heat extracted at each heat exchanger leg is assumed to be equal to that gained by the steam, thereby neglecting directed radiation and losses. The remaining inputs required by the model is the flue gas temperature downstream of SH3 and the overall mass flow rate. These inputs are obtained from indirect measurement as described in [1]. The model solves the remaining thermodynamic states and flow distributions of the flue gas.

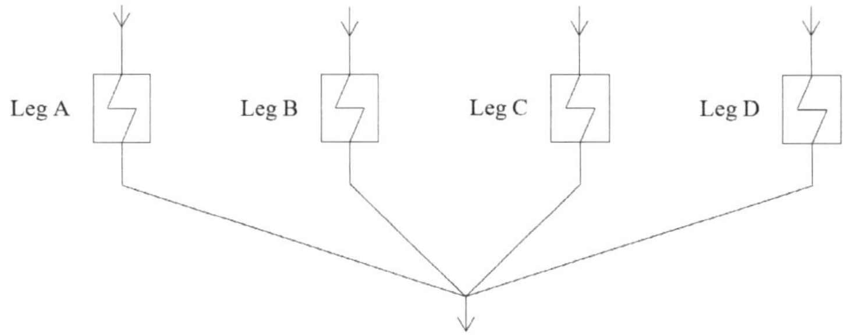

Figure 4: Flue gas model layout.

3.3. Thermal performance evaluation

The thermal performance of a heat exchanger is represented by its total thermal resistance R_{total}. Using the average properties obtained with the soft sensors, the actual value of R_{total} can be determined for each heat exchanger leg using the heat transfer rate, $\dot{Q}$, and the log mean temperature difference, ΔT_{LMTD} [7]. If the actual resistance is baselined against that expected at the corresponding load according to design, conclusions can be drawn about the magnitude of deviations from the design values. However, this approach does not allow one to determine whether this deviation is caused by changes in the physical heat exchanger characteristics or by deviation of process properties from design values. Changes in heat exchanger characteristics relate to increased conduction thermal resistance caused by external ash and internal oxide layer build-up on the tubes. Deviations in process properties lead to changes in the convection and radiation thermal resistance, as well as changes in temperature differences driving heat transfer. These causes are respectively classified as uncontrollable- and controllable losses as discussed in [8], where uncontrollable losses are related to aging and degradation of plant equipment. Using the actual process properties obtained with the soft sensors, the expected convection and radiation thermal resistance can be calculated using suitable correlations. The difference between these quantities and the total thermal resistance yields the conduction thermal resistance. In this study, the Dittus and Boelter correlation [9] and the Hillpert correlation [10] is used to respectively estimate the steam and flue gas convection thermal resistance, while radiation thermal resistance is neglected.

4. METHOD 2 – DATA-BASED APPROACH

The data-based method proposed here is analogous to the reconstruction-based contribution (RBC) method described in [11]. In the RBC method, the normal operating region (NOR) is defined with principal components analysis (PCA) of process history data obtained during normal operation. When a new measurement or sample vector is outside the NOR boundaries, a fault is detected. Isolation is then achieved by removing the effects of possible faults from the sample vector. The fault vector of which the removal reconstructs the sample vector back into the NOR is then the assumed fault. This method requires the fault library, which contains all fault vectors or matrices, to be known [12].

However, the proposed method does not construct a NOR from process history data, but rather defines the normal operating point (NOP) using design data corresponding to the load. Instead of placing the fault library at the sample vector and evaluating which fault vector penetrates the NOR, the proposed method centres the data around the NOP and performs a change of basis from the measurement space using the fault library as the transformation matrix. This produces the contribution of each possible fault to a measurement sample if the fault library is invertible, allowing diagnosis of multiple simultaneous faults. This method also requires the fault library to be known.

The fault library can be constructed using either labelled process history data or model simulation. Labelled data refers to data collected while the fault status of equipment was known. This poses limitations for application to boilers because of, amongst other reasons, the following: (1) Obtaining data corresponding to a known online fault status implies the existence of a means to determine its fault status online. This is difficult for power plant boilers because the harsh environment in which they operate impedes measurements for verifying faults. (2) Imposing faults in an actual plant for training purposes are not feasible as the relevant fault parameters are neither measured nor controlled. (3) Training is further complicated by boilers being prone to several faults being active simultaneously. (4) Service intervals are quite extended, reducing the opportunity for correlating data of the previous operating cycle to the fault status established from offline inspection. (5) Diagnosis is limited to encountered faults.

Because of these limitations, we use a thermo-fluid process model to construct the fault library. The process model is simulated with imposed faults to obtain fault signatures. The accuracy of the fault signatures therefore depends directly on the accuracy of modelling. Nonlinearities in process parameter relations can however cause changes in fault signatures depending on the operating conditions and the magnitude of deviations. It is posited that, for a given load, the state of parameters will not deviate drastically from design values. We therefore generate the fault library as a function of the load only.

4.1. Model configuration

The model is developed in Flownex for steady state operation with the same layout as that of the soft sensor model described above. The input-output configuration differs from that used in the soft sensor model. The soft sensor model receives measurements as inputs to calculate unmeasured parameters of interest. For constructing the fault library, we use unmeasured parameters of interest as inputs to calculate the expected measurements. This allows one to impose changes to the unmeasured parameters and observe the measurement response. Some boundary conditions required to fully determine the model are also measurements, but their response is a function of unmodeled processes. These parameters are therefore also included as inputs and treated as unmeasured parameters.

Heat exchanger legs are modelled to account for convection heat transfer on the flue gas and steam sides with the same correlations used for the soft sensor model. The model also accounts for the effects of fouling on heat transfer resistance by incorporating a cleanliness factor. The attemperator spray water mass flow rates are calculated to achieve the steady state outlet steam temperature setpoint. The model is adjusted to align results with design data.

4.2. Training methodology

The different faults are imposed on the process model to construct the fault library. A fault is any parameter that deviates from the design conditions at the corresponding load.

Without assigning structure to the changes in the flue gas side properties (biasing or deviations from the mean), we cannot distinguish if decreased heat transfer is caused by lower flue gas temperature, lower flue gas mass flow rate, or increased fouling. Similarly, the cause of increased heat transfer being either increased flue gas temperature or flow rate cannot be isolated. This is because these faults ultimately affect the measurement space in the same way, therefore yielding linearly dependent fault vectors for any of these faults. All these parameters are therefore lumped into one fault called the heat transfer index (HI), which represents a deviation from the design heat transfer rate. A HI of less than zero represents either less heat provided by the flue gas, or the increased heat transfer resistance of fouling. A HI of more than zero indicates that the greater than design heat provided by the flue gas has a greater influence on the heat transfer than the resistance caused by fouling.

The table below shows the parameters that are changed respectively for each leg or the whole stage during training, along with the specific magnitude of the deviation from design conditions.

Table 1: Faults imposed as deviations from design values for training.

Parameter	**Size of deviation**
Steam inlet temperature per leg	+5 °C
Heat transfer index (HI)	+0.4 MW
Total steam mass flow rate for stage	+5 kg/s

The changes imposed on these parameters during the training phase do not need to be both in the positive and negative directions, but only in one direction. This is because the fault signature corresponding to a decrease in a specific parameter is the negative of the fault signature for an increase in that same parameter. Further, only one simulation is needed per fault case.

The simulation above is repeated for various loads to construct fault libraries as a function of load. During monitoring, the fault library is interpolated between the simulation cases to match the actual plant load.

5. RESULTS

A soot blowing event is used to compare the two methods described in this paper. The reason for this choice is that the results produced by the two monitoring frameworks are expected to respond in a predictable way during such an event. The expected nett effect is a decrease in the conduction thermal resistance of the heat exchanger in the region where soot blowing is taking place, by the removal of ash from the tubes. Therefore, it should result in an increase in the heat transfer rate and in the HI after soot blowing. However, it is possible for the heat transfer rate to first decrease during the event, if the steam used for soot blowing is at a lower temperature than the flue gas, only to increase afterwards.

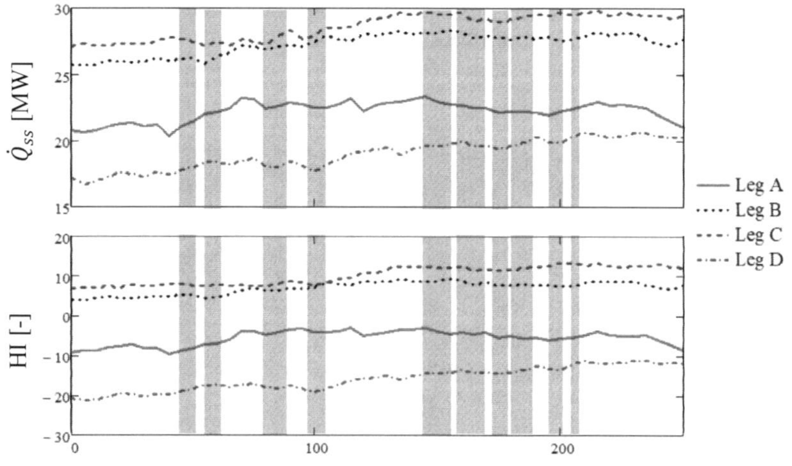

Figure 5: Heat transfer indicators obtained by the soft sensor (top) and data-based (bottom) methods for all legs

Actual plant data was recorded at five-minute intervals over a period of 250 minutes during which soot blowing took place at a constant load of 618MW. The darkened backgrounds in Figure 5 to Figure 8 indicate the activation of tube leak detectors near SH3, which is an indication of active soot blowing. The heat transfer index being monitored by the data-based method, HI, as well as the heat transfer rate calculated with the soft sensors, $\dot{Q}_{SS}$, are plotted for all legs in Figure 5. The HI indicates a deviation from the design heat transfer rate with a value of one corresponding to the trained deviation of +0.4 MW in Table 1.

The results are quite similar for all legs and for brevity only that for leg D are shown in more detail below. The HI and $\dot{Q}_{SS}$ are overlaid in Figure 6.

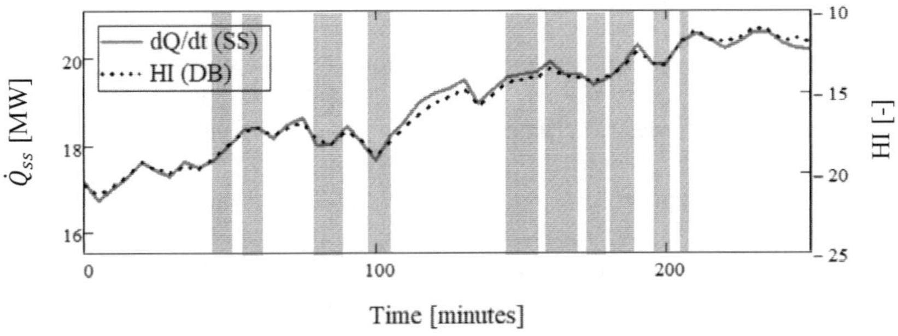

Figure 6: Heat transfer indicators overlaid for leg D

The total thermal resistance obtained with the soft sensors along with its components of conduction and convection are shown in Figure 7.

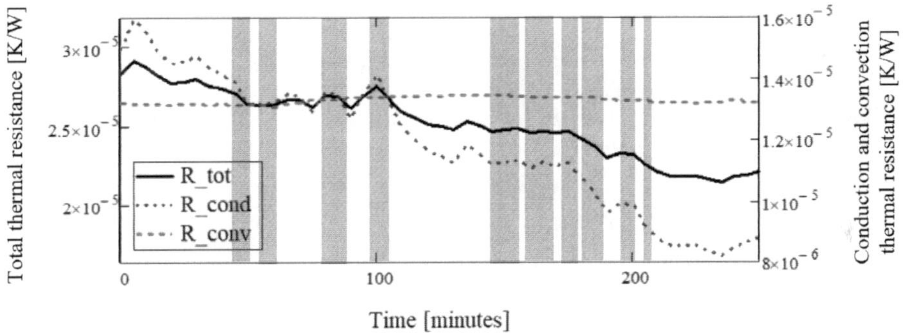

Figure 7: Thermal resistance constituents of leg D obtained with soft sensors

The log mean temperature difference, ΔT_{LMTD}, of the two fluids across the heat exchanger leg, obtained with the soft sensors is plotted in Figure 8.

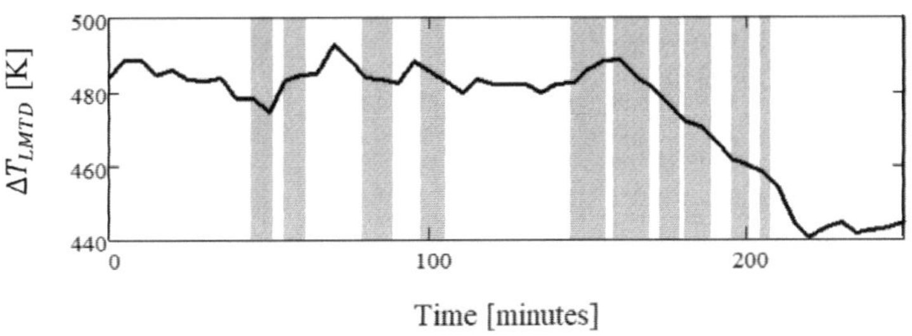

Figure 8: Log mean temperature difference across leg D

6. DISCUSSION AND CONCLUSIONS

The results shown in Figure 5 correspond to what is expected. Legs A and D, which are located at the opposite sidewalls of the boiler, consistently have lower heat transfer rates than the two centre legs as determined with the soft sensor method. This corresponds with the fact that heat is lost through the boiler walls. The indicators of heat transfer rate determined with the separate methods are in excellent agreement, providing confidence in the validity of both methods. Figure 7 shows that the conduction thermal resistance undergoes much more changes during the soot blowing event than the convection thermal resistance, since the ash layer forms part of the conduction resistance. The response in the temperature difference across the heat exchanger shown in Figure 8 can be caused by some combination of changes in process parameters because of soot blowing and controls responding to these changes.

While the data-based method is limited to monitoring deviations from the design heat transfer rate, the soft sensors can further identify the contribution of conduction thermal resistance, convection thermal resistance and temperature differences to this deviation. This allows one to classify losses as being either controllable or uncontrollable. improving root cause analysis and allowing corrective action to be taken more effectively.

The training methodology employed for the data-based method only used one simulation for each type of deviation, but specialised software was required for these simulations. However, its application to online monitoring only requires relatively small sized matrix multiplication, which makes this an attractive method for real-time diagnosis. The soft sensor method required simulation with specialised thermo-fluid software for each monitored sample. These simulations are more computationally expensive and requires manual intervention.

7. ACKNOWLEDGEMENTS

The authors would like to acknowledge the support provided by Eskom through the Eskom Power Plant Engineering Institute (EPPEI) and the University of Cape Town, as well as M-Tech Industrial (Pty) Ltd for allowing the use of Flownex®.

REFERENCES

[1] A. Valero and C. Cortbs, "Ash Fouling in Coal-Fired Utility Boilers. Monitoring and Optimization of On-Load Cleaning," vol. 22, no. 96, pp. 189–200, 1996.

[2] C. Cantrell and S. Idem, "On-Line Performance Model of the Convection Passes of a Pulverized Coal Boiler," *Heat Transf. Eng.*, vol. 31, no. 14, pp. 1173–1183, 2010.

[3] IEA - International Energy Agency, "Key World Energy Statistics 2017," 2017.

[4] J. Taler, M. Trojan, and D. Taler, "Computer System for On-Line Monitoring of Slagging and Fouling and Optimization of Sootblowing in Steam Boilers," pp. 1–12, 2010.

[5] S. A. Idem, "An Integrated Approach to Online Monitoring of Fouling in Coal-Fired Power Plants," *Heat Transf. Eng.*, vol. 28, no. 2, pp. 73–75, 2007.

[6] M. . Nielson, F.S., Danesi, P. and Radhakrishnan, "Modern Boiler Design," *BWE Energy India*, no. January, 2012.

[7] F. P. Incropera, D. P. DeWitt, T. L. Bergman, and A. S. Lavine, *Fundamentals of Heat and Mass Transfer*. 2007.

[8] H. Kim, M. G. Na, and G. Heo, "Application of monitoring, diagnosis, and prognosis in thermal performance analysis for nuclear power plants," *Nucl. Eng. Technol.*, vol. 46, no. 6, pp. 737–752, 2014.

[9] R. H. S. Winterton, "Where did the Dittus and Boelter equation come from?," *Int. J. Heat Mass Transf.*, vol. 41, no. 4–5, pp. 809–810, 1998.

[10] R. Hilpert, "Wärmeabgabe von geheizten Drähten und Rohren im Luftstrom," *Forsch. auf dem Gebiet des Ingenieurwesens A*, vol. 4, no. 5, pp. 215–224, Sep. 1933.

[11] C. F. Alcala and S. J. Qin, "Reconstruction-based contribution for process monitoring," *Automatica*, vol. 45, no. 7, pp. 1593–1600, 2009.

[12] C. F. Alcala and S. J. Qin, "Reconstruction-based contribution for process monitoring with kernel principal component analysis," *Am. Control Conf. (ACC), 2010*, pp. 7022–7027, 2010.

Simulation Based LOX/Methane Rocket Engine Fault Features Analysis

Jingyu xiong, Jianjun Wu, Yuqiang Cheng
College of Aerospace Science and Engineering, National University of Defense Technology, China

ABSTRACT

LOX/Methane Rocket Engine has attracted widely attention due to its advantages, such as economy, nontoxicity and good comprehensive performance. In order to solve the problem of parameter selection and optimization for LOX/methane rocket engine health monitoring, the fault features of reusable LOX/methane rocket engine are investigated and the key parameters are selected for engine health monitoring. Due to the lack of fault sample data, it is difficult to obtain the features of fault modes. So the modular model library was built on the basis of lumped parameter approximation, and the dynamic simulation of the work process for the engine was carried out on the basis of MATLAB Simulink system. To establish fault simulation model, fault factors are added into the balance equations of components. Comparison results show that the simulation results are in good agreement with the real test data of the engine. It can complement important fault sample data in the research of fault detection and diagnosis methods for the engine. Using these data we can analyze the change of parameters in the dynamic transition process from a normal state to a fault state. Finally, the key parameters for typical fault detection and diagnosis are obtained.

Keywords: LOX/Methane Rocket Engine; Modeling and Simulation; Fault Features Analysis; Parameter Selection; Fault Simulation

Corresponding author: Jianjun Wu (jjwu@nudt. edu. cn)

1. INTRODUCTION

LOX/Methane rocket engine is a new type of liquid rocket engine. Its propellant has significant advantages such as low viscosity, low density, high heat capacity, high specific impulse, low carbon deposition during combustion, low price, non-toxicity, and non-pollution. LOX/Methane is called 21st-century rocket propellant, and LOX/Methane rocket engine has broad application prospects[1].

The fault test of rock engine is dangerous and the cost is expensive. So it is difficult to get fault sample data. System modeling and simulation for liquid propulsion systems provides a way to research the fault of rock engine. Since 1970s, lots of contributions has been made for modeling and simulation of Liquid Propellant Rocket Engine. In 1973, Rocketdyne published a report regarding SSME's dynamic characteristics related to Pogo phenomenon, engine components were mathematically modeled using linear differential[2]. In 1995, C. Goertz proposed a modular method for different cycles of engine. Through this method, users can select the basic components to simulate the static performances of engines such as SSME and RD-12[3]. In 2000, K. Liu, Y. Zhang used Microsoft Visual C++ 6.0 to develop a method for modular dynamic simulation of liquid propellant rocket engines[4]. In 2017, M. Naderi, et al. developed a modular simulation software for steady state analysis of Stage Combustion Cycle Liquid Propellant Rocket Engine using MATLAB Simulink. In their work, SSME is studied as a case study and the required input and output data for each module has been described and the algorithm is presented but the coefficients required for the equations are not described and the programming algorithm which also has not been explained in detail[5].

In the current paper, the second chapter gives a brief structure introduction to the LOX/Methane rocket engine. In the third chapter, the fault model of each engine component is established using the lumped parameter approximation modeling theory and modular modeling idea. Then connect each components and form the engine integrated system model in MATLAB Simulink. In the fourth chapter, simulate and analyze the typical faults. Finally the features of fault and the key parameters for typical fault detection and diagnosis can be extracted from the simulation data.

2. STRUCTURE OF ENGINE

LOX/Methane rocket engine adopts liquid oxygen and liquid methane cryogenic propellant. This gas generator cycle for pump-fed liquid rocke engine was chosen. The engine consists of thrust chambers, gas generator, turbopumps, propellant supply systems, valves and regulating components. The two propellants come out of the tank and are pumped into the thrust chamber and the gas generator. The gas produced by gas generator goes into the methane turbine and the oxygen turbine respectively and then excreted. The advantages of using this cycle are: The system is relatively simple, the turbine power is small and the quality of the engine is lighter, so the requirements for production and test are relatively low, and the cost of the engine itself is reduced. However the drawback of this design is that the specific impulse of the engine is lower because of the turbine exhaust loss. The system diagram is shown in figure 1.

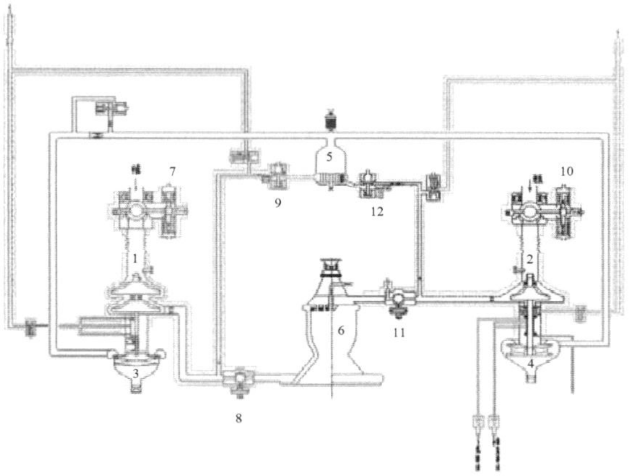

1 - methane pump 2 - oxygen pump 3 - methane turbine 4 - oxygen turbine 5 - gas generator6 - thrust chamber 7 - methane pump front valve
8 - methane main control valve9 - methane secondary control valve 10 - oxygen pump front valve 11 - oxygen main control valve12 - oxygen pair control valve

Figure 1. The system diagram

3. MODEL OF COMPONENTS

The main components of the turbo-pump liquid rocket engine system are fixed, including turbine, pump, thermal assembly (combustion chamber, gas generator, gas conduit), regulator (throttling device) and automatic device. For different types of liquid rocket engines, the dynamic mathematical model has a certain generality in form. Using this feature, build each component separately. After the component model is created, it can be debugged separately, and then combined together to form the entire system model according to its physical connection. In this way, the modular modeling and simulation of the engine system can be performed. Similarly, we use this method to model and analyze LOX/Methane rocket engine[6].

Literature [7] summed up the main fault of pump pressure liquid rocket engine. Turbo pump, combustion chamber and pipeline fault sum more than 70% including pump cavitation, combustion chamber ablation, blocked and leakage.

1. Turbine model

The engine turbine includes an oxidizer turbine and a fuel turbine. The relationship between various parameters in the turbine working process can be described by the following equation.

Turbine power equation:

$$N = v_e^2 / 2 \cdot q \cdot \eta , \tag{1}$$

Turbine efficiency equation:

$$\eta = b_1 + b_2\left(\frac{n}{v_e}\right) + b_3\left(\frac{n}{v_e}\right)^2 , \tag{2}$$

Turbine flow equation:

$$q=\begin{cases}\dfrac{\mu AP_i}{\sqrt{(RT)_i}}\sqrt{k\left(\dfrac{2}{k+1}\right)^{\frac{k+1}{k-1}}} & \dfrac{P_0}{P_i}\le\left(\dfrac{2}{k+1}\right)^{\frac{k}{k-1}}\\ \dfrac{\mu AP_i}{\sqrt{(RT)_i}}\sqrt{\dfrac{2k}{k-1}\left[\left(\dfrac{P_0}{P_i}\right)^{\frac{2}{k}}-\left(\dfrac{P_0}{P_i}\right)^{\frac{k+1}{k}}\right]} & \dfrac{P_0}{P_i}>\left(\dfrac{2}{k+1}\right)^{\frac{k}{k-1}}\end{cases}, \tag{3}$$

Turbine gas theory jet velocity:

$$v_e=\sqrt{\frac{2k}{k-1}(RT)_i\left[1-\left(\frac{P_e}{P_i}\right)^{\frac{k-1}{k}}\right]}, \tag{4}$$

Turbine power balance equation:

$$J\frac{dn}{dt}=\frac{N-\sum N_p}{(\pi/30)^2\cdot n}, \tag{5}$$

Gas pressure between turbine stator and rotor:

$$P_0=P_i\left[\theta+(1-\theta)\left(\frac{P_e}{P_i}\right)^{\frac{k-1}{k}}\right]^{\frac{k}{k-1}}, \tag{6}$$

Where k represents gas adiabatic index, R represents gas constant, T represents gas temperature, b represents coefficient of experience, μ represents flow coefficient of nozzle, A represents turbine nozzle area, N represents turbo power, $\sum N_p$ represents sum of power driven by turbine. θ represents turbine reaction force.

2. Pump model

When simulating the working process of the engine, the pump's static characteristics are used to describe the operation of the pump. Its static characteristics are completely determined based on experience and can be described as follows.

Pump head:

$$\Delta P=P_e-P_i=\left(\mu_{p1}n_p^2+\mu_{p2}n_pq_p+\mu_{p3}q_p^2\right)\cdot F_{p1}, \tag{7}$$

Pump power:

$$N_p=\upsilon_{n1}n_p^3+\upsilon_{n2}n_p^2q_p+\upsilon_{n3}n_pq_p^2, \tag{8}$$

Were ΔP represents pump head, $\mu_{p1},\mu_{p2},\mu_{p3}$ represents the empirical coefficient of the pump head, υ_{n1}, υ_{n2}, υ_{n3} represents the empirical coefficient of the pump power, F_{p1} represents failure factor, when $F_{p1}=1$, represents pump operating in normal state ;when $0\le F_{p1}\le 1$ indicates pump cavitation failure.

3. Thermal Component Model

The engine thrust chamber mathematical model can be described by the following equations:

Mass conservation equation:

$$\frac{dm_g}{dt}=q_{ig}+q_{lo}+q_{lf}-(1+F_{g2})q_{eg}, \tag{9}$$

Variation of gas density in thermal components:

$$\frac{d\rho}{dt}=\frac{1}{V}\cdot\frac{dm_g}{dt},\tag{10}$$

Change rate of gas mixture ratio in thermal components:

$$\frac{dr}{dt}=(1+r)(q_o-r\times q_f)\cdot\frac{RT}{PV},\tag{11}$$

Gas heat value:

$$RT=RT(r),\tag{12}$$

Both sides derivative of Ideal gas equation:

$$\frac{dP}{dt}=\frac{RT}{V}\cdot\frac{dm_g}{dt}+\frac{P}{RT}\cdot\frac{d(RT)}{dt}-\frac{P}{V}\cdot\frac{dV}{dt},\tag{13}$$

Outlet flow equation:

$$q_e=\begin{cases}\dfrac{\zeta F_{g1}AP}{\sqrt{(RT)_i}}\sqrt{k\left(\dfrac{2}{k+1}\right)^{\frac{k+1}{k-1}}} & \dfrac{P_e}{P}\le\left(\dfrac{2}{k+1}\right)^{\frac{k}{k-1}}\\ \dfrac{\zeta F_{g1}AP}{\sqrt{RT}}\sqrt{\dfrac{2k}{k-1}\left[\left(\dfrac{P_e}{P}\right)^{\frac{2}{k}}-\left(\dfrac{P_e}{P}\right)^{\frac{k+1}{k}}\right]} & \dfrac{P_e}{P}>\left(\dfrac{2}{k+1}\right)^{\frac{k}{k-1}}\end{cases},\tag{14}$$

Where m_g represents the quality of high temperature gas in thermal components, ρ represents density of gas, V represents volume of gas, P represents pressure of gas, k represents mixing ratio of gas. q_{gi} 、 q_{lo} and q_{lf} represent the gas mass flow, the liquid oxidant mass flow and the liquid fuel mass flow into the heat module respectively; q_{ge} represents Outlet flow of thermal components; ζ represents the flow coefficient of thermal component throat; A represents the throat area of thermal components, F_{g1} represents throat ablation factor, normal state: $F_{g1}=1$, ablation state, $F_{g1}>1$; F_{g2} represents gas leakage, normal state $F_{g2}=0$, failure state, $F_{g2}>0$.

4. Liquid pipe model

Liquid pipeline plays the role of propellant transmission to components in the engine. Its input is pressure and propellant mass flow, and at the same time, the output pressure and propellant mass flow rate of the next component. In modeling and simulation, it is necessary to consider the inertia, viscosity and compressibility of the liquid.

Considering the inertia and viscosity of fluids, according to the principle of pressure superposition, the equations of motion of the propellant elements in the pipeline can be obtained:

$$L\frac{dq}{dt}=F_{l1}[(P_i-P_e)-F_{l2}\alpha q^2],\tag{15}$$

Considering the compressibility of the fluid, according to the mass conservation equation, the continuous equation of the propellant component in the pipeline can be obtained by derivation:

$$\xi\frac{dP_e}{dt}=(q_i-q_e)/F_{l2},\tag{16}$$

Where ξ represents liquid pipe flow capacity coefficient, it reflects the compressibility of the liquid in the pipeline; L represents the inertial resistance coefficient of the liquid; α represents the resistance coefficient of the pipeline; q_i and q_e represent the inlet and outlet mass flow of the pipeline separately; P_i and P_e represent the pressure at the inlet and outlet of the pipeline separately; V represents line volume, a represents sound speed in liquid, F_{l1} represents factor of leakage, leakage state: $1 > F_{l1} > 0$, normal state: $F_{l1} = 1$; F_{l2} represents factor of blocked, blocked state: $F_{l2} > 1$, normal state $F_{l2} = 1$.

Based on the modular model of the engine system, fault factors are added into the balance equations of components the engine, fault simulation system is designed. We use MATLAB Simulink to model each module. Finally, connect each module according to engine structure, the fault simulation software for LOX/methane rocket engine was obtained, as shown in figure 2. According to the input conditions of the simulated working process, by calling the components coefficients of the engine database and the input coefficients of each component can be set and call the four order Runge-Kutta method to solve.

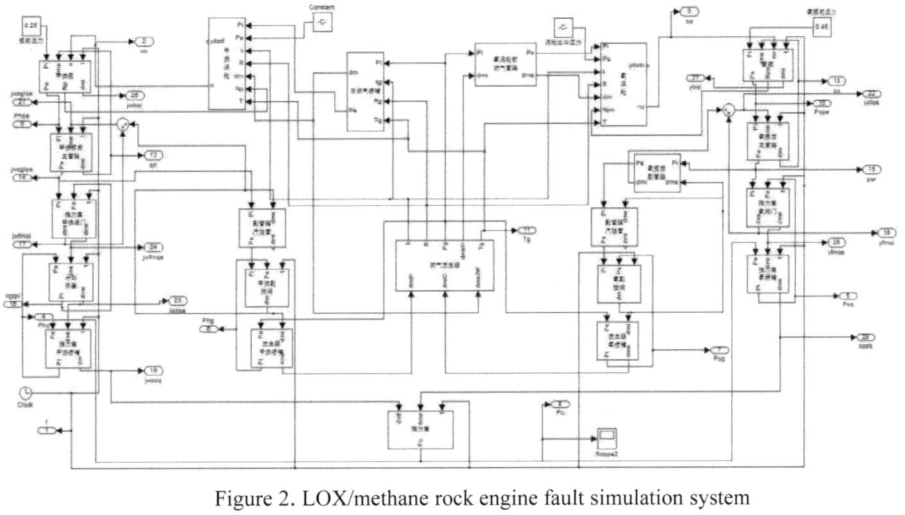

Figure 2. LOX/methane rock engine fault simulation system

4. RESULT ANALYSIS

In steady state the maximum error of actual data and simulation data is less than 12%, so the simulation system is considered to be accurate.

In this paper, rock engine typical failures in steady state (leakage, cavitation, ablation, blocked) are simulated. There are 27 monitoring parameters set in the system. Taking methane pipeline fault as example, the changes of oxygen parameters are much smaller than the methane's, so the figures of the oxygen parameters are ignored

Use the dimensionless parameter D to characterize the change of each parameter after the failure.

Annotate: $D = \dfrac{\text{Fault value} - \text{Rated value}}{\text{Rated value}}$

1 Methane main pipe blocked

The blocked factor of the methane main pipe in the simulation model is set to 2.0, the parameters change with time as shown in figure 3.

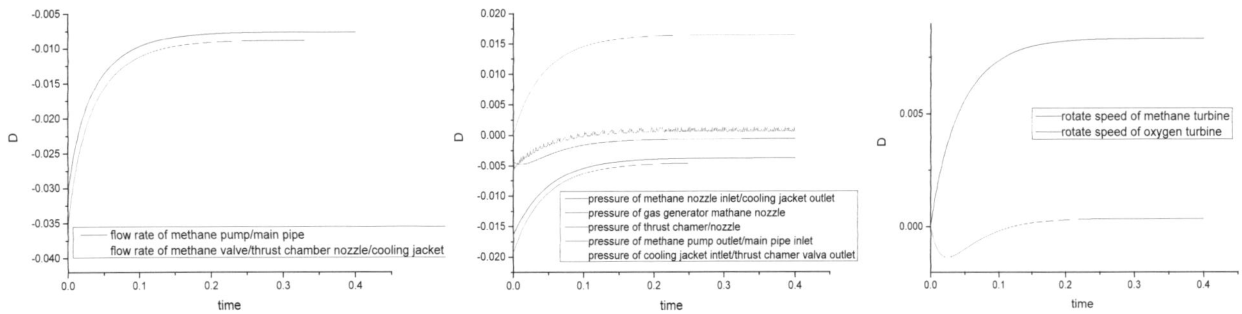

Figure 3. Change of methane parameters with methane main pipe blockage factor 2

When a blocked fault occurs, the equivalent flow resistance coefficient increases, and the value of the blockage failure factor is equivalent to the increase in the flow resistance through the component. When the flow resistance of the methane main pipe is doubled, the methane main flow rate is the same as the methane pump flow rate change. The methane valve flow rate, the methane nozzle flow rate, and the cooling jacket flow rate in the thrust chamber are the same and are reduced by approximately 1%. The inlet pressure of the main methane pipe increased by about 1.5%. The pressure of the remaining components of the methane line decreased in different degrees. After the stable combustion chamber, the inlet pressure of the methane nozzle was restored to the vicinity of the rated value. By comparison, the outlet pressure of the methane main pipe drops by a maximum of 0.45%. The increased rotating speed of the methane turbine is significantly faster than that of the oxygen turbine.

2. Methane main pipe leakage fault

The methane main pipe leakage factor of the simulation model is set to 0.2, the parameters change with time as shown in figure 4.

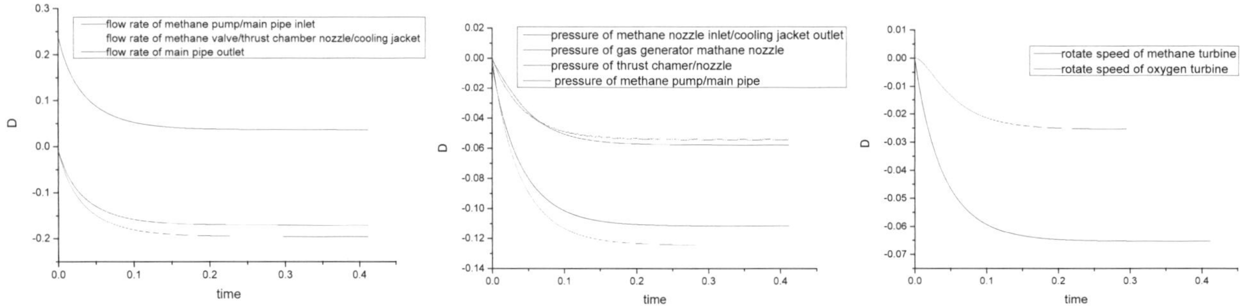

Figure 4. Change of methane parameters with methane main pipe leakage factor 0.2

When the leakage factor of the methane main pipe is 0.2, the flow of the main pipe inlet is the same as that of the methane pump, and the flow rate increases by 5% when reaching the steady state. The flow rate of the methane valve in the thrust chamber, the flow rate of the methane nozzle, and the cooling jacket flow rate are the same, which are reduced by about 20%. The pressure of each components of the methane pipeline decreased in different degrees. Among them, the methane pump outlet, the methane main pipe outlet, and the pressure in the thrust room valve outlet were more similar, their reduction is nearly 12.6%. The speed of the methane turbine and the oxygen turbine is reduced.

3. Ablative fault of thrust chamber

Setting the thrust chamber ablation factor of the simulation model to 2.0, the parameters change with time as shown in figure 5.

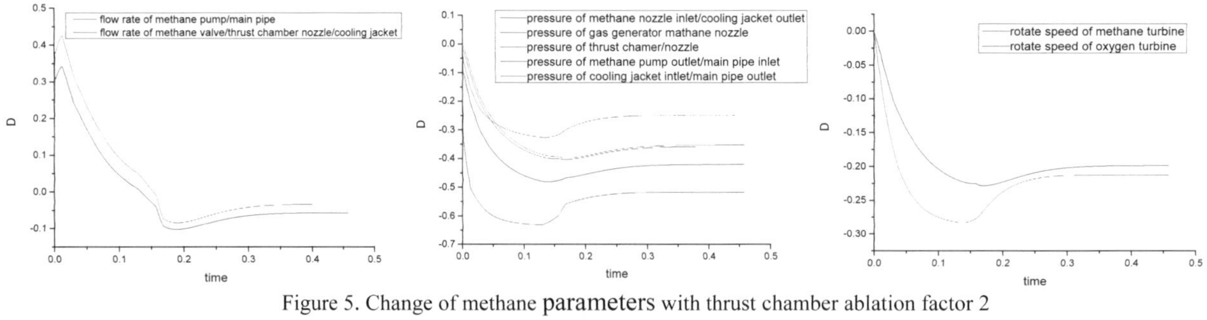

Figure 5. Change of methane parameters with thrust chamber ablation factor 2

When a failure with a thrust chamber ablation coefficient of 2.0 occurs, the flow of each components of the methane and oxygen pipeline will increase suddenly and then decrease with time, finally, after reaching steady state, the flow rate is lower than the rated 3.5%. The pressures of various parts of the methane and oxygen pipelines gradually decreased with time and then rose slightly to reach a steady state, in which the pressure of the thrust chamber changed most significantly. Changes in two turbines are similar to changes in pressure. It is worth mentioning that when the thrust chamber ablation occurs, the pressure changes much more than the flow rate changes.

From the results, we do not look at the specific values of the parameters, but compare their relative sizes to judge the sensitive parameter to fault.

Take into account of the length of the article to omit the other results.

According to the comparison of each parameters' value between the failure and rated condition, the different types of faults can be diagnosed, as shown in table 1:

Table 1: the change of parameters for different fault

Fault	Flow rate of methane pipeline							Pressure of methane pipeline					Rotating speed
	q_{pif}	q_{mif}	q_{mof}	q_{cvif}	q_{cvof}	q_{jof}	q_{cof}	P_{pof}	P_{mof}	P_{cvof}	P_{jof}	P_c	n_f
Leakage of methane pump	+	–	–	–	–	–	–	–	–	–	–	–	–
Methane main pipe blocked	–	–	–	–	–	–	–	+	–	–	–	–	+
Methane pump leakage	+	+	–	–	–	–	–	–	–	–	–	–	–
Cavitation of methane pump	+	+	+	+	+	+	+	+	+	+	+	+	+
Thrust chamber ablated	–	–	–	–	–	–	–	–	–	–	–	–	–

Annotate: + represents value after the fault is greater than the rated value,– represents value after the fault is less than the rated value.

Here only some typical faults of methane pipeline are listed, the same approach can be used for oxygen'.

5. CONCLUSION

In this study, The modular modeling of fault simulation for LOX engine is made by using MATLAB Simulink, some typical faults occurred in steady state are simulated, and the change of each parameter after releasing fault is analyzed. For blocked and leakage faults, the variation of flow rate parameters is more sensitive to other parameters. In the leakage failure, the upstream components' flow rate will increase, while downstream components' flow rate will decrease. Other parameters will decrease for blocked failure, but the entry pressure will increase. For ablative fault, all parameters will decrease, and the change of pressure parameters is more significant. The extraction of these fault features is helpful for studying LOX/methane rocket engine diagnosis. And other faults of LOX/methane rocket engine can also be further studied on this platform.

REFERENCES

[1] Kai Yang, Manrui Cai. The Latest Development of LOX/Methane Rocket Engine. Aerospace China. 2017. 10:14-19.

[2] Wilhelm W. F. SSME Model Dynamic Characteristics Related to POGO, Rocketdyne division Rockwel International, 1973. Report No. : N76-15246.

[3] Goertz C. , A modular method for the analysis of liquid rocket engine cycles. 1995.

[4] Liu K. , Zhang Y. , A study on versatile simulation of liquid propellant rocket engine systems transients. 2000.

[5] Naderi M. , Liang G. , Karimi H. , Modeling and simulation of staged combustion cycle LPRE, in 8th International conference on mechanical and aerospace engineering (ICMAE). 22-25 July, 2017, IEEE Xplore: Prague, Czech Republic.

[6] Yanjun Li, Jianjun Wu. Study on Key Techniques of Fault Detection and Diagnosis for New Generation Large-scale Liquid-propellant Rocket Engines. April 2014.

[7] Yang Erfu, Zhang Zhenpeng. Study on Fault Monitoring and Diagnosis Techniques for Thrust Chamber and Turbo-pump System of Liquid Rock Engines . Journal of Beijing University of Aeronautics and Astronautics. 1999. 10:619-622.

Pipe network leak detection: Sensor placement optimization using Support Vector Machines and a model-based leak detection technique

J.C. van der Walt[1], P.S Heyns[1], D.N Wilke[1]
[1] Centre of Asset Integrity Management, Department of Mechanical and Aeronautical Engineering, University of Pretoria, South Africa

ABSTRACT

This paper investigates a model-based method to detect leaks in water distribution networks. The model is used to generate data on which a Support Vector Machine is trained to locate and estimate the size of leaks on an experimental network. In this paper the leak location and size for an experimental network are shown as well as the effect of pressure sensor placement. The experimental investigation includes model calibration and the estimation of two simultaneous leaks in the network. The estimated solution for the experimental network showed the leaking pipe to be detectable with the leak sizes estimated within 0.7mm. The sensor placement optimization was completed with an exhaustive search algorithm and by removing sensors with the smallest impact on the detection accuracy, where it was found that training time of the SVM can be reduced significantly by using less sensors. Additionally, using less sensors for the estimation of the leaks was found to increase the probability up to 38% in finding the leak.

Keywords: Condition Monitoring, Fault Detection and Localization, Practical Low-Cost Condition Monitoring, Leak Detection

1. INTRODUCTION

In the 2011-2012 National Non-Revenue Water assessment (1) in South Africa the average national non-revenue water consumption was found to be 37%. While the world average in that period was 36.6%. Currently, to reduce non-revenue water consumption in South Africa, pressure management systems are installed, minimum night flows are logged, and water balances are completed. These methods only reduce the amount of leaked water or raise water waste awareness. The primary method of locating leaks comes from consumers. Contractors use listening sticks, geophones, ground penetrating radar, and noise loggers to locate potential leaks. More advanced techniques such as the Smartball, the Saharah system, and JD7 systems are also used (2) to detect potential leaks.

In recent years research to find leaks using on-line machine learning techniques with simulated time data has intensified with studies by Mashford et al. (3), Mandal et al. (4), Kayaalp et al. (5), ect. That showed the use of support vector machines (SVM) to be effective in finding leaks in a variety of networks. Mashford et al. (3) optimized the cost and kernel function parameters of the SVM on simple networks to find the best classification of the leaks. They could achieve a classification accuracy of 76.8%, with 77.5% of the leak cases within 100m of the actual leak. They went on to apply this technique on a water network in South Eastern Melbourne (6) where they investigated the accuracy of the SVM when classifying the emitter coefficients for each leak. They also investigated the application of this method for low leak rates, ranging between 50-100l/h. They found that all the leaks could be classified within a 500m range from the actual leak location with the network having a size of 1000m ×1100m.

Mandal et al. (4) investigated different training algorithms for SVMs on the classification of leaks in water networks. They compared Particle Swarm Optimization (PSO) and Enhanced Particle Swarm (EPSO) Optimization with Artificial Bee Colony (ABC) optimization and found that the SVM trained by ABC could classify the leaks nearly 10% more accurately than PSO or EPSO. Kayaalp et al. (5) used an experimental network to compare three algorithms for leak identification. The three algorithms were the RAkEL$_d$ (Random k-Labelsets), BRkNN (Binary Relevance k-Nearest Neighbors) and a SVM. They found that RAkEL$_d$ performed the best between the three algorithms, with the SVM performing better than BRkNN. Salam et al. (7) investigated the use of SVMs for leak identification on a network in Makassar, Indonesia. They found that the leak size could be correctly predicted with an accuracy of 86.68%, while the leak location could be predicted with an accuracy of 76.14% on the dataset.

Rosich et al. (8) investigated optimum sensor placement on a district metered area (DMA) in the Barcelona water distribution network. Their methodology is based on a depth-first search by choosing the nodes with the lowest cost and back-tracking to other un-explored nodes. Sanz and Pèrez (9) investigated the optimal placement between pressure and flow sensors on the Nova Icaria DMA in Spain. Their method estimates pressure and flow sensitivity using Singular Value Decomposition. They tested different sensor configurations with the use of three to six sensors and found that the demand calibration error decreases with increased number of sensors. They also found that pressure sensors are more beneficial in the calibration of demand components. Meseguer et al. (10) used a Genetic Algorithm (GA) for the optimization of the sensor positions on a DMA in Barcelona. They compared an optimal and non-optimal pressure sensor distribution and showed that the sum of the five largest non-isolable leaks could be reduced from 2976 nodes to 1357 nodes. Steffelbauer et al. (11) considered uncertainties with optimum sensor placement. They applied their methodology to a real-word DMA and found that the number of sensors to the cost followed an extended power law equation. Their methodology included a GA to optimize the sensor placement. Jung et al. (12) investigated a multi-objective optimal meter placement model which maximized the detection probability and robustness of a metered network, while minimizing the amount of false alarms. They tested this methodology on the Austin network and found that a linear relationship exists between the detection probability and the amount of false alarms.

Cugueró-Escofet et al. (13) completed sensor placement optimization by considering topological issues on the network which lead to mislabeling of leaks. Their methodology was applied on two DMA within the Barcelona water distribution network and they could suggest optimum number of sensors and their placement from their results. Fuchs-Hanusch et al. (14) compared six optimal sensor placement algorithms by applying them on real-world networks. They evaluated these algorithms on their robustness and their ability to localize small leaks. They found that the localization results could find leaks up to 0.5l/s if sensors were placed close to them. Soldevila et al. (15) proposed a sensor placement approach for classifier-based leak localization. The proposed method is based on a hybrid feature selection algorithm that combines a filter based on relevancy and redundancy with a wrapper based on genetic algorithms. They applied the proposed methodology on the Hanoi and Limassol water networks.

In this paper a linear kernel function SVM is trained on a calibrated experimental network to localize and classify leaking pipes. The calibration of the EPANET model is completed by optimizing pipe roughness, pump efficiency and minor loss coefficients. Two leak cases are classified by the SVM to find a baseline accuracy of the model. The experimental measurements included 12 pressure sensors allowing for optimum sensor selection for the training of the SVM. The optimization is completed with an exhaustive search algorithm and an algorithm removing sensors with the lowest cost to the detection probability.

2. METHODOLOGY

This methodology uses SVMs to calculate probabilities for the classification of leaks, as well as predictions for the size and location of the leaks. The SVM uses vectors, referred to as the kernel function, to generate decision boundaries between different data sets. The trained kernel functions can then be used to give probabilities and regressive estimations for the leaking pipe, the leak size and leak location.

The SVMs are generated using scikit-learn (16), which is a Python package used for machine learning. The kernel functions were left as default, which were linear kernel functions. The methodology is model-based to monitor pressure-flow deviations and require a dataset of possible leaks to train the SVMs on.

The dataset contained 1000 data samples randomly chosen depending on three variables. The leak size was randomly selected between 1-4mm with the leak location between 0-3m. These two variables were then fixed to a specific pipe which were randomly chosen. Three SVMs were created, each solving a single output. These three SVMs consisted of one for classification of the leaking pipe, one for regression on the leak size and, one for regression on the leak location. From the randomly chosen data an EPANET model could be simulated to find pressure values. These pressure values are used as inputs for the SVMs.

No artificial noise was added to the model measurements, and the classification of the leaks depended on the error between the EPANET model and the experimental measurements. This was done to disentangle error contributions for solving the inverse problem.

3. EXPERIMENTAL INVESTIGATION

Figure 1 shows the experimental network under investigation in this study. This network consists of six pipes serving as a distribution network. Each pipe in this network is installed with two pressure sensors, one at the start and one at the end of each pipe. These pressure sensors are marked as $P_1 - P_{12}$ in Figure 1. Additional flow sensors are installed to measure the demands or outflows for each pipe, marked as $F_1 - F_6$. The flow measurements are used to help with the calibration of the network. The network is fed by pumping water from a reservoir, with the demands and leaks feeding back into the reservoir.

The pipes in this network have a diameter of 10mm and have lengths of 3m each with 10cm of pipe before the pressure is measured. This is to reduce the turbulence after the T-pieces and L-bows before measuring the pressure. The 3m pipe sections are spaced 30cm apart. The demands in this network are created by drilling 3mm holes into the end of the pipes.

The experimental network contains three leaks, which can be switched on and off individually. The leaks can be found on pipe 1, 3, and 5. The first leak have a diameter of 3mm while the other two have a diameter of 2mm. Figure 2 shows an example of the application of a 2mm leak on this network. In this paper leaks on pipe 3 and 5 are tested and can be found 1m and 1.5m from the start of their respective pipes. The leaks and demands in this network were created following the approach illustrated in Figure 2.

An EPANET model of this network was created and calibrated to simulate the experimental network. The calibration process comprised of the optimization of 16 variables. They include one pump efficiency coefficient, minor loss coefficients, and roughness coefficients to minimize the error between the measurements. In Figure 1 red squares indicate the location of 12 minor loss coefficients added to the EPANET model to simulate the losses due to the connections. Two additional minor loss coefficients were added: one to the demand return pipe and one to the leak return pipe. Two additional calibration factors were added to the optimization algorithm, the pump efficiency and the roughness coefficient for all the pipes in the model.

The optimization was completed by minimizing a mean squared error between the 12 pressures and 6 flow measurements for the experimental measurements and the simulated model measurements. An error was calculated over all the leaking and non-leaking cases for this model calibration. Table 1 shows the results found for the 16 optimization parameters. It is evident that the minor loss coefficient depends on the flow rate within a pipe.

From this table the roughness coefficients of the pipes can be seen to be 245.7, while the pump efficiency was found to be 80.1%. As expected, the minor loss coefficients for L-bows and T-pieces differ. This can be seen where minor loss 1 is much smaller than minor losses 2-6. Minor loss coefficients 7-12 can be seen to be much smaller, this is because of an entering flow condition at these T-Pieces.

The minor losses in the demand return pipe can be seen to be much higher than that of the minor losses in the leak return pipe. This is due to the flow being much higher in the demand return pipe. The errors after optimizing these coefficients were calculated, and for the case where no leak exists, a mean squared error of 2.52% between the model and experimental measurements. For the first leak case an error of 44.7% was found and for the second leak case an error of 28.8% was found.

To find the leaks for these two leak cases, a dataset for the SVM were created using the calibrated model. The training methodology described in Section 2 was used, and the results achieved is given in Section 4. The two leak cases estimated on the experimental had sizes of 2mm and was located 1m and 1.5m from the start of each pipe. The leaking pipes were pipe 3 and 5 respectively. All 12 pressure sensors were used for the results shown in the next section.

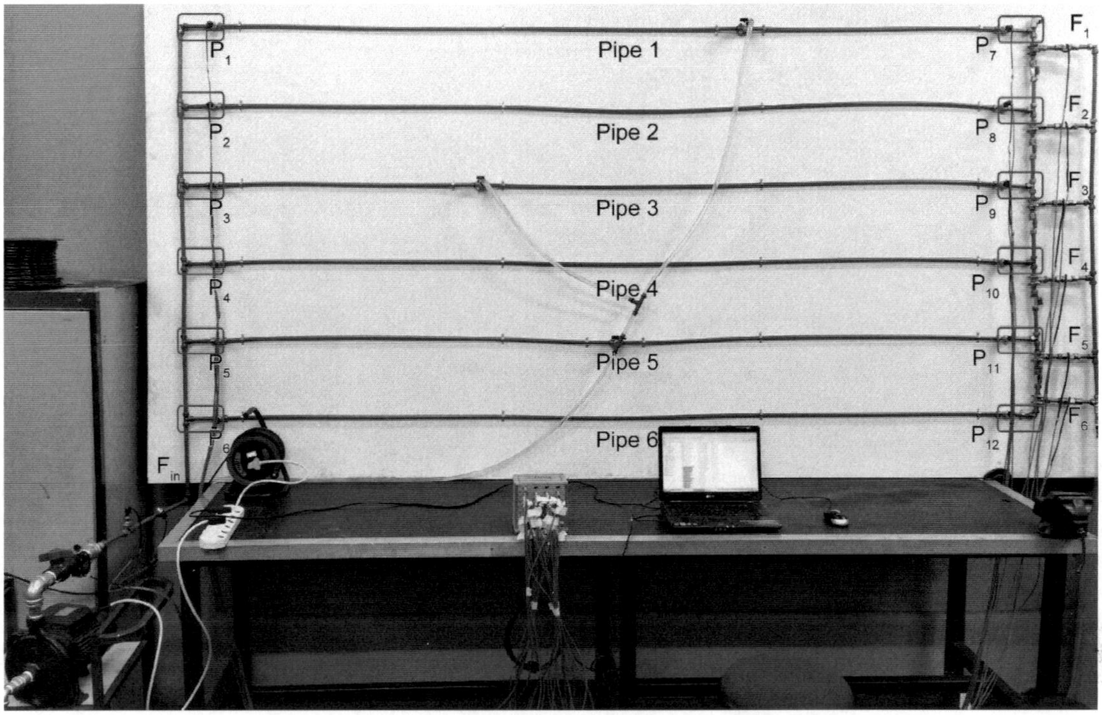

Figure 1: Photo of the Experimental Network

Figure 2: Photo of the 2mm leak applied to the Experimental Network

Table 1: Optimized Parameters for Experimental Network

Optimized Parameter	***Value***
Roughness Coefficient	245.7
Pump Efficiency	80.1%
Minor Loss 1	11.0
Minor Loss 2-6	28.6-40.8
Minor Loss 7-12	0.1-2.4
Minor Loss Demands	287.0
Minor Loss Leaks	14.2

4. RESULTS

The results found for the first leak case are depicted in Figure 3. The estimated leak using the simulated experimental measurement (ideal result) and actual experimental (measured) results for each pipe section are shown. The ideal results show the results if there was no model error, the leaking pipe could be estimated with a 100% probability. Additionally, it can be seen that the leak size and location could be estimated correctly with a size of 2mm and at a location of 1m. The results from the experimental measurements, estimated from the SVM trained on a model with a 44.7% error, could estimate the leaking pipe with a probability of 60.5%. The leak size was estimated to be 1.34mm and the leak location was found at 2.22m. This results in a leak size estimation error of 0.66mm and a location estimation error of 1.22m.

For the second leak case the results found by the SVM can be seen in Figure 4. For the ideal model, the leak could be located with a probability of 100%, with the leak size and location estimated perfectly as 2mm and 1.5m. With the experimental measurements, having a model error of 28.8%, the leaking pipe could be located with a probability of 93.6%. The leak size using the experimental measurements was found to be 1.29mm at a location of 1.28m. This resulted in a leak size estimation error of 0.71mm and a leak location estimation error of 0.22m.

5. SENSOR PLACEMENT OPTIMIZATION

To find the optimum sensors to be used in the experimental network, an exhaustive search algorithm is used with the SVM to find the highest probability of each leak for every sensor combination. Due to the amount of possible combinations, the search algorithm is limited to a depth of four sensor combinations.

This allows two test scenarios: the first test scenario up to the four most important sensors while the other test scenario up to the four least important sensors for each leak case. Table 2 shows the results found by this search algorithm for the first leak case on the experimental network. It can be seen that if a single sensor is used, where the ninth pressure sensor gave the best results, the leak could only be found with a probability of 33.45%. The exhaustive optimization is repeated by adding up to three more sensors. The highest probability of the leak being found increased to 99.87% with the addition of two sensors for a total of three sensors.

Initially, using all twelve pressure sensors, the leak could be located with a probability of 60.5%. If only eleven pressure sensors were used, removing the fifth pressure sensor, the leak could be located with a probability of 82.59%. Removing an additional three sensors results in the leak being located with a probability of 98%.

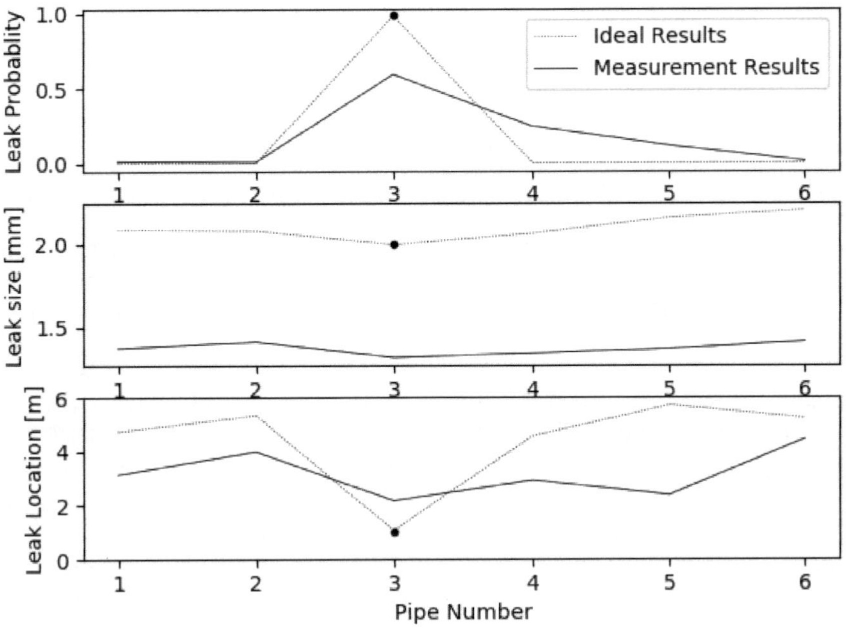

Figure 3: Results for the first leak on the experimental network

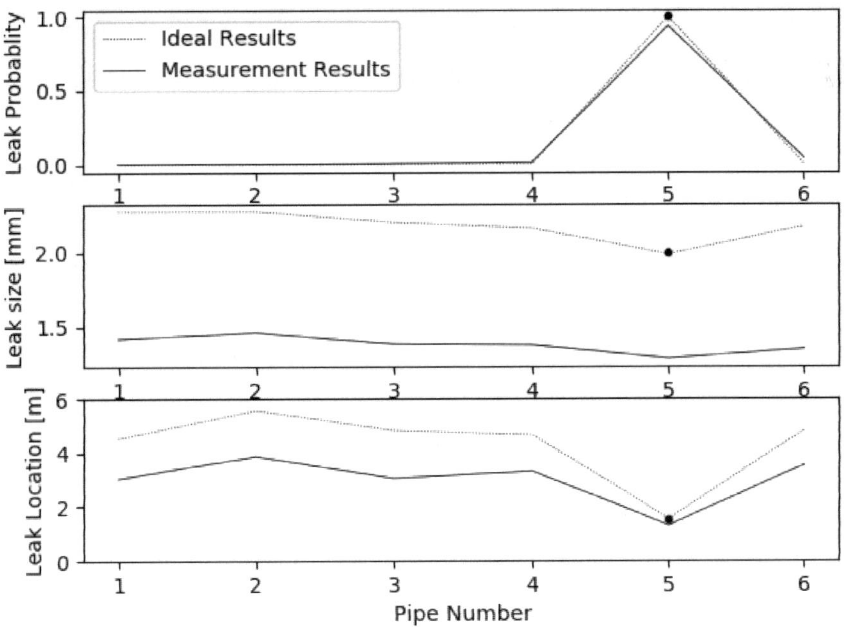

Figure 4: Results for the second leak on the experimental network

Table 2: Pressure sensor usage optimization for the first leak in the experimental network

Sensor Order	*Sensors Used*				*Sensors Not Used*				
	One Sensor	Two Sensors	Three Sensors	Four Sensors	Eight Sensors	Nine Sensors	Ten Sensors	Eleven Sensors	Twelve Sensors
1st	P9	P9	P9	P9	P5	P5	P5	P5	
2nd		P2	P2	P2	P3	P3	P3		
3rd			P11	P11	P2	P2			
4th				P10	P4				
Probability	33.45	92.73	99.87	99.34	98.0	97.11	95.40	82.59	60.50

Table 3 shows the results found by the search algorithm for the second leak case tested on the experimental network. For this leak case it can be seen that if a single pressure sensor was used, the fifth sensor, the leak could only be located with a probability of 12.25%. Adding an additional pressure sensor, pressure sensor 10, the leak could be located with a probability of 99.97% which was found to be the highest for the second leak case.

For this leak case the leak could be located with a probability of 93.6% using all twelve pressure sensors. Removing pressure sensors for this leak case bears smaller effect on the leak locating probability compared to the previous leak case, since this leak could already be located with a high probability. Removing the eleventh pressure sensor from the experimental measurements increases the leak locating probability with only 3.34%.

Table 3: Pressure sensor usage optimization for the second leak in the experimental network

Sensor Order	*Sensors Used*				*Sensors Not Used*				
	One Sensor	Two Sensors	Three Sensors	Four Sensors	Eight Sensors	Nine Sensors	Ten Sensors	Eleven Sensors	Twelve Sensors
1st	P5	P5	P5	P5	P11	P11	P11	P11	
2nd		P10	P10	P10	P6	P6	P6		
3rd			P2	P4	P12	P12			
4th				P7	P1				
Probability	12.25	99.97	99.86	99.82	99.53	99.22	98.54	96.94	93.6

The leak probability can be plotted over the number of sensors used and can be seen in Figure 5. For these two leak cases, using only 5, 6, and 7 pressure sensors, the results were unavailable and were linearly interpolated between the calculated results. This is shown with dashed lines in this figure. From the figure there can be seen that a peak leak locating probability was found for both leak cases when only 2-3 pressure sensors were used, with a gradual decline in detection probability as more sensors are added. This gradual decline in probability increases significantly after 10 sensors being used.

This behavior of the optimum sensor placement when removing sensors is due to uncertainty and noise from the additional sensors, with little additional information. This increases the detection probability up to 8 sensors to be used. When removing more sensors, the ratio between information and noise decreases and the problem becomes ill posed with too little information.

To evaluate the optimum sensor placements for both the leak cases, an additional approach for the optimization of the sensor placements were followed. The approach started with all the sensors and removed the sensor with the smallest cost on the detection probability. This could be completed until a single sensor was left. The results for this technique can be seen in Figure 6.

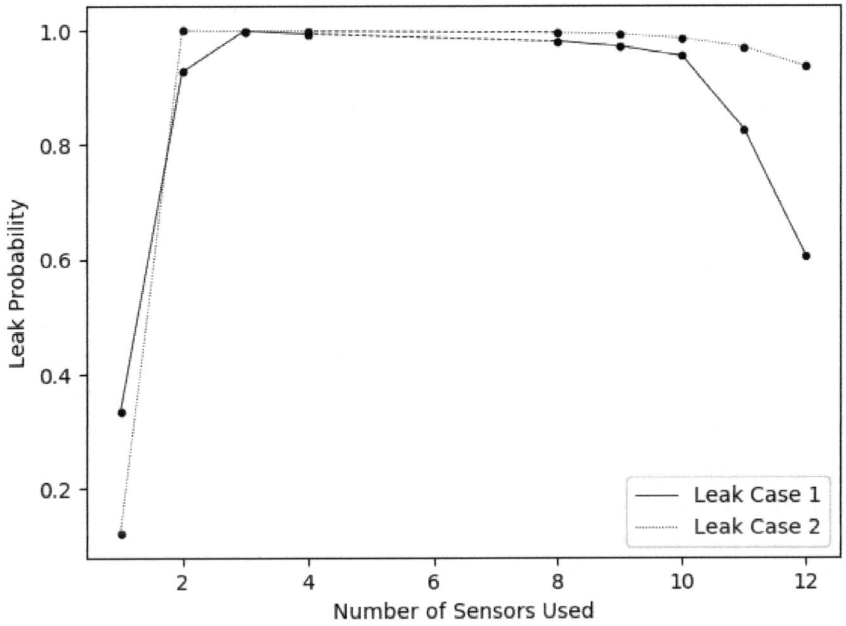

Figure 5: Summary of the leak estimation probability for the number of sensors used

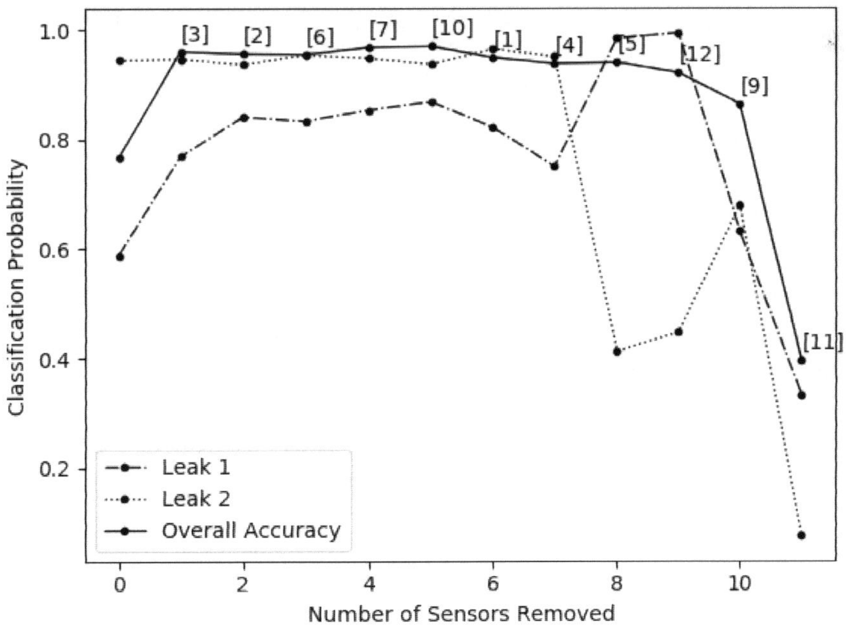

Figure 6 Optimal Sensors used for the two leak cases. Number of sensor removed indicated in square bracket

In this figure there can be seen that the leak detection probability was plotted against the number of sensors used. Three cases were plotted, the first leak case, the second leak case and the overall accuracy over the training dataset. From the results there can be seen that removing up to 5 sensors gradually increases the overall average for detecting the two leaks. Labeled in the square brackets are the sensor number that was removed from the SVM prediction. From the figure there can be seen that sensor 8 has the largest contribution for the detection of the leaks, which is similar to the results found in Table 2. Using only three sensors, sensor 8, 11 and 9, results in leak 1 and 2 being detected with a probability of 51.3% and 79.2% respectively. Overall the detection accuracy for the complete dataset is shown to be 87.3%.

6. DISCUSSION OF RESUTLS

In the experimental investigation, model calibration was completed on the EPANET model to fit it to the experimental measurements. Sixteen calibration factors were used and it could clearly be seen that the amount of flow within a pipe effects the minor loss coefficient. Additionally, it was shown that L-bows and T-pieces have different minor loss coefficients and that flow into these fittings generate nearly no loss coefficients.

The SVM combined with this model-based method could find the two leaks in the actual experimental leaking pipes with 60.5% and 93.6% probabilities respectively. This was found with calibration model errors for the two leak cases of 44.7% and 28.8%. For the first leak case the leak size and location were calculated with an error of 0.66mm and 1.22m. This result is impressive considering the model error. For the second leak case the leak size and location were calculated with an error of 0.71mm and 0.22m. For this network it was shown that this method of leak detection can solve the problem using all 12 pressure sensor measurements and a zero-model error with 100% accuracy.

With the exhaustive search algorithm, the sensor placement optimization was completed. It was seen that using between 2-3 sensors for each leak case can increase the leak detection probability to 99.86%. This increases the leak detection probability to ideal model conditions. The sensors to be used was found as the two sensors connected to the pipe under consideration for possible leaks. Furthermore, it was found that using more than 10 pressure sensors starts reducing the leak detection probability drastically for the experimental network. This is due to the noise to pressure information ratio from the experimental measurements.

7. CONCLUSION

This paper investigated the usage of a SVM with a model-based methodology. An experimental network was under consideration where after model calibration was completed, the model errors were calculated to be 44.7% and 28.8% for the two leak cases. Using the SVM the leaks could be located accurately with the leak sizes within 0.7mm and the leak location within 1.22m

Performing sensor placement optimization, it was found that the SVM can find the leaks with higher probabilities using less pressure sensors. The pressure sensors with the best locating probability was found to be the sensors connected to the pipe under consideration. An increase of 32.23% probability in leak location confidence was found by using only two sensors instead of twelve.

This model-based method shows promise in leak detection although certain limitations with this method exist. This paper shows the importance of selecting sensors when using the pressure flow deviation method.

8. REFERENCES

1. **R. Mckenzie, ZN. Siqalaba, WA. Wegelin.** *The state of non-revenue water in South Africa.* s.l. : Water Research Commission, 2012.

2. **Mckenzie, RS.** *Guidelines for reducing water losses in South African Municipalities.* s.l. : Water Research Commission, 2014.

3. *An approach to leak detection in pipe networks using analysis of monitored pressure values by support vector machines.* **J. Mashford, D. De Silva, D. Marney, S. Burn.** s.l. : International Conference on Network and System Security, 2009, Vol. 3.

4. *Leak detection of pipeline: An integrated approach of rough set theory and artificial bee colony trained SVM.* **S.K Madal, F.T.S Chan, M.K Tiwari.** s.l. : Expert Systems with Applications, 2012, Vol. 39.

5. *Leak detection and localization on water trasportation pipelines: a multi-label classification approach.* **F. Kayaalp, A. Zengin, R. Kara, S. Zavrak.** 10, s.l. : Neural Computing Applications Forum, 2017, Vol. 28.

6. *Leak detection in simulated water pipe networks using SVM.* **J. Mashford, D. De Silva, S. Burn, D. Marney.** 26, s.l. : Applied Artificial Intelegence, 2012, Vol. 5.

7. *A leakage detection system on the Water Pipe Network through Support Vector Machine method.* **A.E.U Salam, M. Tola, M. Selintung, F. Maricar.** s.l. : Makassar International Conference on Electrical Engineering and Informatics, 2014.

8. *Optimal sensor placement for leakage detection and isolation in water distribution networks.* **A. Rosich, R. Sarrate, F. Nejjari.** s.l. : IFAC Symposium on Fault Detection, 2012, Vol. 8.

9. *Comparison of demand calibration in water distribution networks using pressure and flow sensors.* **G. Sanz, R. Pèrez.** s.l. : Computer Control for Water Industry Conference, 2015, Vol. 13.

10. *Model-based monitoring techniques for leakage localisaiton in distribution water networks.* **J. Meseguer, JM. Mirats-Tur, G. Cembrano, V. Puig.** s.l. : Computer Control for Water Industry Conference, 2015, Vol. 13.

11. *Effiecient sensor placement for leak localization considering uncertainties.* **DB. SteffelBauer, D. Fuchs-Hasusch.** s.l. : Water Resource Management, 2016, Vol. 30.

12. *Robust meter network for water distribution pipe burst detection.* **D. Jung, JH. Kim.** s.l. : Water, 2017, Vol. 9.
13. *Optimal pressure sensor placement and assessment for leak location using a relaxed isolation index: Application to the Barcelona water network.* **MA. Cugueró-Escofet, V. Puig, J. Quevedo.** s.l. : Control Engineering Practice, 2017, Vol. 63.

14. *Real-world comparison of sensor placement algorithms for leakage localization.* **D. Fuchs-Hanusch, D. Steffelbauer.** s.l. : International Conference on Water Distribuion System Analysis, 2017, Vol. 18.

15. *Sensor placement for classifier-based leak localization in water distribuion networks using hybrid feature selection.* **A. Soldevila, J. Blesa, S. Tornil-Sin, RM. Fernandez-Canti, V. Puig.** s.l. : Computers and Chemical Engineering, 2018, Vol. 108.

16. *Scikit-learn: Machine Learning in Python.* **F. Varoquaux, G Gramfort, A. Michel, V. Thirion, B. Grisel, O. Blondel, M. Prettenhofer, P. Weiss, R. Dubourg, V. Vanderplas, J. Passos, A. Cournapeau, D. Burcher, M. Perrot, M. Duchesnay.** 2011, Journal of Machine Learning Research, pp. 2825-2830.

Performance visualisation of a transcritical CO_2 heat pump under fault conditions

J.J.A. de Bruin[1], K.R. Uren[1], G. van Schoor[2], M. van Eldik[3]
[1] School of Electrical, Electronic and Computer Engineering, North-West University, South Africa
[2] Unit for Energy and Technology Systems, Faculty of Engineering, North-West University, South Africa
[3] School of Mechanical and Nuclear Engineering, North-West University, South Africa

ABSTRACT

Heat pump systems have gained acceptance and appeal as an energy efficient alternative to electrical geysers for the purpose of water heating. This paper investigates a next-generation transcritical heat pump system using carbon dioxide as its working fluid. The water-to-water heat pump system used in this study simultaneously produces cooled and heated water. The heat pump system consists of four components. An evaporator is used to extract heat from a stream of water in order to evaporate and then superheat the working fluid. Chronologically after the evaporator comes the reciprocating compressor that raises the pressure and temperature of the working fluid to enable its circulation through the system. The gas cooler follows the compressor and is responsible for transferring heat from the working fluid to the second stream of interacting water. The expansion valve causes the working fluid, that leaves the gas cooler component, to undergo a large pressure drop and phase change. The working fluid that leaves the expansion valve enters the evaporator component for evaporation and superheating. The continuous conversion between different forms of energy, as enabled by the heat pump's components, make heat pumps susceptible to many different types of fault conditions. In this paper, the identification of faults that can occur during the operation of a heat pump and the degrading effects thereof on system performance were investigated. Fault detection and monitoring of the heat pump system via various visual representations are proposed. Specifically; fouling, working fluid leakage and coinciding water pump failure as system faults were investigated. The visual representations could uniquely identify and distinguish between the investigated fault conditions. The graphs were also able to monitor the severity and progression of system faults and their contribution to performance degradation.

Keywords: Transcritical heat pump, carbon dioxide, diagnosis, system performance, performance visualisation

1. INTRODUCTION

In the residential and domestic sectors, electrical geysers are traditionally used to provide warm water on demand. Carré [1] states that space and water heating accounts for approximately 70% of the IEA19 country group domestic sector energy consumption. Heat pump systems have gained notoriety in recent times due to their high efficiency when applied as alternative space heating and fluid heating solutions.

Heat pumps contain different system components in which energy is converted back and forth between different forms. Energy can be present in the form of thermal-, mechanical- and electrical energy in heat pump systems. The relative complexity of heat pump systems and the continuous conversion between different forms of energy by components that are present in heat pump systems make such a system susceptible to many different types of fault conditions [2].

Li *et al.* [3] did a theoretical and numerical study on an air-source heat pump system in the high-altitude region of Tibet, China. The study was performed to assess the performance of a specific air-source heat pump (ASHP) system operating at various ambient conditions. The study benchmarked and represented the ASHP system's performance using various derived two-dimensional relationships. Parameters like system power consumption, heat pump heating capacity and the coefficient of performance (COP) were related to ambient air temperature at various produced hot water temperatures. Li *et al.* focused only on the influence of ambient factors on system performance. Fault conditions were not considered in the study. Another study by Casteleiro-Roca *et al.* [4] considered a geothermal heat pump system used to heat a bioclimatic house in Galicia, Spain. The study used a novel intelligent approach to detect possible fault conditions in the system's geothermal heat exchanger. The fault detection approach implemented classification techniques and temperature probe date from two temperature sensors, located on the system's geothermal heat exchanger, to identify the presence of fault conditions and their effect on the system. The approach proved to have high accuracy, but only for the identification of fault conditions which stemmed from problems associated with the system's geothermal heat exchanger. Mehrabi and Yuill [5] did a study on the generalised effects of faults on the normalised performance of air conditioners and heat pumps. Fault intensity (FI) factors were used to model the drift of various defined parameters, which occur due to faults, from their nominal values. Fault conditions which incorporated the effects of change in refrigerant system mass, COP, input power, working fluid mass flow and pressure drop due to liquid line restrictions were considered in the study. Different operational scenarios were presented by depicting the heat pump cycle on temperature-entropy (T-s) and pressure-enthalpy (P-h) diagrams. Each depicted cycle diagram corresponded to various unique FI factor values.

In this paper, the use of unique two-dimensional and three-dimensional visual representations which correlate system performance parameters to the progressive effects of faults are investigated. The approach uses the interpretation of the proposed graphs as a means to identify specific faults and quantify their magnitude and progression in the considered heat pump system.

In Section 2 a system description is given of the investigated system and its component arrangement. Section 2 also discusses the holistic system level modelling approach used to solve for the system parameters in the selected simulation environment. After that, Section 3 presents and discusses the indices used to benchmark the overall system performance of the investigated heat pump. In Section 4 the unique visual representation to be used to visualise the degradation impacts of faults on the heat pump system's performance is presented. The paper concludes with Section 5 where the summary of key findings of the study and recommendations for future work on the topic are discussed.

2. SYSTEM DESCRIPTION AND MODEL

2.1. Overview

The heat pump system considered in this paper is a thermal-fluid device that uses carbon dioxide as an environmentally friendly working fluid. The secondary interacting fluid in the system is water. Two different streams of water are circulated through the system. The first stream that flows through the evaporator component serves as an energy source and is therefore cooled during system operation. The second stream flowing through the gas cooler serves as an energy sink and is therefore heated during system operation. Additionally, the heat pump system consists of a compressor which circulates the carbon dioxide and an expansion valve which regulates pressure drop and initiates the evaporation process in the evaporator. The heat pump system thus transfers heat energy from the one water stream to the other via the circulated carbon dioxide gas. The heat pump system's component layout is illustrated in Figure 1.

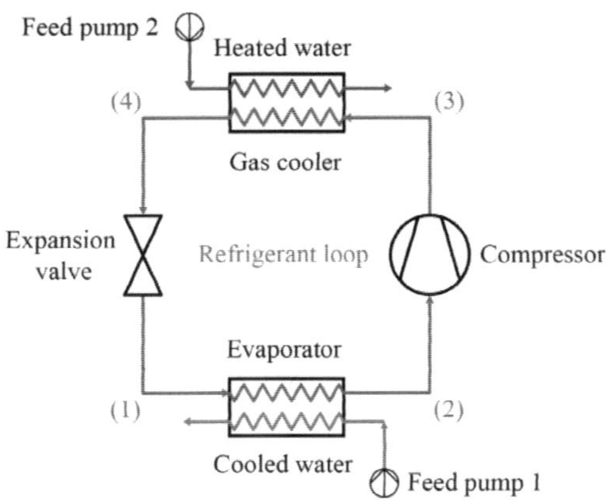

Figure 1. Diagrammatic representation of heat pump system layout

The first feed pump ensures circulation of water through the evaporator in which heat is transferred from the water to the carbon dioxide. The water is thus cooled, and the carbon dioxide absorbs heat to evaporate to the saturated gas state and heat even further into the superheated gas state. The second feed pump circulates the water to be heated through the gas cooler component. The water absorbs heat from the supercritical carbon dioxide. The supercritical gas is thus cooled, and heated water is delivered at the gas cooler outlet during system operation.

2.2. Refrigerant cycle representation

The thermodynamic behaviour of the system's refrigerant may be illustrated via a graphical representation that uses critical cycle points. Cycle points are graphed points that are representative of refrigerant thermophysical properties at specific locations in the system. Cycle point (1) to (2) represents heat absorption by the refrigerant in the evaporator component. The compressor does work on the refrigerant to circulate it through the system. The action of the compressor is depicted by cycle point (2) to (3). After compression, the refrigerant transfers heat to water in the gas cooler component. The heat rejection from the refrigerant is represented by cycle point (3) to (4). After heat rejection, the expansion valve component decreases the refrigerant pressure and enables the process of evaporation. The expansion valve isenthalpic pressure drop process occurs from point (4) to (1). The refrigerant behaviour may be visualised on a temperature-entropy diagram as shown in Figure 2.

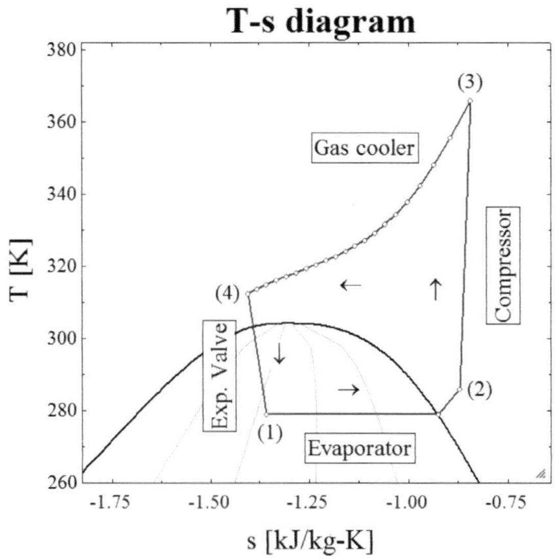

Figure 2. Heat pump refrigerant loop depicted on a T-s diagram [6]

2.3. System model

The system model was developed in the *Engineering Equation Solver* [7] (EES®) environment. The simulation methodology used a system level approach. Semi-theoretical models were used with the implementation of conservation laws to solve for the required parameters which quantified the behavioural properties of each component in the system. The laws of conservation of mass, momentum and energy (for steady incompressible flow) were each applied to every component in the system. The use of the incompressible form of the conservation of momentum law is valid due to the Mach number of the refrigerant never exceeding 0.1 throughout the system within the range of simulated system conditions. At flow Mach numbers below 0.3, the compressible and incompressible form of the law of conservation of momentum both yield the same thermo-physical property results. The law of conservation of mass for steady incompressible flow through an individual component's control volume region (CV) is expressed mathematically as [8]:

$$\sum_{i=1}^{n_1} \dot{m}_{in} \mid_i = \sum_{j=1}^{n_2} \dot{m}_{out} , \tag{1}$$

where the first term is the summation of all the mass flow streams entering the CV and the second term is the summation of all the mass flow streams flowing out of the CV. The refrigerant and interacting water all flow through pipes with no other fluid flowing in or out of the piping. The conservation of mass for every system component may thus be simplified to:

$$\dot{m}_{in} = \dot{m}_{out} . \tag{2}$$

Equation (2) holds true for each component and for the refrigerant side of the heat exchanger as well as the water side. The law of conservation of momentum for steady (incompressible low Mach number) flow through an individual component's CR is expressed as [8]:

$$\sum_{i=1}^{n_1} (P_{tot})_{in} \mid_i + \sum_{j=1}^{n_2} (\rho g z)_{in} \mid_j = \sum_{k=1}^{n_3} (P_{tot})_{out} \mid_k + \sum_{l=1}^{n_4} (\rho g z)_{out} \mid_l + \Delta P_{tot}, \tag{3}$$

where the first term is the total pressure of the fluid(s) entering the CV, the second term is the hydrostatic pressure of fluid(s) entering the CV, the third and fourth term represents the total- and hydrostatic pressure of fluid(s) leaving the CV and the fifth term represents the pressure drop which occurs as the various interacting fluids flow through the CV. Equation (3) may be simplified by only considering the interacting fluids for each component. For the refrigerant side of the evaporator, gas cooler and expansion valve components are:

$$(P_{tot|r})_{in} = (P_{tot|r})_{out} + \Delta P_{tot|r} . \tag{4}$$

The conservation of momentum for the water side of the evaporator and gas cooler components is:

$$(P_{tot|w})_{in} = (P_{tot|w})_{out} + \Delta P_{tot|w} . \tag{5}$$

The pressure ratio of the compressor is used to find the discharge pressure of the refrigerant leaving the compressor. The ratio of pressure for a compressor is defined as [9]:

$$r_p = \frac{P_{tot|outlet}}{P_{tot|inlet}} . \tag{6}$$

Finally, the law of conservation of energy (First Law of Thermodynamics) for a steady flow process of an individual component's CV is given by [8]:

$$\dot{q}_{net} + \dot{W}_{net} + \sum_{i=1}^{n_1} (\dot{m} h_{tot})_{in} \mid_i + \sum_{j=1}^{n_2} (\dot{m} g z)_{in} \mid_j = \sum_{k=1}^{n_3} (\dot{m} h_{tot})_{out} \mid_k + \sum_{l=1}^{n_4} (\dot{m} g z)_{out} \mid_l , \tag{7}$$

where the first term is the net heat transfer into the CV, the second term is the net work absorbed by the CV, the third term is the total flow energy entering the CV, the fourth term is the gravitational potential energy of the flow(s) entering the system, the fifth term represents the flow energy leaving the CV, and the sixth term represents the gravitational potential energy of the flow(s) departing the CV. The conservation of energy for each component was compiled under the assumption that the elevation head difference throughout the system is negligible. The conservation of energy for the evaporator and gas cooler component is

$$(\dot{m} h_{tot|r})_{in} + (\dot{m} h_{tot|w})_{in} = (\dot{m} h_{tot|r})_{out} + (\dot{m} h_{tot|w})_{out} . \tag{8}$$

The conservation of energy for the compressor component is

$$(\dot{m} h_{tot|r})_{in} + \dot{W}_C = (\dot{m} h_{tot|r})_{out} . \tag{9}$$

The expansion valve component's throttling process is assumed to be isenthalpic. Under the isenthalpic assumption, the conservation of energy for the expansion valve simplifies to:

$$(\dot{m} h_{tot|r})_{in} = (\dot{m} h_{tot|r})_{out} . \tag{10}$$

In addition to the traditional conservation laws, an exergy balance was also applied to each component to solve for its corresponding internal irreversibility rate within each system component. An exergy balance for an individual component's CV has the form [10]:

$$\dot{E}_{CR}^{\dot{q}} + \dot{W}_{net} + \sum_{i=1}^{n_1} (\dot{E})_{in} \mid_i = \sum_{j=1}^{n_2} (\dot{E})_{out} \mid_j + \dot{I}_R, \tag{11}$$

where the first term is the exergy transfer to the CV due to heat transfer, the second term is the net work absorbed by the CV, the third term is the exergy entering the CV, the fourth term is exergy leaving the CV and the fifth term is the rate at which exergy is irreversibility destroyed inside of the CV. The evaporator and gas cooler component do not have associated work terms in their respective exergy balance equations. The exergy balance for the evaporator is:

$$\left(\dot{E}_g\right)_{in} + \left(\dot{E}_w\right)_{in} = \left(\dot{E}_g\right)_{out} + \left(\dot{E}_w\right)_{out} + \dot{I}_E \,. \tag{12}$$

The exergy balance for the gas cooler component is identical to the exergy balance for the evaporator component, as in (12), with the only difference being that the irreversibility rate's subscript change to GC. The exergy balance for the compressor component is:

$$\left(\dot{E}_g\right)_{in} + \dot{W}_C = \left(\dot{E}_g\right)_{out} + \dot{I}_C \,. \tag{13}$$

The exergy balance for the expansion valve component is:

$$\left(\dot{E}_g\right)_{in} = \left(\dot{E}_g\right)_{out} + \dot{I}_{EV} \,. \tag{14}$$

2.4. Validation of the system model

The EES® system simulation model from which results were generated for this paper has been validated by de Bruin *et al.* [11] using the comparison of simulation data to experimental results overlaid on a cycle T-s diagram. This paper generates results from an updated and improved version of the previously validated simulation model.

3. SYSTEM PERFORMANCE MEASURES

Thermal-fluid systems can be modelled and evaluated by usage of their energy as well as exergy characteristics. The application of system energy indices is considered first. The most well-known benchmark of heat pump performance is the *coefficient of performance* [9]. The COP relates how much useful energy leaves the refrigerant loop compared to the energy invested in driving the heat pump's compressor. The formula for the coefficient of performance of a heat pump assigned to a heating objective is:

$$COP = \frac{\dot{q}_{heat}}{\dot{E}_{elec}}, \tag{15}$$

where $\dot{q}_{heat}$ is the heat transferred to the water being heated and $\dot{E}_{elec}$ is the electrical energy required by the system's compressor. The COP does not give an indication of how well the system is performing compared to ideal energy transfer system conditions. Stene [12] recommends the *Lorentz efficiency* index in order to benchmark a heat pump system's COP to a maximum COP which could be achieved under theoretical ideal circumstances. The Lorentz efficiency for a transcritical heat pump has the form:

$$\eta_{LZ} = \frac{COP}{COP_{max}} 100, \tag{16}$$

where COP_{max} is expressed as

$$COP_{max} = \frac{T_{\overline{GC}}}{T_{\overline{GC}} - T_{evap}}. \tag{17}$$

$T_{\overline{GC}}$ is the logarithmic mean heat rejection temperature in the gas cooler component and T_{evap} is the evaporation temperature respectively found in the evaporator component. Kotas [10] argues that it is more rational to use exergy analysis to evaluate the performance of thermal-fluid systems. Kotas purposes the formulation of rational exergy domain efficiency indices which compare process operation to that of a process which is completely reversible. The rational efficiency of a process can be defined in two different manners. The first way to define rational efficiency is by relating how much irreversibility is deducted from the driving exergy of a system due to effects of intrinsic and avoidable process inefficiencies. Mathematically the first form of rational efficiency for a heat pump is expressed as:

$$Rat_1 = \frac{\dot{E}_{elec} - \sum_{i=1}^{n} \dot{I}_n}{\dot{E}_{elec}} 100, \tag{18}$$

where $\dot{I}_n$ is irreversibility rate inside a sub-system component which must always have a positive and greater than zero value. The second rational efficiency has equivalent logic to the COP because it is defined as the ratio of useful desired exergy delivered by the system to the exergy expenditure used to drive the system. The second form of rational efficiency is expressed as

$$Rat_2 = \frac{\Delta \dot{X}_{GC}}{\dot{E}_{elec}} 100, \tag{19}$$

where $\Delta \dot{X}_{GC}$ is the exergy absorbed by the water that is heated by the heat transfer process in the gas cooler component.

4. VISUALISATION OF FAULT CONDITIONS

4.1. Heat pump system faults

There are multiple faults which can occur in the considered heat pump system. A fault with a fast-enough onset could be considered so severe that it may cause the system to cut out completely and to stop working instantly. A severe system fault could, for example, be a loss of electrical power to the system's compressor. If the heat pump system's compressor cuts out during operation, then the heat pump system will stop transferring heat between the two water streams almost instantaneously. Other system faults manifest and progress slowly with time. These faults can slowly reduce the performance of the system over time without an operator noticing due to the slow and incremental, steady decrease in system performance.

This paper considers three faults that may develop slowly over time in a carbon dioxide heat pump system namely: the build-up of unwanted material inside the heat exchanger tubes, gas leakage from the refrigerant loop and finally external water pump failure. The build-up of unwanted material inside heat exchangers is formally called fouling [13]. Fouling build-up occurs when water with impurities is circulated through heat exchangers. The impurities precipitate onto the inside of the tube walls with time. The refrigerant may leak from the primary loop in a heat pump system due to the culmination of erosive effects of particulate matter that is carried with the flow of a fluid through system tubing. The erosion of the tubing may ultimately lead to a hole in the piping from which refrigerant may leak. Pump failure may occur due to problems with the pump's electric motor or due to pump impeller damage that has occurred over time due to insufficient pump net positive suction head (NPSH) at the pump's inlet [14]. Fouling may also build-up with time inside water pumps just like inside the tubing of components.

4.2. Visualisation of gas cooler fouling build-up

Fouling in the system was simulated by varying the gas cooler's inner tube's outer fouling factor (FF_{io}), which effects the gas cooler component's thermal resistance levels, from 0.0 to 2.0 [m^2-K/kW]. Fouling has the effect of introducing extra thermal resistance in the path of heat transfer inside a heat exchanger. The effect of fouling is the same as with the introduction of extra electrical resistance in an electrical circuit. The analogy between thermal and electrical resistance is depicted in Figure 3.

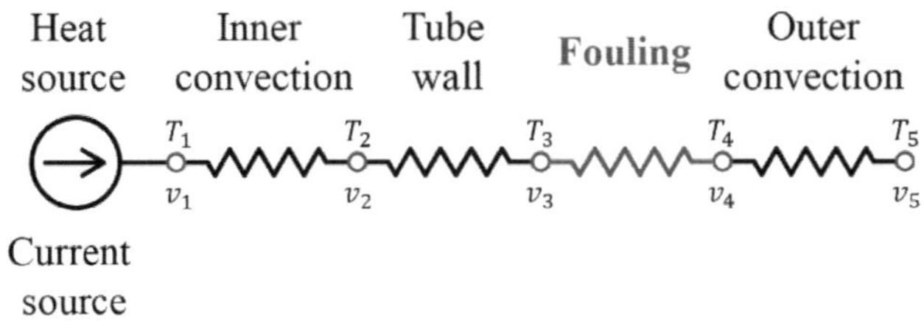

Figure 3. Circuit analogy that shows the similarities between thermal and electrical resistance

An increasing amount of fouling presence in the transcritical heat pump system's gas cooler was found to have the effect on the magnitude of the four performance indices used to benchmark the refrigerant cycle as summarised in Table 1:

Table 1: The effect of the fouling fault condition on the magnitude of the heat pump's performance indices

FAULT	COP	η_{LZ}	Rat_1	Rat_2
Gas cooler fouling	Decreases	Decreases	Decreases	Increases

A decrease in the heat pump system's performance may be due to the progression of any number of faults in the system. To specifically visualise the impact and the progression of fouling a logarithmic pressure – flow energy (log. P-h) diagram is suitable. The log P-h diagram that illustrates the heat pump's working fluid under various gas cooler fouling conditions is depicted in Figure 4.

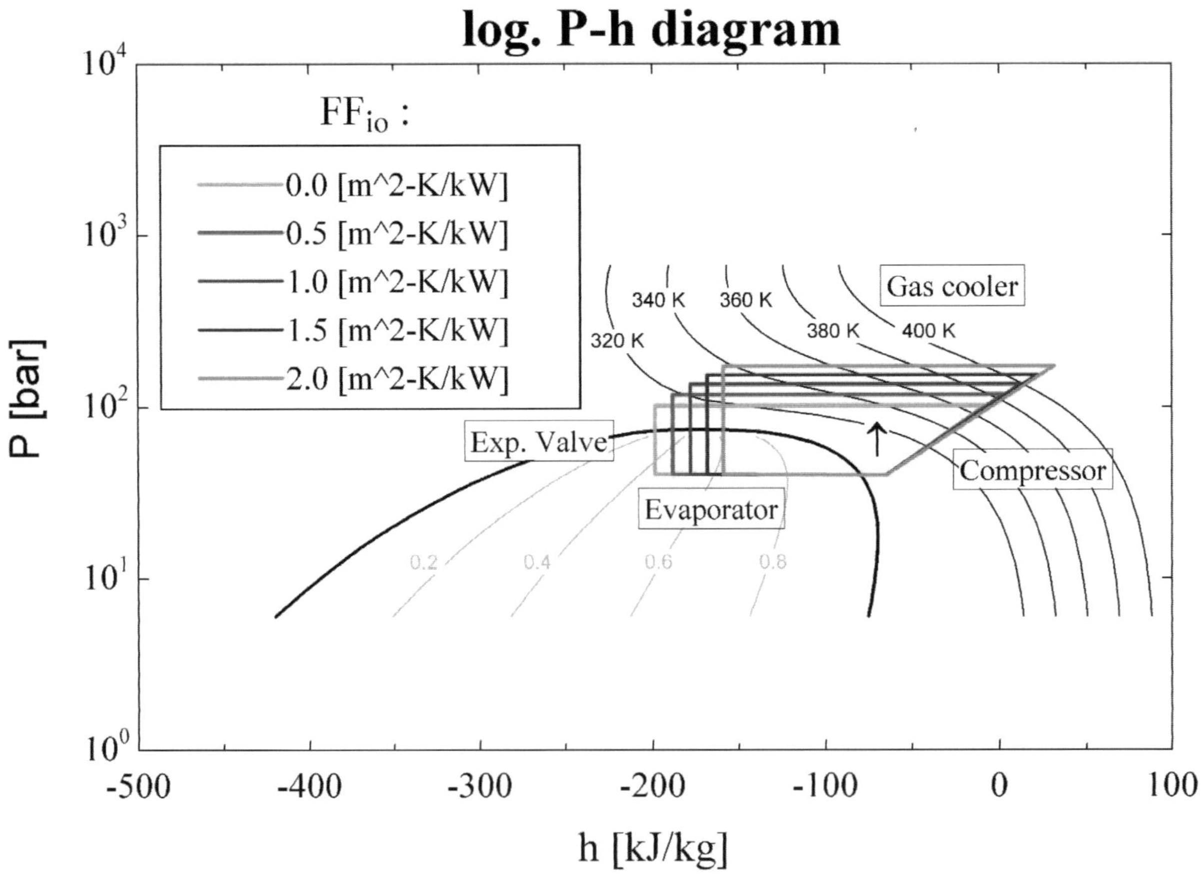

Figure 4. Fouling fault condition visualised on a log. P-h diagram

Figure 4 represents the thermo-physical condition of the heat pump's working fluid at various physical locations throughout the heat pump system's constituent components. Visual inspection of Figure 4 shows that in order to maintain the required heat transfer rate in the gas cooler component – the compressor's discharge pressure and temperature trends to increase with an increase in the fouling build-up in the gas cooler.

4.3. Visualisation of working fluid leakage

Working fluid leakage was simulated using a parametric study in which the mass flow rate value for the reference condition was used as the initial entry and with subsequent decreasing values for the mass flow rate onwards in the parametric table. Table 2 summarises the effect that an increasing amount of working fluid leakage has on the four performance indices.

Table 2: The effect of the working fluid fault condition on the magnitude of the heat pump's performance indices

FAULT	COP	η_{LZ}	Rat_1	Rat_2
Refrigerant leakage	Decreases	Decreases	Increases	Decreases

A discretised temperature-position (T-pos) diagram has some interesting properties suited for the visual investigation of working fluid leakage. Figure 5 illustrates the lengthwise thermal interaction between the cooled working fluid stream and the heated water stream, that respectively interact in the gas cooler component, before and after working fluid leakage has occurred.

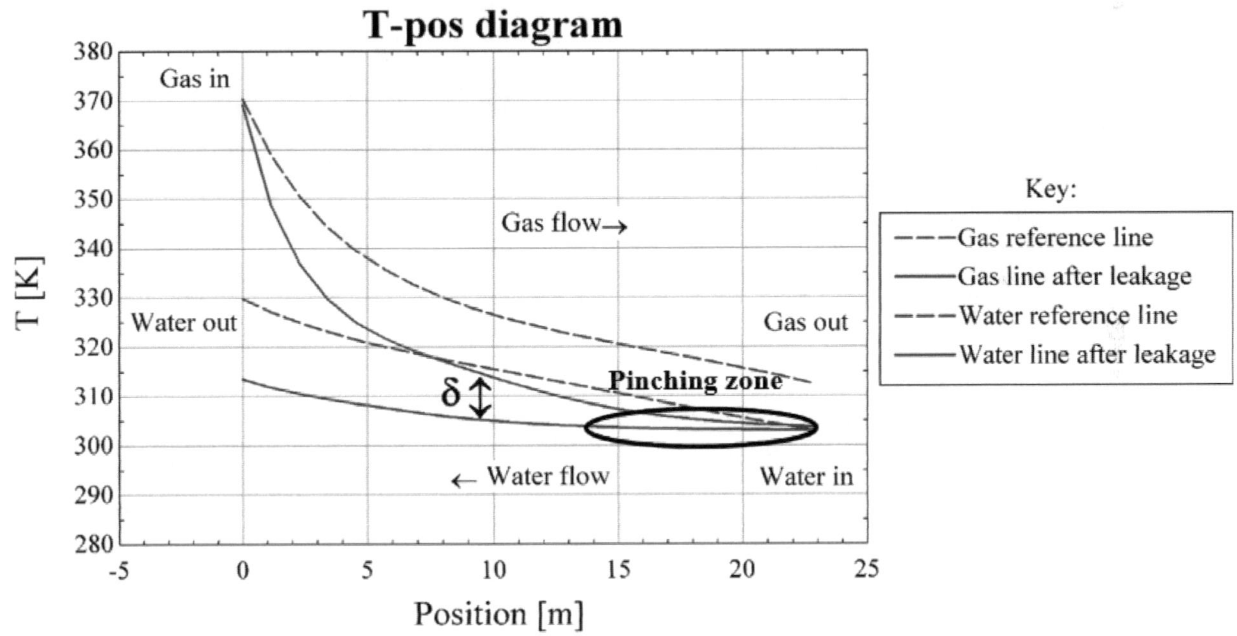

Figure 5. Working fluid leakage fault condition visualised on a T-pos diagram

From Figure 5 it should be noted that a *pinching zone* may be identified on the diagram. It can be seen that in the interval between 0 [m] and 5 [m] that the working fluid undergoes a rapid temperature decrease when compared to the reference condition in the same length interval. The pinching zone occurs due to the decreased ability of the working fluid to carry and store thermal energy as it flows through the gas cooler. The pinching zone occurs between the interacting fluid streams after working fluid leakage has occurred and the region is indicated by the black ellipse. Due to working fluid leakage the temperature difference between the interaction fluid streams in the pinching zone, as represented by the symbol – $\boldsymbol{\delta}$, trends increasingly to zero.

4.4. Visualisation of simultaneous water pump failure

Simultaneous water pump failure was simulated by varying the delivered volumetric flow rate of each pump of the two water pumps simultaneously between 16 and 6.5 [l/min] in a parametric table. Simultaneous failure was simulated to investigate a system fault condition that may stem from improper pump installation/ maintenance of both system heat exchangers. A decreased flow rate delivery from a pump can be thought of as a forerunner effect that may be a warning sign in the lead up to complete pump failure. The Lorentz efficiency was selected as the z-axis parameters for the compilation of a three-dimensional graph. The x-axis represents the volumetric flow rate of feed pump one and the y-axis represents the volumetric flow rate of feed pump two. Figure 6 depicts the relationship between heat pump Lorentz efficiency and the volumetric flow rates delivered by the heat pump's feed pumps.

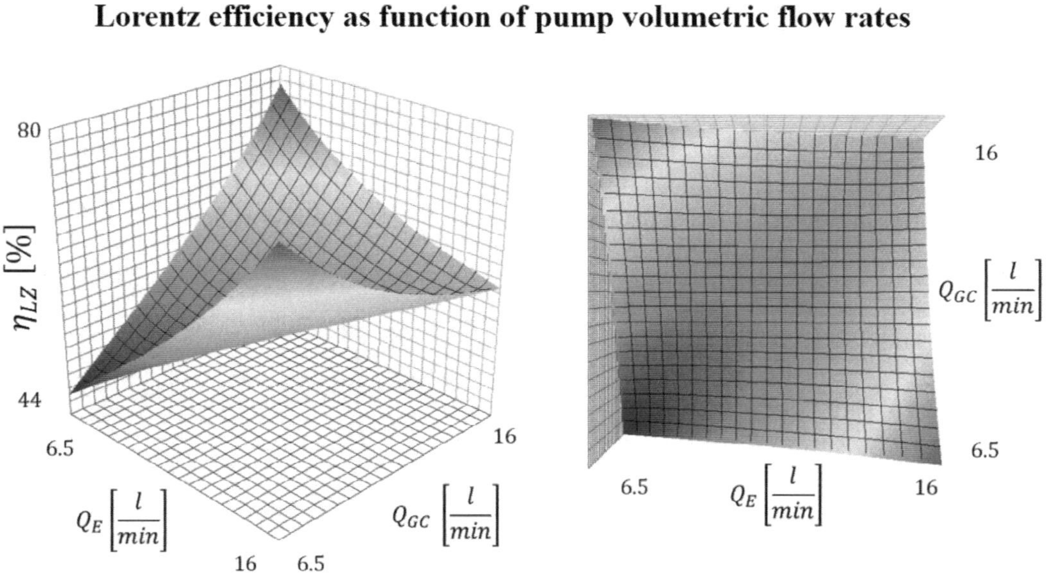

Figure 6. Lorentz efficiency three-dimensional representation as a function of feed pump volumetric flow rates

5. CONCLUSION

Figure 4 shows a pattern where the representative expansion valve line moves from the left to the right on the graph. The line that is representative of the heat transfer in the gas cooler shows a pattern of moving from a lower pressure to a higher pressure. A transcritical heat pump of which the associated refrigerant shows the shift in cyclic representation as illustrated in Figure 4 may be diagnosed as having fouling present as a fault condition in it. The presence of a pinching zone on a T-pos diagram may be used as a means to identify and isolate a working fluid leakage fault condition. The comparison between representative fluid lines on T-pos diagrams may be used as a means to track the progression of working fluid leakage within a heat pump system. The results presented in Table 1 are distinguishable from the results present in Table 2. Gas cooler fouling and refrigerant leakage are thus distinguishable fault conditions when only using the behaviour of the performance indices as a fault detection procedure.

From Figure 6 it can be seen that when the volumetric flow rate of the evaporator's water pump, Q_E, and the volumetric flow rate of the gas cooler's water pump, Q_{GC}, are decreased then the Lorentz efficiency, η_{LZ}, of the associated heat pump trends to a minimum value. The Lorentz efficiency is at its lowest value in the blue 'cold spot' region in Figure 6. The cold spot region coincides with the lowest values of the simulated volumetric flow delivery rates by the heat pump system's two water pumps.

Future research should include an investigation of whether the same patterns and visual representations, as suggested in this paper, could be used to identify faults in larger simulated data ranges. Various combinations of fault conditions also need to be investigated and such an investigation may result in the development of improved residual definitions.

ACKNOWLEDGEMENTS

The authors would like to thank the National Research Foundation (NRF) of South Africa (Grant numbers 91093 and 103392). The opinions expressed in this paper are those of the authors and are not necessarily to be attributed to the NRF.

REFERENCES

[1] J.-B. Carré, "Experimental investigation of electrical domestic heat pumps equipped with a twin-stage oil-free radial compressor," Ecole Polytechnique Fédérale de Lausanne, Lausanne, 2015.

[2] H. Madani and E. Roccatello, "A comprehensive study on the important faults in heat pump system during the warranty period," *International Journal of Refrigeration,* no. 48, pp. 19-25, 2014.

[3] Y. Li, W. Li, J. Liu, J. Lu, L. Zeng, L. Yang and L. Xie, "Theoretical and numerical study on performance of the air-source heat pump system in Tibet," *Renewable Energy,* no. 144, pp. 489-501, 2017.

[4] J. Casteleiro-Roca, H. Quintián, J. Calvo-Rolle, E. Corchado, M. d. C. Meizoso-López and A. Piñón-Pazos, "An intelligent fault detection system for a heat pump installation based on a geothermal heat exchanger," *Journal of Applied Logic,* no. 17, pp. 36-47, 2016.

[5] M. Mehrabi and D. Yuill, "International Journal of Refrigeration," 11 October 2017. [Online]. Available: https://doi.org/doi:10.1016/j.ijrefrig.2017.10.017.. [Accessed March 2018].

[6] Y. A. Cengel and M. A. Boles, "Thermodynamics an engineering approach. Eighth ed.," Mc Graw Hill Education, New York, 2015.

[7] F-Chart Software, "EES Engineering Equation Solver," F-Chart Software, 2018. [Online]. Available: http://www.fchart.com/ees/. [Accessed March 2018].

[8] Y. A. Çengel and R. H. Turner, Fundamentals of thermal-fluid sciences, 2nd ed., McGraw-Hill Higher Education, 2004.

[9] Borgnakke and Sonntag, Fundamentals of thermodynamics, 8th ed., Don Fowley, 2013.

[10] T. Kotas, in *The exergy method of thermal plant analysis*, Malabar, Krieger Publishing Company, 1995, pp. 99-103.

[11] J. de Bruin, K. Uren, G. van Schoor and M. van Eldik, "A thermodynamic cycle model of a transcritical CO2 heat pump for energy-visualisation," Stellenbosch, 2017.

[12] J. Stene, "Residential CO2 heat pump system for combined space heating and hot water heating," Norwegian University of Technology and Science, Trondheim, 2004.

[13] Y. A. Çengel and A. J. Ghajar, in *Heat and mass transfer*, New York, McGraw-Hill Education, 2015, p. 656.

[14] A. Sayers, Hydraulic and compressible flow turbomachines, London: McGraw-Hill, 1990.

North-West University
Potchefstroom Campus
Private Bag X6001
Potchefstroom
2520
South Africa

ISBN 978-1-5108-7038-3

9 781510 870383

CURRAN ASSOCIATES INC.
proceedings
.com